RUSSIA AND THE POST-SOVIET SCENE

RUSSIA AND THE POST-SOVIET SCENE

A Geographical Perspective

James H Bater

Professor of Geography, University of Waterloo, Canada

A member of the Hodder Headline Group
LONDON • NEW YORK • SYDNEY • AUCKLAND

To
Linda, Lisa, Steven, Kevin and Ian

Published in Great Britain in 1996 by
Arnold, a member of the Hodder Headline Group
338 Euston Road, London NW1 3BH

Copublished in the US, Central and South America by
John Wiley & Sons, Inc., 605 Third Avenue
New York, NY 10158-0012

British Library Cataloguing in Publication Data
A catalogue entry for this book is available from the British Library

Library of Congress Cataloging-in-Publication Data
A catalog entry for this book is available from the Library of Congress

ISBN 0 340 60149 3 (Pb)
ISBN 0 340 23685 X (Wiley)

ISBN 0 340 67710 4 (Hb)
ISBN 0 340 23636 1 (Wiley)

Typeset in 10/12 Palatino and produced by Gray Publishing, Tunbridge Wells, Kent
Printed and bound in Great Britain by Bookcraft (Bath) Ltd, Midsomer Norton, Bath

CONTENTS

LIST OF TABLES

LIST OF FIGURES

PREFACE

In the few short years since the collapse of the Soviet Union in 1991 there has been great change in the geography of the region. Fifteen new states have appeared where one existed before. Others may yet emerge from the ashes of the former Soviet empire, for there are still disputes over control of territory within Russia and elsewhere in the former Soviet Union. It is the purpose of this book to serve as an introduction to the geography of this vast region of the world, a challenge of no small order given the fundamental nature of the political, economic and social changes which the disintegration of the Soviet system has precipitated. Two basic themes integrate the content of the book. The first is that an historical perspective is essential if the present is to be understood. It is only a short time since the Soviet socialist system collapsed under its own weight of internal contradictions and economic inefficiencies. It is necessary to understand the nature of this system, and what it has bequeathed to its successor states. Put simply, the Soviet past has left both an indelible impress on the landscape, and helped to shape human values. The second major integrating theme is that of decision-making. For more than seven decades the principles of Soviet socialism provided the framework for decision-making. Ideology shaped human values, and this legacy of the past has not yet been entirely erased. Privatization and market reform are now the bywords of the new economies across the post-Soviet scene. Not all states have moved at the same speed in reforming their economies in the years since independence, however. Thus, one of the tasks of this book is to provide some sense of both the tempo and direction of the changes which are taking place.

This is not a regional geography in the traditional sense of providing a detailed description and discussion of the major peoples and their lands. It is systematically organized, and is intended to serve as a point of departure for more study. In an introductory survey such as this it is not possible to cover all topics, nor cover those selected in equal depth for all post-Soviet states. Because Russia is the most important of all post-Soviet states in political, economic and military terms as well as the largest in terms of population and area, the explicit emphasis of the discussion in the ensuing chapters is on Russia. Comparing and contrasting developments in Russia with those across the whole post-Soviet region provides the basic frame of reference for understanding the nature of the changes underway. If what is presented here serves to stimulate further interest and enquiry, then its purpose will have been served.

The information provided in this book has been drawn from a variety of sources, but recent statistical handbooks, newspapers and the journal literature figure prominently. While there is original research which underpins some of the discussion herein, the vast majority of what follows is necessarily a distillation and synthesis of the research and writing of other people. The text has been intentionally kept free of references and footnotes. However, each chapter contains a list of recommended readings which the interested reader may wish to consult. Books only rather than journal articles are listed since these are more likely to be available in libraries than journals in this era of cut-backs and underfunding of such institutions. However, some journals are essential, and for this region *Post-Soviet Geography and Economics* and *The Current Digest of the Post-Soviet Press* are without doubt the most valuable. For those readers who can access the internet, there are now some extremely useful sources, two of which are noted below. Place names have changed across the post-Soviet scene as vestiges of the Soviet system have been purged and old place names restored or, less often, new ones more compatible with the present era introduced. In the following chapters new names are used. For less well-known places, the old name is included in brackets since most atlases are hard pressed to keep abreast of the changing place names.

Over the course of writing this book, parts of which are drawn from *The Soviet Scene* (1989), a great deal of assistance has been provided by a number of people. Colleagues in Moscow all have been most helpful, but special thanks are owing to Andrei Degtyarev and Vladimir Amelin. In Washington, Tim Heliniak of the World Bank and Stephen Rapawy of the US Bureau of Census provided me with much appreciated recent statistical and other materials. In Toronto, Mark Opgenorth assisted with library research. At Waterloo, Barry Levely drew the many maps, while Angela Cuthbert and Rose Turvey gave some timely help with diagrams and index respectively.

The Social Science and Humanities Research Council has provided financial assistance over the years, and for it I am most grateful.

Some basic geographies of the region

Baransky, N.N. 1956 *Economic Geography of the USSR*. Moscow, Foreign Language Publishing House.

Bater, J.H. 1989 *The Soviet Scene. A Geographical Perspective*. London, Edward Arnold.

Bradshaw, M. (ed) 1991 *The Soviet Union. A New Regional Geography?* London, Belhaven Press.

Cole, J.P. 1983 *Geography of the Soviet Union*. London, Butterworth.

Demko, G.J., Fuchs, R.J. (eds) 1974 *Geographical Perspective in the Soviet Union*. Columbus, Ohio State University Press.

Dewdney, J.C. 1970 *A Geography of the Soviet Union*, 2nd edn. Oxford, Pergamon.

Gregory, J.S. 1968 *A Geography of the Soviet Union*. London, Harrap.

Gregory, J.S. (ed) 1975 *Geography of the USSR: An Outline*. London, Collet's.

Hooson, D.J.M. 1966 *The Soviet Union*. London, University of London Press.

Howe, G.M. 1983 *The Soviet Union: A Geographical Survey*, 2nd edn. Plymouth, Macdonald and Evans.

Jorre, G. 1967 *The Soviet Union: The Land and its People*, 3rd edn. London, Longman.

Lydolph, P.E. 1990 *Geography of the USSR*, 5th edn. Elkhart Lake, WI, Misty Valley Publishers.

Mathieson, R.S. 1975 *The Soviet Union: An Economic Geography*. London, Heinemann.

Mellor, R.E.H. 1982 *The Soviet Union and its Geographical Problems*, 2nd edn. London, Macmillan.

Shaw, D.J.B. (ed) 1995 *The Post-Soviet Republics*. London, Longman.

Symons, L., et al. 1983 *The Soviet Union. A Systematic Geography*. Totawa, NJ, Barnes and Noble.

Some useful addresses on the internet

1. The WWW Virtual Library: Russian and East European Studies

   http://www.pitt.edu/~cjp/rees.html

   The most comprehensive, up to date listing of all related sites. All other sites, servers, lists and use-groups are listed at this site.

2. Russian News Navigator

   http://www.hooked.net/users/igort/news.html

   A home-page set up by the San Francisco Russian Community. It has an almost-complete list of news-sources on the net, something which the REES page does not have as a special category.

Some basic references for the region

Brown, A., Kaser, M., Smith, G.S. (eds) 1993 *The Cambridge Encyclopedia of Russia and the Former Soviet Union*. Cambridge, Cambridge University Press.

Dewdney, J.J. 1982 *USSR in Maps*. New York, Holmes & Meier.

Heleniak, T. (compiler) 1994 *Economic Geography of the Russian Federation*, 2 Vols. Washington, Country Department III, Europe and Central Asia Region, The World Bank.

Horak, S.M. 1985 *The Soviet Union and Eastern Europe: A Bibliographic Guide to Recommended Books for Small and Medium-Sized Libraries and School Media Centers*. Littleton, CO, Libraries Unlimited.

Mote, V. (ed) 1994 *An Industrial Atlas of the Soviet Successor States*. Houston, Industrial Information Resources Inc.

Sanchez, J. 1985 *Bibliography for Soviet Geography: with Special Reference to Cultural, Historical and Economic Geography*. Chicago, Council of Planning Librarians.

1

AN INTRODUCTION TO RUSSIA AND THE POST-SOVIET SCENE

'The Soviet Union was not defeated by the United States in the cold war, we defeated ourselves'
(A. Dobrynin, former Soviet Ambassador to the United States, in a television interview, October 1995)

The Russian revolution of 1917 heralded the rise of the world's first socialist state, the Union of Soviet Socialist Republics, or the Soviet Union. From the beginning, the new Soviet state actively promulgated the basic values of socialism in all corners of the globe. Socialism, it was contended, not only is morally superior to capitalism, it is a far more effective system than capitalism in improving the material circumstances of daily life and labour for all people. By the end of World War II, the Soviet Union had emerged as one of the world's leading military-industrial powers. Having forcibly pushed back its borders during the war, the Soviet Union controlled a territory that rivalled that of its predecessor state, the imperial Russian Empire. Soviet political hegemony now extended over all of Eastern Europe as well. Small wonder that western leaders in the immediate post-World War II years viewed with alarm the activities of the Soviet Union. While the theoretical benefits of Soviet socialism were aggressively promoted everywhere, they in fact little resembled the reality of daily life endured by the Soviet peoples.

From the beginning, the Soviet state entered into what might be described as a social contract with its citizens. In exchange for compliance with the system, people's housing, education and health care were to be provided by the government at little or no cost. In exchange for cradle to grave social security, employment for life and a guaranteed minimum standard of living, all citizens were expected to conform, and if necessary to accept hardship in their lives with the assurance that the material circumstances of daily life and labour for subsequent generations would improve. In the early years of Soviet power this promise of a better future was sufficient for many tens of millions of people, since what was then being provided by the state was in any event already better than life in the Russian Empire. However, during the oppressive years from the late 1920s until 1953 when Josef Stalin was the Soviet leader, any citizen who had the temerity, indeed courage, to express publicly dissatisfaction with the economic, social or political circumstances of Soviet life, ran the certain risk of retribution. Even criticism voiced to presumed friends could find its way to the omnipresent police. The consequence of such an indiscretion was customarily brutal. The state controlled virtually every aspect of the lives of the citizenry, from where they could live, work or relax. The harsh consequences of not complying with authority were not lost on the vast majority of state's citizenry.

Soviet claims of rapid economic growth in the early years were mostly valid. During the Stalin era, annual increases in gross national product were customarily high. And during the 1930s, substantial annual increases in economic growth were especially remarkable because most of the capitalist world was mired in a deep recession. But high rates of economic growth in the Soviet Union came at an exceedingly high price. For example, the forcible collectivization of agriculture initiated by Stalin in the late 1920s resulted in staggering losses of human life. Draconian measures introduced at the same time to facilitate rapid urban-industrialization dictated where one could live and work. Stalin died in 1953. As his successors tolerated more public criticism of economic and social conditions, ever greater demands were placed on the system. Improvements in the material conditions of daily life were forthcoming, but never fast enough to keep pace with rising expectations. Ultimately, the cost to the state of attempting to meet demands for an ever higher standard of living could not be sustained. By the late 1970s, annual rates of economic growth had dropped precipitously from those attained half a century earlier. An era of stagnation slowly enveloped the whole Soviet system. Productivity of labour was unacceptably low, partly because of technological backwardness in most sectors of the non-military economy, partly because of prevailing attitudes toward work itself. By the 1980s, economic growth to all intents and purposes had ground to a halt. The Union of Soviet Socialist Republics, the world's largest state in terms of territory, and with nearly 290 million citizens the third largest after China and India, had reached a crossroads. The Soviet system would either be reformed, or it would atrophy. That the Soviet Union, including its hegemony over Eastern Europe, simply collapsed is one of the most remarkable events of the twentieth century.

Problems were apparent early on. To make the Soviet system work even at a minimal level necessitated a growing deviation between ideal and reality, between theory and practice. What emerged was a system riven with inconsistency, if not hypocrisy. Instead of the interests of the collective taking precedence over individual self-interest, from the beginning privatism and privilege were nurtured by the state for an elite who, while usually unobtrusive, were nonetheless an essential part of the Soviet scene. For most people the revolution of 1917 portended a level of personal freedom beyond reach in autocratic, imperial Russia. Yet throughout the Soviet era there remained firmly in place various bureaucratic controls over the movement and activities of the citizenry. The Russian revolution resulted in the nationalization of all resources and the substitution of central planning by the government for the market place. Yet a second, or shadow, economy of private production, distribution and sale persisted simply because without it the system would never have functioned as well as it did. The Soviet Union had clearly demonstrated its technological capabilities with its manned space programme and production of some of the world's most sophisticated military hardware. Yet in the countryside millions of peasants worked with decidedly primitive farm equipment in an increasingly vain effort to wrest centrally planned levels of agricultural production from a generally begrudging soil. What originally was perceived as a system in which each member's basic needs would be met in abundance, was from the beginning a society in which most members' daily lives were shaped by the values of a 'deficit' culture. Access to scarce goods defined the elites of Soviet society at least as much as did income level. Socialist, egalitarian ideals were publicly promulgated on a daily basis. But these very ideals were remote from the reality of daily life and labour amongst the masses, because getting ahead usually involved working around the system. It is little wonder that so much seemed so overtly hypocritical at the end of the Soviet era. Paradox and contradiction were the essence of things Soviet.

Contradiction and paradox may be detected in differing degrees in all societies, of course. What made the Soviet Union so fascinating was in large part related to the fact that it began as a unique experiment in the management and development of society, an experiment in which the solutions to many practical tasks still had to be worked out in practice. Contradiction between ideal and reality was inevitable under such circumstances. That the system itself would ultimately collapse may have been inevitable, but few observers predicted its actual demise in 1991. What is emerging this time from the ashes of an empire is much less clear. In 1917, revolution brought a small, highly disciplined socialist political organization into power, and it soon extended its control over most of the territory and the diverse peoples which had been caught up in the expansionist Russian Empire. The Soviet Union is no more. Fifteen independent countries have emerged from its ashes, of which Russia is in all respects the largest and most important. Figure 1.1 locates each state, and the capital city of each. But what is replacing Soviet socialism as the basis of these newly emergent political economies? What is the nature, indeed the direction of the changes being introduced?

Market reform has been widely touted as the panacea for the problems associated with the Soviet system, but public proclamation of new ideas is always simpler than altering basic values and attitudes which shape decisions on a day-to-day basis. Thus, dismantling the apparatus for central planning is one thing, accepting the consequences of introducing a market-based economy is quite another. After seven decades of socialist ideology and practice, old habits in all parts of the post-Soviet scene are proving extremely hard to break, even when there is the political will to change. And change is not always welcome, especially when it makes an already hard life more unpredictable and uncertain. Revolutionary beginnings notwithstanding, the Soviet system and Soviet society had a distinctly conservative hue. Notwithstanding the collapse of the Soviet Union, prompted by popularly supported democratic movements demanding change, this legacy of the past conditions much of what is happening across the post-Soviet scene today.

Politics reflect popular sentiments. And popular sentiments in the mid-1990s post-Soviet scene rarely provide clear majority support for fundamental change. To be sure, in most states citizens now have a wide array of political parties and blocs to choose from when casting their votes, a stark contrast to the days when voting for the only party of choice, the Communist Party of the Soviet Union, was not just a responsibility but a legal requirement. Indeed, in some states there are so many parties that exercising an informed choice amongst platforms that run from one extreme of the political spectrum to the other is far from simple. Reformist parties and blocs are defined by, amongst other things, a commitment to dismantle the old command economy structures. But introducing a market economy into a Soviet-type command economy inevitably leads to profound social change, not least of which is growing income disparity, widespread hardship, and the new reality of unemployment. In many post-Soviet states, the public has come to appreciate that market reform is not simple, that it is not an easy and fast road to a higher standard of living, views that were commonplace just a few years ago. Thus, reformist political parties when in power have had to adjust the timetable for introducing a market-based economic system. In some post-Soviet states, reformist parties elected to govern in the first flush of grass-roots democratic movements have been ousted in subsequent elections. Invariably, they have been replaced by more conservative parties, led quite often by former Communists whose mandate usually is to slow down the tempo of change. In short, across the former Soviet Union democratic elections do not always result in clear mandates for fundamental change. Economic and political reform proceeds at different tempos, and in some cases in different directions, in the post-Soviet scene.

Political platforms and economic reform aside, political changes have profoundly impacted the lives of millions of former Soviet citizens. The very nature of the Soviet federal political-administrative system, which recognized the rights and privileges of national minorities in terms of a hierarchy of territorial units (15 Soviet

FIGURE I.1 *The post-Soviet scene*
(*Source*: Adapted from *Atlas SSSR* (Moscow: Glavnoye Upravleniye Geodezii i Kartografii pri Sovete Ministov SSSR, 1969), 2–3)

Socialist Republics, 20 Autonomous Republics, eight Autonomous Oblasts, and ten Autonomous Okrugs, see Fig. 2.5, p. 47) within the Union of Soviet Socialist Republics, made the transition from being part of the Soviet empire to post-Soviet political independence relatively straightforward. The map of Soviet-era political administrative territories for national minorities infers a measure of tidiness in distribution, but on the ground reality was seldom so neat. There were few political-administrative regions in which there were not substantial numbers of other minority groups. Indeed, in two of the 15 Soviet Socialist Republics, Kazakhstan and Kyrgyzstan, the titular group in fact was as an absolute minority. In some post-Soviet independent states, 'indigenization', or replacement of 'foreigners' by members of the titular nationality in all administrative and economic structures, is being systematically pursued as official policy of the new governments. This has resulted in substantial out-migration of non-indigenous peoples, most notably Russians, who now feel the effects of discriminatory policies.

Refugees and ethnic diasporas add immeasurably to the collective hardship which many peoples across the post-Soviet scene now endure. For example, for Russians, who traditionally dominated the decision-making process throughout the former Soviet Union, and the Russian Empire before it, whose language always assured them easy passage since it was the *lingua franca* of all important discussions, and who willingly relocated to distant regions in the quest for personal advancement and a better standard of living, for them the emergence of politically independent countries has created an entirely new situation. Previously they were part of the dominant culture, politically and economically, even though ethnic Russians had been reduced to a bare majority of the total Soviet population on the eve of the system's collapse in 1991.

Russia's population is the largest by far of all the newly independent post-Soviet states. And ethnic Russians now account for more than 80 per cent of the total population of 148 million. But about 25 million Russians were living outside the Russian Republic at the time the Soviet Union imploded. Already close to two million Russians have fled regions where their families may have settled generations ago, but which are now parts of new states in which their continuing presence is at best uncomfortable, or possibly at risk because of civil war. This situation is scarcely unique to Russians, however. Indeed, some members of virtually all 111 national minority groups identified in the 1989 Soviet census are in a broadly similar position, for in addition to 25 million Russians, another 18 million people were left outside their home territory at the time the Soviet Union collapsed. Many had departed their home regions in search of better opportunities elsewhere, and like Russians now living in the so-called 'near abroad', face similar discomfort, or risk to their personal safety. If this were not problem enough, hundreds of thousands of people have fled their ancestral homes as inter-ethnic or territorial disputes have boiled over into outright civil war. During the Stalin era, several million ethnic German, Tatar, and other national minority members were deemed politically suspect for one reason or another and forcibly relocated to distant, often harsh, frontiers where they have settled and adapted to local conditions. Official Soviet policy of *sliyaniye* encouraged the so-called 'merging' of ethnic minorities. Inter-marriage therefore adds yet another dimension to the minority group problem in the post-Soviet scene, one which means that simply pulling up and returning 'home' is hardly simple. There are still other legacies of the former Soviet Union which constitute different, but equally formidable, challenges to countries struggling with how best to deal with the present. Principal among these is the legacy of the Soviet centrally planned economy.

Soviet central planning fostered scale economies and regional economic inter-dependencies. Thus, a mere handful of huge production facilities would supply particular products for the whole country. Now essential supplies of manufactured components, semi-processed inputs, or basic industrial raw material supplies for these plants must now be obtained from foreign countries. Literally thousands of such specialized production facilities still exist, but without the sources of supply or markets provided by Soviet central planning. Restructuring such a system of industrial production is in itself a major challenge for the leadership of the 15 newly independent states. But the task is further complicated by the appearance of new, non-convertible, currencies which have replaced everywhere the former Soviet rouble, varying rates of inflation, and enterprise, if not state, bankruptcy. Given these problems, the urgent need to do something about another legacy of the Soviet era – extreme environmental degradation – goes largely unanswered. From the lingering effects of the nuclear meltdown of Chernobyl' in Ukraine in 1986 which devastated a vast area, to the drying up of the Aral Sea and resultant climate change in much of Middle Asia, to pollution of the rivers and the atmosphere in industrial centres everywhere, serious problems abound in virtually all parts of the post-Soviet scene. They have perceptible consequences in terms of public health. Since many environmental degradation issues are now trans-national in scope, even if one country could do something it is often beyond its jurisdiction to implement ameliorative programmes.

THE SCOPE OF OUR ENQUIRY

It is the purpose of this book to introduce students to the geography of the post-Soviet scene. The challenge is to provide a coherent overview of the new geography which the collapse of the Soviet Union has precipitated. In place of a single political entity there are now 15 independent and internationally recognized states, each of which was previously a Republic in the Union (Fig. 1.1). Russia clearly dominates in size of territory and population, but as the data in Table 1.1 indicate there are several other post-Soviet states with substantial populations and territories. Even the smallest post-Soviet state, Estonia, is larger in population and area than many of the world's nation states.

Russia is not only the largest of the post-Soviet states, it has assumed the mantle of successor state to the Soviet Union in the international economic and political arenas. It is Russia which commands most attention amongst western business interests, in the world's capital cities, and in the United Nations. And, of course, Russia's military capabilities remain formidable. Indeed, it is to Russia that some new post-Soviet states have turned in time of trouble. Russian military forces have been dispatched to a number of former Soviet regions to quell civil war or to protect international borders. In the new political and economic reality which is arising from the ashes of the former Soviet empire, Russia is clearly pre-eminent, and therefore the discussion in this book concentrates on Russia. To facilitate a comparative perspective on developments across the post-Soviet scene, much of the material presented in the ensuing chapters of this book is organized in such a manner as to highlight the situation in Russia in relation to what are labelled 'western' and 'southern' tier states (Table 1.1). The western tier states include the Baltic states of Estonia, Latvia and Lithuania, plus Ukraine, Belarus' and Moldova. The southern tier group includes the Caucasian states, Georgia, Armenia and Azerbaijan, and the Middle Asian states, Turkmenistan, Uzbekistan, Tajikistan, Kyrgyzstan, and Kazakhstan.

Aside from a focus on regional issues, which constitutes the cornerstone of geographic enquiry, two overriding themes figure prominently in the organization and content of what follows. The first is that of historical continuity and change. There is a conscious attempt in the ensuing chapters to convey an appreciation of the past, for without such a perspective there is little chance of even coming close to understanding the present. The second integrating theme is that of political structures and decision-making. The transition from the Soviet political and economic systems, which were characterized by totalitarianism, state ownership of all resources and central planning, to systems based on democratic processes and institutions, private property and a market economy, is fraught with difficulty. Reconciling rhetoric with reality in assessing change was a challenge in the Soviet era; it remains a significant one in the post-Soviet era as

TABLE 1.1 *Post-Soviet states population and area – 1995*

	Population (millions)			Area	Population
	Total	Urban	Rural	(1000 sq. km.)	per sq. km.
Russia	148.3	108.3	40.0	17075.4	8.7
WESTERN TIER					
Ukraine	51.7	35.1	16.6	603.7	85.7
Belarus'	10.4	7.1	3.3	207.6	49.8
Estonia	1.5	0.8	0.7	45.2	33.2
Latvia	2.5	1.7	0.8	64.6	38.7
Lithuania	3.7	2.5	1.2	65.3	56.9
Moldova	4.3	2.0	2.3	33.7	129.0
SOUTHERN TIER					
Georgia	5.4	3.0	2.4	69.7	77.6
Armenia	3.8	2.6	1.2	29.8	126.0
Azerbaijan	7.5	4.0	3.5	86.6	86.5
Kazakhstan	16.7	9.3	7.4	2717.3	6.1
Kyrgyzstan	4.5	1.6	2.9	199.9	22.4
Tajikistan	5.8	1.6	4.2	143.1	40.4
Uzbekistan	22.6	8.7	13.8	447.4	50.4
Turkmenistan	4.5	2.0	2.5	488.1	9.1

Sources: Demograficheskiy Yezhegodnik 1994 (Moscow: Statkomitet SNG, 1995) 11; *Demographic Yearbook, 1994* (Vil'nius, Lithuanian Department of Statistics, 1995), 11–12; *Macroeconomic Indicators of Latvia* (Riga: Central Statistical Bureau of Latvia, 1995), 6

well. Two other themes are developed in several of the ensuing chapters. The role of minority ethnic groups and territories is one. This highly charged issue has already precipitated bloody civil war over who will control territory and govern. The other theme which is developed in several places in this book is that of environmental degradation. Soviet development priorities for too long gave precedence to short-term benefits. Degradation, if not destruction, of the environment was a commonplace consequence. This legacy of the Soviet era afflicts in some degree all parts of the post-Soviet scene.

In order to provide a benchmark for assessing the change in the post-Soviet era, the results of more than seven decades of Soviet development must be established. And this in turn requires some appreciation of the Russian Empire before the revolution. The Russian revolution of 1917 obviously produced a major transformation of society, but it should not be seen as a hard and fast watershed. The Bolshevik-led revolution certainly gave rise to new political, social and economic structures, but these were based on the traditional Russian traits of collectivism and authoritarianism. Thus, the next chapter will first sketch in broad brush fashion some of the principal features of the Russian Empire's agrarian system, its urban-industrial economy and the currents of social and political change in the late imperial era which culminated in the Russian revolution of 1917. Secondly, the Soviet political and administrative structure which the revolution spawned will be briefly described. Thirdly, some features of the official Soviet social structure will be outlined, and then contrasted with the reality of the social system and values which defined the Soviet era. Finally, the principal events leading up to the collapse of the Soviet system will be briefly described.

Chapter 3 describes the essential features of the Soviet planned economy, the nature of the decision-making process and regional development priorities. The Soviet approach to economic development traditionally generated high rates of growth, but as the economy both modernized and diversified central planning foundered, labour productivity flagged and regional development priorities shifted. Part of the reality of the Soviet economic system which evolved was periodic and unpredictable shortage. It is in this context that the aforementioned second, or shadow, economy came into play, and its role will be described as well. To maintain momentum, there were periodic campaigns to reform the economy, but right up until the end of the Soviet era always within the confines of a revamped socialist system. Nonetheless, the final years were characterized by the most far reaching economic reform initiatives in the history of the state. Principally associated with then President Mikhail Gorbachev, they proved too little, too late. The Gorbachev reform initiatives provoked debates over fundamental issues related to private property, debates which resonate still right across the post-Soviet scene. The collapse of the Soviet system precipitated fundamental economic change. Market reform is the catch phrase everywhere, although progress in implementation is not everywhere the same. The basic features of the post-Soviet market reform process will be highlighted in the final section of this chapter.

Chapter 4 examines the geography of population during both the Soviet and post-Soviet eras. Events which occurred decades ago have a perceptible impact on the age/sex structure of the population today. In this respect the legacy of the Soviet period in terms of population losses resulting from world war, civil war, and draconian domestic policies remains indelibly impressed on the basic population structure. While the Soviet era produced its own population problems for planners concerned with future labour supply, the collapse of the system and the resultant hardship and turmoil have brought their own very serious demographic consequences. Across most of the post-Soviet scene, birth rates have dropped, death rates have escalated, and in a number of countries the customary population growth owing to a surplus of births over deaths has evaporated. The concept of demographic momentum simply means that the consequences of these changes in vital statistics will be part of the population profile of the post-Soviet scene for decades to come. Indeed, not a few post-Soviet states have actually experienced an absolute decrease in their total population since independence in 1991. And in most instances this is not simply a function of net out-migration, although this phenomenon in fact has exacerbated the demographic crisis in some countries. The new geo-political realties have precipitated substantial migration of people across the borders of the new states. Most of this migration is forced, as will be described in Chapter 4.

That most people live in cities is a commonplace feature in developed countries across the world. The urbanization of the Soviet population was controlled as will be discussed in Chapter 5. Not only was migration to the city restricted, but the Soviet city itself was supposed to be planned and built according to socialist principles. The idea that the built environment of the city could help to shape human values, could play a part in shaping human behaviour patterns, played a major role in the design and development of the Soviet socialist city. But as with so much else in the Soviet era, theory and practice were customarily rather far apart. In Chapter 5, the urban growth process, the evolution of the city, and the relationship between ideal and reality in urban development will provide a benchmark for assessing the impact of market reform on the city in the post-Soviet era. Privatization plays a central role in the market reform process, and perhaps nowhere else are the consequences so palpable as in the city. After examining the changes in the tempo of the urban growth process, and some of the most important privatization initiatives such as the privatization of housing across the post-Soviet scene, the discussion in the final section of Chapter 5 will turn to a more detailed examination of urban developments in Russia in general, and in Moscow in particular.

Some of the new post-Soviet states have significant natural resource endowments, and development of them is often central to economic growth. But the challenge is not simply one of finding and exploiting natural resources, it includes jettisoning resource management practices of the Soviet era. This legacy of the Soviet period will be briefly outlined in Chapter 6. Since independence and the advent of market reform, fundamental changes in land and natural resource ownership have been initiated in most new states. Management of resources is also changing so as to comply with the requirements of the marketplace. But progress is variable. There remains still great resistance to the idea that individuals, or companies, should be able to own, buy and sell land and natural resources. Such a concept is fundamental to the market reform process in the opinion of many people in government. But in some new states this is not always a widely shared opinion. In the second part of Chapter 6, how the privatization of land and resource ownership is being handled across the post-Soviet scene will be examined.

Chapter 7 explores the management of the land for food production. Early in the Soviet period agriculture was collectivized with colossal loss of life, and lasting negative impact on the state's ability to produce sufficient food to meet domestic demand. By the early 1960s, the Soviet Union was obliged to import grain for the first time in its history, a dependence which continued right down to the demise of the system, indeed beyond for countries like Russia. The organizational structure which was imposed on the agricultural sector was wasteful and inefficient, and was recognized to be so in comparison with the more or less market-oriented personal sector which was tolerated, but on a small scale, during the Soviet period. Since independence there have been numerous attempts to break the collectivized structure inherited from the Soviet era, and these are discussed in the second part of Chapter 7. In some post-Soviet states progress in privatizing agriculture has been rapid. In others it has hardly begun, notwithstanding the changes in the ownership of farming units which have been introduced. The ability of each new state to meet domestic food demand is important. Few are yet doing so very successfully, as will be described in the final section of this chapter.

Chapter 8 examines the energy scene. Patterns of energy production, distribution and consumption are presently in a state of flux. But many of the post-Soviet states have enormous energy reserves as yet untapped, sufficient inducement for foreign investment on the part of multi-national corporations. In the Soviet period, energy resource development was significant to be sure, but invariably lagged behind western industrial countries in terms of the relative importance of particular sources of energy in the national fuel balance. In order to have some appreciation of the evolution of Soviet energy policies and their consequences for the present, the first part of this chapter will examine the geography of Soviet energy resource development. The discussion then turns to a detailed examination of energy resource developments in the 1990s. As not all post-Soviet states are equally endowed with energy resources, the discussion will focus on Russia which possesses the richest energy resource base. In a geo-political sense, Russia also figures prominently in the energy resource development strategies of other states seeking to export their energy surplus.

Chapter 9 examines the past and present situation in terms of the relationship between industrial raw materials and markets. Much of the legacy of the Soviet era in terms of industrial infrastructure, transportation infrastructure, and movements of raw materials and finished products is simply a burden for post-Soviet economies struggling to adapt to a market economy. The attempts to restructure the industrial sector has been especially difficult, and new patterns of industrial raw material flows are emerging. Not only were specific factories designed to supply large parts of the total domestic demand for consumer goods during the Soviet period, there were whole cities created to serve the needs of the particular industrial sectors, both civilian and military. Adapting parts of the military industrial complex to the civilian market is challenging in the extreme.

The fundamental issue of the quality of life in states in transition from being part of a single planned economy to independent market economies is the subject of Chapter 10. Not only are unemployment and inflation impacting on substantial numbers of people who are now living in poverty, the social safety net which had provided Soviet citizens at least a basic standard of living is now no longer guaranteed. Post-Soviet states cannot afford the systems which had developed during the Soviet era, even if they wanted to maintain them. Security of life is also a major problem for many people. Political instability, civil war or threatened civil war, and discrimination of one kind or another have turned the lives of hundreds of thousands of people upside down since independence. Huge numbers are forced migrants, or refugees. For those spared such trauma, life is still far from easy. Everywhere life expectancies are in general decline. Partly this is a function of economic collapse. But in a great many places environmental pollution figures prominently in shortening life expectancy. All told, the current situation in quality of life terms, however defined, is less than auspicious for the average citizen.

The final chapter focuses on some aspects of political geography which characterize Russia and the post-Soviet scene. Particular attention is accorded nationalism and core-periphery relations in Russia, both of which have important implications for the emerging geography of this vast and fascinating region.

The physical setting

The post-Soviet scene comprises new nation states, but the territories over which they have jurisdiction were integral parts of two empires, Russian and Soviet. If this book is to convey something of a sense of place, it is clearly essential to have an appreciation of the physical setting which bears the impress of Soviet development, and Russian before it, and which is now beginning to bear mute testimony to the ideas and actions of the post-Soviet era market reform. In order to provide a general frame of reference, the basic features of the physical environment of all the former Soviet territory are described in Figs 1.2–1.5. The physical environment of Russia and the western and southern tier states will be described separately. In subsequent chapters, where some aspect of the physical environment figures prominently in the discussion of a particular post-Soviet state, it will be dealt with in greater detail. For present purposes, to facilitate comprehension of the broad similarities and differences in the physical environment of post-Soviet states, an outline of their borders from Fig. 1.1 has been superimposed on Figs 1.2–1.5.

Perhaps the most striking feature of Fig. 1.2 is the northerliness of most post-Soviet states. While Canada is also a northerly country, 90 per cent of its 28 million people live within an hour's drive of the border with the United States. Figure 1.2 also conveys clearly the enormous size of the territory of the former Soviet Union. This huge size has direct bearing on the climatic regimes of the successor states, which in most cases are primarily continental. Winters tend to be long and not infrequently extremely cold; summers are warm and all too short. Russia not only is the largest in area and population of all post-Soviet states, it is also the most northerly. Thus, its people have the most to contend with in terms of adapting to, or overcoming, distance and a harsh climate.

The Russian realm

The sheer size of Russia is hard to comprehend. It stretches more than 10,000 kilometres from west to east, embracing more than eleven time zones. It is more than 2500 kilometres from its most northerly point to its most southerly. To travel by train from St Petersburg in the northwest to Vladivostok in the far east, for example, takes a full week. For many people such a transcontinental rail journey holds enormous appeal. But few likely would want to make the journey more than once. The customary fare offered in the dining car usually pales long before the journey's end, and this combined with the seemingly permanent line up to get in, and the prices actually charged once seated, are sufficient reasons for the tourist to adopt the locals' habit of taking as much food on board at the outset as possible. Fortunately for the novice traveller, with some practice it is possible to purchase supplementary foodstuffs from peasant vendors on the station platforms at scheduled stops. These hasty transactions may be the first encounter with trade in the era of privatization. Peasants, they will soon discern, have little to learn from their city cousins in terms of quickly

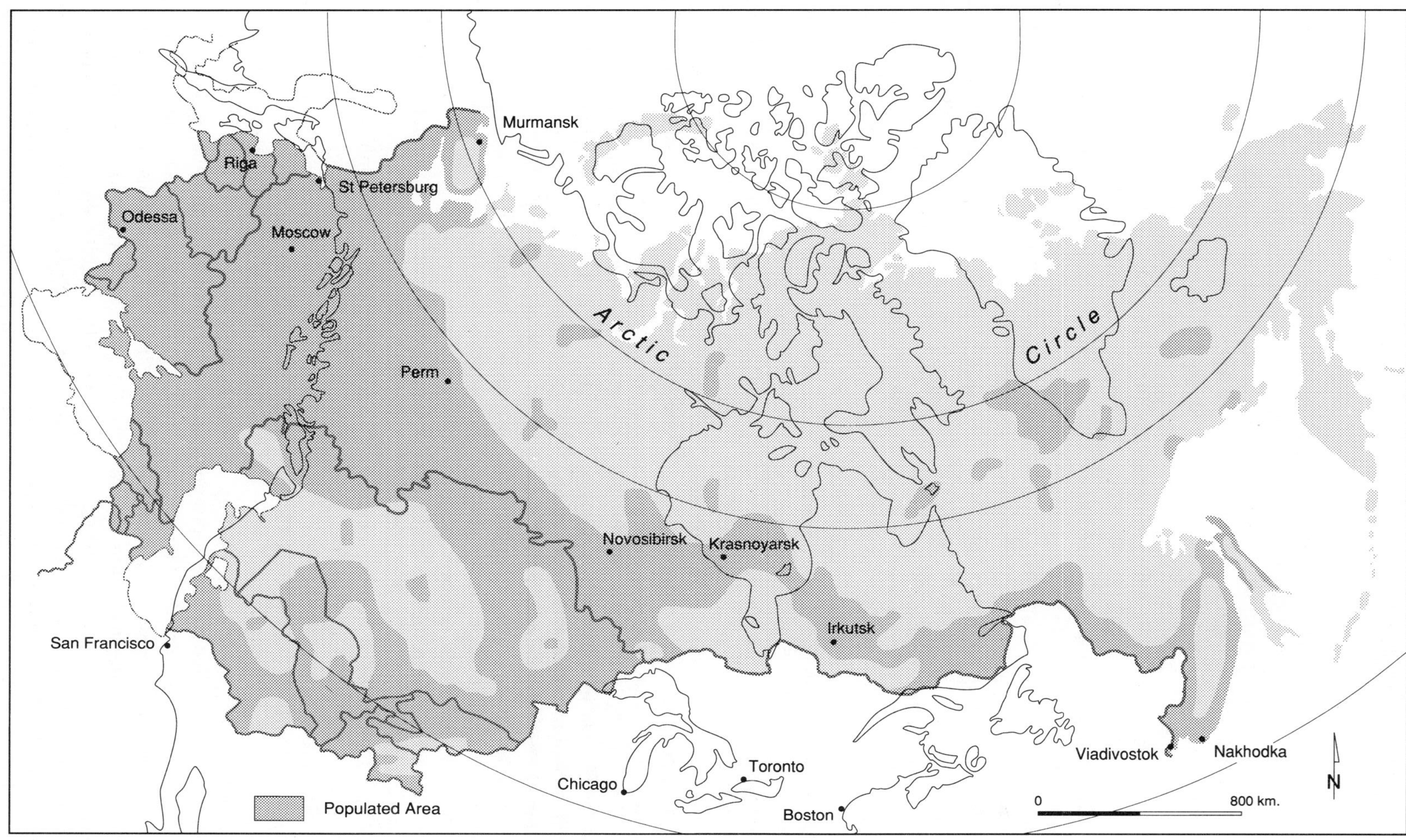

FIGURE 1.2 *Relative location of post-Soviet states and North America*
(*Source*: Adapted from A. Brown *et al.* (eds), *The Cambridge Encyclopedia of Russia and the Soviet Union* (Cambridge: CUP, 1982), 339)

gauging what price the market will bear. Given the time required to travel across Russia by train, the greatest volume of trans-continental passenger traffic is handled by Aeroflot, the domestic air service. In the Soviet era, domestic travel for foreigners was decidedly inexpensive. It is no longer such a bargain. Indeed, even getting to and from airports is notoriously expensive nowadays, since the drivers of privatized taxis often demand, and get, a substantial portion of what would be a government employee's monthly salary from a single fare! Once on the plane, problems in meeting regular maintenance schedules, shortages of spare parts, and the general unease over airline safety standards in the post-Soviet era, are all best put out of one's mind. Whether people or goods are moved by air, land or water transport, the huge distances involved impose costs which ultimately must be absorbed by the state. However, for Russia it would be short-sighted to think of its vast territory purely in terms of cost. Within its lengthy borders are found in some quantity virtually all resources required by a modern industrial economy.

High latitude means that in Russia most ports are ice-bound during the long winter. Murmansk on the Kola peninsula in the north is a major exception. It benefits from the tail end of the warm Gulf Stream and is open year round. Some ports, such as Russia's western enclave of Kaliningrad on the Baltic Sea, or Vladivostok on the Sea of Japan in the far east, can be kept open by ice-breakers. But if year round access to the open seas is a problem for Russia, it at least has access. As Fig. 1.1 indicates, a number of post-Soviet states have no access to the open seas whatsoever. This problem of limited access to the open sea stands in marked contrast to the United States or, for that matter, Canada.

Figure 1.3 describes in general terms the physical geography of Russia. What this map does not show is the considerable uniformity of landscape across Russia. To be sure, there is considerable variety of natural landscapes including tundra, tayga, steppe, alpine, and semi-desert, some bearing ample evidence of human intervention, others hardly at all. But owing to the basic geological structure, and the huge distances covered, for those travelling across Russia by rail the landscape visible through the lace-curtained window of a soft class compartment is characterized by one feature, uniformity. A bed of hard crystalline rock of igneous and metamorphic origin underlies much of the country. In European Russia, that is, Russia west of the Ural mountains, glacially-eroded outcrops occur most notably in the northwest near the border with Finland. Over the rest of European Russia, the bed of crystalline rock is buried under several score metres of sedimentary deposits, the physical forms of which bear mute testimony to the last continental ice sheet. To the south of the ice sheet's furthest extent, fine dustlike yellowish material, called loess, was carried by the winds off the moraines and deposited to considerable depth, further muffling ground relief. Where these loess deposits occur in Russia, they provide a rich base for agriculture. Less beneficial are those glacial deposits which have disrupted drainage. Where this occurs, extensive swamp and bog and scrub forest frequently dominate the landscape. In general, the European Russian landscape is mostly gently rolling. However, throughout this broad region, where rivers have eroded sedimentary deposits and loess, the results on occasion can be spectacular. But for most journeys, it is the uniformity of landform which is the dominant impression. Indeed, after two days on an eastbound train out of St Petersburg, the string of grey villages, scrub forest, patches of agriculture and rolling vistas seems quite without end. But the trip across Russia has only just begun.

Proceeding east, the first physiographic exception to this vast expanse of plain is the Ural mountain range. Much eroded by the succession of ice sheets during the Pleistocene period, rarely do the Urals exceed 2000 metres in elevation. In most places they are only 600–700 metres above sea level, and in age and general appearance are similar to the Appalachians of the eastern United States. In short, the Urals do not constitute a major barrier to transport. Numerous broad valleys in the middle of the range facilitate links with Siberia. Passengers on the trans-Siberian railroad follow a northerly route across the Urals, reaching a maximum elevation of 400 metres

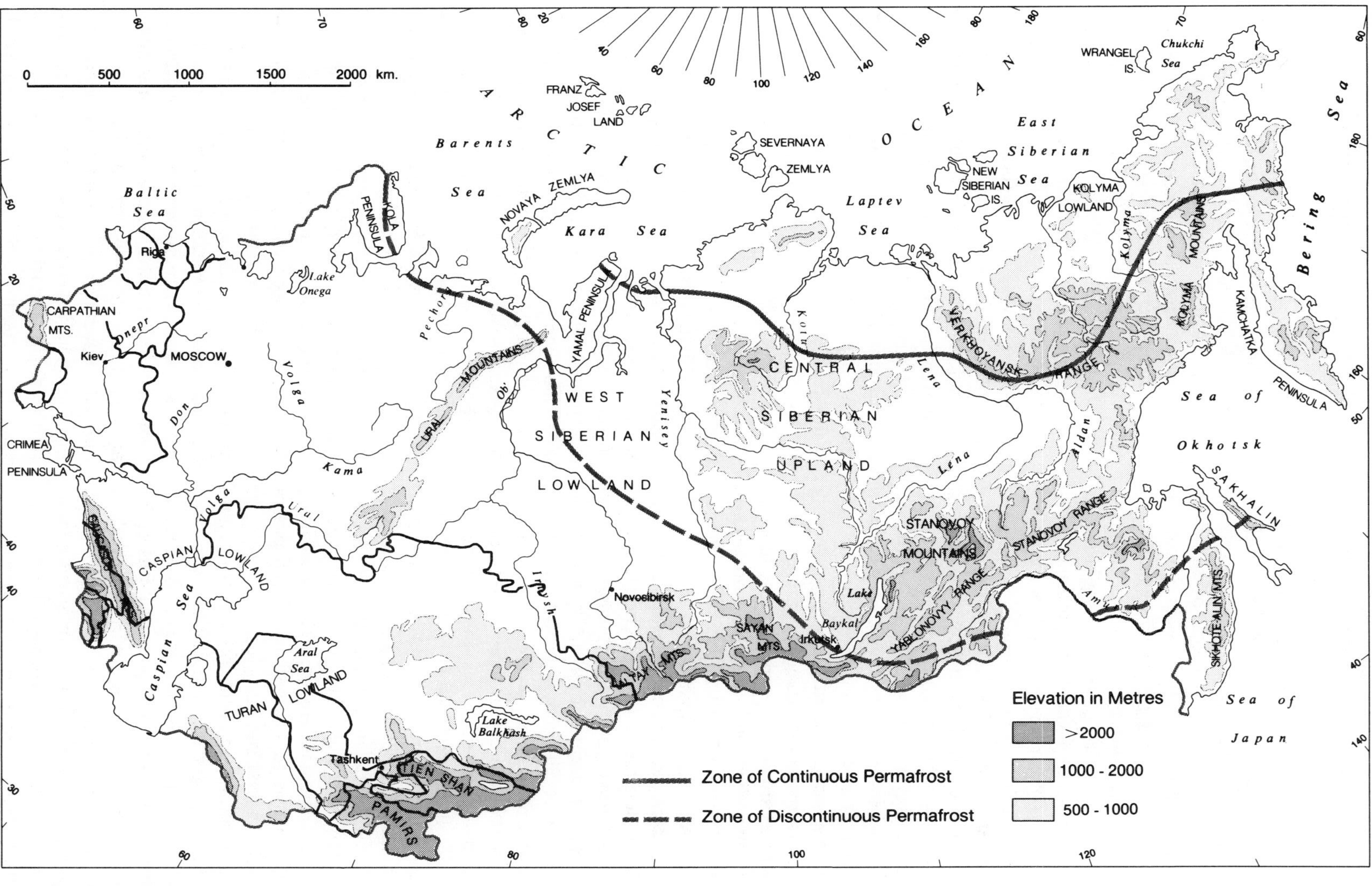

FIGURE 1.3 *Physical geography*
(*Source*: Adapted from P.E. Lydolph, *Geography of the USSR* (New York: Wiley, 1970, 2nd edn), 12)

above sea level, before descending the eastern flank of the range to the city of Yekaterinburg. After watching kilometre after kilometre of gently rolling plain and plateau, dissected by ravine and valley, but all dressed in similar vegetation, little relief, figuratively or literally, is occasioned by the journey across the Urals. For the next thousand kilometres there is more of the same. To the east of the Ural mountains the underlying bed of hard crystalline rock continues, giving physical form to the West Siberian Lowland and the Central Siberian Upland. To the south, it finds expression in the plateaus and lowlands of Middle Asia (Fig. 1.3).

Leaving the Urals behind, the trans-Siberian traveller crosses the Tobol' river and passes on through the Ishim steppe before reaching the city of Omsk. The grasslands of the steppe are only periodically disturbed by stands of pine or birch, or by marshes supporting aspen and willow – a landscape, like that before it, always dominated by the sky. To the north lies the West Siberian Lowland. This vast expanse of sedimentary strata laid down during Tertiary and Quaternary times when the Arctic Ocean reached deep into the middle of the Eurasian land mass, is remarkably flat. Poorly drained by the Ob' and Irtysh rivers, the West Siberian Lowland presents a panorama of forest, bog, swamp and meadow. Only in the post-World War II period was its vast potential as a source of oil and natural gas recognized. It was developed rapidly by the Soviets, with customary disregard for maximizing output in the long term, and the environment. Production has fallen significantly in recent years, and much ecological disruption remains to be mitigated. After crossing the Irtysh river at the city of Omsk, the trans-Siberian railroad passes through the Baraba Steppe, which because of its multitude of swamps and shallow lakes, occasionally fresh, more often brackish, is visually more of a waterscape than a landscape. Eventually Novosibirsk is reached, and once past this major city of 1.4 million people the southwestern extremity of the rugged Siberian Upland, and then the tayga itself, slowly come into view. Evidence of attempts to wrest a living from cultivating the soil diminishes accordingly.

The Central Siberian Upland covers most of the territory between the Yenisey and Lena rivers (Fig. 1.3). Plateaus of about 700 metres elevation, heavily dissected by several major drainage systems, are the principal elements in the landscape. Only to the east of the Lena river, and along Russia's southern border with Mongolia and China, do mountains prevail. As first Krasnoyarsk, and then Irkutsk, are left behind, the terrain becomes rougher. But it is only east of Lake Baykal that one is aware of being in a mountainous region, an event perhaps all the more striking by virtue of having waited so long. Once over the Yablonovyy range, which lies a few hundred kilometres east of Lake Baykal, all rivers drain to the Pacific. However, from Chita, the first major city on this eastern flank of the watershed, it is still 3093 kilometres to Vladivostok. Put another way, the train trip will last another two and a half days! Eventually the tayga is left behind and deciduous forest of the eastern region takes over. By now the enormous size of Russia will have registered on even the most unobservant of travellers. For those who began the trip in St Petersburg and journeyed straight through, disembarking at the eastern terminus of the trans-Siberian railroad is usually a most welcome event.

The combination of limited relief, huge territorial expanse and high latitude, as noted earlier, clearly influences the climatic regime of Russia. Dominant air streams are westerly. While weather systems move across European Russia largely unimpeded, vast regions further east are simply too distant to benefit much from the moderate air masses originating over the Gulf Stream. Few countries have such a uniform climate as Russia. No place is excused frost and snow during the winter months. Indeed, most of the country experiences average temperatures of 0 degrees Celsius, or lower, for at least six months. Arctic air masses impact virtually all regions of Russia at some time during the long winter. Over Siberia and the far eastern reaches of Russia, a pronounced high-pressure system develops during the winter, which draws cold Arctic air from the north into the maritime zone of the far east, thereby further inhibiting any moderating influence of the Pacific Ocean and ice-free eastern seas. In the centre of this very intense high-pressure system clear skies and calm air prevail. This facilitates heat loss and record low temperatures result. Conditions in Verkhoyansk, a mining centre in the middle of this high-pressure area, are often cited to illustrate the severity of the winter. Average January temperatures are usually in the order of –50 degrees Celsius. Winter temperatures gradually moderate from northeast to southwest. And so too does the duration of winter. It lasts ten or more months in the extreme northeast. Another measure of the severity of the climate there is that continuous permafrost is common (Fig. 1.3). However, about two thirds of the whole country is influenced to some degree by permafrost conditions. For the substantial number of people living on the northern margin of the ecumene, or permanently settled zone, continentality means extreme cold, while the high latitude means there is little daylight during the long winter months. On the Russian shores of the Black sea, winters are benign in comparison with the northeast and last perhaps three months, but the risk of frost is ever present. Winter is the dry season in Russia, and it seldom brings more than an average of 13 centimetres of precipitation.

Summer comes quickly over most of Russia, spring being a brief spell of dirty snow and much mud. Historically, spring was the time of *bezdorozh'ye* or roadlessness. Indeed, as many rural roads are still unpaved they remain largely impassable for a period in the spring, and sometimes again during heavy autumn rains. Sun or frost is still required to transform an impenetrable sea of mud into a more or less negotiable maze of ruts. During the summer it is generally warm and frequently very hot. Even in northeast Siberia where January record low temperatures are recorded, the short summer offers a reasonable number of days in the upper 20 degree Celsius range (and on occasion higher still). As the land mass warms, a low-pressure system replaces the winter's dominant high over Siberia. This situation helps to draw moist air masses off the Atlantic into European Russia. But Russia's vast west to east expanse limits the moderating influence of these weather systems. Therefore, the western region receives the bulk of the rainfall (Fig. 1.4). Summer is the wet season in most of Russia. In European Russia and the western parts of Siberia, there are usually 25 centimetres of precipitation, much of it produced by thunderstorms. The Chinese monsoon brings heavy rains to the southeastern corner of the Russian far east, especially during August and September, and in this region precipitation is often as much as 60 centimetres over the summer. Although summer is the wet season, for agriculture rainfall is rarely abundant. Moreover, precipitation amounts during the summer months are highly variable. Thus, as will be discussed in more detail in a subsequent chapter, few areas have anything like an ideal balance between adequate heat and moisture for growing crops.

The distribution of natural vegetation is portrayed by Fig. 1.5, and like climatic zones it is broadly latitudinal. The tayga, or boreal forest, covers most of Russia and constitutes an enormous reserve of softwood. Larch and spruce prevail amongst the conifers, the former on the poorly drained podzolic soils of Siberia and the far east, the latter in European Russia. Forestry operations, however, are largely confined to the more accessible western and southern regions. To the north, climatic severity increases until tree growth can no longer be sustained and tundra dominates the landscape. Human occupancy there is mostly tied to mineral exploitation. Compared to the northern parts of other polar states, such as Canada, Russia has a far more extensive permanent claim. Noril'sk, a city of 163,000, is the largest settlement north of the Arctic Circle anywhere, but is just one of a number of major northern Russian cities which are the legacy of Soviet regional development strategy. To the west and south the tayga gives way to a zone of mixed forest. Mixed forest in turn blends into the wooded steppe and steppe. Each of these ecosystems has been substantially altered over the centuries. For example, more than half of the mixed forest has been axed. Since their grey-brown podzolic soils offered some potential for agriculture, there was considerable clearing of the mixed forest long before the end of the nine-

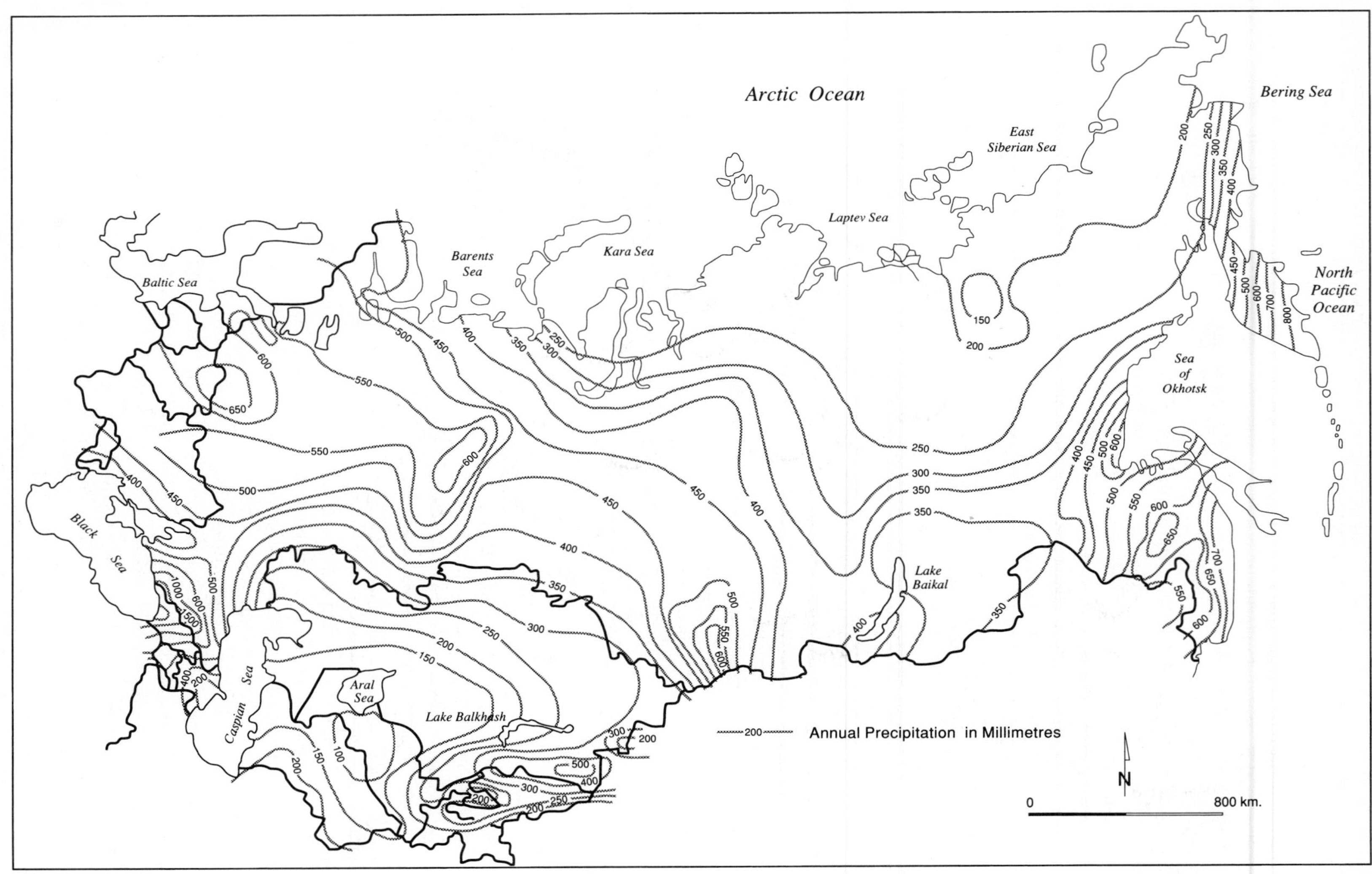

FIGURE 1.4 *Annual precipitation*
(*Source*: L.S. Berg, *Natural Regions of the U.S.S.R.* (New York: Macmillan, 1950), 360)

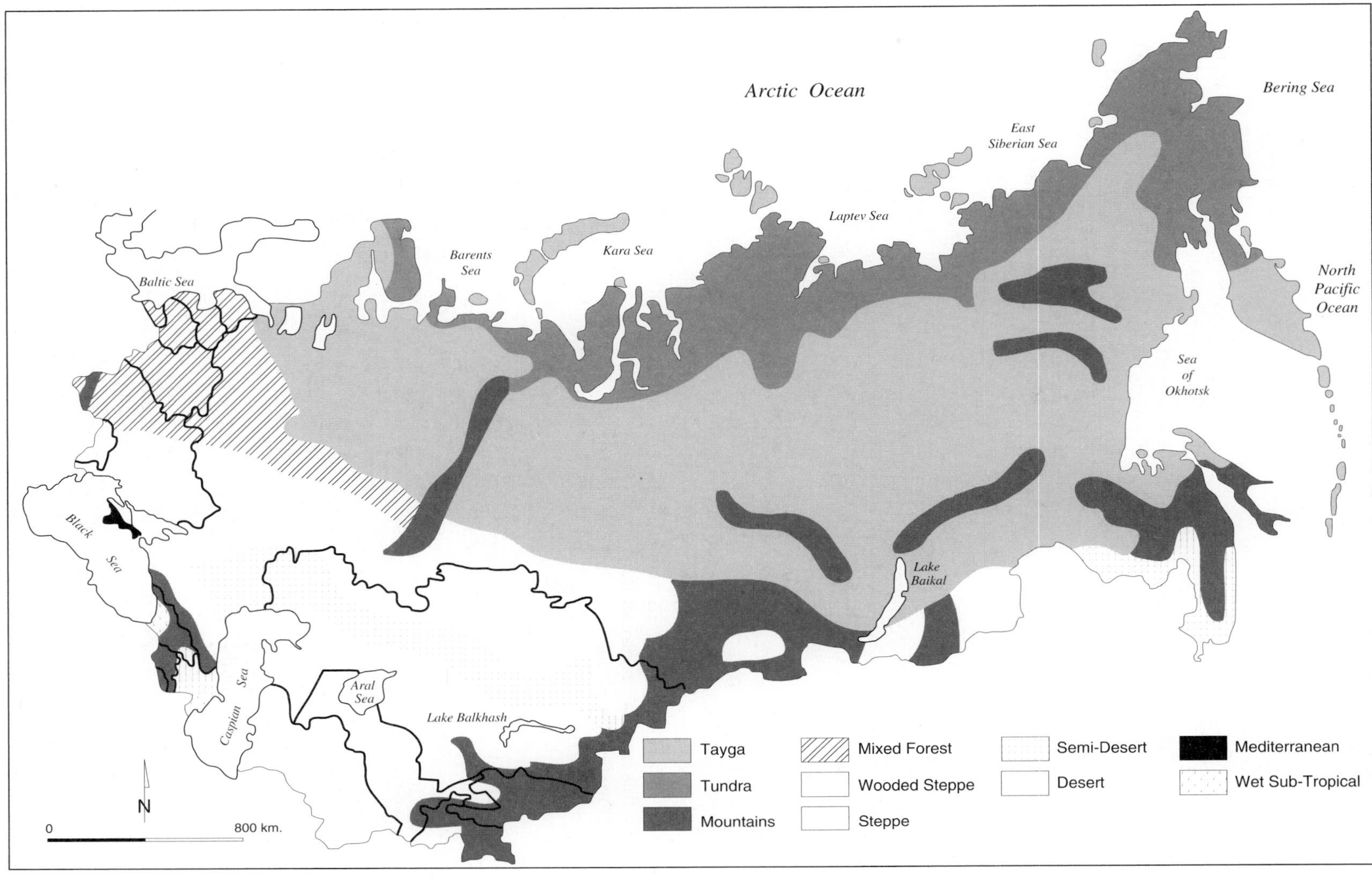

Figure 1.5 *Vegetative zones*
(*Source*: D.J.M. Hooson, *The Soviet Union* (London: University of London Press, 1966), 43)

teenth century. During the twentieth century agriculture has made more inroads into the zone of mixed forest. The wooded steppe with its river-bottom tree growth amidst open grassland, and the steppe proper, are largely coincident with chernozem, or black earth soils. These soils have long been valued for their natural fertility. Notwithstanding the natural fertility of the rich chernozem soils, the climate which helped to produce them is fickle, and bountiful harvests cannot be taken for granted.

The western tier states

As a close inspection of Figs 1.2–1.5 reveals, there are many common elements between the physical geography of Russia and the physical geography of the western tier states – Ukraine, Belarus', Moldova, Estonia, Latvia and Lithuania. Most of these states are characterized by relatively uniform relief. Only the Carpathian mountains in the western region of Ukraine and the mountains and upland region of the Crimean peninsula offer very much by way of variety in landscape. Elsewhere in this region, gently rolling lands predominate. Quaternary period glacial deposits are common, and are complemented by Baltic Sea coastal zone marine deposits, which in some parts of Estonia, for example, reach impressive scale. The western part of Ukraine and southwestern Moldova in particular have benefited from the thick layer of loess deposited during the last ice age, but so too have some parts of most of the western tier states. These deposits have softened the surface features of the underlying bed of crystalline rocks, the much eroded surface examples of which are a rare occurrence in this region. Glacial deposition not only has helped to soften the landscape's appearance, it has disrupted drainage. The combination of disrupted drainage systems, and surplus moisture which the westerly storm systems emanating over the Gulf Stream bring to the western tier states (Fig. 1.4), results in numerous bogs and swamps. They are especially commonplace in Belarus' and the Baltic states. The Pripyat Marsh of southeastern Belarus' is certainly the largest area of disrupted drainage in the region, and probably the best known. Its sandy soil and coniferous vegetation stand in marked contrast to the flora of the surrounding territory, indeed, in contrast to that of most of the western tier states. As Fig. 1.5 indicates, the Baltic states, the Russian enclave of Kaliningrad, and Belarus' are all characterized by mostly deciduous species mixed forest. Throughout this mixed forest vegetative zone podzolic soils predominate, grading off in the south to the black earth of the wooded steppe and then to the steppe proper. In Ukraine and Moldova are extensive areas of wooded steppe and steppe (Fig. 1.5). Owing to the productivity of the soil, virtually all of the wooded steppe and steppe have long been under the plough. The southern fringe of the Crimean peninsula offers the advantages of a Mediterranean climatic regime. Its predominantly limestone base is porous and relatively easily eroded, and its vegetation is sparse. But the combination of sun and sea at places such as Yalta have made the whole of the southern coastal strip of the Crimean peninsula a popular winter retreat since the nineteenth century. For Russians, however, the Crimean resort area is now part of a foreign country, Ukraine. Though Crimea's population is dominated by ethnic Russians, in the 1990s it has become a place of considerable political and social tension. Its once bustling tourist industry has been decimated in consequence.

The Mediterranean temperature regime and vegetation of the Crimea comprise only a tiny part of the territory of the western tier states, and stands in distinct contrast to the climate of most of the region. Weather conditions in the winter are associated mostly with the movement of air masses from over the Atlantic Ocean, interspersed with colder, drier weather systems from the eastern interior regions. Most of the western tier states have snow in the winter. Estonia, Latvia, Lithuania, and the Kaliningrad enclave have the mildest winters. Average January temperatures there are just a shade under freezing. Winter season lasts about three months, the days being frequently overcast, cold and damp. To the east and south in Belarus', Ukraine and Moldova, the winter cold is deeper, and lasts about a month longer than along the Baltic coast.

Summer conditions are again affected by the predominantly westerly air mass movements off the Atlantic. Along the Baltic coast, July average temperatures are a shade above 20 degrees Celsius, but weather conditions are comparatively humid. Again, to the south and east, the summer average temperatures increase with distance from the Baltic coast, and humidity drops. In much of Moldova and Ukraine, summer temperatures average in the mid-twenty degree range, and in the Mediterranean parts of the Crimea considerably higher still, of course. Summer, as in Russia, is the wet season. Rain comes principally as a result of weather systems off the Atlantic, but there is also considerable cyclonic activity along the few areas of higher elevation. As Fig. 1.4 indicates, annual precipitation falls off markedly from the Baltic coastal region to southern Ukraine.

The southern tier states

As is conveyed clearly by Figs 1.2–1.5, the Caucasian southern tier states of Georgia, Armenia and Azerbaijan possess some very distinctive landscapes. Mountains, upland plateaus, patches of semi-desert, and wet sub-tropical lowlands are just some of the ecological environments which characterize this region. The Greater Caucasus mountain range, which comprises mostly Precambrian granite and gneiss, dominates the northern part of this region, while the Lesser Caucasus range does the same along the southern borders with Turkey. The Surami range links the Greater and Lesser Caucasus, separating in the process the western lowland basin and the Rioni river system from the eastern lowland basin and the Kura river system. The highest peak in the Greater Caucasus is in excess of 5500 metres, and many others exceed 5000. The southern approaches to this range present the greatest obstacle for communication with Russia as they are much steeper than the northern approaches. The main routes over the Greater Caucasus all include passes which are about 3000 metres above sea level. Put simply, the topography of the region does little to facilitate ground transport. In both the Black and Caspian Sea coastal zones between the two major mountain ranges are wet sub-tropical lowlands. But in both these areas and the interior plateaus, a rugged, alpine vista everywhere dominates the skyline of the Caucasus.

Climatically, the Caucasus is influenced both by the moist westerly air masses that track in from the Atlantic, and the dry, warm air mass movements that originate over the desert and semi-desert regions to the east across the Caspian Sea. The presence of the massive mountain ranges to the north and to the south, and substantial water bodies to the east and west, also impact on the movement of, and nature of, the weather systems, and thus the climate (Fig. 1.3). Copious amounts of precipitation fall on the windward sides of the mountain ranges. In the western wet sub-tropical lowland region precipitation frequently exceeds 120 centimetres annually (Fig. 1.4). At high elevations precipitation is often in the order of 150 or more centimetres, some of it being snow, of course. On the leeside upland plateaus, precipitation is limited, often making agriculture irrigation dependent. As Fig. 1.4 indicates, rainfall decreases from the west to the east, but as already implied, there are many, often quite significant, local variations on this basic pattern.

Vegetation in the Caucasus runs the gamut from high alpine eco-systems to wet sub-tropical, with many others in between. Many of the upland plateaus have prairie grassland eco-systems, others are characteristically semi-desert. On the Black Sea coastal zone where in places the Greater Caucasus range runs right to the coast, there are pockets of eastern Mediterranean vegetation. At higher elevations where there is sufficient precipitation, a broadleaf forest belt takes over. In keeping with the precipitation gradient from west to east, the predominant species in this forest belt also change accordingly. Moving from the west to the east, spruce and fir forests phase out, and are replaced by species such as elm, ash, beech and linden, all more suited to the increasingly arid conditions. Soils are commonly lateritic, not ideally suited to most agricultural crops, but with appropriate chemical fertilizers certainly reasonably productive. At higher altitudes still, the forest belt gives way to alpine meadows.

Winters can be harsh indeed, especially in the high alpine valleys. On the plateaus, frost is a commonplace occurrence, but the winter months are certainly more benign than in most of Russia. And in the lowlands, frost occurs only rarely. Indeed, along the Black Sea coast mean January temperatures are about 7 degrees Celsius. On the interior plateau of central Armenia, for example, the mean January temperature is about –3 degrees Celsius (Fig. 1.1). Summer temperatures are generally high throughout the region. On the Armenian plateau, the July mean temperature is around 26 degrees Celsius, and conditions are arid. In the coastal lowland areas, temperatures in July are higher still, and conditions are invariably quite humid.

The southern tier states which occupy the vast area east of the Caspian Sea are much more uniform in terms of landscape, climate and vegetation than their Caucasian counterparts (Figs 1.1 and 1.3). Partly this is the result of the fact that deserts comprise most of Turkmenistan and Uzbekistan, and a substantial share of Kazakhstan's territory. Tajikistan and Kyrgyzstan both have desert areas, but are mostly characterized by mountainous terrain (Figs 1.3 and 1.5). Kazakhstan is dominated in terms of physical geography by a huge geosyncline, which extends south from the Arctic Ocean and lies between the Ural mountains to the west and the Siberian Upland region and the mountain ranges along the border with China to the east. Sedimentary deposits lie deep over much of this natural geological depression which covers a substantial part of Kazakhstan, and fortunately so, for it is from them that coal, oil and natural gas already have been extracted in abundance and recent exploration has determined that there is much more still to be tapped. The northern reaches of Kazakhstan are characterized by a belt of uplands and plateaus which are from 1000 to 1500 metres above sea level. To the southeast, is the Ustyurt Plateau, a vast expanse of sand desert and occasional clay soil. Its average elevation is about 120 metres. And further south still lies an even larger expanse of low-lying desert. To the east of the Ustyurt Plateau is a series of ancient geological depressions, all of limited elevation and all mostly desert. Moving further east across the Kazakh Uplands, eventually the foothills of the Western Sayan mountains come into view. Along most of the eastern and southern borders of these southern tier states are mountain ranges – the Sayan, Altai, Tien-Shan, Pamirs and Kopet Dag, to name only a few of them.

The vast expanse of deserts such as the Kyzylkum and Karakum, which cover large parts of Uzbekistan and Turkmenistan, are a clear indication of the nature of the climatic regime in much of Middle Asia. Winter is the dry season, with typically about ten days with some snow or rainfall. Total precipitation rarely exceeds two or three centimetres. Winter temperatures are often extremely variable. Arctic air masses sometimes reach as far south as Middle Asia's mountain perimeter, and warm air masses from Iran and Afghanistan occasionally penetrate the region as well. As a rule, January mean temperatures are a shade above freezing, though Arctic air masses can drive down the temperatures to –20 degrees Celsius. Conversely, if sub-tropical systems penetrate the region, daily maximum temperatures in January can reach 20 degrees Celsius. By spring, convectional rainstorms occur as a result of the convergence of warm and cold frontal systems, but the amounts of precipitation are not large. Summer season temperatures are brutally hot. July daytime temperatures rarely dip below 25 degrees Celsius, and are frequently well in excess of 40. Convectional air currents are commonplace owing to the extremely high ground temperatures, but as the air is customarily dry little precipitation occurs. Over the lengthy summer period, precipitation amounts to only a few centimetres. Late summer and autumn bring peak amounts of rainfall to the desert owing to frontal storm activity, but in total it is obviously quite limited. In the uplands and foothill regions precipitation amounts increase rather substantially, and if offered any protection from the Arctic air masses, temperatures during the winter months are usually quite moderate. In the foothills, annual rainfall is customarily between 30 to 40 centimetres, certainly adequate to grow crops. In the desert areas, agriculture is restricted to the historic oases, or to irrigated areas. At higher altitudes, precipitation increases dramatically, and seasonal average temperature regimes are lower. Indeed, during the winter months conditions are often very severe.

Throughout most of Middle Asia the vegetation varies according to elevation. As noted already, huge areas are desert eco-systems. The desert soils are capable of supporting crops in some places, but only under irrigation. The many natural depressions in the desert usually offer sufficient natural vegetation for grazing sheep and goats. More bountiful for grazing are the prairie grassland eco-systems of the steppe, but save for Kazakhstan these areas are limited in Middle Asia. At higher elevations, more arid steppe vegetation predominates. Between 1200 and 2250 metres, the grassland and meadow vegetation is much more lush as a result of the greater rainfall. At higher elevations, there used to be a zone of upland forest, albeit not an especially extensive one. Over the centuries, most of the forest has been axed. Above 3000 metres are extensive alpine meadows, and occasional tracts of high desert and semi-desert eco-systems.

Further reading

Berg, L.S. 1950 *Natural Regions of the USSR*. London, Macmillan.

Borisov, A.A. 1965 *Climates of the USSR*. Edinburgh, Oliver and Boyd.

Farson, D. 1994 *A Dry Ship to the Mountains: Down the Volga and Across the Caucasus – In my Father's Footsteps*. New York, Viking Penguin.

Lydolph, P.E. 1977 *Climates of the Soviet Union*, Vol. 7 in *World Survey of Climatology*. Oxford, Elsevier Science.

Nalivkin, D.V. 1973 *Geology of the USSR*. Edinburgh, Oliver and Boyd.

Newby, E. 1984 *The Big Red Train Ride*. London, Penguin.

Parker, W.H. 1969 *The World's Landscapes: 3, The Soviet Union*. London, Longman.

2

FROM TSAR TO TOTALITARIANISM: THE CREATION AND COLLAPSE OF THE RUSSIAN AND SOVIET EMPIRES

'political revolutions are preparing for the world a common destiny and solution of the problems affecting the life of nations'
(Elisee Reclus, *The Universal Geography*, 1890, Vol. 6, 32)

The collapse of the Soviet system in 1991 changed fundamentally the world map. Fifteen independent states replaced the single Union of Soviet Socialist Republics, a political geographic entity that had accounted for one-sixth of the earth's land area and embraced 290 million people. For more than seven decades the Soviet Union had struggled to retain, or in some cases regain, territories and peoples accumulated during several centuries of territorial expansion of its predecessor state, the Russian Empire. In imperial Russia, the ultimate authority was the ruling Tsar, who stood next to God in the minds of the majority of the Empire's subjects. For Russians in particular, the relationship between the Tsar and subject was usually characterized by reverence and obedience. Amongst the growing number of non-Russians caught in the maw of expansionist imperial Russia however, the relationship between Tsar and subject was usually far less subservient. But whatever imperial dicta of the Tsar failed to elicit by way of appropriate response from any of the 160 million subjects living within the borders of the Empire by 1914, the army, police and courts were more than capable of compensating for. The rule of the Tsar was brought to an end by the Revolution of 1917, which was spawned by the internal contradictions in Russian society and political economy. The new Soviet state which emerged from the ashes of the Empire faced both new challenges and the old problem of maintaining control. While its territory had been reduced by successful, though mostly impermanent, declarations of political independence on the part of some non-Russian minorities, its extensive borders still embraced scores of different ethnic groups. The forging of the first ever socialist society out of this heterogeneous population necessitated a massive campaign of social engineering. The leader of the Communist Party replaced the Tsar as the ultimate symbol of authority. The new state demanded of its citizens compliance with, if not enthusiastic endorsement of, a multitude of decrees. In the early years of Soviet power, those who failed to comply, or who had the temerity to publicly question the new ideology, were cruelly purged. Ultimately, as in the case of the Russian Empire in the late imperial era, Soviet central authority waned, publicly promulgated values were increasingly divorced from personal behaviour, and contradiction and paradox came to characterize society and political economy. The ultimate demise of the Soviet system must be seen in historical context. Thus, it is the purpose of this chapter to first of all sketch in broad brush fashion a portrait of the Russian Empire in its final years, highlighting some of the major forces of change which first challenged, and then toppled, the status quo. The political and administrative structures which characterized the Soviet system which emerged in the aftermath of revolution will then be briefly described. These structures were in hindsight flawed, and their principal deficiencies will be outlined. Finally, the basic structure of Soviet society, both official and unofficial, will be described and the essential contradictions highlighted.

Emergence of the Russian Empire

The process of modernization in Russia was intimately bound up with urban-industrialization, with the penetration of capitalism into a society which had evolved under conditions of an absolute aristocracy. But modernization was not easily accommodated by institutional systems designed to perpetuate the authority of the Tsar, by a society whose elites were bent upon maintenance of the status quo, by patterns of behaviour which were quintessentially rural and not urban. At the turn of the nineteenth century, Russia was already the largest country in the world in terms of area, and the largest European state in terms of population. In many ways, it was also one of the more backward of European countries. Outstanding achievements in science, in the humanities, in the world of culture in the broadest sense, combined with the pomp and splendour of the imperial court were to belie the harsh and impoverished conditions of daily life and labour for so many of the Empire's inhabitants. The trappings of modernization lay like a thin veneer over much that was Russian. On the eve of the World War I, the vast majority of the Empire's more than 160 million subjects were peasants. Many could recall from personal experience the reality of serfdom, which was formally abolished only in 1861. Prior to that date, a large proportion of the population of the Russian Empire, which then numbered about 60 million, was the personal property of a small elite, or of the state itself. Indeed, wealth was usually measured in terms of number of male peasants, or souls as they were called, that one owned. Thus, compared to countries like Britain, France or Germany, modernization had come late to Russia. Its impact was no less profound for the delayed start, however. Before examining the process of modernization in the context of the agrarian and urban-industrial economies, and describing some of the social and political changes set in motion, it is appropriate that the territorial expansion of the Russian Empire from the early Middle Ages to the eve of the World War I, and the ethnic composition of the population which came to be embraced by its ever-expanding borders, be briefly surveyed.

The expansion of the Russian Empire depicted in Fig. 2.1 suggests that frontier and colonization played an important role in Russian history. And indeed they did. From the first settlement of Slavic peoples in the territory around the upper Volga river and its tributary the Oka in the twelfth century, the process of territorial expansion continued through the nineteenth century. The earliest Slavic settlements developed as quasi-independent princedoms in the mixed forest of the upper Volga region. Independence ended with the invasion of the Tatars, or Mongols, in the early 1200s. With the notable exception of Novgorod, all of these early princedoms were brought under Tatar control. The princes of Moscow eventually gained the favour of the Great Khan, and while serving him as a tax gatherer they at the same time were able to extend their hold over other princedoms. As the wealth and prestige of the Principality of Muscovy increased, so too did the so-called Tatar Yoke begin to chafe. From the thirteenth to the fifteenth centuries, the princes of Muscovy steadily expanded their territory, the subjugation of Novgorod in the late 1300s adding substantially to their domain to the north and northeast – the traditional hinterland of Novgorod's merchantry (Fig. 2.1). During this period of territorial expansion, confrontation with the Tatars steadily escalated. It was not until the mid-sixteenth century, however, that Russia,

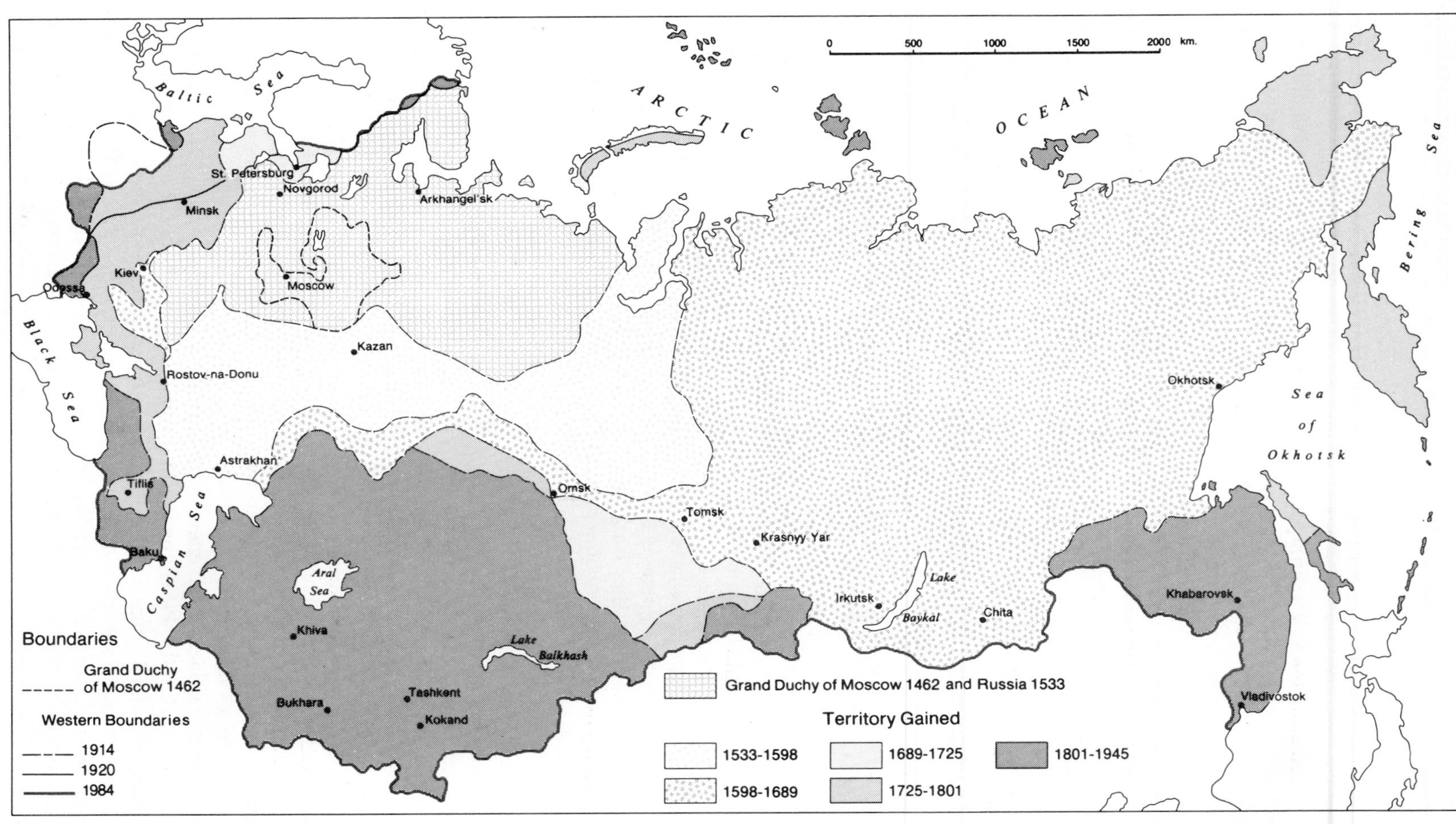

FIGURE 2.1 *Expansion of the Russian Empire*
(*Source*: Adapted from P.E. Lydolph, *Geography of the USSR* (New York: Wiley, 1970, 2nd edn), 6)

as the Principality of Muscovy eventually became known, finally put paid to the last vestiges of external control by defeating the Tatars in the battle of Kazan' (1552). This victory was celebrated by Ivan IV through his commissioning the construction of St Basil's Cathedral in Moscow's Red Square. The Tatars remained a considerable threat, however. For example, in the late 1580s they were able to muster sufficient forces to sack Moscow yet again. Still, the defeat of the Tatar Khanate based at Kazan' was significant. The Tatars no longer wielded influence as before, and a major obstacle to the exploration and colonization of Siberia was removed.

Those peoples who inhabited the region between the Volga river and the Urals offered minimal resistance to the small bands of cossacks, soldiers and hunters who comprised the first wave of explorers, and who sought not to colonize so much as to collect fur. Sable, prized in Europe, and thus high in price, was the principal attraction. The vast expanse of Siberia across the easily traversed Ural mountains beckoned. Moving by river through the tayga, Russians reached the Sea of Okhotsk by 1649 and established there another colonial outpost (Fig. 2.1). Scattered in the forest behind them were numerous *ostrogi*, or fortresses. These were customarily small since the indigenous peoples had proven relatively easy to subjugate, and thus a token garrison force was usually sufficient to administer the collection of fur tribute, or *yassak*. The fur trade brought trappers and traders, but few settlers. The Russian realm was limited to the forest, since in the steppe to the south nomads still held sway. Russian claim to this vast tract of Siberian tayga was formally acknowledged by the Treaty of Nerchinsk concluded with China in 1689. The Stanovoy mountains served as the boundary between the Russian and Chinese spheres of influence in the east. The wealth of the fur trade had drawn Russia to the Pacific. Potential profits from trade in seal skins would soon lure Russians onto North American soil as well. Expansion of the Russian Empire during the seventeenth century was by no means confined to the eastern regions, although as Fig. 2.1 clearly reveals, territorial gains in Siberia greatly exceeded all others.

The consolidation of Russian control over the steppe region of European Russia, and to a lesser extent in the Siberian steppe, was one of the major developments during the eighteenth century (Fig. 2.1). For centuries, permanent settlement had been confined to the poorer soils of the forest zone of European Russia. Save for a few fortified cossack settlements, the rich soils of the steppe could not be cultivated because of the constant threat of attack by the nomads. With the defeat of the Crimean Tatars in the late eighteenth century, the historic grip by a war-like nomadic people over the Ukrainian steppe was finally broken. But this did not result in widespread colonization by Slavs. By the late eighteenth century, serfdom had been legally entrenched in Russia for more than a century. Free colonization, the authorities feared, would simply undermine the existing social, economic and political order. Thus, foreigners, including Mennonites, Hutterites and other ethnic and religious minorities from Europe were invited to settle these new lands. Settlement of the steppe by subjects of the Empire remained closely monitored. Meanwhile, on the western frontier of the Empire, European powers steadily gave way to Russian expansion, most notably in terms of the creation of the Grand Duchy of Warsaw (the Russian spoils from the late eighteenth century process of partitioning Poland with Prussia and Austro-Hungary) and control over the Baltic region.

Nineteenth-century territorial acquisitions brought large numbers of non-Slavic peoples into the Empire. Finland was acquired from Sweden in 1809 and also given the status of Grand Duchy. The Georgians and Armenians of the Caucasus were ostensibly rescued from the Turks and Persians during the first quarter of the century. In extending control over the Caucasus region, the Russian Empire had to absorb a vast array of peoples differentiated by race, religion and language. However, the rugged, mountainous terrain afforded numerous opportunities for ethnic minorities to seclude themselves from external influence. Thus, many groups were able to perpetuate a cultural identity despite having been incorporated into an Empire noted for Great Russian chauvinism. The rugged mountainous environment also facilitated armed resistance to Russian suzerainty by some minorities, something which took years to eradicate. That the mountain environment and resolute resistance to Russian authority have again come to haunt Russia in the 1990s in the effort to curtail the Chechen independence movement is one of the many ironies of history.

In the second half of the nineteenth century, Russian military penetration into Middle Asia began in earnest. This first of all opened the way for colonization in Siberia of the southern reaches of the steppe and the northern margins of the semi-desert, a zone previously controlled by nomadic peoples. Across the deserts of Middle Asia were the Moslem Khanates and Emirates centred on the oasis and montane valley. One by one they fell under Russian control – the city of Tashkent in 1865, the city of Samarkand in 1868, the Emirate of Bukhara in 1868, the Khanate of Khiva in 1873. The mountain perimeter of Middle Asia, and British India beyond, forestalled further Russian advance.

In the far east, the weakening authority of the Manchu Empire prompted Russian incursion into Chinese territory after mid-century. Russian settlements on the Pacific shores, to say nothing of those in Russian Alaska or northern California, had long proven to be extremely difficult, and costly, to supply. Thus, the presumed agricultural potential of the Amur and Ussuri river valleys was especially attractive. But these lands were beyond the Stanovoy mountains, the boundary between Russia and China established in 1689. The region proved too enticing to forego, and therefore was simply annexed. Permanent settlement was required and colonization was promoted by the state in consequence. The formal abolition of serfdom in 1861 had created a vast pool of potential colonists, and the construction of the trans-Siberian railroad in the last years of the nineteenth century opened the door to the lands of Siberia and the far east. Large numbers of Ukrainian peasants, amongst others, came to farm in the Amur region. Access to the far east was greatly facilitated when Russia gained the right in 1894, through negotiations with a further weakened China, to build the last section of the trans-Siberian railway across Manchuria to the Russian port of Vladivostok. Indeed, for a short period, the whole of Manchuria was in Russian hands.

The Russo-Japanese War of 1904 brought to a close Russian territorial expansion in the far east. Defeat by Japan resulted in Manchuria passing out of Russian hands. As well, the southern half of Sakhalin Island, acquired from Japan in 1875 in exchange for the Kuril Islands, was ceded to Japan. It was not until World War II that control over all of Sakhalin was regained.

The zenith of Russian territorial expansion had been reached by the early 1900s (Fig. 2.1). The partitioning of Poland and the acquisition of Finland pushed Russian borders to their furthest western extent. The Lesser Caucasus mountains separated Russia from the domain of the Turks and Persians to the south. In Middle Asia, mountains separated British colonial India from Russian colonial Middle Asia. The abandonment of the California outposts and the sale of Alaska to the United States in 1867 signalled the end of Russian territorial interests across the Pacific. The Arctic realm, of course, had been under the Russian flag for centuries. The pattern of territorial expansion depicted by Fig. 2.1 has on more than one occasion been reduced to single, simple historical explanations – the search for warm water ports, the felt need to unite all Slavic peoples being two examples. Historical reality, as usual, defies such simplistic interpretations.

Of the more than 160 million people living within the extensive borders of the Russian Empire on the eve of the World War I, a considerable number were neither Russian, nor indeed even Slavic, as the data drawn from the census of 1897 presented in Table 2.1 indicate. Aside from the East Slavs, that is, Russians, Ukrainians and Belorussians, who comprised about three-quarters of the total population, there were more than 170 other distinct ethnic groups. Some of these national minorities could claim only a few thousand members. The indigenous peoples of Siberia and the north, such as the Chukchi, Koryaks or Kamchadals, are examples. Other groups were both sizeable in number and restive subjects of the Tsar. For example, more than 12 million

Table 2.1 *Nationalities of the Russian Empire: 1897 (by language group)*

Nationality	Number (thousands)	% of total
SLAVS		
Russian	55,600	44.8
Ukrainian	22,800	18.4
Belorussian	5800	4.7
BALTS		
Estonian	1000	0.8
Latvian	1430	1.2
Lithuanian	1660	1.3
MOLDOVAN	1120	0.9
TRANSCAUCASIAN		
Armenian	1170	0.9
Azeri	1470	1.2
Georgian	1310	1.1
MUSLIM		
Kazakh & Kyrgyz	4280	3.4
Tajik	850	0.7
Turkmen	630	0.5
Uzbek	2790	2.2
TATAR	2230	1.8
JEWS	3770	3.0
Region, total	124,200	100.0

Source: Basile Kerblay, *Modern Soviet Society* (New York: Pantheon Books, 1983), Tables 10, 40. Translated by Rupert Swyer

people lived in the Grand Duchy of Warsaw. Attempts to Russify the Poles during the latter part of the nineteenth century prompted concerted resistance to Russian authority. The obligatory substitution of Russian for Polish in Polish schools was especially offensive, but was just one example of Russification. Resistance increased, and so too did the stream of Polish nationals exiled to Siberia. In Finland, Russian rule was much more tolerant of local interests. Greater sensitivity in dealing with Finland did not preclude the emergence of a Finnish nationalist movement, however. Estonians, Latvians, and Lithuanians did not have the separate status of the Poles and Finns, but by virtue of language and religion they were equally separate and distinguishable culture groups. Across the vast expanse of the Empire there were many other such minorities.

While most minorities suffered in one way or another from Slavic, or more commonly Russian, chauvinism, it is doubtful if any were more oppressed than the Jews. Living for the most part in a region straddling the western reaches of Ukraine and European Russia and spilling over into part of the Baltic provinces, the geographical and social mobility of the Empire's Jewish population was limited by statute. Denied the right to live wherever they wanted, denied the freedom to pursue certain professions, denied in any sense the rights of self-determination, Jews were certainly amongst the most persecuted of the minority groups. Notwithstanding the many legal shackles imposed upon them and the widespread anti-Semitism, Jews still became a major force in industry and commerce, and figured prominently in a number of professions. Not all ethnic minorities caught up in the historical expansion of the Russian Empire had much contact with Russians or with the Empire's colonial administration. In the Caucasus, for example, most Georgians, Armenians and Azeris carried on very much as they had always done. Indeed, Russian culture was not always regarded as being of a higher order, and the intellectual elites of some national minorities perceived Russians as inferior. In Middle Asia, the ten million or so Muslims vastly outnumber the emigrant Slav colonists who had arrived during the last decades of the imperial era. In the few cities, Russian colonial administrators created a built environment based on the European model, segregated from the hotch-potch of adobe structures which made up the traditional Middle Asian urban settlement, much as the British did in India. In the countryside, strict segregation of colonists from the indigenous population was also the rule.

The Emancipation of the Serfs in 1861, the construction of the trans-Siberian railroad, and the Stolypin Agrarian Reform, about which we will say more later, all served to spur migration from the overpopulated lands of European Russia to Siberia. Migration to Siberia, of course, was not a new phenomenon. In earlier times, hundreds of thousands of peasants fled to escape the injustices of serfdom. Scores of thousands more sought religious freedom, seeing in Siberia a possible refuge from the intolerance of the Eastern Orthodox religious principles and practice. And always Siberia had been the place of exile. Tens of thousands of people of all classes and nationalities had been forced to trek across the Empire to a place of internment somewhere in the distant reaches of the Siberian realm. Some exiles never did return, choosing instead to stay in Siberia upon the completion of their sentence. For them, and many others, life in a region known then (and now) for an independence of spirit and action was preferable to the surveillance and oppression associated with the aristocracy and European Russia. By 1914, more than ten million people lived in the enormous area stretching from the Ural mountains to the Pacific. The indigenous populations, always small in number, were now completely submerged by recent migrants, most of whom were Russians and Ukrainians. Notwithstanding the enormous scale of the late nineteenth, early twentieth century exodus to the east, many regions in European Russia remained acutely overpopulated. Save for the oasis-based settlements in Middle Asia, population density east of the Urals was seldom very great. Most of the Empire's inhabitants lived west of the Urals and earned a livelihood from the land. Unfortunately, the soil there provided but a bare subsistence existence for the majority, little changed from decades, if not centuries, before.

THE AGRARIAN SCENE IN LATE NINETEENTH CENTURY RUSSIA

Serfdom was legally entrenched in Russia in a code of laws adopted in 1649. It lasted until 1861. From that year on, peasants could no longer be bought, sold or mortgaged. They could not be forced to labour free of charge on the landlord's estate or serve as domestic servants. They no longer had to pay for the privilege of departing the village, temporarily, to seek work elsewhere in the countryside or in the city. Put simply, trade in human capital ended with the Emancipation Act of 1861. But the peasantry paid a heavy price for their newly acquired freedoms. Serf owners had to be recompensed by the state for their loss, and for future income foregone. The peasantry was thus saddled with an additional annual tax, or redemption payment as it was called, to continue for 49 years to compensate for the state having made a financial settlement with the serf-owning gentry. Furthermore, the Emancipation Act required that peasant lands be formally separated from those of the former landlord. Since tradition, rather than legal documentation, often differentiated the areal extent of the peasants' agricultural operations from the landlord's, there was certainly scope for manipulation in the land allocation process. And, indeed, it appears that this occurred on a fairly wide scale. So far as available data are accurate, they indicate that the average peasant land holding was about 4 per cent less after the Emancipation than before it. In those regions where the soil productivity was high, and therefore where estate owners had a vested interest in acquiring land customarily cultivated by the peasants for their own needs, peasant landholdings were as much as 20 per cent less. Thus, not only was there the new burden of redemption payments after the Emancipation, but peasant households had less land on average from which to generate the income to pay them. For a great many peasants, the burden was such as to promote rural–urban migration in search of wage employment. The Emancipation produced a major transformation of the legal underpinnings of the rural social structure, but this did little to change the appearance of the countryside.

Although by the time of the Great War some peasants had acquired their own farms, and the large estates of the land-owning gentry were still an important feature in the rural landscape, most rural inhabitants were villagers and members of a rural commune, or *mir*. Thus, the village remained the customary settlement form in rural Russia and the commune the primary form of social organization outside the family. A sense of group solidarity no doubt had been nurtured by the peculiar demands of serfdom. However, even before becoming enserfed, there appears to have been a sizeable element of collective decision-making within the village. The Emancipation did not alter this time-honoured pattern of behaviour. Within the village commune, or *mir*, an elected group of elders took decisions on behalf of all members. For example, they allocated land to spring crops, winter crops and fallow in accordance with the principles of the three-field system. Collective actions like these were taken in hundreds of thousands of villages across Russia, though it should be noted that such expressions of collectivism were more common in the central Russian core than in the peripheral regions of the Empire. The allocation of the individual strips of land within the cultivated 'fields' was also governed by the collective decision-making. The number of strips allocated to each household was usually determined by the number of mouths to feed. Strips were distributed in as equitable a fashion as possible. Each household was to share the good, bad or indifferent soils and would travel roughly similar distances to work them. The actual distribution of one household's allocation of strips is portrayed in Fig. 2.2. In some parts of Russia, strips of land were periodically reallocated in order to reflect changes in household membership. In the Baltic region, in the south Ukraine and in the steppe region of Siberia, however, strips were customarily hereditary allotments, that is, they were not periodically redistributed. Cultivating scattered parcels of land was clearly inefficient. The Emancipation, however, did not improve agricultural efficiency in the peasant domain. Indeed, the government had a vested interest in preserving the status quo.

The communal *mir* had administrative functions in addition to determining land allocations. As a rule, it administered the selection of recruits for the army and was responsible for the collection of taxes from all households in the village. Thus, should one household default, the village as a whole was responsible for making up the tax obligation. Consequently, the village community had a vested interest in superintending the migratory habits of individual village members. Prior to the Emancipation, peasants could leave the village only with the necessary authorization of the *mir* and the consent of the landlord. After the Emancipation, permits issued by the *mir* were still required. To be sure, from time to time some peasants left the village without the approval of the *mir*; however, because of tax obligations, and tradition, there was considerable peer group pressure to preserve the village community structure.

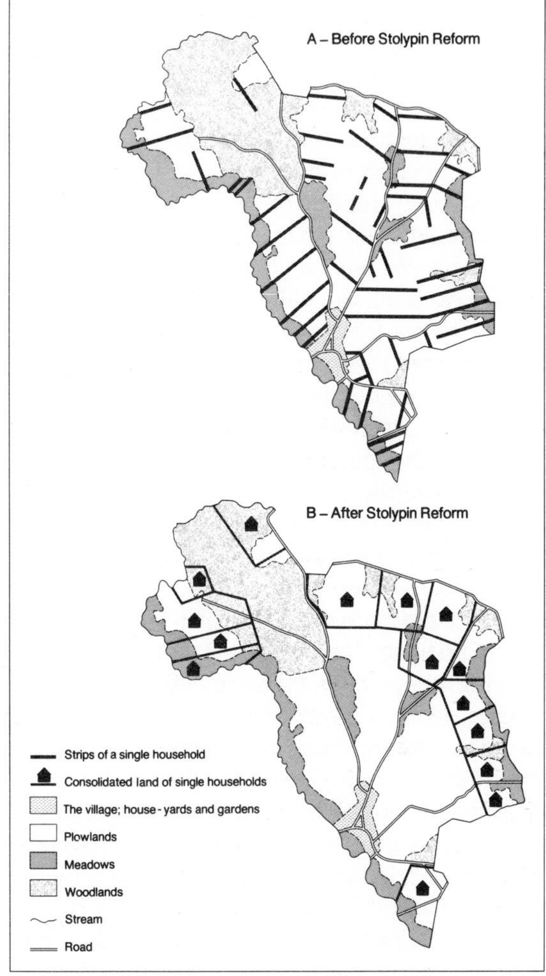

FIGURE 2.2 *Rural transformation at the time of the Stolypin Agrarian Reform*
(*Source*: Adapted from G.R. Robinson, *Rural Russia Under The Old Regime* (New York: Macmillan, 1967, sixth printing), 217)

Rural–urban migration increased in volume in the decades following the Emancipation. Most peasants sought out work in the city on a seasonal basis, or as their permits dictated. However, growing numbers of peasants, mostly males, managed to stay in the city for periods of years, and slowly acquired some characteristics of a working-class culture. But ties with the village, both legal and familial, remained strong. The numerous religious holidays facilitated journeys back to the village at regular intervals where distances, or cost of travel, were not too great. Even amongst the growing number of peasants who had become part of the permanent workforce, many retained more than a passing interest in village affairs. After all, the village was potential refuge in times of hardship, where tradition ensured that those in need would be looked after. The sense of collective was deeply embedded in the peasant mentality, perpetuating rural customs on the one hand, and because of the nature of the rural–urban migration process, helping to shape the urban milieu itself on the other.

Traditional peasant agriculture was largely subsistence. Therefore, it could contribute relatively little to grain exports, which were so important in financing industrial development. Peasant agriculture had helped to meet domestic and foreign demand for grain by expanding the area under cultivation, and in particular by bringing the new lands of Siberia and Middle Asia into production. However, the technology employed, and the organization of agriculture itself, dictated that yields remained low. Of course, not all Russian agriculture was in the hands of peasant households. From the estates of the land-owning gentry substantial amounts of grain moved to both domestic and foreign markets. Following the Emancipation, some estate owners adopted technological innovations and substituted capital for the labour that was no longer free. With sufficient acreage, economies of scale were also possible. In the Baltic region and the south Ukraine, capitalist agriculture took firm root on many estates, the very best comparing favourably in terms of management, technology and production with advanced agricultural operations in Europe and America. But the dead weight of the peasant agriculture economy ensured that average yields for the Empire as a whole remained comparatively low.

By the turn of the nineteenth century, growing population pressure made even more difficult the task of wresting a living from the soil. The occasional natural calamity meant that life itself was sometimes at risk. On occasion, the *mir* was able to extend village lands by purchase or lease, often from former landlords. In most cases, however, it was confronted with the task of sharing out a fixed amount of land amongst more mouths. The average size of peasant land allocation steadily declined, incomes shrivelled, poverty intensified. A growing number of peasants showed physical manifestations of an impoverished existence. For example, one-fifth of the military conscripts called up between 1899 and 1901 were rejected as physically unfit. The impoverishment of the peasantry was in no sense a universal phenomenon, however; indeed, a class of wealthy peasants emerged. Some peasants prospered by virtue of greater ability as farmers, other were able to augment their income from non-agricultural sources, other simply at the expense of the growing number of households with subsistence-size allotments or less. Thus, the peasantry itself was increasingly stratified according to wealth, creating in the process further grounds for grievance.

Peasant dissatisfaction reached breaking point by the turn of the century. An economic depression prevailed throughout most of the 1880s, and from 1898 the Russian economy was caught up in yet another. From 1900 to 1904, hundreds of peasant disturbances, revolts, and uprisings took place in Russia. Some were minor, but a good number were serious. In 1905, peasant disturbances combined with widespread strike activity in the cities. The catalyst in sparking what is customarily labelled the '1905 revolution' was the Russo-Japanese War. Initiated with much popular support in 1904, the incompetent management of the war effort culminated in an ignominious settlement which was forced on Russia in 1905. This brought the grievances of peasants and workers alike to a head. Tsar Nicholas II was obliged to introduce some major reforms in order to settle the tide of unrest. In 1906, a constitutional government, or *Duma*, was created, legitimizing for the first time the existence of political parties in Russia. Workers in the factories were given the right to strike. In the countryside, the pressure for change was every bit as acute as in the city. Inadequate allotments, restrictions on access to woodlot and pasture imposed by landowners on what before the Emancipation had been part of the peasant agricultural realm, and the continuing authority of the *mir*, all served to perpetuate subsistence agriculture, and thus inhibit development of market-oriented agricultural production. Propagandists and agitators advocating change found more and more receptive ears amongst the villagers. Conscripts' first-hand accounts of the debacle on the far eastern front simply fired further the discontent. That preservation of the status quo in the countryside was weakening, not bolstering, the basis for the aristocracy was now conventional wisdom amongst many of the Tsar's key advisors. Thus, beginning in 1906, Tsar Nicholas II signed into law a series of measures which are commonly referred to as the Stolypin Agrarian Reform.

In recognition of the impoverished state of the mass peasantry, the Tsar already had issued a decree in late 1905 cancelling all further redemption payments to the state. The main thrust of the agrarian reform, however, was still to come. The Emancipation of 1861, as we have emphasized, perpetuated traditional – and highly inefficient – agricultural practices by reaffirming the authority of the communal *mir*. In decrees passed in 1906, however, the right of the individual peasant to leave the village at will, to consolidate the household's allotment of strips, and to receive title to this land, was confirmed in law. What the state intended to create by the Stolypin Agrarian Reform was a new class of independent, landowning peasant farmers. The vested interest of this class would entail the support of the Tsar and government to whom it would owe its independence, if not its prosperity. The reform's attraction was greatest for the roughly nine million peasant households belonging to communes in which strips of land were not hereditary holdings, but were periodically redistributed, or repartitioned, according to local custom. As this group comprised about three-quarters of all peasant households, the potential impact of the reform was clearly enormous.

Available data suggest that between 1906 and 1915, some 2.5 million households gained title to land previously held by the commune. That represents more than one-quarter of all households involved in repartitional communes. Such conversions of title, however, did not necessarily result in the consolidation of strips into a single holding. Indeed, while some conversion of titles brought both consolidation of strips and the creation of new, independent farms as indicated by Fig. 2.2B, this was an exception, although an important exception to be sure. While the communal *mir* no longer had the legal authority to direct the actions of peasant members, the fact is there was not a wholesale conversion of landholdings of the type depicted in Fig. 2.2A to that described by Fig. 2.2B. Such consolidation, of course, afforded those participating households the opportunity to decide for themselves which crops to sow, how much land should be left fallow, and so on. Consolidation occurred most often in the peripheral regions of the Empire where the communal system was not as well established. Consolidation also offered the possibility of greater efficiency. No longer was there weed-infested waste land separating strips under the plough. No longer was it necessary to travel to a score or more individual strips. No longer was there a disincentive to improve soil quality through fertilization because the improved strips might be assigned to another household at the next repartition. Even in the absence of financial capital and knowledge of such agrarian innovations as improved seeds, consolidation helped to raise productivity. There were limits to what could be achieved, however.

The difficulty of deciding how to consolidate strips of differing soil qualities and locations no doubt persuaded some peasants to persist with the old ways even when they could see advantage to consolidation. For others, indeed, perhaps the majority, their daily lives were so deeply embedded in traditional ways that they simply did not want change. There was after all a measure of security in belonging to the commune. Thus, over much of Russia, agriculture carried on after the Stolypin Reform much as it had before. Each dawn witnessed millions of peasants trudging off to tend individual strips of land, a scene transcending the revolution of 1917, continuing, in fact, down to the collectivization drive initiated by Stalin in the late 1920s. Still, to dismiss the Stolypin Agrarian Reform as having had little or no impact would be incorrect. Peasants had much more personal freedom after the reform than before it. Free now to stay in the village or to leave, huge numbers decided to migrate to the frontiers of settlement, to colonize. Even larger numbers abandoned the village for the factory and the city. After the Stolypin Reform there occurred a period of economic growth unparalleled in Russian history. Perceived opportunities for a better life outside the village unleashed an unprecedented wave of rural–urban migration. In the following pages we will discuss some of the features of the urban-industrialization process which helped to stimulate migration and outline some of its consequences.

Industrialization in late nineteenth century Russia

At the turn of the century, three-quarters of the Empire's labour force were employed in agriculture. Manufacturing engaged perhaps a tenth of the total. However, its share was growing quickly and in any event greatly understated the importance of industrialization. While the development of industry was late in comparison with the major European powers and the United States, this was not because of inadequate industrial resource base. All of the basic materials necessary for industrialization in the nineteenth century were available within the Russian Empire. For example, iron ore, coking coal, even oil deposits were being tapped – with profit – at the turn of the century. Indeed, in the case of oil, production from the Baku region of what is now the state of Azerbaijan in the eastern Caucasus was the largest in the world. While the enormous resource potential of Siberia was still largely unrecognized, within European Russia there were sufficient mineral and energy resources to sustain industrialization. Russia also had a huge handicraft production, albeit largely peasant in origin and therefore closely tied to the village. There was therefore a sizeable population familiar with some facets of the manufacturing process. Still, industrial development in Russia was not just comparatively late, it was also distinctive because of the degree of state involvement, the scale of enterprise, the location, and the role of foreign capital.

State involvement in industry was of long-standing importance despite some notable efforts by Tsar Peter I (1682–1725) to foster private enterprise, both Russian and foreign. But in a state in which the few people capable of assuming responsibility for industrial development more often than not spurned industry (and sometimes commerce as well), for a position in the state bureaucracy, both entrepreneurial talent and financial resources were at a premium. Thus, from the outset, industrial development necessitated the purchase of foreign technology and administrative personnel. State orders, guaranteed profits, protective tariffs, conscription of state-owned serfs to labour in factory or mine, and encouragement of education, science and technology, all were intended to nurture industrialization. However, as each new military encounter with European powers served to illustrate, Russian industry and armaments remained weak links in the defence of the Empire. The ability of the state to defend its borders was further compromised by the policy of Tsar Nicholas I (1825–1855) not to modernize the internal transport system. At mid-nineteenth century, for example, the Empire had but a single railroad, that joining the capital St Petersburg with Moscow. A vast network of public highways and post-roads served the needs of internal transport at least during those seasons when movement was possible. When spring brought *rasputitsa*, or the time of mud, land transport came to a halt. Autumn rains produced the same state of *bezdorozh'ye*, or roadlessness, in Russia, a condition that lasted from several days to several weeks depending on prevailing climate and soil characteristics. River and canal transport were even more limited by the seasons than was the largely unpaved dirt track road system. The consequences were more than just economic; the Crimean War of 1854–1956 was resolved unsatisfactorily for Russia at least in part because road and river transport were inadequate to the task of supplying the forces on the Crimean peninsula. In war, technological innovation was now just as important as the ability to muster men. Backwardness, not modernity, had characterized much of the Empire's industrial structure, its military equipment, and the internal transport system. But following the Crimean War, industrial development was promoted as never before, especially in terms of the construction of rail and rolling stock in Russia. The Emancipation of the Serfs played an obvious role in the state's plans for modernization.

By the middle of the nineteenth century, there were about 10,000 manufacturing establishments with just over one-half million workers. In the most important sectors of industry, however, private enterprise played a decidedly minor role. As before, the state was not only a major purchaser of industrial goods, it was a major producer as well. The changes set in motion by the post-Crimean War development policies culminated in a period of frenetic industrialization around the turn of the century. Although the figures are only approximations, the number of factories and employees increased threefold between 1850 and 1890. Nearly 1.5 million people were now working in over 30,000 factories. By 1913, about 2.6 million workers were employed in the Empire's manufacturing establishments, nearly twice the figure of 1890. However, the number of factories had increased by only 50 per cent in the same period; the concentration of workers in large establishments, already a distinctive feature of Russian industrialization in 1890, was therefore accentuated. In 1910, for instance, more than half of the industrial workforce was employed in factories with more than 500 workers. In the United States, where economy of scale was the byword, barely one-third of the factory workforce was employed in such large enterprises. Amongst European countries, the share was smaller still. Of course, modernization is not necessarily to be equated with the number of workers in a factory; compared to both Europe and America, the quality, and cost, of Russian labour was lower. Low wages, of course, reflected low productivity. But labour could be, and was, substituted in some production processes for more expensive machinery, and this was one reason for such large individual factory workforces. It has also been suggested that because managerial and technical personnel were scarce in late imperial Russia, concentration of production in a single enterprise made more efficient use of the limited available talent. Whatever the reasons involved, the fact is that there were many very large factories. Thousands of workers in a single factory was not uncommon. Indeed, in 1913 a number of factories employed more than 10,000 workers each. State policies concerning industrialization in the nineteenth century were somewhat ambivalent. Positive endorsement appeared, for example, in the form of tariff protection (Russia had the highest average import duties of any major European industrial country in 1914), in the form of government purchase of domestic production at prices high enough to ensure large profits, and in the form of state involvement in industrial production, especially in the armaments sector. Yet there was no wish to have industrialization proceed without some measure of control. The experience in some European cities where a large 'lumpen-proletariat' had proven socially and politically disruptive was sufficient to worry those who sought to preserve the aristocracy. Thus, after the Emancipation of 1861, the state attempted to keep in place the means for controlling the movement of the peasants. Endorsement of the communal *mir* and its authority to issue permits to peasants wishing to depart the village temporarily went hand in hand with the maintenance of bureaucracy in the cities to monitor the 'coming and going' of such peasants. Factory workers in Russian cities consequently were slow to develop a sense of solidarity based upon long association with industrial production and permanent ties with the city. To be sure, when unions were legalized after 1905, the situation began to change, and quickly. But even so, many peasant factory workers retained close ties with the village.

The state had long encouraged factories to locate in the countryside. To some extent a rural orientation of industry was reinforced by handicraft activities, since these were mostly found in the villages, especially those of Central European Russia where only marginal prospects for earning a livelihood from the land had long ago fostered the development of alternative, non-agrarian occupations. Thus, a potential labour supply could be tapped. But perhaps a more important reason for promoting rural industry was that many government officials reckoned it would be potentially less disruptive than urban industry. As it turned out, this hope was not always realized, since strikes and labour unrest were by no means confined to the city. In encouraging industry to locate in the countryside, state policy was in fact reasonably successful. By 1902, for example, only 41 per cent of the one million factory hands lived in cities. In 1914, there were few urban- industrial regions in Russia, St Petersburg, Moscow, and Riga being the most notable exceptions. Together these three areas accounted for about 18 per cent of the Empire's factory workforce.

As Fig. 2.3 indicates, there was considerable regional variation in the distribution of industrial activity in Russia in the early 1900s. The area around Moscow was the most heavily industrialized, yet even here the vast majority of people still worked the land. The Baltic ports of St Petersburg and Riga were major industrial centres, more spatially concentrated than the Moscow region, and more modern. The central industrial district, focused on the city of Moscow, was still dominated by textile manufacture, while in St Petersburg and Riga, metallurgy and engineering accounted for most industrial jobs. Metallurgy, and food product industries, notably sugar beet processing, characterized the industrial structure of Ukraine. In the Baku region, the petroleum industry was particularly important, while in the Urals, mineral resource exploitation was the source of most industrial employment.

As Table 2.2 reveals, by the early 1900s the Empire's industrial structure remained dominated by the traditional sectors – textiles and foodstuffs. Still, there were clear signs of modernization. The metallurgical industries, particularly steel and engineering, had developed quickly and now commanded a reasonably sizeable share of total industrial employment. The expansion of the metalworking and machinery sector was very much associated with the programme of railroad construction initiated after the abortive Crimean War. At mid-nineteenth century, Russia could boast of barely 1600 kilometres of railroad; by 1910, there were more than 66,000. The expansion of the rail network drew whole regions into the world of national, and indeed international, markets. The demand for rails and rolling stock helped to shift the traditional emphasis on production of consumer goods, such as textiles and clothing, to producer goods, that is, items which are not in themselves end-products, but which are used to

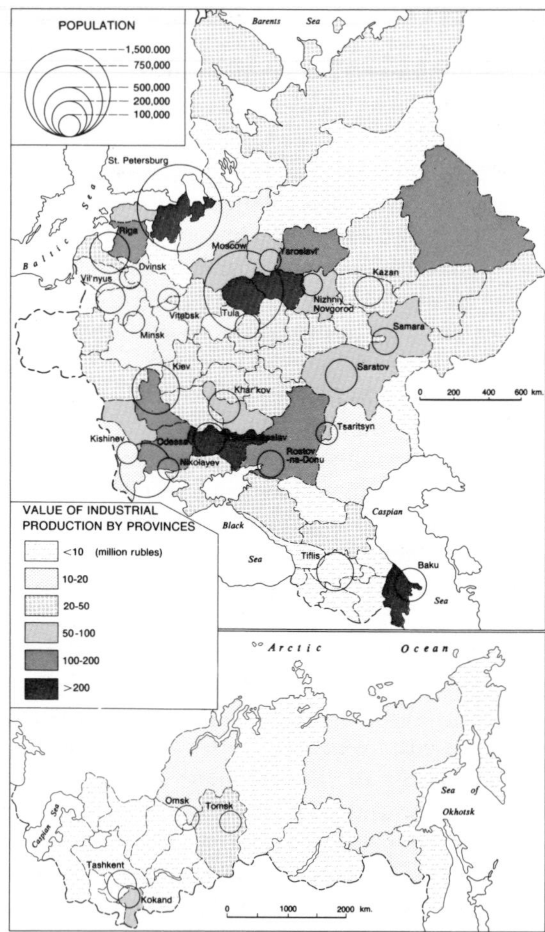

FIGURE 2.3 *Industrial regions in the Russian Empire, 1914* (*Sources*: Based on W.H. Parker, *An Historical Geography of Russia* (London: University of London Press, 1968), 316; P.I. Lyashchenko, *History of the National Economy of Russia to the 1917 Revolution* (New York: Macmillan, 1949), 595)

Table 2.2 *Industry in the Russian Empire: 1914*

<table>
<tr><th rowspan="2">Industrial group</th><th colspan="2">Workers</th><th colspan="2">Gross value of production</th></tr>
<tr><th>1897</th><th>1908</th><th>1897</th><th>1908</th></tr>
<tr><td>Textiles</td><td>30.6</td><td>36.5</td><td>33.3</td><td>29.8</td></tr>
<tr><td>Food products</td><td>12.2</td><td>17.1</td><td>22.8</td><td>33.9</td></tr>
<tr><td>Mining and metallurgy</td><td>25.9</td><td rowspan="2">24.5</td><td>13.9</td><td rowspan="2">16.4</td></tr>
<tr><td>Metalworking and machinery</td><td>10.2</td><td>10.9</td></tr>
<tr><td>Timber processing</td><td>4.1</td><td>4.1</td><td>3.6</td><td>3.7</td></tr>
<tr><td>Livestock products</td><td>3.1</td><td>2.8</td><td>4.7</td><td>3.5</td></tr>
<tr><td>Ceramics & building materials</td><td>6.8</td><td>–</td><td>2.9</td><td>–</td></tr>
<tr><td>Chemical</td><td>1.7</td><td>2.9</td><td>2.1</td><td>3.8</td></tr>
<tr><td>Paper</td><td>2.2</td><td>3.9</td><td>1.6</td><td>2.8</td></tr>
<tr><td>Other</td><td>3.2</td><td>8.2</td><td>4.2</td><td>6.1</td></tr>
<tr><td>Total</td><td>100.0</td><td>100.0</td><td>100.0</td><td>100.0</td></tr>
</table>

Source: Olga Crisp, 'Labour and industrialization in Russia,' in Mathias, Peter and Postan, M. M. (eds), *The Cambridge Economic History of Europe; The Industrial Economies, Capital, Labour and Enterprise; The United States, Japan and Russia* (Cambridge: Cambridge University Press, 1978), Vol. VII, Part 2, 354

manufacture other commodities. Consumer goods from rural handicraft, or *kustar'*, industries continued to find a market. Of course, when factory technology was applied to the manufacture of the same commodities, *kustar'* production was rarely able to compete in terms of price and suffered accordingly.

The modernization of the industrial structure suggested by the growth of the metalworking, machinery, and chemicals industries in Table 2.2 was made possible by a massive influx of foreign capital, especially from the late 1890s on. Foreign investment flowed in largest volumes into those industries where growth potential and profits were greatest. Mining, electrical engineering, rubber, chemicals, petroleum production – all are examples of industries dominated by foreign capital. In many instances, government support through tariff protection or lucrative contracts further boosted profits for foreign investors. In the early 1900s, capital investment on the part of Russian entrepreneurs was still concentrated in the traditional areas of industrial enterprise – textiles, clothing, woodworking, tobacco, and food products.

Only a small share of the Empire's total workforce actually toiled amidst the din and dust of the factory. But in absolute terms the numbers involved were still huge, and given the transient nature of at least a part of the factory labour force, many more people than the 2.6 million operatives counted in manufacturing establishments in 1913 would have had first-hand acquaintance with factory production. Of course, industrialization was much more than just factories and workers; for the recently arrived peasant whose life had been governed by the seasons, the daily regularity of habit dictated by the factory whistle and clock was often an entirely new, and not necessarily welcome, experience. The factory changed the conditions of employment for those who had some work experience in the city in trade, commerce, or domestic service, from the close personal relationships, good or bad, to impersonal, bureaucratic ones, thereby creating a void between owner or manager and employee. Industrialization required new financial structures; it required more transport facilities, if not new modes; it destroyed some handicraft activities and spawned others. Put simply, industrialization transformed in both subtle and obvious ways the whole of the national economy. It was both an example and an agent of modernization, and there were few parts of the Empire which did not experience some of the changes set in motion. But it was in the Russian city that the impact of industrialization was most evident.

URBANIZATION IN LATE NINETEENTH CENTURY RUSSIA

Russia's urban population was always small in relative terms. At mid-nineteenth century, about 3.4 million people were classified as urban, barely 5 per cent of the total population. By 1914, there were more than 28 million people living in urban places, but this was still less than one-fifth of the total population. In England, France, Germany, and the United States, for instance, urbanites now comprised 50 per cent or more of the total population. What is more, in Europe and America, urban growth was more nearly synonymous with urbanization; those who departed the countryside for the city usually did not return. Such migrants helped to change the city, and, in turn, their behaviour and attitudes were modified by the environment in which they now lived. In short, they became urbanized. In Russia, on the other hand, the development of cities and life within them was very much under the thumb of officialdom from at least the middle of the seventeenth century until well into the nineteenth. Peasant rural–urban migration was customarily monitored through permits and passes, and thus a peasant's sojourn into the city was intended to be short-term. The seasonal nature of residence in the city was by no means restricted to the peasantry, as we shall see shortly.

Urbanization, like industrialization, was both an example and an agent of modernization. The trappings of modern urban-industrialism, however, had little to do with the reality of the urban experience for perhaps the majority of the Russian Empire's urban inhabitants. On the eve of the World War I, for example, peasants comprised almost three-quarters of the total population in the Empire's two largest cities, St Petersburg and Moscow. This was nearly twice the share fifty years earlier, and resulted from the surge of peasant migration following the Stolypin Agrarian Reforms. The official *sosloviya*, or legal estates, were certainly not an accurate reflection of the urban class structure. But for three-quarters of the inhabitants of St Petersburg and Moscow who belonged to the peasant *sosloviya*, or estate, it is not unreasonable to suggest that they were little urbanized and even less urbane. St Petersburg and Moscow were dual economies in the sense that international industrial enterprises, joint-stock banks, companies with world-wide business dealings, and so on, lay like a thin veneer over the peasant economy which more often than not involved the itinerant pedlar and bazaar. In the city, there existed different social worlds as well as business ones, with prince and peasant representing opposite ends of the social class continuum. But in terms of sheer numbers, the world of the peasant was absolutely dominant. Moving down the urban hierarchy from capital city to provincial centre to country town, the social class structure of the population changed. Peasants comprised a smaller share of the population, while the *meshchane*, or petty trading estate, increased in proportion. But the links with the countryside were still palpable.

The proportion of the Russian Empire's population living in towns was never very great. The role of the city as a centre of cultural change was therefore different from its European and American counterparts. But who acquainted with the broad canvas of Russian history would assign to the city an importance commensurate with its share of the Empire's population? After all, the city was the scene of revolution. Indeed, it was in the city where the fundamental contradictions between the old and new, between the traditional values of Russian society and the forces of modernization, were to be seen in boldest relief. In the next few pages we will explore, albeit briefly, a few of these contradictions and some of their consequences.

After a long period of comparative neglect, the eighteenth century witnessed a renewed interest in the city on the part of the aristocracy. Peter I (1682–1725) and Catherine II (1762–1796), in particular, busied themselves with the creation of an ordered, and orderly, urban environment. The founding of St Petersburg in 1703, and its development after 1712 as the planned capital of Russia, served as a model for new town development elsewhere in the Empire. While what was accomplished there was rarely achieved in other settings, it nonetheless reflected the prevailing values of the aristocracy.

Throughout the eighteenth and early nineteenth centuries, the principal emphasis in urban development was on external form. However, the fascination with the geometry and symmetry of the town plan tended to mask the fact that it was more than simply a technical working document. Within the city, land-use segregation was a first priority. This went far beyond simply banishing polluting industry to peripheral locations, or dispatching industry to the countryside. The plan usually included provisions for the enforcement of social class segregation. While attempts to separate the various constituents met with only limited success owing to the costs involved and the less than perfect enforcement, the important point is that the town plan was used to reinforce the existing social order.

Despite the hundreds of plans created to guide the development of new towns and the reconstruction of existing ones, by the middle of the nineteenth century, the Russian city rarely gave much evidence of close adherence of ideal with reality. The fabric of the typical Russian town was still largely wooden. Even in the imperial capital, St Petersburg, wooden buildings prevailed, notwithstanding requirements to build with stone, bricks and mortar. Countless imperial edicts stressing the need for architectural consistency and harmony in the design of public and private buildings alike did little to alter the appearance of ancient cities like Kiev, Novgorod, or Moscow, where the main impression remained one of a rich confusion of design and construction materials. The smaller, more remote the town, the less did such regulations have any real meaning. Moreover, as the frontiers of the Empire came to embrace growing numbers of non-Slavic peoples, regional differences in urban form and fabric were accentuated. Cities of the Baltic region, such as Reval' (Tallinn), Riga, and Vil'na (Vilnius) had their own distinctive architecture. So, too, did more recently acquired Caucasian cities like Tiflis (Tbilisi), Yerevan, and Baku. Still to be brought into the Empire were the ancient urban settlements of Middle Asia, but at mid-century, the frontier was fast approaching them (Figs 2.1 and 2.3). While few cities were actually developed with close adherence to town-planning principles, they, like St Petersburg itself, often symbolized the aims and aspirations of the aristocracy.

The essence of planned development of cities was controlled, predictable growth. As urban-industrialization took hold in the latter part of the nineteenth century, the bureaucratic machinery for supervising rural–urban migration was severely tested. During the period from mid-nineteenth century to the Great War, the increasing tempo of urban growth demanded more attention be given by the authorities to the material conditions under which people lived, and that less attention be given to the plan and external appearance of the Russian city. But despite a conscious shift in emphasis, it came too late. Cities lacked autonomy, lacked finances, and consequently lacked sufficient municipal services to meet demands. From sewage systems to water supply networks to public transport to primary education, the urban infrastructure was inadequate, antediluvian, or both. When the Stolypin Agrarian Reform unleashed an unprecedented flood of rural migrants, any remaining possibility of an orderly development of the Russian city simply disappeared. Instead, there ensued a crisis of numbers. The urban population of Russia may not have been very large in relative terms on the eve of World War I, but the rate of increase, and the sheer mass of humanity it represented, simply outstripped the urban system's capacity to absorb it.

In the Russian city, males were numerically dominant, unlike European and American cities where females customarily outnumbered males. The principal reason for this demographic pattern was the Russian practice of permitting peasants to depart the village only for limited periods. Even after the Stolypin Agrarian Reform, when families could migrate to the city at will, the acute housing shortage helped to perpetuate sex-selective rural–urban migration. To be sure, the demographic profile of the Russian city was slowly changing, as more single females, and indeed wives and families, left the countryside for the town. But on the eve of World War I, there were many married men with reasonably secure city jobs paying higher than average wages who still maintained a family in the countryside. Even when tradition no longer governed behaviour, the crisis in housing forestalled the creation of a normal family life in the city for a great many working men.

Throughout Russia, the signs of an impending housing crisis were evident already in the 1860s. The rapid conversion of previously uninhabited cellars, the year-round occupancy of suburban summer *dachi*, or cottages, the constant carving up of the existing housing stock into smaller and smaller rental units, until in all too many instances, people were forced to live, literally, in just a corner of a room, all of this and more testified to the deterioration of the housing scene. Rents escalated, and in their wake appeared more and more jerry-built tenements. In the early 1900s, there were three times as many people per apartment in St Petersburg and Moscow as in Vienna, Paris, Berlin, or London.

In addition to the hardships produced by the acute housing shortage, there was disease, the spread and persistence of which were causally linked to deficient municipal services, especially water supply and sewage networks. Throughout urban Russia, the onset of spring was regarded as a mixed blessing, for the easing of the icy grip of winter was sure to bring infectious disease. Maintaining satisfactory standards of public health was perhaps the most important task confronting municipal government. Cities large and small alike, however, generally failed to fulfil this particular mandate. Ports like Odessa, Baku, Astrakhan, and even the capital, St Petersburg, were notorious bastions of infectious disease. But serious epidemics rarely spared cities further inland. For example, in the short period from 1883 to 1917, the citizens of Moscow were ravaged by no less than 32 outbreaks of smallpox, cholera, and typhus. In the smaller provincial towns, municipal services such as centralized water supply and sewage systems were usually quite limited. This situation had a pronounced negative influence on rates of mortality. In few other countries were cities so hazardous to live in. The brunt of epidemic disease was of course borne by the masses who were descending upon the Russian city in ever greater number in the early 1900s. Infectious disease regularly accounted for a third or more of all deaths in cities. While births exceeded deaths on a more regular basis in the late nineteenth century, even in the early 1900s it was not unusual for cities to register more deaths than births in some years.

Migration was the principal source of urban growth throughout most of the imperial era. In light of the abysmal state of public health, those who came to the city had little reason to stay very long. Indeed, for most members of urban society who had some choice in the matter, residence in the city was often very much a seasonal affair. For the social elites, especially the nobility, winter was the time to sojourn in the city, since it was during that period that the social season was in full swing. With the arrival of spring, many of those who could, departed for the *dacha*, or estate. Peasants, too, had limited attachment to the city, though as we have noted, for rather different reasons than the elites. But whether because of social custom, expiry of permit, inadequate or too-costly housing, or the legitimate fear of being caught up in the regular outbreak of epidemic disease, transience was an ingrained feature of urban life in the Russian Empire. Indeed, even those who ostensibly had some claim to permanent urban residence were very much on the move. Frequent changes of address (made easier owing to the very high proportion of rental accommodation in the housing stock), were common for those people listed in the city directory. High levels of transience meant many things, and among them were less commitment to the city, less concern with its management, and its resultant environmental quality.

Change usually associated with rapid urban-industrialization was in Russia inhibited by technological backwardness and by ingrained social custom. Technological backwardness was manifested in many ways in the Russian city. For example, the limited development of public transport meant that the voluntary residential segregation of particular groups, especially the elites and emerging middle classes, in socially homogeneous suburbs was very severely restricted. In contrast to many European and North American cities, where the process of residential segregation had long since left an indelible stamp on urban form, in some of the larger Russian cities a kind of three-dimensional residential segregation occurred. In the multi-storeyed built environment of central St Petersburg, for example, the lower classes were often forced to live in cellars and garrets (Fig. 2.4). On the floors in between lived a heterogeneous mix of social groups. Toward the periphery of the city, land-use intensity fell off, and the proportion of the lower classes amongst the population increased correspondingly. Ability to pay was as often reflected in the size of the apartment and the manner in which it was furnished as it was through an exclusive claim to territory. In effect, the lavish decoration of living space might help to insulate social elites from the poor who might be housed above and below in the same building. But once on the street, the social elites were dependent on the time-honoured personal symbols of rank and status, that is, upon uniform or dress, to proclaim their place amidst the seeming confusion of classes and activities. However, on the eve of World War I, such symbols no longer guaranteed a deferential response from the burgeoning masses. Indeed, at a time when the status quo in Russian society was increasingly being called into question, such personalized symbols of rank and class might well have heightened resentment. Deficient municipal services and hypercongestion had already made the Russian city uncommonly hazardous in terms of public health. For the social elites and the leading government bureaucrats, residence in the city was becoming increasingly hazardous in other ways as well.

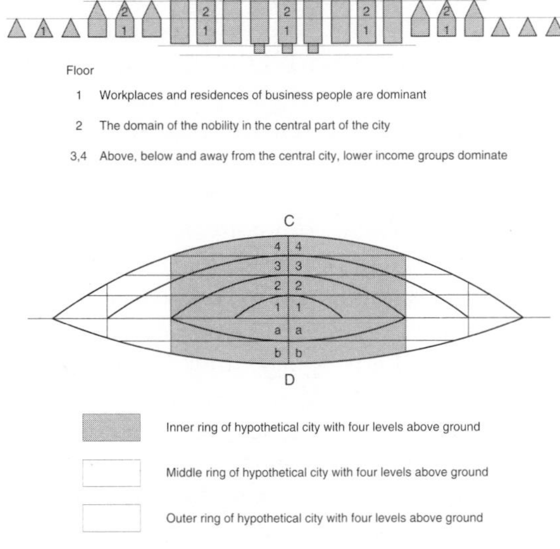

FIGURE 2.4 *St Petersburg: model of social geography* (*Source*: Based on J.H. Bater, *St Petersburg. Industrialization and Change* (London: Edward Arnold, 1976), 380)

REFORM AND REVOLUTION

Alexander II, the Tsar who had 'liberated' the serfs by passing the Emancipation Act of 1861, was assassinated in 1881. Thereafter, despite a vastly expanded network of police, spies and agent-provocateurs, the government was unable to stem the anarchist and revolutionary movements. Of course, many legitimate grievances served to generate support for groups advocating social, economic, and political reform, whether by peaceful or violent means. As was noted earlier, at the turn of the century, Russia was in the grip of a severe depression. Famines, and then riots, spread across the countryside. In the cities, the cost of living was rapidly outstripping the rise in wages. Unions were illegal, strikes were illegal. There was no real mechanism for legitimately voicing the concerns and aspirations of the majority of the population. In the city, there was elected municipal government. However, under the highly restrictive franchise imposed by the state, usually less than one percent of a city's population was entitled to vote. Few bothered. Thus, the potential for social reform on the part of municipal government was severely limited. From time to time, strikes occurred, invariably to be put down by brute force. Resentment festered. After the turn of the century, each year witnessed more instances of resistance to authority. A growing list of prominent people fell victim to assassination. Years of economic depression and growing social unrest culminated in 1905 in widespread strikes and mass demonstrations to publicize perceived injustices and inequities, as we have already seen. The debacle of the Russo-Japanese War had brought matters to a head. The social, economic, and political upheaval which occurred in 1905 was in every sense a revolution. It was not the first revolution in Russia, nor was it to be the last.

The Stolypin Agrarian Reform was one manifestation of Tsar Nicholas II's attempt to mollify public opinion and accommodate change. An equally significant gesture was the creation in 1906 of a *Duma*, or Consultative Assembly, to which members would be elected. The existence of a state *Duma* meant that political parties were now legal in the Empire. Still, the authority of the Tsar was not entirely emasculated. On issues vital to national interests, the Tsar had veto power over legislation passed by the *Duma*. Indeed, the very existence of a *Duma* was dependent upon imperial goodwill. For example, after elections to the first *Duma* revealed a strong liberal sentiment amongst its members, it was disbanded by the Tsar and the election procedures were changed. Manipulation of the franchise resulted in a political complexion more or less satisfactory to the Tsar and his advisors. All parties, including the leader of the revolutionary Bolsheviks, Vladimir Lenin, now had a legal, and public, forum in which issues could be raised and debated.

Through a combination of reforms, conciliatory gesture, and, wherever necessary, brute force, the upheaval of 1905 was finally subdued by the Russian government forces. In the process, grass-roots organizations of workers called 'soviets' were abolished. These had appeared spontaneously during 1905. Aside from coordinating strikes, the soviets had on occasion helped to maintain public order where local government and police had fallen into disarray. During 1905, the formation of soviets became a more formal democratic process. All factories and trades, for instance, were encouraged to elect delegates to local soviets on the basis of one delegate for every 100 workers. The network of soviets was centred on the capital, St Petersburg. The spread of an overtly socialist, grass-roots organization had been rapid. Clearly it could not be tolerated under an aristocracy. But it proved simpler to suppress the formal trappings of this organization than it did the idea.

In the years following the revolution of 1905, the process of urban-industrialization quickened perceptibly. But the fundamental contradictions inherent in an aristocracy persisted. And so, too, did the privation experienced both by large numbers of urban workers and by millions of peasants dependent for a living upon marginal land. As the demand for labour increased, vast numbers departed the countryside for the town, only to be confronted by an inflationary spiral in which wages lagged behind the already high costs of food, fuel, and housing. Social unrest was always a potential threat to the stability of the Empire.

The outbreak of World War I in 1914 fired the nationalism of political leaders and general populace alike. Amidst the initial outburst of enthusiasm to defend the interests of Slavs everywhere, the name of the capital itself was changed from St Petersburg to Petrograd in deference to anti-German sentiment. However, as the war effort dragged on, as casualties mounted, eventually reaching the millions, as management of the armed forces, domestic economy, and government became increasingly inept, dissatisfaction rose. From 1914 to 1917, the annual number of strikes increased from fewer than 100 to 1300, and this was just in those factories subject to inspection. As the domestic economy deteriorated, food shortages grew worse, and food riots in the cities became commonplace. The cost of living soared. Hours of work in factories increased as overtime was demanded of the labour force, a growing proportion of which was made up of women and children. The mobilization of 14 million men left the factory and farm short of labour in general, and it seriously depleted the ranks of skilled workers in particular. In the growing chaos, political activism by socialist and revolutionary groups increased. Meanwhile, from the state *Duma* came appeals to maintain, indeed augment, the war effort. These increasingly fell upon deaf ears. Government was ineffectual. The personal involvement of Tsar

Nicholas II in military strategy was a failure. Surrounded by sycophants, poorly advised, uncertain in action, and dependent upon the advice of his wife Alexandria, who in turn was manipulated by the infamous Gregory Rasputin, Nicholas II and all he stood for was discredited. He abdicated in March 1917. Failure to find a successor to Nicholas II created an unprecedented crisis. Just a few days before abdicating, Nicholas dissolved the fourth, and last, state *Duma*. The Empire now had neither Tsar nor government. As an interim measure, a Provisional Government was formed. But the Provisional Government was handicapped by the very circumstances giving rise to it – world war, a national economy in near ruin, and political conflict. Though committed to the task of bringing into existence a democratically elected Constituent Assembly, the exigencies of the times and the absence of any tradition of consensual politics made a daunting task all the more difficult. Election of the new Constituent Assembly was eventually set for the autumn of 1917. However, a revolutionary solution to the domestic chaos in Russia had already been set in motion.

From Tsar to soviets

The economic chaos during the war and the abdication of the Tsar were sufficient conditions for the re-emergence of the system of soviets. Indeed, during the eight months of Provisional Government, real political power rapidly built up in the network of soviets, once again centred in the capital. Vladimir Lenin returned to Russia in the spring of 1917 after a long period of exile in Europe. While abroad, he shaped the policies and programmes of the Bolsheviks, a faction of the Russian Social Democratic Workers Party he had led since its formation in 1903. Having recognized the political significance of the fast-expanding soviets, Lenin added the slogan 'All power to the soviets' to the Bolshevik calls for an end to the war and a redistribution of agricultural land. Most probably the call to end the war and to redistribute the land would have been more comprehensible to the masses than 'All power to the soviets'. But as the soviet system spread from factory to armed forces to the countryside, its significance began to be appreciated by a growing number of people, including those in the Provisional Government. Indeed, before long, important policy decisions taken by the Provisional Government had to be approved by the soviets in Petrograd.

The Bolsheviks twice attempted to overthrow the Provisional Government in 1917. The first effort in July was a failure and resulted in Lenin fleeing Petrograd to take refuge in nearby Finnish territory. In early autumn, he secretly returned to Petrograd. With a growing Bolshevik presence in the key soviets in the factories and garrisons of the capital, Lenin was able to plan another attempt to overthrow a weakened and increasingly demoralized Provisional Government, a government with no real popular support, a government in name only. This occurred on 25 October 1917, on the eve of the Second All Russian Congress of Workers and Soldiers Deputies. The Bolshevik forces met with scarcely any resistance. With the Provisional Government now gone, the Congress, comprising delegates from the soviets which had spread across much of the country, was confronted with a successful *coup d'état*. A motion was adopted to the effect that the Executive Committee of the Congress should administer the country until the election of the Constituent Assembly. This had been set for 25 November by the now defunct Provisional Government. The 'revolution' of 25 October 1917 did not immediately give power to the Bolsheviks. Indeed, within the soviets across the country Bolsheviks were customarily a minority. But in the Congress's Executive there were a number of Bolsheviks, including Lenin himself. Thus, political power was determined not so much by the number of delegates, but by control over the vital centres of decision-making. It was in this context that the Executive of the Congress was so singularly important.

The election of a Constituent Assembly took place as planned, but it demonstrated that, throughout Russia, support for the Bolsheviks was limited. Still, as the real political power came increasingly into the hands of the soviets, and especially their executive committees, this was not a major problem. By early 1918, Lenin had mustered support for a decree to dissolve the newly elected Assembly. Political authority now resided in the network of soviets, over which the Bolsheviks exercised control. By the summer of that year, a new constitution was adopted in which the soviets became the key element in the new political reality. Peace had been declared on 25 October 1917, but it was only in March 1918 that the war with Germany was formally concluded by the Treaty of Brest-Litovsk. Peasants needed little encouragement to displace estate owners and distribute the land following the decree of 26 October 1917 to nationalize land. Thus, in a matter of a few short months, the Bolsheviks had realized three key objectives – giving political power to the soviets, redistributing land, and ending Russia's involvement in World War I.

The revolution of 1917 confirmed the demise of autocracy and ensured the same fate for capitalism. It also triggered nationalist aspirations amongst a number of the minorities who had been incorporated into the Russian Empire during the centuries-long process of territorial expansion, because the Bolsheviks, led by Lenin, decreed the right of all nationalities to self-determination on 25 October. Finland, Latvia, Lithuania, and Estonia won their independence after the war, while republics were declared, but not sustained, in Ukraine and the Caucasus. Other territories on the western flank of the old empire, most notably Poland, were lost in the process of creating the new states of Eastern Europe following the Treaty of Versailles. Within the new Soviet state, however, the economy was in chaos. Even political control was questionable as civil war erupted in the summer of 1918.

The Bolsheviks had finessed their way to power, rather than winning outright majority support. Coalitions with other socialist parties had helped, but these were fragile arrangements and soon crumbled. Meanwhile, resistance to socialists in general, and Bolsheviks in particular, mounted. The anti-Bolshevik forces were varied in their political complexion. The Mensheviks (created at the time of the factionalization of the Russian Social Democratic Workers' Party in 1903, which produced the Bolshevik, or literally if not accurately, the majority element of the same party) and the Socialist Revolutionaries criticized, but were not counter-revolutionary in action. Parties of the centre and right, especially those supportive of the monarchist and now dispossessed landed gentry, were in forcible opposition. These anti-Bolshevik, or White, coalitions and groups under such leaders as Wrangel, Denekin, and Kolchak, reduced the Bolshevik-controlled territories to a core area of European Russia focused on Moscow. But they lacked popular support, especially amongst the peasantry, who had been only too willing to 'redistribute' estate lands. As the Red Army was reorganized, largely through the efforts of Trotsky, the anti-Bolshevik forces were slowly pushed back. The direct intervention of some Allied countries during the summer of 1918 further complicated matters. Ostensibly, Allied intervention was intended to prevent the German forces from acquiring munitions supplies located in Russia. But some moral support was also provided to the Whites. Certainly the reconstitution of an eastern front was a widely shared objective amongst the Allied countries, but their intervention also played into the hands of the Bolsheviks, who quickly capitalized on the opportunity to embark on a propaganda war against Allies and Whites alike. The civil war probably would have been concluded earlier were it not for Allied intervention. In any event, it was not until the autumn of 1920 that the last of the anti-Bolshevik forces was defeated.

In the years of civil war following the revolution, the fabric of society and economy alike was pulled apart. Many Bolshevik policies simply exacerbated the disintegration of the domestic economy. In the attempt to remove capitalist structures and replace them with socialist forms, a number of irrational elements were introduced into the economy. It was not until the mid-1920s that conditions in the new Soviet state assumed a measure of normalcy. The old social order of course was no more; but what socialism actually meant in terms of a new socialist society had still to be determined. For the old elites of Russian society the fall from grace was precipitous. Fashioning a new socialist society, however, was not to prove very simple. Egalitarianism and social justice for all proved far easier to proclaim than achieve. Indeed, privilege continued to play a role in fashioning the new Soviet society, albeit in new and usually more subtle ways than in the past.

The structure of Soviet socialist society – rhetoric and reality

In imperial Russia people were differentiated by nationality, culture, tradition and degree of urbanization. As well, there were various bureaucratic methods of pigeon-holing the constituents of society, numerous barriers to social mobility, and a well-ensconced system for defining the niceties of inter-personal and inter-class behaviour. Put simply, everyone had their place and was presumed to know it. The Revolution of 1917 really did portend a new social structure, even if the exact nature of a society under conditions of socialism was little understood. The population was now out from under the thumb of patrimonial, autocratic Russia, but there was a long road between the reality of day-to-day life in 1917 and the Utopian existence portended by the revolution. For many people the prospects were exciting, but they were also uncertain, since there was no existing socialist system to use as a model, and only vague guidelines in the theory available on socialism. In the early years of Soviet power truly egalitarian precepts influenced the laws and decrees issued by the new government. With the abolition of all forms of privatism the ultimate objective, a new society, was in the making.

In the Marxist analysis of capitalist society, class positions are determined by the relationship to the means of production. Thus, the owners of the means of production are the ruling class who exploit the proletariat, who in turn are obliged to sell their labour under disadvantageous circumstances. Under capitalism the state sanctions a system in which one class exploits another, and in which power and status of the industrial bourgeoisie mediate all the social relationships. Thus the abolition of capitalism, by definition, would remove political and social inequality. That in short was the theory.

In 1936 the Soviet Union was officially designated a socialist society and a new system of pigeon-holing members of society was legally ensconced. At this date there was still a small contingent of individual peasant farmers and *kustari*, or handicraftsmen, as Table 2.3 indicates. Still, the official transformation of Soviet society was clearly underway as the abolition of the bourgeoisie, traders and *kulaki* (or wealthy peasant class), category between 1926 and 1939 suggests. Officially the new social order comprised two social classes and one social stratum – workers, collective farmers, and the *intelligentsia*. As the working class under conditions of socialism is no longer exploited by the owners of the means of production, which are now owned by the state, that is by the people themselves, it was therefore not a proletariat. Collective farmers were deemed a separate social class because they were engaged in a form of production in which they, in theory at least, owned the output, a situation discussed more fully in a later chapter. The third official group, or stratum, was the *intelligentsia*, which was distinguished from the working class by virtue of being involved in intellectual as opposed to manual labour. Within each of these three official components of the new social order was a plethora of occupations. For example, within the *intelligentsia* the occupations ran the gamut from clerks, to authors, to senior bureaucrats of the ruling Communist Party (as the Bolshevik faction of the Russian Social Democratic Workers' Party came to be known after March 1918). As the data in Table 2.3 make plain, during the Soviet era there were significant changes in the official social class structure. Most notable was the steady increase in the relative importance of the *intelligentsia*. The concept of social mobility, that is, people moving up some type of social hierarchy which conveyed a rise in material possessions, prestige, privilege, and therefore status, was traditionally down-played during most of the Soviet period. However, there were in fact preferred occupations which did bring different social status, income and privilege. The changes in the official social class categories, depicted by the data presented in Table 2.3, in fact paled by comparison with the societal changes which actually occurred in day-to-day life.

The key to social class mobility in the real world was higher education. By the end of the Soviet era more than 20 million people had some type of higher education. The true measure of this facet of the transformation of Soviet society is that at the time of the Revolution of 1917 about 70 per cent of the total population was illiterate. To be sure, the state had an obvious vested interest in educating the population. But the opportunities provided were pursued with zeal. There quickly emerged an internal pecking order within the system of higher education such that, as in other countries, a graduate of an institution such as Moscow State University was more highly regarded than one from a provincial institution. A higher education opened the door to occupational mobility. And this in the nature of things produced a social class fabric far more intricate in its detail, and nuanced in terms of relationships between social groups, than is conveyed by the 'official statistics' on the two official classes and one stratum (Table 2.3). Soviet policies on wages, salaries and accumulation of wealth, all of which have an obvious bearing on the material measures of status, furthered the process of social class differentiation in the real world.

Under the Soviet leader, Josef Stalin, egalitarianism was in practice abandoned, public proclamations to the contrary notwithstanding. Stalin came to power after the death of Lenin in 1924 and was the main architect of the Soviet industrialization drive which began in 1928. The success of Stalin's industrialization drive was dependent upon the willing participation of the then rather small portion of the labour force which was educationally and technically competent to run industry. Thus, after the government had squeezed wage differentials to a minimum ratio between lowest and highest of around 1.75 in the first years of the Soviet era as part of its promotion of egalitarianism, under Stalin they were widened to reflect skill requirements throughout all sectors of the economy. Salaries of managers were raised so as to nurture incentive. Piecework became commonplace in the factories as well, thereby linking higher production with higher remuneration. Moreover, laws pertaining to inheritance were changed to permit the transfer from one generation to another of accumulated wealth in the form of, for example, money, household furnishings, and a *dacha,* or summer cottage. These and other developments in the 1930s were intended to co-opt the participation of the *intelligentsia* in the drive to industrialize. The legal underpinnings for the emergence of a middle class, materialistic in spirit and upwardly mobile in ambition, were thus put in place. Similar chan-

Table 2.3 *Soviet-era social class structure: percentage change, 1928–1989*

	1928	1939	1959	1979	1985	1987	1989
Working class of which:	17.6	50.2	68.3	85.1	87.6	88.0	88.1
Manual	12.4	33.7	50.2	60.0	61.6	61.8	58.8
Intelligentsia	5.2	16.5	18.1	25.1	26.0	26.2	29.3
Collective farmers	2.9	47.2	31.4	14.9	12.4	12.0	11.7
Individual peasant farmers and *kustari*	74.9	2.6	0.3	–	–	–	
Bourgeoisie, traders, *kulaki*, and others	4.6	–	–	–	–	–	0.2
Total	100.0	100.0	100.0	100.0	100.0	100.0	100.0

Source: Strana Sovetov za 50 Let (Moscow: Finansy i Statistika, 1967), 3; *Narodnoye Khozyaystvo SSSR v 1984g* (Moscow: Finansy i Statistika, 1985), 7; *Narodnoye Khozyaystvo SSSR za 70 Let* (Moscow: Finansy i Statistika, 1987), 11; *Narodnoye Khozyaystvo SSSR v 1989g* (Moscow: Finansy i Statistika, 1990), 33

ges occurred in the military. Ranks were restored in the 1930s after having been abolished in the first flurry of egalitarian decrees issued by the Soviet government after the revolution. Across the whole of Soviet society, state practice, if not official policy, served to enhance social differentiation rather than diminish it. And this at the same time that the Soviet Union was officially proclaimed to be the world's first socialist state!

In any system some jobs are more desired than others, and generally speaking, those people successful in acquiring them benefit accordingly. With the most highly valued jobs comes high social status. The Soviet system was no exception. Indeed, for those occupations deemed especially important, there quickly emerged a wide range of perquisites. These included priority access to better quality state housing, better quality health care, access to special food stores, possibly even a car, chauffeur, and domestic help. In an economic system in which shortage of all consumer goods, and near the end of the Soviet era food stuffs as well, was endemic, such perks were clearly very important. Privileges such as these, and others besides, created a very privileged elite, which by the end of the Soviet period ranged in size from a few hundred thousand to several million people depending on the definition used. Entry into this elite strata of society was controlled by the state of course. A key component of this elite group were the officials of the Communist Party. While joining the Communist Party was by no means the only route to upward occupational and social mobility, it clearly more often than not facilitated both. By the end of the Soviet era there was a very sizeable middle class as well.

The emergence of real-world social class groups, distinctions and relationships lay outside the official, and largely theoretical, purview of Soviet socialist society. The idealized harmonious relationship between manual workers, collective farmers and the *intelligentsia* which official doctrine touted was divorced from the far more complex social class relationships which actually characterized Soviet society. In all societies, reality differs from official census or other bureaucratically determined social categories. The stated goal of the Soviet system, however, was to break down barriers between social class groups, to promote interaction between them, to create a new, egalitarian social order. But over time society became ever more stratified, aided by legislation introduced by the state itself. Social interaction tended, not surprisingly, to be conditioned more by peer group interests than by who happened to live in the apartment next door. Social interaction was most certainly not much influenced by proclamation and exhortation on the part of the state. Perhaps the most notable continuity in the social structure during the Soviet era was the position in it of the *kolkhozniki*, or peasants. Recently arrived members of this huge segment of society were generally as readily identifiable in the cities at the end of the Soviet era as they were at the outset. They occupied the bottom rung of the social class ladder throughout.

THE NATURE OF THE SOVIET POLITICAL-ADMINISTRATIVE SYSTEM

Shortly after the Bolsheviks seized power all resources, including land, were nationalized. The state also assumed control over many sectors of the economy, including banking and foreign trade. Some elements of privatism were immediately abolished; others were soon to follow. In the place of privatism, the ethos of collectivism was to be cultivated instead. To usher in the new order required that plans be formulated, decisions taken, results monitored. All of this demanded new bureaucracies and countless officials to staff them. In this process the Communist Party and the soviets were destined to play central roles. Fundamental changes were required. The first task entailed putting in place a system of government.

The adoption of a constitution for the Russian Soviet Federative Socialist Republic in July 1918 enshrined the Soviet in the new political reality. This served as a model for the constitution ratified for the Union of Soviet Socialist Republics in January 1924. As a federal system, the USSR on the eve of its dissolution comprised 15 Soviet

Socialist republics, which in turn contained 20 Autonomous Soviet Socialist Republics, eight Autonomous Oblasts and ten Autonomous Okrugs (see Fig. 2.5). The highest-order unit was the Soviet Socialist Republic. According to the Soviet constitution, the nationalities so represented had all the rights of a sovereign state, including the right to secede. For that reason, all such republics were peripheral to the Russian Soviet Federative Socialist Republic, as a careful inspection of Fig. 2.5 reveals. A number of nationalities residing within the Russian Soviet Federative Socialist Republic had Autonomous Soviet Socialist Republic status, although they outnumbered several of the minorities having full republic status. Theoretically, the titular nationality of a republic comprised the majority of the population. This was true save for the Kazakh and Kyrgyz peoples. Given the reality of the Soviet system, however, the *de jure* constitutional right to secede, to establish separate foreign relations and so on, was entirely academic.

Government authority until late in the Soviet era resided in the Supreme Soviet (Fig. 2.6). This bicameral legislature comprised the Soviet of the Union, to which members were elected on the basis of one delegate for every 300,000 people, and the Soviet of Nationalities, to which members were elected according to the status of the political-administrative unit. Those nationalities represented in Fig. 2.5 had the right to send delegates in the following numbers: 32 for each Soviet Socialist Republic; 11 for each Autonomous Soviet Socialist Republic; five for each Autonomous Oblast; and one for each Autonomous Okrug. There were many other nationalities in the Soviet Union, but owing to their lower political-administrative status, they were not directly represented in the Soviet of Nationalities.

Administrative functions during most of the Soviet period were carried out mainly by the Council of Ministers, a body elected from the Supreme Soviet and responsible to it, or the Presidium. The latter, which was also elected from the Supreme Soviet, was like an inner cabinet and exercised legislative authority between sessions. In theory, the Supreme Soviet, which met only occasionally, could revoke measures adopted by the Council of Ministers. In practice it did not. Thus, the day-to-day business of government at the national level fell very much on the shoulders of executive committees of one kind or another. The Council of Ministers had about 100 members and met four times a year. It included not just those responsible for various ministries, but beginning in the late 1970s, the chairmen of the State Committees as well. State Committees dealt with a great many concerns. For example, there was Gosplan, the key state planning agency; the KGB, or internal police; science and technology; environmental protection; and the protection of mother and child. The administrative structure of the Supreme Soviet, Council of Ministers, and so on, was replicated in each Soviet Socialist Republic.

Up until the reforms in governance introduced in the late 1980s, the delegates sent to the Supreme Soviet were invariably Communist Party nominees, and elections, while displaying near-perfect turnouts, were, from the standpoint of western democratic procedures, politically uncontested. A modification of the Supreme Soviet, Council of Ministers, and Presidium structure was a central element in the reforms initiated by the last leader of the Soviet Union, Mikhail Gorbachev (1985–1991). Instead of an arrangement in which the Supreme Soviet met twice a year, a new bicameral assembly was created as a result of elections held in the spring of 1989. Comprising 2250 members elected by secret ballot from slates of candidates, the new Congress of Peoples' Deputies was to convene once a year. Of its total membership, 750 representatives would be elected from various civic organizations, such as professional groups or trade unions, and therefore did not have to be members of the Communist Party as was previously the case. By definition, the remaining representatives would be, and therefore Party members still comprised the majority. Each elected representative was limited to two five-year terms. The principal distinction of the Gorbachev era political reform was that some 542 members of this bicameral assembly would be elected to a new permanent parliament, or Supreme Soviet, which would debate and enact both domestic and

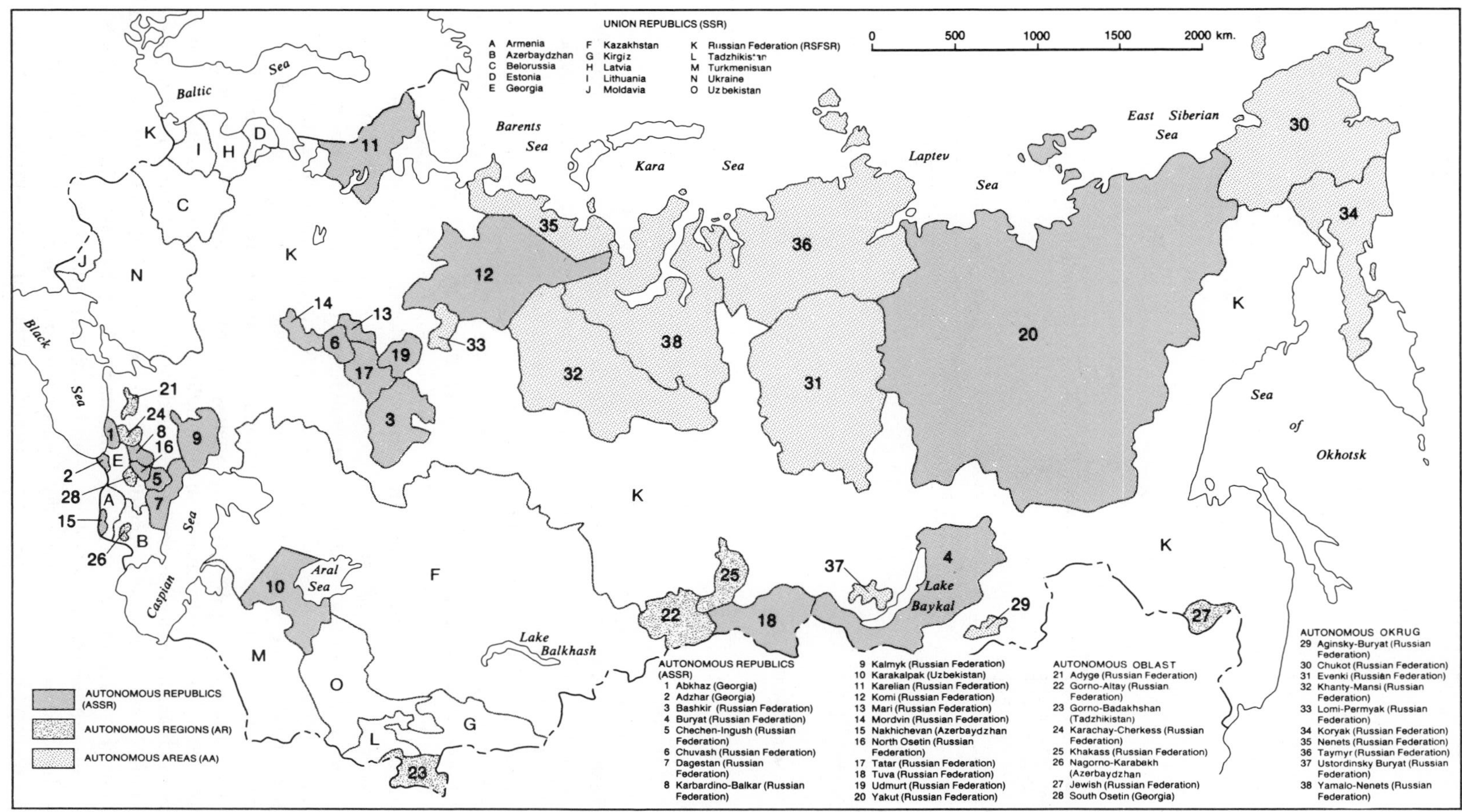

FIGURE 2.5 *Political administrative regions of USSR*
(*Source*: A. Brown *et al.* (eds), *The Cambridge Encyclopedia of Russia and the Soviet Union* (Cambridge: CUP, 1982), 306–7)

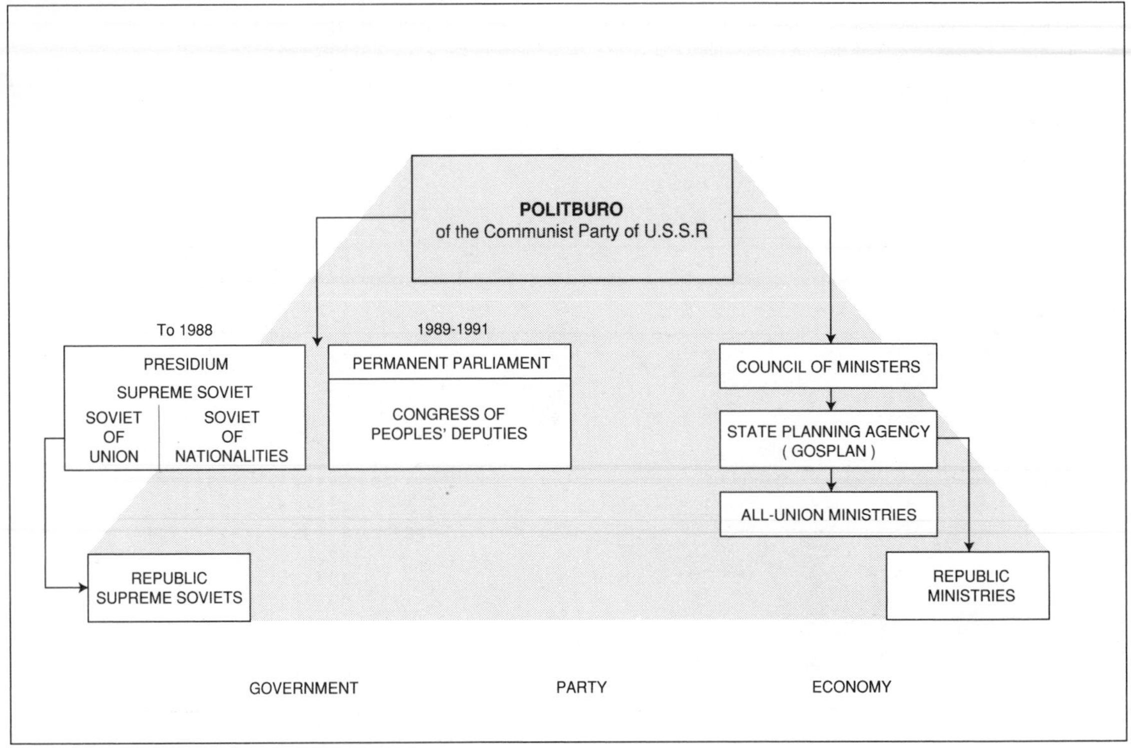

FIGURE 2.6 *Simplified structure of decision-making in the USSR*
(*Source*: J.H. Bater, *The Soviet City. Ideal and Reality* (London: Edward Arnold, 1980), 38)

foreign-policy legislation. The Council of Ministers and Presidium were to disappear, and direct Party involvement in the management of domestic affairs would wane. All of this was to be facilitated by a legal reform which would provide the necessary independence of state from Party. In keeping with these changes at the national level, the apparatus of government at the Republic and lower levels would similarly change. Modification of government structures and responsibilities during the Gorbachev regime was intended to revive the Soviet system, not destroy it. But by the time these changes were implemented in 1989, the Soviet system was in deep and severe crisis. While structures of government were modified, the most important decision-making body was rarely the target of reform. The Communist Party played a central role in all spheres of life throughout the Soviet era. It is important to have some sense of how the Communist Party functioned.

The clandestine nature of the Bolshevik activities in the early years of its existence gave credence to the need for strict discipline and a sense of hierarchy. But that did not preclude serious debate over policy alternatives. It was only after Lenin's death in 1924 and the assumption of power by Josef Stalin that the character of the Party, and debate within it, changed. It was not long after the arrival of Stalin that the typically free-wheeling, vigorous debate over ideological and policy matters within the Party was forcibly curtailed. Under Stalin, policies were increasingly determined at the top, while the mandate of those below was to implement, not to question. Greater discipline was achieved not just by squelching dissident opinion, but by purging those who held dissenting views. Mass arrests, particularly during the 1934–1938 period, political trials, exile, and execution on an unprecedented scale ensured a measure of compliance amongst Party members in particular and within Soviet society in general. Some attributes of totalitarianism were certainly present at this time, including the monopolization of power based on the ever more inflexible ideology underpinning the Communist Party, the bureaucratization of society, the obliteration of dissenting views, mass mobilization of society to attain common goals, usually, but not exclusively, related to production targets, and so on. The Communist Party that emerged out of this process of change was one in which the predispositions, indeed, the whims, of one person, Josef Stalin, often shaped policy and practice. Stalin's manipulation of Party affairs continued until his death in 1953.

The Communist Party increased its membership and shifted from the dominance by a single individual to what is usually described as a collective leadership. But at no time during the Soviet era was the membership of the Communist Party very large, at least in terms of the relative proportion of the total population. On the eve of the collapse of the Soviet system in 1991, the nearly 19 million Communist Party members, including those in a probationary status, comprised less than 7 per cent of the total population. Indeed, there were in 1991 probably more full-time Party officials than there were party members at the time of the revolution in November 1917! The basic structure of the Communist Party is outlined in Fig. 2.7. The hierarchical framework corresponded to the political-administrative organization of the state. At the top of the Party hierarchy was the Politburo, the key decision-making body. Its membership was small, sometimes no more than a baker's dozen in number, but it nonetheless comprised several different interest groups. The Communist Party Central Committee after Stalin's time actually emerged as an influential forum for debating policy alternatives. With 300-odd full members, a much broader range of viewpoints was likely to be found on any specific issue than was the case in the Politburo, if only because the median age of the Politburo members was for so long so much higher than in the Central Committee, and therefore the shared experiences of the past tended to play a more important role in shaping perceptions. The arrival of Mikhail Gorbachev as leader

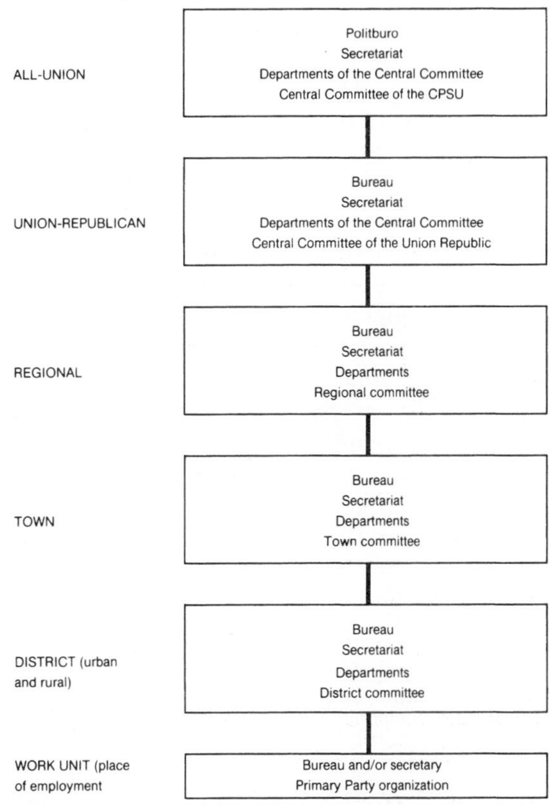

FIGURE 2.7 *Simplified structure of the Communist Party (CPSU)*
(*Source*: A. Brown *et al.* (eds), *The Cambridge Encyclopedia of Russia and the Soviet Union* (Cambridge: CUP, 1982), 296)

in 1984 brought change, however. He was the first youthful leader in decades. In attempting to consolidate his political position, he appointed a number of younger men to the Politburo and retired a number of the most elderly. As noted earlier, the Party bureaucracy itself was enormous. Within it, an elite had emerged with a voice in Party affairs.

Despite the demise of the 'cult of personality' associated with Josef Stalin, the rise of a collective leadership, and the appearance of various interest groups and forums for discussion, an essentially top-down decision-making process characterized Party affairs right to the end. The Politburo convened weekly, while the Central Committee which, in theory, elected it, met twice a year. The Party Congress in turn elected the Central Committee, but the roughly 5000 delegates to the Congress met only once every five years. Thus, the Politburo remained the key element in the decision-making process, even if diversity of opinion and consultation with other groups, including the middle class echelon of the Party membership, was increasingly tolerated during the final years of Soviet power. The role of the Party was all pervasive as Fig. 2.6 implies.

To become a Party member required more than simply filling out a form and paying monthly dues. Although the total membership was around 19 million by the late 1980s, it was by definition something of an elite. To join required approval, and this was usually the culmination of a long process of moving from youth group to the *Komsomol'*, (the All-Union Leninist Communist Union of Youth), to 'probationary' Party status. However, success in some fields of endeavour was often recognized through an invitation to join. Once a member of the Party, public and private behaviour was expected to conform to established norms, and each year witnessed the expulsion of some people for various transgressions. Ostensibly Party membership was monitored to ensure that all strata of Soviet society were adequately represented. As in most societies, however, participation in elitist organizations was usually sought by those wishing to get ahead, and over the years, joining the Communist Party was widely perceived as facilitating this objective.

Clearly, the Communist Party controlled the government (see Fig. 2.6). In less obvious ways, it intruded itself into every other facet of the Soviet daily life as well. To this end, the Party maintained a register of politically dependable people, a component to be sure of the Soviet social class elite, and a list of positions in which political reliability was deemed essential. The *nomenklatura*, as this list was known, was just one way through which compliance with central planning and other directives was exercised. Within the Party itself were organizations which paralleled, for example, those of the government, economy, culture, education, military, and police. Thus, Party members were found in key positions in factories, state farms, ballet companies, and so on. The responsibility of party members so employed was to monitor, to cajole, to manipulate as necessary in order to fulfil directives initiated above. However, as Soviet society became more sophisticated, as consumer expectations rose, the task of making the system work effectively became more difficult.

While the Party and its extensive bureaucracy were mandated to facilitate the realization of national objectives, there was growing, irrefutable evidence that it was becoming an obstacle to change itself. Under Gorbachev, radical remedies were proposed. Instead of a life-time sinecure, Party officials were to have limited terms of office and were to be democratically elected to them rather than just appointed. This was to apply to the office of General Secretary of the Communist Party, Gorbachev's own position, down to the 300-odd members of the Central Committee. After nearly three-quarters of a century of sinecured positions, save for the Stalin-era purges, the unease that such a proposed change caused amongst the Party's officialdom might be easily imagined. However, amongst the public at large, the proposal found much favour, since the privileges of Party officials were the object of considerable envy, if not criticism. Gorbachev's pursuit of greater democracy, under the banner of *glasnost'*, or openness, unleashed enormous public debate on many aspects of the reality of the Soviet Union which were previously taboo subjects. The privileges and sinecured positions of Party functionaries were certainly amongst them. Truly democratic elections was another.

From Party, to academia, to government, to the shop floor, the election of officials from a slate of candidates became possible in the Gorbachev era. Since democratic processes were to a large extent a lost art in the Soviet Union, the relatively free elections held in 1989 were the first opportunity at the national scale for voters to choose amongst candidates and from political factions other than the Communist Party. Thus, democratic processes were slowly being resuscitated during the final years of Soviet power. But not with a view to replacing the basic elements of the Soviet socialist system. The process of democratization initiated by Gorbachev was intended to resuscitate the Soviet economy by providing citizens with a greater say in how the system was to be managed.

What was intended was a more explicit leadership role for the Party and its bureaucracy, and less interference in the management of the economic system in particular and in affairs of day-to-day life in general. The bywords of the time were *glasnost'*, *demokratizatsiya*, *perestroyka*, or restructuring, and *uskoreniye*, or acceleration.

Although the March 1989 election was decisive in terms of offering alternative candidates to those from the Communist Party, barely 400 of the 2250 members of the new assembly, the Congress of People's Deputies, could be regarded as democrats. Still change had been initiated, and the demand for more became both more vocal, more grass-roots in origin and therefore more compelling. Gorbachev was elected Chairman of the 542 member Supreme Soviet of the Soviet Union. He also assumed the role of President of the Soviet Union, although he had not been popularly elected to the position. Gorbachev continued to hold the position of General Secretary of the Communist Party, the top position. In the summer of 1989 discontent amongst workers over the deteriorating state of the economy in general and food shortages in particular gave rise to workers' movements and strikes, most notably amongst miners. Gorbachev continued in his increasingly vain efforts both to reform the economy from the top down, and retain the leading role in society of the Communist Party. As Communist-led governments throughout eastern Europe fell in the face of democratic, revolutionary grass-root forces throughout 1989 and early 1990, Gorbachev, in order to stave off the spread of such movements into the Soviet Union, orchestrated the modification of the Soviet Constitution in February 1990 so that the Communist Party's monopoly on political power was formally abolished. But as with his attempts to reform the Soviet economy, described in the ensuing chapter, this was too little, too late.

The seeds sown by Gorbachev's initiatives to make the Soviet socialist system work better gave rise to movements which ultimately proved its undoing. Gorbachev attempted to decentralize decision-making authority, much as Khrushchev did in the mid-1950s. But in terms of decentralization Gorbachev was not intending to devolve real political power, rather he was in favour of deconcentrating it, that is, breaking the stranglehold which Moscow bureaucrats had over the detailed management of the economy. Like Khrushchev before him, Gorbachev wanted to make the socialist system more efficient and in the 1980s the way to achieve that objective was widely acknowledged to require giving some real authority to grass-roots level managers. In Gorbachev's mind that did not mean the abandonment of socialism, and certainly not the complete removal of the Communist Party from a pre-eminent position in Soviet society. Gorbachev seemingly believed that by deconcentrating decision-making authority, this objective could be realized and at the same time the essence of a socialist state could both be maintained and strengthened. What Gorbachev did not appreciate was the degree to which national minorities in the Republics and regions would fasten upon such an initiative to promote their own interests, interests which included rejection of the whole Soviet system. Very soon there were coalitions of interest groups springing up across the Soviet Union. Elections to Republic and local level governments occurred in early 1990. In many Republics, but most notably those of the Caucasus and the Baltic regions, nationalist independence movements gathered momentum, supported by a wide array of interest groups, including a very strong environmental movement. While Chernobyl' had served as something of a catalyst in bringing environmental degradation and public health-related issues to the fore, there had been a ground swell of resistance to the decision-making processes which had given rise to horrific levels of pollution in many parts of the country over the years. Now there was an opportunity to do something about them. But not as Gorbachev had intended through improvement of the existing system, rather through its outright rejection. Gorbachev put in place the mechanism for deconcentration of decision-making. By so doing he opened a 'Pandora's Box'. People perceived an opportunity to achieve a real devolution of authority, to achieve independence. They took it.

The local and Republic level elections of March 1990 returned substantial numbers of democrats and reformers, although there was a significant regional variation. In the major cities of Moscow and St Petersburg (formally Leningrad, renamed once again after the local elections in 1990, but this time in deference to anti-Soviet popular sentiment) democrats gained substantial absolute majorities. Elsewhere across the northern reaches of the Soviet Union (roughly north of the latitude of Moscow), reformers made substantial inroads, whereas in the southern regions the old guard, Communist Party establishment candidates dominated the election results. But everywhere people grew increasingly fed up with the continuing deterioration of the economy. Public resentment often manifested itself in mass demonstrations. In many regions, greater autonomy in economic affairs was perceived as necessary. In the largest Republic, the Russian Federation, its newly elected Congress formally moved adoption of a decree of State Sovereignty in June 1990. Within months even more extremist decrees were adopted in many other Republics. Nationalist movements gained momentum in many Republics of the Soviet Union, and calls for independence from some of them became louder. The constitutional ties that had bound the Union of Soviet Socialist Republics together for decades were fast coming undone.

Confrontation between the increasingly ineffectual central government in Moscow led by Gorbachev and independence-minded reformist Republican governments escalated in late 1990 and early 1991. The former was wracked by internal factions and disputes. Conservative Communist Party members and reformist elements could not agree on how best to accommodate change. The semi-democratic central government which had legislative responsibilities continually clashed with senior bureaucrats and Party officials, to say nothing of some members of the Congress of People's Deputies itself whose concept of decision-making was conditioned by years of those in positions of executive responsibility simply issuing orders. At all levels of government, there was tension between executive and legislative roles in the new era of democracy. Meanwhile, independence-minded Republican leaders became even more resolute in their efforts to leave the Union. The central government responded to overt manifestations of political independence by Republic governments in some instances with military intervention, in others with economic sanctions or even blockades. In March 1991, Gorbachev initiated a nationwide referendum which sought voters' opinion concerning the future of the Union. However, vague wording in the referendum gave no conclusive result. At the same time, plans were being put in place to draft a new Union Treaty between the Republics, one which would embrace the democratic principle of Republican sovereignty. Negotiations proceeded. A draft agreement was scheduled to be formally adopted August 20, 1991. But not all Republics participated in the negotiations. Including the Russian Republic, only ten of the 15 Republics in the Union were involved. The remaining five – Lithuania, Latvia, Estonia, Georgia and Armenia – refused to participate, indication enough of their common objective of complete political independence. Meanwhile, Gorbachev continued to push for economic reform and the creation of a market economy. A radical plan for a 500-day transition from a centrally-planned to a market economy was proposed, debated, and ultimately rejected by Gorbachev. The economic situation continued to deteriorate while debate over how to make the transition continued to rage in central government. Not surprisingly, there were many people who viewed the evolving political situation with alarm. Many more were shocked by the visible collapse of the economy.

Presidential elections in Russia in June 1991 confirmed Boris Yeltsin's rising stature as a populist leader in the government of the Russian Republic. Running against five other candidates, Yeltsin was elected with a convincing level of support in most regions across Russia (Fig. 2.8). The core of his support lay in the industrial heartland of European Russia, in the industrial belt of the Urals and West Siberia, in the Far East economic region, and in parts of the north Caucasus (see also Figs 9.13 and 9.14, pp. 280, 282) Unlike Gorbachev who had climbed up the Communist Party ladder to the top position, Yeltsin had faced an electorate and won. Within

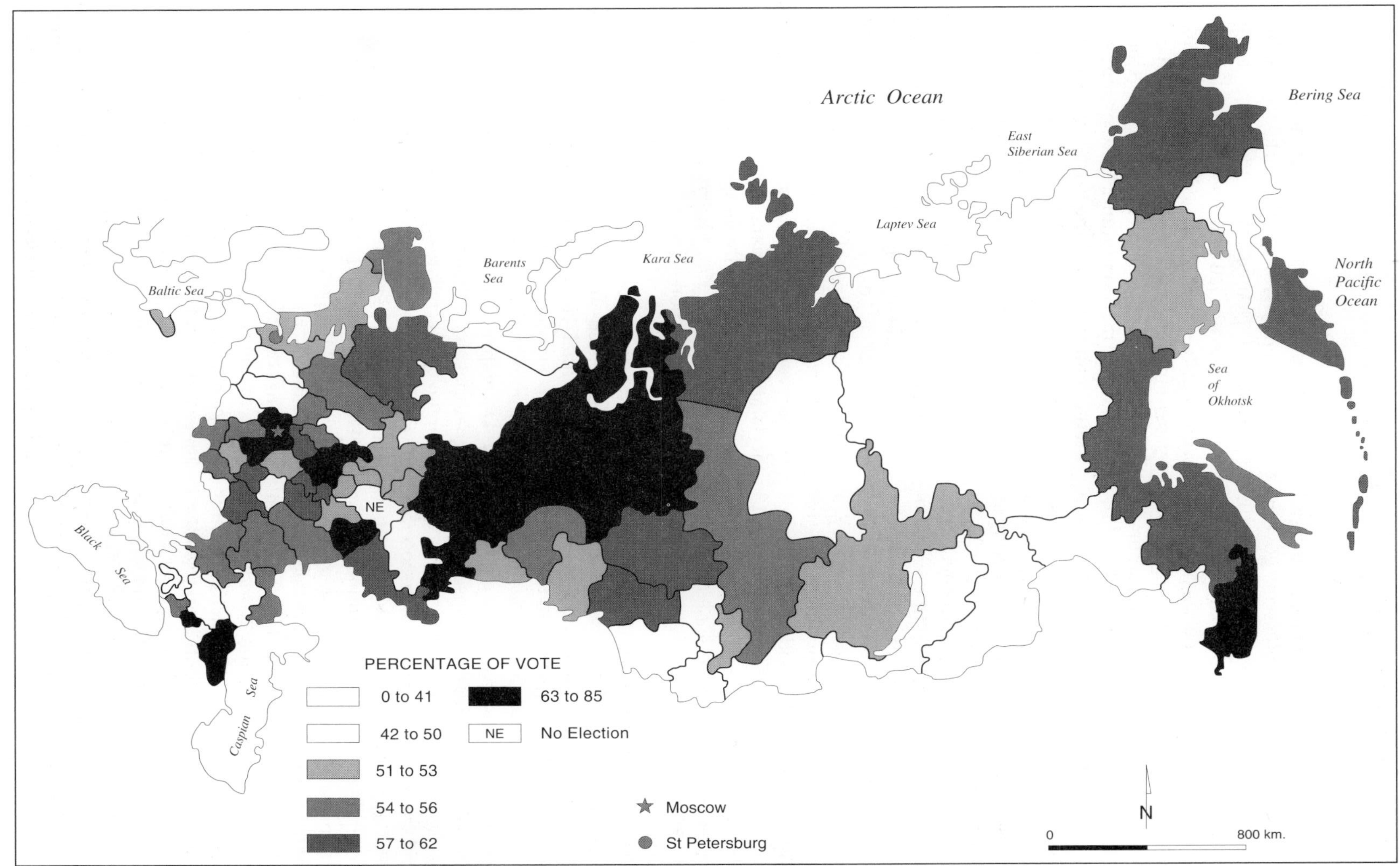

FIGURE 2.8 *Percentage of the vote for Yeltsin, 1991 Russia presidential election*
(*Source*: V. Kolossov, 'The electoral geography of the former Soviet Union 1989–91: retrospective comparisons and theoretical issues', in O'Loughlin, J. and van der Wusten, H. (eds), *The New Political Geography of Eastern Europe* (London: Belhaven Press, 1993), 199)

Russia the ground swell of support for the increasingly independence-minded Russian government headed by Yeltsin grew. Support for the Congress of People's Deputies waned. Faced with disaffected Republican governments in many parts of the country, conservative forces in the central government in Moscow were finally mobilized to action in the summer of 1991. In July, a manifesto calling for grass-roots resistance to the proposed dismantling of the Union of Soviet Socialist Republics was issued by a small group of leading conservatives. Just before the official signing of the new Union Treaty on August 20, an act which would effectively signal the demise of the Union, an even smaller group in the central government orchestrated a coup. Gorbachev, at this time on a holiday in the Crimea, was apprised of the takeover of the government by a delegation sent from Moscow by the coup leaders. He condemned the coup and its leaders, but was otherwise non-resistant. Monday, August 19 witnessed a nationwide declaration of a state of emergency signed by the coup leaders in Moscow. Troops were dispatched to Moscow, ostensibly to ensure compliance with the coup. But in Moscow popular support was quickly marshalled, not for the coup but in resistance to it. Thousands of ordinary citizens flocked to central Moscow, in order to defend the White House, the seat of the Russian government headed by Boris Yeltsin, who was quick to spearhead a defence of democracy. As it turned out, the troops dispatched to Moscow had little instruction from the coup leaders, and were in any event more disposed to support the forces of democracy than they were the coup leaders in the Kremlin and their attempt to preserve the Soviet system. Boris Yeltsin emerged as a national hero within hours, and was seen on television across the Soviet Union and, indeed, around the world, as the defender of democracy in the Soviet Union. By August 21, just three days after the coup leaders' public declaration of a national state of emergency, they were routed. Gorbachev returned to Moscow, but it was Boris Yeltsin, President of the Russian Republic, not Mikhail Gorbachev, President of the Soviet Union and General Secretary of the Communist Party, who was feted.

Yeltsin in his capacity as President of the Russian Republic moved quickly to initiate fundamental change. With the coup leaders now detained in jail and the Communist Party widely discredited, he signed a decree formally suspending the Party in the Russian Republic. On August 24 Gorbachev resigned as General Secretary of the Communist Party of the Soviet Union. A few days later the Party was suspended everywhere in the Soviet Union by the now widely discredited central government, the Congress of People's Deputies formed in 1989 and led by Gorbachev. This first-ever elected body to permit other than Communist Party candidates to stand, and one of Gorbachev's real, even if ephemeral, achievements, voted itself out of existence in early September. Meanwhile, formal declarations of political independence were forthcoming from Republics across the Union even during the coup, joining the earlier declarations of Lithuania (March 1991) and Georgia (April 1991). By the end of October 1991, only Russia and Kazakhstan remained in the old Soviet Union. Efforts by Gorbachev to negotiate a new Union Treaty which would recognize the new, post-coup political realities and be based upon international standards of individual rights and freedoms eventually brought eleven former Soviet republics into a Commonwealth of Independent States (CIS) by December 1991. Four countries were not signatories – Lithuania, Latvia, Estonia and Georgia. Gorbachev resigned as President of the Soviet Union on December 25. The Union of Soviet Socialist Republics was formally abolished on December 31, 1991.

In each of the post-Soviet successor states new governments have been formed as a result of elections contested by a vast array of parties and blocs of all possible political stripes. But not everywhere does governance proceed smoothly. For example, in Russia, Yeltsin asked for, and was given, considerable Presidential powers by the Russian Parliament in early 1992. But after the first flush of excitement over achieving independence, the reality of governing a newly independent country amidst economic collapse, high inflation and internal bickering between political and regional vested interests, soon descended upon everyone. Divisions between the executive authority, represented by President Yeltsin, and legislative authority, represented by the Russian Parliament, quickly deepened, and working relationships deteriorated. Alliances forged amongst parties and interests groups in the late Soviet era soon fell apart in the new political reality of post-Soviet Russia. Yeltsin relied increasingly on Presidential Decrees in order to continue the economic reform momentum. But resistance to the economic reform initiatives on the part of the executive authority rose amongst those who had been elected to the Russian Parliament in 1990. Eventually relations deteriorated to the point that the seat of government, the White House in Moscow, became the focus of armed conflict in early October 1993.

Conservative forces within the Russian Parliament, and some democrats as well, united in armed resistance to a September 1993 Presidential decree which dissolved the Russian Parliament and called for elections to a new constituent assembly, the State Duma. Elections to the State Duma were to take place in early December 1993, with elections to regional and municipal governments early in 1994. Anti-Yeltsin forces in the Russian Parliament called upon the public and the armed forces to support democracy, which they argued had been overturned by President Yeltsin's unilateral and unconstitutional actions. On the streets of Moscow armed conflict broke out. The military forces sided with the President. Despite the Parliamentary opposition force of civilians and former military veterans storming the Moscow Mayoralty building and the main television station, the Presidential authority prevailed. The military responded by turning tank guns on the White House, and then storming it. Resistance was forcibly broken. In the process the White House building was nearly destroyed. Well over 100 civilians were killed in the three days of armed conflict which occurred in specific parts of Moscow. Russia came perilously close to civil war. Parliament was thus forcibly dissolved and elections at all levels of government followed in December 1993 and early 1994.

Figure 2.9 outlines the current system of governance in Russia. The mediating position of the judicial branch as embodied in the Constitutional Court was in abeyance from December 1993, when the new Constitution was adopted, until February 1995, when at long last it was finally re-constituted. The State Duma, with only 450 Deputies in contrast to the 2250 members of the Congress of People's Deputies which was abolished in September 1993, is in theory a more manageable body simply because of the smaller number of elected representatives. But Russian voters sent to the new State Duma representatives of all of the factions which had previously confounded many of President Yeltsin's market reform initiatives, and some others besides. Put simply, elections to the State Duma in 1993 did not produced a legislature more amenable to Presidential authority, and neither did the election in December 1995. An additional problem with the current legislative structure outlined in Fig. 2.9 is that the composition of the Federation Council has not been precisely specified in the new Russian Constitution. Presently it has 178 members, two from each of Russia's 89 regions, one to represent the legislative authority of the region, one to represent the executive (see Fig. 9.13 p. 280). But there is no specified maximum number of members for the Federation Council, as there is for the State Duma. Moreover, there is no specific procedure laid out in the new Constitution regarding how to form the Federation Council. Are members to be elected or appointed, and by whom? Should members be elected by regional legislatures, or directly by the populations of each of the 89 regions? The lack of clarity in this regard is further complicated by the fact that in late 1995 about a quarter of the 89 regions did not yet have an elected local legislature. As part of President Yeltsin's strategy to out-manoeuvre disaffected regions, he appointed 60 governors to them. In theory, they are to be elected to these positions. Governors are the heads of the executive authorities in these regions. Having them represent the regions in the Federation Council is anathema to many of Yeltsin's opponents in the State Duma, as well as in the regions. Yeltsin has continued to support

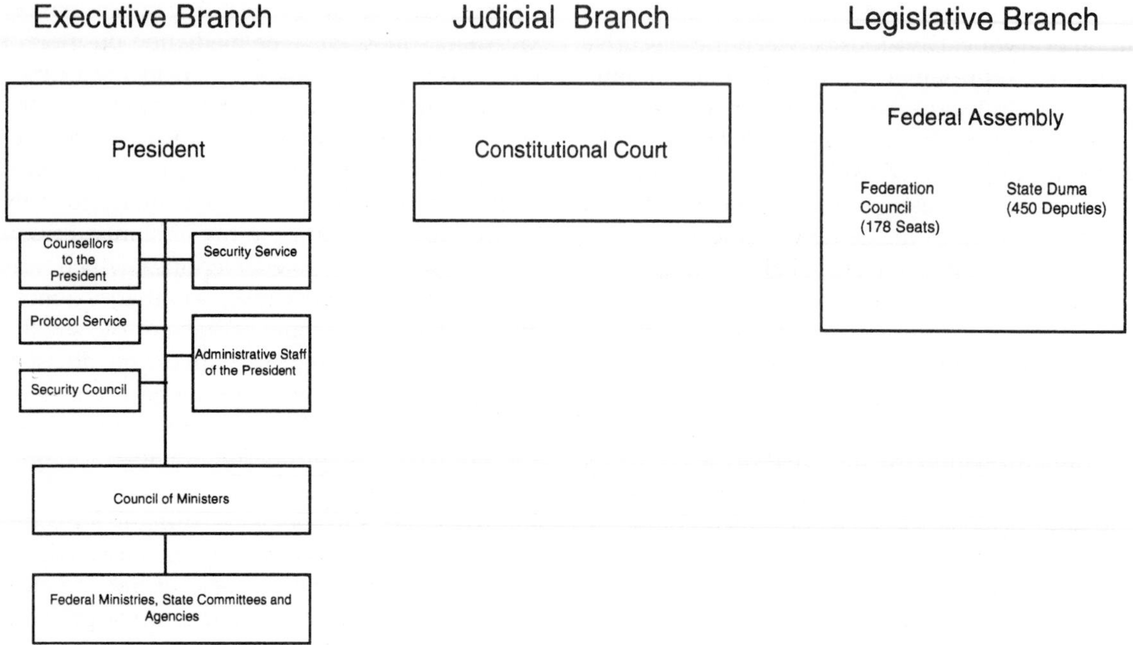

FIGURE 2.9 *Current system for governance in Russia*
(*Source*: Based on Russian Constitution, adopted December 1993)

the concept of an 'ex officio' Federation Council in his Presidential decrees. Most members of the State Duma support an elected Council. The relationship between the State Duma and the Federation Council is also important. Having passed a law, the State Duma must then despatch it to the Federation Council. The Federation Council is to receive all such bills within five days and make a decision on them within two weeks. As the Council sits irregularly, the two-week time line creates problems. Should the deadline be met but debate on the bill result in its rejection, a joint commission comprising Council and Duma representatives is formed to search for a consensus position. Failing agreement, the bill is returned to the State Duma where a two-thirds vote there is sufficient to override the Federation Council objection. Because the regions have a direct link to the Federation Council there is a considerable amount of lobbying. Important national priorities are frequently submerged by provincial ones in the business of the Council. Given all of this, it is small wonder that tensions continue to characterize relationships between the State Duma and Federation Council, and between the legislative and executive branches of central government. Yeltsin's position as President clearly gives him great authority, including the right to veto legislation passed by the Federal Assembly. His presidency was not contested until June 1996. Meanwhile Yeltsin's public support, like Gorbachev's before him, has plummeted.

Throughout the western tier states, that is, Lithuania, Latvia, Estonia, Ukraine, Belarus' and Moldova, the political situation has been far more stable. While there certainly has been heated argument as to the measures to be implemented to reform the economy, society and political structures, the kind of violence Moscow experienced has been avoided. Plans and governments have changed through the normal electoral process. In the Baltic states, for example, new elections have been conducted without incident. They resulted in ousting from power some nationalist parties which had spearheaded the independence movement, but which once in power could not satisfy the public's expectations.

It would seem that these parties could not keep the confidence and support of the people when it came to management of the domestic economy after independence. In Lithuania, for example, former Communists have been elected to lead the government. In the southern tier states of the Caucasus and Middle Asia, the situation is more complex politically. Since independence in 1991, civil war has engulfed regions in both Georgia and Azerbaijan in the Caucasus. The Russian military has been 'invited' back to assist in keeping the peace. Armenia and Azerbaijan have been engaged in an ongoing dispute over territorial claims to the economic detriment of both countries. In Middle Asia, Tajikistan has been beset with internal civil war between competing Tajik and Uzbek 'patriots'. The ruling government is supported by Russia, and relies on the Russian army to defend its borders. Since independence it has had frequent military engagements with anti-government forces based in Afghanistan. The other Middle Asian states have been spared such internal upheaval and disruption. In all of them, former Communists figure prominently in government, and participatory democratic processes tend to take second place to the public's apparent preference for strong leadership, political stability and economic development. In several of the Middle Asian states, the Presidents have breached their newly adopted constitutions by using referenda to extend their terms of office without election. Invariably, they have been given overwhelming endorsement by the voters to do so.

Political processes in the post-Soviet scene are still evolving. This is scarcely a surprising development given the fact that for most of the former Soviet territory and peoples there is no history of democracy, and considerable uncertainty as to what it means in practice. During the 1990s, a stable economy with some prospect of improvement will likely be of greater importance to most people throughout the post-Soviet scene than will the niceties of political protocol commonplace in western democracies.

The collapse of the Soviet Union may be attributed to many failings. By the late 1980s there were deep divisions in society. These included divisions between social classes, between regions and the central government, between the food-consuming cities and food-producing countryside. These divisions were exacerbated by the rise of nationalist movements and the call for political independence. Gorbachev's attempts to reform the economic system by promoting the transition to a market economy and facilitating the deconcentration of responsibility and authority to the regions were assumed to be compatible with the preservation of a socialist system of benefits. He permitted public criticism of the status quo, while presuming that somehow the leading role in society of the Communist Party could somehow be preserved. In many ways these initiatives fuelled the independence movements. In a very real sense, Gorbachev's many initiatives ended up confounding his basic objective of reviving the Soviet socialist system so as to prepare the country for the next century. Economic issues initially were his principal concern. Their resolution required him to accommodate to political change. Like Pandora's Box, once it was opened it proved impossible to stifle the forces for change which were unleashed. In the next chapter, we will briefly examine the nature of the Soviet economic system, its achievements and failures. We will then turn to the nature of the post-Soviet economy, focusing principally on Russia, but drawing comparisons and contrasts with other post-Soviet states as well.

FURTHER READING

Anderson, B. 1980 *Internal Migration During Modernization in Late Nineteenth Century Russia.* Princeton, Princeton University Press.

Andrle, V. 1994 *A Social History of Twentieth-Century Russia.* London, Edward Arnold.

Batalden, S.K., Sandra, L. 1993 *The Newly Independent States of Eurasia: Handbook of Former Soviet Republics.* Phoenix, AZ, The Oryx Press.

Bater, J.H. 1976 *St Petersburg: Industrialization and Change.* London, Edward Arnold.

Bater, J.H., French R.A. (eds) 1983 *Studies in Russian Historical Geography*, 2 Vols. London, Academic Press.

Becker, S. 1985 *Nobility and Privilege in Late Imperial Russia.* Dekalb, Northern Illinois University Press.

Blackwell, W.L. 1994 *The Industrialization of Russia: A Historical Perspective*, 3rd edn. Arlington Heights, IL, H. Davidson.

Bonnell, V., Cooper, A., Freidlin, G. (eds) 1994 *Russia at the Barricades. Eyewitness Accounts of the August 1991 Coup*. Armonk, NY, M.E. Sharpe.

Brower, D.R. 1990 *The Russian City Between Tradition and Modernity, 1850–1900*. Berkeley, University of California Press.

Crisp, O. 1976 *Studies in the Russian Economy Before 1914*. London, Macmillan.

Dawisha, K., Parrott, B. 1994 *Russia and the New States of Eurasia. The Politics of Upheaval*. Cambridge, Cambridge University Press.

Fedor, T. 1975 *Patterns of Urban Growth in the Russian Empire During the Nineteenth Century*, Research Paper No. 163. Chicago, University of Chicago Department of Geography.

Ferdinand, P. 1994 *The New States of Central Asia and Their Neighbours*. New York, Council on Foreign Relations Press.

Fitzmaurice, J. 1992 *The Baltic—A Regional Future?* New York, St. Martin's Press Inc.

Fitzpatrick, S. 1994 *The Russian Revolution*, 2nd edn. Oxford, Oxford University Press.

Friedberg, M. and Isham, H. (eds) 1987 *Soviet Society under Gorbachev. Current Trends and the Prospects for Reform*. Armonk, NY, M.E. Sharpe.

Goldenberg, S. 1994 *Pride of Small Nations: The Caucasus and Post Soviet Disorder*, Politics of Contemporary Asia Series, Vol. 9. London, Zed Books Ltd.

Hamm, M.F. (ed) 1976 *The City in Russian History*. Lexington, The University Press of Kentucky.

Hamm, M.F. (ed) 1986 *The City in Late Imperial Russia*. Bloomington, Indiana University Press.

Hosking, G.A. 1993 *The First Socialist Society: A History of the Soviet Union From Within*. Cambridge, MA, Harvard University Press.

Hunter, S.T. 1994 *The Transcaucasus in Transition. Nation Building and Conflict*. Washington, DC, Center for Strategic and International Studies.

Kaiser, D.H. 1987 *The Workers' Revolution in Russia, 1917. The View from Below*. Cambridge, Cambridge University Press.

Keep, J. 1976 *The Russian Revolution: A Study in Mass Mobilization*. New York, Norton.

Kerblay, B. 1983 *Modern Soviet Society*. New York, Pantheon Books. Translated by Rupert Swyer.

Kingston-Mann, E., Mixter, T. (eds) 1991 *Peasant Economy, Culture, and Politics of European Russia, 1800–1921*. Princeton, NJ, Princeton University Press.

Kort, M. 1993 *The Soviet Colossus. The Rise and Fall of the USSR*. Armonk, New York, M.E. Sharpe.

Kuzio, T., Wilson, A. 1994 *Ukraine: Perestroika to Independence*. London, Macmillan.

Lyashchenko, P.I. 1949 *History of the National Economy of Russia to the 1917 Revolution*. New York, Macmillan.

Mandelbaum, M. (ed) 1994 *Central Asia and the World: Kazakhstan, Uzbekistan, Tajikistan, Kyrgizstan, Turkmenistan*. New York, Council on Foreign Relations.

Manz, B.F. (ed) 1994 *Central Asia in Historical Perspective*. Oxford, Westview Press.

Motyl, A.J. 1993 *Dilemmas of Independence. Ukraine after Totalitarianism*. New York, Council on Foreign Relations.

Pares, B. 1964 *A History of Russia*. New York, Alfred A. Knopf.

Parker, W.H. 1968 *An Historical Geography of Russia*. London, University of London Press.

Pitcher, H.J. 1994 *Witnesses of the Russian Revolution*. London, John Murray.

Ransel, D. (ed) 1978 *The Family in Imperial Russia*. Urbana, University of Illinois Press.

Rieber, A. 1982 *Merchants and Entrepreneurs in Imperial Russia*. Chapel Hill, University of North Carolina Press.

Rywkin, M. 1994 *Moscow's Lost Empire*. Armonk, NY, M.E. Sharpe.

Sacks, M.P., Parkurst, J.G. (ed) *Understanding Soviet Society*. Winchester, Unwin Hyman.

Slezkine,Y. 1994 *Arctic Mirrors: Russia and the Small Peoples of the North*. Ithaca, NY, Cornell University Press.

Stephan, J.J. 1994 *The Russian Far East: A History*. Stanford, CA, Stanford University Press.

Yanowitch, M. (ed) 1986 *The Social Structure of the USSR: Recent Soviet Studies*. Armonk, NY, M.E. Sharpe.

3

FROM SOVIET PLANNED ECONOMY TO POST-SOVIET MARKET ECONOMIES

'In more than three years of reforms, the results of economic development never before inspired such hope for getting out of the crisis'
(*Izvestiya*, April 15, 1995, 1)

The rise of the Soviet Union as one of the world's major economic and military-industrial powers and then its decline and collapse, was an economic and political-geographic transformation of enormous proportion. It had a profound impact on the lives of millions of people. A whole way of life, indeed, a people's history and the official ideology upon which it was based, was found wanting. It is still far from certain what will replace the Soviet economic system. While the average standard of living never did match that of many western industrialized states, all Soviet citizens at least were assured a minimal standard of living as part of a 'social contract' with the state. Those who had been successful in the Soviet system often lived quite well, even if below the customary standard enjoyed by comparable elites in the west. However, since the collapse of the Soviet system in 1991, and the declarations of independence which were so widely and enthusiastically applauded, the economic situation has deteriorated dramatically in all of the newly created states. It was reckoned that 30 per cent of Russia's total population of 148 million lived below the official poverty line in 1995. In a number of post-Soviet states, the proportion living below the official poverty line is even higher. Across the post-Soviet scene each year witnesses a growing gap between the newly rich and the poor.

In order to have some appreciation of the economic development challenges confronting post-Soviet states, it is necessary to understand the legacy of Soviet economic development policies and practices. In this chapter, we will first examine the nature of the Soviet economy, describe its specific geographic characteristics, outline its achievements and describe its systemic weaknesses. Soviet officials were not oblivious to the system's inherent problems, and a number of economic reforms were initiated over the years. None succeeded. The last attempt to reform the Soviet economy is associated with former President, Mikhail Gorbachev. He is credited or castigated, depending on the point of view, for having unleashed a series of reforms in the mid-1980s which culminated in the collapse of the state itself. While the need to introduce a market economy is now publicly acknowledged everywhere, the debate over how best to do this, and how quickly, resonates still across the post-Soviet scene. The essential elements of the market reform process will be outlined in the final section of this chapter.

THE SOVIET ECONOMIC DEVELOPMENT SYSTEM

The national economy was devastated by World War I and the years of civil war which followed. Only in 1921 did internal political conditions stabilize to the point that Vladimir Lenin could introduce his New Economic Policy. This pragmatic accommodation with reality encouraged small-scale private enterprise in both town and countryside, a necessary measure if the Soviet Union was to regain the levels of domestic production which existed in 1914. Even with the contribution of small-scale, private sector activity, such levels of output were generally not reached until the mid-1920s. Lenin's death in 1924 precipitated an internal power struggle in the Soviet Union out of which Josef Stalin emerged the central political figure. Stalin introduced the Five Year Plan in 1928, and along with it the initial apparatus of the ministerial system. This signalled the end of Lenin's relatively pragmatic New Economic Policy period and the official tolerance of small-scale private enterprise in selected sectors of the national economy. Stalin's centralization of control over decision-making in economic matters paralleled his centralization of control over the Communist Party. Henceforth, economic planning was administered through a hierarchy of ministries, committees and departments.

The basic components of the decision-making framework for managing the Soviet economy are

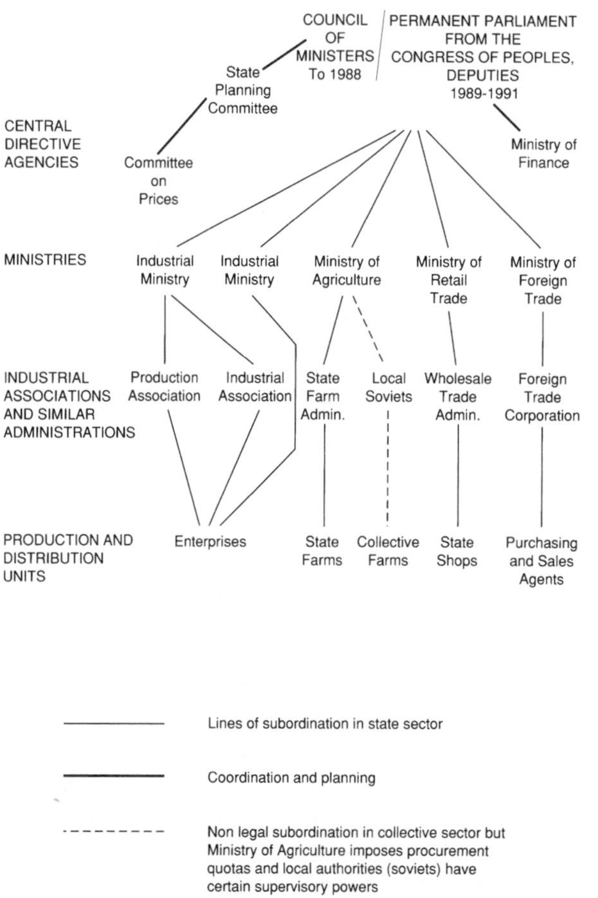

FIGURE 3.1 *Simplified decision-making framework for managing the Soviet economy*
(*Source*: A. Brown *et al.* (eds), *The Cambridge Encyclopedia of Russia and the Soviet Union* (Cambridge: CUP, 1982), 356)

outlined in Fig. 3.1. While this structure dates from the late 1980s, and not the early Stalin era, its basic features were the same. Over time the structure was embellished as opposed to radically restructured. As already noted in the previous chapter, fundamental policy decisions were the prerogative of the Politburo. The major policies set by this small, inner group of the Communist Party hierarchy usually bore the personal stamp of approval of the leader. Certainly during Stalin's time this was so. Toward the end of the Soviet era more discussion and consensus seems to have been common, but in any event the actions taken by the Politburo were not accountable to the government, but to the Party Congress. In the realm of executing the economic policies of the Politburo, the State Planning Committee, or Gosplan, played a key role throughout the Soviet era. Chaired by a deputy premier, it was assured a high profile in the decision-making system. Once the basic development priorities were established, the huge Gosplan bureaucracy translated them into detailed plan-instructions for ministries representing the various sectors of the economy. As is only partially suggested by Fig. 3.1, economic planning was a multi-faceted process involving a host of organizations located from one end of the country to the other. Theoretically, information required for decision-making passed up through the administrative hierarchy, while directives were sent down. The reality of the process was rarely so simple. The following discussion describes the nature of the planning process up until the mid-1980s, after which Gorbachev altered established procedures.

Ministries operated at three levels – All-Union, Union-Republic, and Republic – were numerous in number, and were in a constant state of flux. In the mid-1980s, around 60 were concerned with economic matters. Plan-instructions were customarily issued for five-year time periods, and theoretically determined for each year what was to be produced, how and by what means, when and to whom it was to be delivered, and at what price it was to be sold. Prices, both 'wholesale' and 'retail', in the absence of a market were established by the State Committee on Prices. The price structure throughout the Soviet era was characterized by extreme rigidity. It was generally based upon estimated costs of production rather than resource scarcity, and therefore was of limited use in resource allocation. In some cases, prices at the consumer level were maintained at unrealistically low levels as part of the unwritten 'social contract' with the citizenry.

In practice central planning authorities did not make most of the detailed decisions regarding management of the economy. If they had done so, the system would not have worked as well as it did. What evolved over the years was something of a balance of real decision-making authority between ministries and central planners. Central planners specified goals, set production targets, and the ministerial bureaucrats did whatever was deemed necessary to meet them. This decision-making process has been called 'centralized pluralism'. What this means is that ministerial bureaucracies inevitably became 'interest groups', and informally lobbied and sought benefits from the leadership. Some ministries, such as those in the military-industrial complex, were economically and politically more important than others. Nevertheless, within any particular ministry all bureaucrats had a vested interest in at least appearing successful. Hence, in order to meet output targets, managers tended to bring under their control as many inputs as possible. Where they were not able to control the essential components for increasing production themselves, systems of mutually beneficial personal contacts were developed to take up the slack. These networks continue to play a very important role in the post-Soviet economic scene, as will be discussed in subsequent chapters. Where semi-legal bartering or exchange between Soviet era-production units failed to garner sufficient supplies, production targets were not fulfilled. This shortfall usually had a cascading, multiplier effect within the system, since output from one production facility was frequently an input for another. Production bottlenecks were thus endemic to the planning system. Depending on the political or economic importance of the ministry, failure to meet output targets specified for annual and five-year-plan periods could make positions of bureaucratic responsibility exceedingly tenuous. Even those in charge of important ministries were sometimes dismissed for perceived failures. Therefore, during the Soviet period all participants had a real stake in making things work. As added incentive, bonuses for managers were often the reward for successful plan fulfilment, in addition perhaps to being selected to move up the administrative hierarchy. Despite the inherent weaknesses, the economy did function. The military-industrial complex which received the highest priority often performed exceedingly well indeed.

In terms of economic growth, some impressive gains were registered in the early years of Soviet power. In the 1930s, for example, annual rates of growth in industrial production frequently exceeded 10 per cent. During the world-wide depression of the 1930s the Soviet economic experiment attracted much favourable attention, and many otherwise unemployed European and North American technical and professional personnel took jobs in the Soviet Union. The 1940s witnessed the colossal destruction wrought by World War II. Population losses exceeded 25 million, and thousands of villages and cities were heavily damaged or destroyed. A vast post-war programme of reconstruction was undertaken during the Fourth Five Year Plan, 1946–1950. By the early 1950s, real economic growth resumed. But as was the case before World War II, it was based on a narrow spectrum of activities, notably those related to the military-industrial complex, including such basic industries as iron and steel. Indeed, in many instances the Soviet Union lagged behind technological developments in the West. The energy sector, for instance, remained heavily dependent upon coal well into the 1950s, whereas in Europe and America oil and natural gas were increasingly important sources of energy. Agriculture and the consumer goods sector were intentionally given short shrift in the allocation of scarce capital, just as they had been before the war. In short, the Soviet economy registered rapid annual growth, but it was locked into a pattern established in the early 1930s. Modernization of some sectors was much needed, and so too was better integration of development plans framed in an explicitly sectoral mode. Regional economic inequality was just one of the legacies of imperial Russia that the new socialist state was committed to eradicating.

Soviet Regional Economic Development Strategies

Differences between the developed core of European Russia and peripheral regions such as Middle Asia were staggering in the early years of Soviet power. No matter which measure was used – level of literacy, life expectancy, infant mortality, income, industrial output per capita – each highlighted a national economy with significant regional development disparities. The economic and social needs were obvious enough, the regional development strategies to ameliorate them less so. However, policies and programmes of action were soon put in place to effect the desired fundamental change in the distribution of economic activity.

To the extent that governments influence the location of economic activity, they are all confronted with the dilemma of encouraging either national efficiency and most likely more rapid economic growth, or regional equity and the real prospect of less rapid economic growth. In the Soviet Union, three broad categories of criteria, or guidelines, for allocation of investment funds were used. The first group of guidelines for planners focused on technical optimization. The minimization of transport costs by reducing or eliminating long hauls of industrial raw materials or final products was one such guideline. Thus, there was long-standing advocacy of locating industry either near markets, or sources of raw materials, depending on the nature of the production process involved. Hauling long distances weight-losing industrial raw materials, such as low metal-content ore, makes little economic sense no matter whether the resources are publicly or privately owned. Related to this concern was the promotion of regional industrial specialization. The use of local raw materials, even if the quality was lower, and hence the cost of recovery higher, was viewed positively by Soviet decision-makers. Thus, if specialization in industrial output on a regional basis could be combined with some degree of regional self-sufficiency owing to utilization of local or regional resources, in the Soviet scheme of things the whole national economy was assumed to benefit. This led to the creation of a number of sub-optimal production facilities being developed in non-Slavic areas especially, facilities that were in effect heavily subsidized. While the main emphasis in this context tended to be on utilization of industrial raw materials, the same principle applied in terms of utilization of regional labour resources.

The second group of guidelines in locational decision-making had to do with spreading the benefits of socialism. A more even distribution of economic activity was desired. This was far from a novel, national objective since it was shared by many western countries in the post-war era. In the Soviet case, reduction of regional disparity certainly involved fostering urban-industrialization in economically less developed, non-Slavic areas, such as Middle Asia and the Caucasus. The objective of creating a more even distribution of economic activity also entailed eradicating the differences between city and countryside. A central tenant of Marxist ideology is that under advanced socialism there is minimal difference in the quality of life between town and village. Greater participation in the economic development process in the peripheral, non-Slavic realm would result in an increase in the political and cultural sophistication of the population, and presumably by extension, support for the Soviet system. In other words, by ensuring minority groups opportunities to participate in the development of the national economy, socialist, if not communist, values would be more easily inculcated.

A third set of guidelines dealt with strategic factors in industrial location. Their particular impact over the years is difficult to gauge. While the general nature of the locational criteria, or guidelines, precludes determining how any one of them might have shaped the distribution of industry over the years, it would seem that some industrial development in the eastern regions of the country was related to the need for security from external attack. So too was the creation of scores of so-called secret cities located in various regions throughout the Soviet Union. These were in effect closed cities, places where even the inhabitants' contacts with relatives and friends were strictly controlled owing to their classified, highly strategic military-industrial research and production facilities.

In the Soviet economy, it was clearly possible to direct investment to specific regions or places. It was similarly possible to issue decrees restricting the further development of particular places. Along with these methods for directing investment, the Soviet state also used a system of wage differentials to encourage internal migration to regions where economic development was desired. Regional wage coefficients played an important role in promoting migration to Siberia, especially to the eastern and northern regions, and were used from the late 1930s to the end. Figure 3.2 shows the regions where wage coefficients were applicable in the early 1980s. It is apparent that wage coefficients of one kind or another were used over most of the Soviet Union. Benefits were always greatest in the Far North, and in areas deemed to be equivalent. Typically, wage rates were about double those that were paid for the same work in areas where wage coefficients did not apply. Additionally, each year served in such a harsh regime counted twofold toward retirement and a pension. Notwithstanding the attraction of high wages, early retirement possibilities, generous holidays and so on, the fact that the system was still in place at the end of the Soviet era speaks to the problems of attracting and holding labour in climatically harsh regions. The Soviet state had other means at its disposal, however.

Various forms of directed labour were also utilized. The most benign form was probably the assignment of graduates of university or technical school programmes to periods of employment in a remote region. As higher education was free, indeed, students typically received a small monthly stipend from the state, this was not necessarily an unreasonable arrangement. But by the late 1930s, labour was often simply assigned to specific locations, there to stay until permission to depart was granted. During World War II the substantial number of people involved, willingly or otherwise, in this programme was much

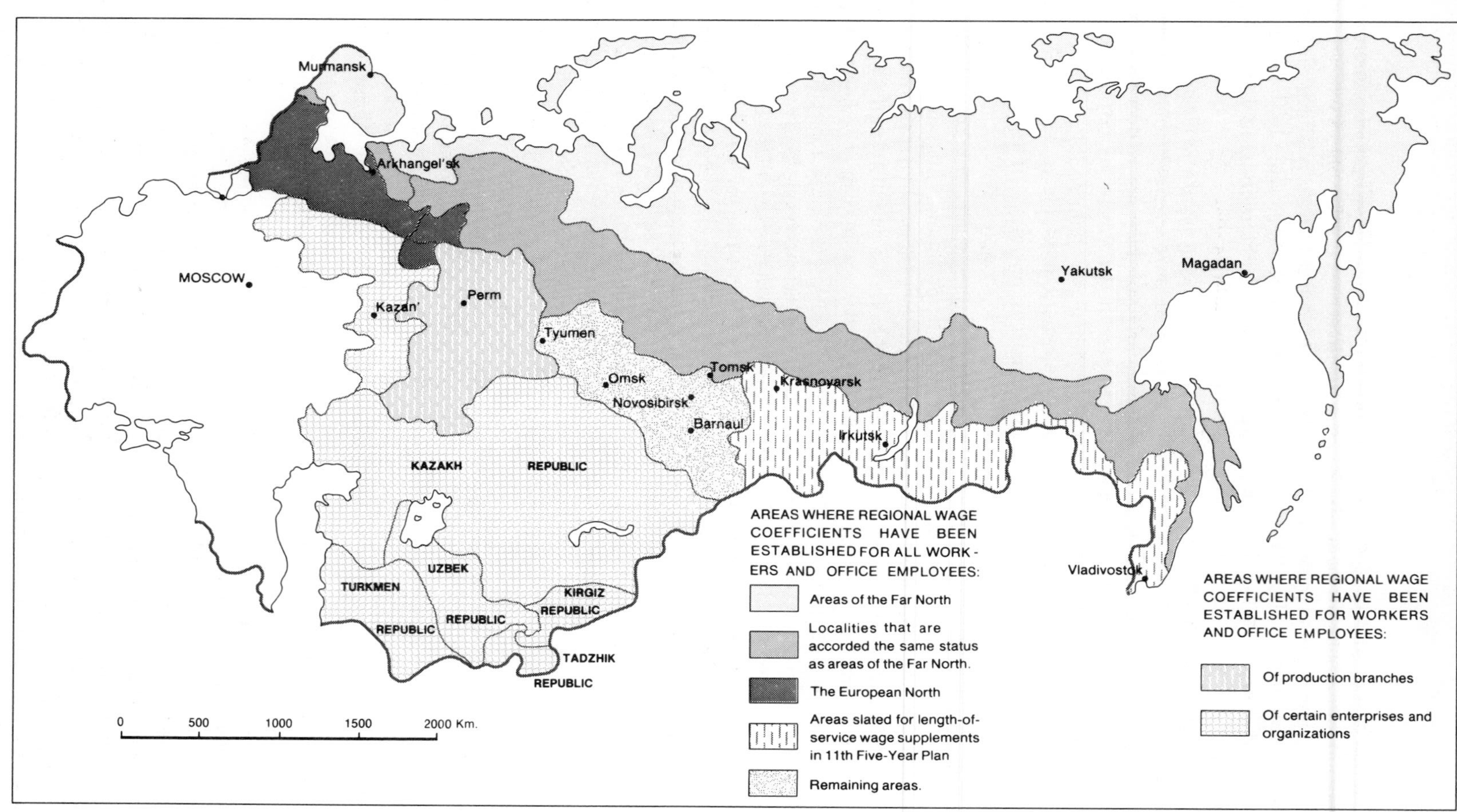

FIGURE 3.2 *Regional wage coefficients during the Soviet era*
(*Source*: Adapted from *The Current Digest of the Soviet Press*, Vol. 24, No. 23, 7 July 1982, 20)

augmented. Firstly, whole groups of Soviet peoples fell into disfavour because their loyalty to the state was suspect. Thus, Crimean Tatars and the sizeable number of Germans from Ukraine and the Volga region were relocated to the eastern and northern parts of the country where labour was needed, and where possible collaboration with the invading German forces was not possible. Prisoners of war were yet another source of labour for the frontier regions. The most notorious component of what developed into a vast system of camps was that known as the 'gulag'. During the Stalin era especially (1928–1953), several million Soviet citizens, accused of one thing or another, were summarily dispatched to the 'gulag' prison camps which multiplied across the eastern and northern regions of the country. Whether or not this pool of forced labour ever really contributed more to the regional development of the economy than it cost to supervise and supply is a moot point. The diet was appalling, the physical condition of the incarcerated population was seldom very good and output per person was therefore low. A vast organization emerged just to supervise the system. While the post-Stalin years witnessed a steady dismantling of the 'gulag', camps continued to operated right up until the end of the Soviet system. Only in the final years of the Soviet era were small numbers of Crimean Tatars, Germans and some other minorities successful in petitioning the government for permission to return to those regions from which they had been forcibly removed. A good many other Crimean Tatars simply returned from various places throughout Middle Asia to the Crimea without authorization, however. What they returned to was not what they had been taken from, however, as their homes and villages had long been occupied by Russian and Ukrainian migrants. Clearly, to make such a perverse system work at all required that there be some means of control over internal migration. Thus, to assist the state's bureaucracy to monitor and direct the movement of all Soviet citizens, internal passports and residence permits, or *propiski,* were instituted in 1932. More will be said about the vestiges of this system in the post-Soviet era in a later chapter.

Within this framework of state control over the spatial allocation of investment and movement of labour, there were several specific approaches to regional economic development which in fact did change the distribution of economic activity. The earliest and perhaps most controversial of them was the decision to create the so-called Urals–Kuznetsk Combine. This regional economic development scheme involved the functional integration of the vast iron-ore reserves of the southern Urals with the substantial, high-quality coking coal deposits of the Kuznetsk Basin (KUZBAS) in southeastern Siberia. The Urals region lacked coking coal; the KUZBAS lacked iron ore. The proposal to integrate the two complementary resources had appeared at least as early as 1916. Envisioned were modern iron and steel complexes at either end. But hauling weight-losing materials more than 1600 kilometres flew in the face of conventional economic logic and industrial location theory. Some Soviet economists attacked the proposal on the grounds that the resultant transport costs would greatly exceed the presumed benefits. Better to invest, they argued, in the existing iron and steel industry of the south Ukraine. A greater return on investment would ensue, and much sooner. But invoking western industrial location theory to buttress the argument against proceeding with the Urals–Kuznetsk Combine simply invited the obvious criticism that under socialism things could be different, and better. The project offered an opportunity to alter quite fundamentally the economic geography of the state, and in a manner which conformed to the notion that under socialism there would be a more equitable distribution of economic opportunity. To invest only in the Ukrainian steel industry would just reinforce the pre-Soviet pattern of economic development. In any event, some observers have suggested that the economics of the project were largely irrelevant because it made good sense from the standpoint of national security. Whatever the real reasons, the Urals–Kuznetsk Combine went ahead notwithstanding a cost which is reputed to be at least equal to one-third of the total investment capital available during the First Five Year Plan alone!

By the mid-1930s large scale integrated iron and steel complexes had been constructed in the southern Urals at a new city called Magnitogorsk, and in the KUZBAS at the city of Novokuznetsk. However, as the opponents of the Combine had predicted, the costs of hauling weight-losing materials over such vast distances soon began to take their toll. Little had been done to bolster the trans-Siberian railroad, and thus the operational efficiency of that part of the rail system linking the two ends of the Combine was greatly reduced. Plans for resource substitution soon emerged. For example, the coking-coal reserves which had been discovered in what is now the state of Kazakhstan near the city of Karaganda were already being exploited by the early 1930s in order to supply part of the demand of the southern Urals mills. Located about half way between the Urals and the KUZBAS, Karaganda coking coal helped to relieve the growing pressure on the trans-Siberian railroad. Eventually, iron ore was discovered at Atasu in Kazakhstan, and in combination with Karaganda coking coal and water brought several hundred kilometres across the steppe by canal from the Irtysh river to the east, the basis for yet another large-scale integrated iron and steel complex was created in a peripheral, less developed region. Furthermore, the growing realization that it would be necessary to replace at least some of the ore transported from the Urals to the KUZBAS, 1600 kilometre to the east, spurred geological exploration in the latter region. By the end of the 1930s, iron ore was being mined in the southern part of the Kuznetsk Basin. Even though it was of lower iron content and difficult to use in the smelters, it nonetheless helped to reduce some of the pressure on the rail system. Some observers reckon that in the absence of the metallurgical complexes in the southern Urals and in the KUZBAS, the Soviet Union's capability to mount an effective military offensive to counter the German invasion of 1941 would have been seriously hampered. In any event, principles upon which the Combine were originally based were exceedingly costly to underwrite in other places. Theoretical discussions of alternative regional development strategies under the conditions of socialism continued. The work of N.N. Kolosovskiy published during the 1940s is one example.

Kolosovskiy is closely associated with the concept of territorial production complexes (TPC). In Kolosovskiy's scheme, functional linkages based on flows of raw materials and energy were the basis of what he labelled energy-production cycles. Kolosovskiy identified eight such energy-production cycles, and later proponents of this concept have added to the roster. The cycles range from ferrous metallurgy, to petro-chemicals, to agricultural production. In many ways the Kolosovskiy energy-production cycle concept is similar to the notion of growth poles as developed by the French economist, F. Perroux, in the 1950s. Both concepts focus on the delineation of sets of inter-related, if not directly interdependent, production processes, which theoretically could be anticipated to generate economic growth. Both concepts, in the first instance, are non-spatial. Within the context of a centrally planned economy, Kolosovskiy's scheme had some obvious attractions for Soviet authorities. By concentrating on production cycles, which by their very nature would embrace activities across several sectors of the economy (managed by separate ministries, of course), some obvious efficiencies of integration might occur. The conscious selection of production cycles to match the resource base of a particular region gives rise to the territorial production complex, or TPC. Clearly, within a planned economy it was possible to develop inter-related production facilities so as to maximize economies of scale, minimize waste products, and ensure some balance between the demands for male and female labour.

Territorial production complexes were assumed to exist at different scales and to evolve through different stages. Something of the hierarchy of the component parts of a regional production complex is portrayed by Fig. 3.3. In theory, a regional production complex was conceived as comprising several TPCs, each TPC comprising several functional and spatially integrated industrial nodes, that is, communities with an industrial economic. Stages of possible TPC development ran the gamut from regions with latent resource potential for future development,

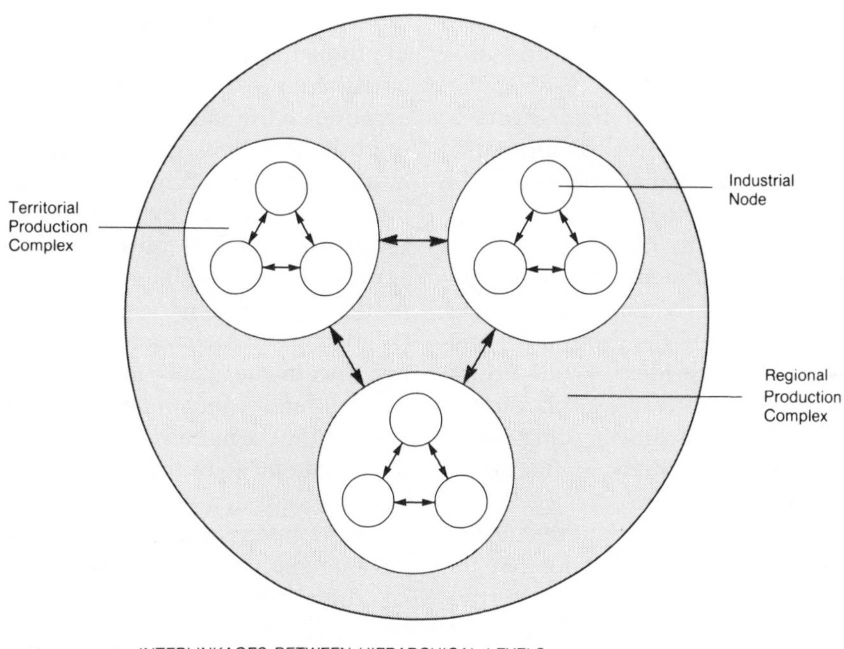

FIGURE 3.3 *Soviet era territorial production complex (TPC)*
(*Source*: V.L. Mote, 'The Baikal-Amur Mainline and its Implications for the Pacific Basin', in Jensen, R.G., Shabad, T., Wright, A.W. (eds), *Soviet Natural Resources in the World Economy* (Chicago: University of Chicago Press, 1983), 171)

to developing 'pioneer' regions, to regions in which TPCs and industrial nodes were already formed, to regions of advanced development. Since by definition the energy-production cycle upon which a TPC was based spanned several different sectors of the economy, each managed by a separate ministry, effective coordination of the functional and spatial development of TPCs proved difficult.

In the absence of a regional decision-making body with sufficient economic and political clout to challenge ministries, the planned development of TPCs often did not proceed smoothly, and not infrequently resulted in quite unsatisfactory situations. It was frequently the case that production facilities were built according to TPC requirements, but the housing, social and consumer service needs of the workers were not. What occurred in a frontier TPC complex in central Siberia in the 1960s was all too typical. The TPC approach was used in the Angara river region to plan the development of energy-intensive production processes to utilize the large amounts of seemingly inexpensive hydroelectric power from the huge Bratsk and Ust-Ilimsk power plants. At Bratsk, an industrial node within the Angara TPC, the production of large amounts of aluminium and wood products was based on natural resources found in the region. A single new town was to be built. In actuality, the individual ministries responsible for different components of the industrial node ended up building their own housing facilities, each physically separate from the other! There were considerable differences in the provision of basic social and consumer services between the settlements which made up the single administrative authority of Bratsk. The experience in Bratsk is typical of that elsewhere in the former Soviet Union in terms of translating a good idea in theory into reality.

For most of the post-war period, the TPC was generally regarded to be the principal mechanism for planning regional economic development. Perroux's growth-pole concept was similarly extended to regional development planning in western countries, though in the absence of direct government intervention the creation of so-called growth centres has been far from successful. Indeed, even where there have been substantial government incentives to encourage firms to locate in designated growth centres, there are few examples of permanent spatial change in the distribution of economic activities resulting from the adoption of a growth centre strategy in regional economic development. The nationalized resources and planned economy of the Soviet Union were much more fertile ground for successful application of such ideas. While not all plans were executed properly, the application of the TPC concept did leave an indelible impress on the pattern of economic activity in the Soviet Union and in its successor states.

What had brought the need for more effective regional economic planning to the fore in the 1950s, and hence gave impetus to the further development of Kolosovskiy's territorial production complex idea, was the widespread recognition that the existing sectorally based, decision-making structure was no longer working satisfactorily. The many criticisms of the ministerial system included the following: it resulted in production bottlenecks and shortages of goods; it contributed to unacceptably high levels of cross-hauling of material supplies; it tended to emphasize quantitative, rather than qualitative, aspects of production; and it fostered what was called ministerial autarchy.

Ministerial autarchy, or self-sufficiency, however, was a logical response to two features of the Soviet decision-making environment: the constantly changing (usually higher) production targets set by central planners, and the uncertainty associated with dependence upon other ministries for critical material supplies. Thus, the more control a ministry had over inputs for the production process, including labour, the greater the potential for a successful, adaptive response to changing production targets set by central planners. While striving for self-sufficiency was perfectly rational behaviour at the ministerial and sub-ministerial level, this pattern of behaviour was frequently objectively irrational, and very costly, when viewed from the perspective of the state. It not only led to ministries producing goods properly the mandate of other ministries (which reduced the potential economies of scale in production for the latter), it also contributed to cross-hauling. The Soviet press was thick with accounts of the same goods being transported in opposite directions, often between two quite distant places, each origin and destination of movement determined by separate ministries. A broader perspective on demand and source of supply would highlight the complementary nature of each ministry's requirement. The compelling logic of each ministry meeting the need of the other by means of a local exchange is obvious. But logic and reality in decision-making were not always the same, as the example of the TPC development at Bratsk illustrated. The burden that unnecessary cross-hauling placed on the Soviet railroad system was enormous. Another obvious diseconomy associated with the ministerial drive for self-sufficiency was 'hoarding' labour. This resulted in the under-employment of an important resource in many production facilities. The long-standing preoccupation with meeting those targets set in physical output terms in the easiest possible manner inevitably led to poor quality. The widespread habit of managers consciously underestimating what their particular factory was capable of producing when submitting information about plant output potential to higher level planning authorities, and the latter's anticipation of this behaviour by way of even higher targets for production being imposed, frequently resulted in critical shortages when the targets set in this game of bluff really did exceed what was physically possible to produce. All of these problems, and more besides, dictated that the decision-making process be re-examined.

SOVIET ECONOMIC REFORM INITIATIVES

The death of Stalin in 1953 provided an opportunity for serious discussion of fundamental reform of the Soviet economy. The problems were apparent; the solutions were less so. Following the assumption of control by Nikita Khrushchev, Soviet leader until 1964, several potentially significant changes in the management and structure of the economy were instigated. In 1957, despite being well into the Sixth Five Year Plan, the plan was aborted. A Seven Year Plan, designed to carry development strategies through until 1965, was adopted instead. With the strong endorsement of Khrushchev, the Seven Year Plan witnessed the dismantling of all but a handful of the key All-Union ministries and the creation of over 100 regional economic councils, or *sovnarkhozy*, to take their place. Each *sovnarkhoz* was responsible for planning, managing, and budgeting for virtually all economic activities in the region. Certainly there had long been a clear need for better integration of sectoral planning in a regional context. With so much authority vested in ministerial offices located in Moscow, there was often legitimate reason for contending that local requirements, or potentials, were being overlooked. The fact that the political importance of ministers heading up powerful All-Union Ministries would be emasculated in this restructuring of the decision-making system was perhaps not an insignificant benefit to someone like Khrushchev. He had barely survived the political in-fighting which followed Stalin's death, as opposed to having been immediately and enthusiastically acclaimed as leader.

The radical transformation of the decision-making structure initiated by Khrushchev was not without some political pitfalls, however. The decision-making responsibility for achieving targets was widely decentralized geographically, and therefore the relationship between plan and reality was even more difficult to monitor. The basic industrial structure of the Soviet Union was also altered in several important ways at this time. Such changes usually prove problematic. We need only note at this point a couple of examples. In the energy sector, the long-standing emphasis on coal production was changed; instead, the development of oil, and especially natural gas, which until this date had largely gone unused as an energy source, was promoted. Consumer goods, and related items like housing, municipal services, and so on, were given a substantial fillip. Finally, putting authority into the hands of decision-makers at the regional level risked feeding nationalist aspirations, because the regions used as *sovnarkhozy* were political-administrative in origin, and therefore reflected the basic distribution of the various national minorities. Some minority groups needed little encouragement to put their interests ahead of those of the state as a whole. The Baltic Republics, which had been independent in the inter-World War period, and the peoples of the Caucasus, for example, were well-known for their separatist tendencies. Thus, the *sovnarkhoz* experiment offered considerable scope for putting local interests ahead of the state's, notwithstanding the fact that the Party apparatus existed at all levels of the political-administrative cum regional system and ought to have been able to squelch such nationalist tendencies as came to the fore. In fact, some regional Party bureaucracies even succumbed to nationalist tendencies! For a variety of reasons, the *sovnarkhoz* experiment soon came upon rocky ground.

Within a few years, the clumsiness of decision-making vested in more than 100 regional economic councils took its toll. The number of *sovnarkhozy* was steadily reduced, and finally consolidated into 47 regions. Centralized state committees were created in an effort to establish control from 'above', and discussion of other reform options proceeded apace. One such option involved redirecting emphasis from the quantitative fulfilment of planned targets, which was the essence of the Material Balance Planning method used by the Soviet authorities since the 1930s, to meeting contracts with 'customers', thereby focusing more attention on the quality of goods produced. The measure of success would be the volume of sales and associated surplus of revenue over costs, or what might in the Soviet context be seen as profit. To encourage participation in this new approach to economic management, the economist E.G. Liberman, with whom it is most frequently associated, proposed that a share of the profit be used in a discretionary system of bonuses for workers and managers alike. Toward the end of the Seven Year Plan, some limited experimentation with the Liberman method was approved.

All of this rather radical, at least by Soviet standards, discussion and experimentation took place during a period of general liberalization of the Soviet system introduced by Khrushchev. Not only were economists free to debate the merits of various schemes for improving the performance of the economy, but artists, writers, and others were permitted freedom of expression not seen for decades. For instance, Solzhenitsyn's *One Day in the Life of Ivan Denisovich* was published in the premier literary journal *Novy Mir* in 1962 and signalled the state's tolerance of public assessment of the Stalin era. Khrushchev had earlier set in motion a widespread programme of de-Stalinization, which included rewriting history and the geography of place names. Across the Soviet Union, public recognition of Stalin was purged, the renaming of countless streets, squares, and cities being one manifestation of the process. For many thousands of Soviet citizens, the easing of restrictions on internal migration provided a new lease on life. Thus, the Khrushchev period is notable for considerable experimentation, a general relaxation of control over individual behaviour, and a frank assessment of the Stalin era. But many of the schemes with which he was closely associated were not successful. In consequence, he was ousted in 1964. The assumption of power by L. Brezhnev and A. Kosygin brought management of the economy nearly full circle.

In 1965, another major economic reform was announced, though given the tenor of the debate which characterized the last years of the Khrushchev period, it could scarcely be regarded as radical. The now discredited experiment with regional economic councils (*sovnarkhozy*) was shelved. The ministerial system was put back in place along with an expanded system of State Committees – but with some potentially important differences. The number of plan indicators that central authorities used to monitor the performance of individual enterprises was to be reduced drastically. More autonomy was given to decision-makers at the grass-roots level to manage their affairs so as to best satisfy national priorities as set by central planners. Of course, the five-year plans continued to serve medium-term development strategies, while fifteen- or twenty-year planning scenarios were intended to keep the national economy abreast of technological changes and social needs. The annual plan, frequently altered, almost always unpredictable, remained the principal management tool. The 1965 reform, which the Eighth Five Year Plan beginning in 1966 officially introduced, included some concepts advocated by the economist Liberman. These included an emphasis on sales, profitability, and discretionary incentive funds, all of which were intended to foster innovation and produce higher quality goods more efficiently.

Given the importance attached to enterprise profitability in the 1965 reform, a major price revision was required if there was to be anything like a rational accounting of profit. This occurred in 1967. As noted earlier, both retail and wholesale prices were established by the State Committee on Prices. While scarcity did play some part in price determination, most prices were set with state policy and state needs in mind. In other words, price was not regarded as an autonomous factor in resource allocation. For example, despite huge increases in the cost of providing public transport, the fares charged at the close of the Soviet era were essentially the same as they were a quarter century earlier. In such an environment, a realistic assessment of cost, and profit, was bound to be elusive. The 1965 reform also introduced a charge on investment capital advanced by state banking institutions to some sectors of the economy. This charge, perhaps most readily likened to our notion of interest, was intended to bring about a more efficient use of financial capital. It was often the case that huge sums were allocated 'free of charge' to capital projects such as hydroelectric power stations. Most produced no substantial amount of electricity until years after they were scheduled to begin operation. Meanwhile, state investment capital drew no return. In a similar vein, the assignment of industrial materials supply planning to the State Supply Committee was supposed to offset the autarchic tendencies of the ministerial system. By giving this committee a higher profile, material flows were supposed to be more predictable, and from the national perspective, less wasteful.

Some features of the 1965 economic reform implied a loosening of control from above. Perhaps not too surprising is the fact that such initiatives did not always meet with widespread enthusiasm, since they were often direct challenges to well-ensconced vested interests in the Soviet system. For nearly 40 years, the locus of decision-making had been based in the upper echelons of the Communist Party, government, and economic planning committee (excepting in this latter case, of course, the Khrushchev-initiated *sovnarkhozy* experiment). After some initial success with the 'new' ministerial structure, several of the traditional problems reappeared. The number of performance indicators steadily grew, and thus eroded such local autonomy as had existed. Ministries resisted materials supply planning by the State Supply Committee, preferring instead to achieve a measure of self-sufficiency with all of the attendant inefficiencies already described. The revision of industrial wholesale prices in 1967 did not bring all the changes desired either. Given the rigidity of the price structure, innovation continued to be frustrated. It was frequently more 'profitable' for a plant to manufacture an old product than to introduce new lines, where the margin between cost and sale price was less. Indeed, the distribution of enterprise profits soon became less a discretionary bonus to workers for higher productivity than an expected regular supplement to the monthly wage packets. Put simply, the introduction of profitability did not radically improve the propensity to innovate, nor did the creation of bonus funds bring about a fundamental change in the attitude toward work as evidenced by the various measures of labour productivity. The 1965 economic reform brought some improvements, but it did not mark a divergent trend away from centralized decision-making in economic planning. By the early 1970s, it was recognized that something else had to be done.

In a complex economy, many important development opportunities lie at the interface of traditional industrial sectors, which in the Soviet Union were customarily set up as separate ministries. The Soviet decision-making structure announced in 1965 did little to facilitate inter-ministerial initiatives. Indeed, as the economy expanded, the sheer number of individual enterprises, be they factories, warehouses, or scientific research institutes, grew correspondingly, and so too did the task of centrally coordinating them all. Not infrequently, by-products of one manufacturing process which could be used as input in another simply ended up as waste, owing to the lack of integration of enterprises belonging to different ministries. The loss to the national economy was sizeable, and thus in 1973, a new organizational unit was created – the *obedineniye*, or association. Intended to replace the individual enterprise (*predpriyatiye*) as the basic administrative unit in the industrial and resource development sectors, the association conceivably could include design, production, and distribution functions previously under the jurisdiction of several ministries. There were two types, production associations and industrial associations, but their common purpose was to promote greater efficiency through economies of scale and, because of their legal status, to minimize bureaucratic interference in production. But even here, the apparent benefits were not sufficient to alter the behaviour of institutions and individuals alike. Ministries continued to protect their turf, and attempts to transfer real decision-making to the *obedineniye* were successfully resisted.

The 1965 economic reform and the creation of the *obedineniye* in 1973 were both intended to provide greater scope for grass-roots initiative, while at the same time facilitating central planning. Fostering grass-roots initiative is one thing; however, weakening the central planning function is quite another. The dilemma for the state was that to the extent that real control from the centre was given up, central planning became more problematic. In any event, the inertia, or resistance to change, within the Soviet system was considerable. The ministerial apparatus, the Party bureaucracy, and probably even a segment of the population itself, was disinclined to readily accommodate major change. To do so perhaps raised the possibility of losing power, working harder for less certain gain, or both. Thus, economic policy changes consistently affirmed the role of central planning and were intended to facilitate it, not to undermine it. In 1978, Brezhnev endorsed the need for central planning by arguing for more, not less, authority for Gosplan. A year later, a formal decree confirmed the sanctity of central planning. After the 1979 decree, other measures were introduced to further improve the central planning function.

Another revision of basic industrial wholesale prices was undertaken in 1982. Some retail prices were also increased substantially at this time. The retail cost of gasoline, for example, was about doubled. But for the growing number of people who owned cars, this was less traumatic than might initially appear, since it was estimated that close to half of all gasoline consumed by the private sector came from other than 'official' channels. In other words, it was acquired illegally through purchase, barter, or outright theft. The increase in gasoline prices at the pump was but one instance of the attempt by the state to facilitate central planning of resource development and allocation through a more realistic price structure intended to encourage conservation.

With the death of A. Kosygin in 1980, then of L. Brezhnev in 1982, new opportunities for change arose. The selection of Yuri Andropov as Soviet leader portended a campaign against corruption and slack labour discipline. And, indeed, the former head of the internal security department, the KGB, was responsible for both decree and action in this area. The forcible assault on drunkenness, 'parasitism' (or voluntary unemployment), graft, and the widespread casualness concerning hours of work did produce some positive results. Labour productivity increased dramatically during the early months of the Andropov era (from November 1982 to February 1984), but failing health precluded a sustained effort on Andropov's part. Upon his death, Konstantin Chernenko assumed control. Aged and infirm, Chernenko continued most of the Andropov initiatives, but he possessed few of the attributes of the true reformer. His passing away in March 1985 brought Mikhail Gorbachev to the fore. As noted in the preceding chapter, he quickly consolidated his position and introduced the prospect of wide-reaching change throughout the Soviet system. As chairman of the Politburo and of the Secretariat of the Party Central Committee, he was well placed to shape events. The call for improved management of the economy through *perestroyka*, *glasnost'*, and *uskoreniye* was espoused with more vigour than consequence, as a review of economic performance in the final years of the Soviet system makes clears.

In speech after speech, Gorbachev made the relationship between improvement of the material conditions of the Soviet population and economic performance his central theme. The following excerpts from a May 1985 report on the upcoming 27th Communist Party Congress were typical:

> The development of Soviet society will be determined, to a decisive extent, by qualitative changes in the economy, by its switch onto the tracks of intensive growth and by an all-out increase in efficiency...
>
> It is known that, along with the successes that have been achieved in the economic development of the country, in the past few years unfavourable tendencies have intensified, and a good many difficulties have arisen.
>
> ... Life is making even higher demands on planning, which is the core of management. It should become an active lever for the intensification of production and the implementation of progressive economic decisions, and it should ensure the balanced and dynamic growth of the economy. At the same time, the plans of association and enterprise must drop a great many indices and make broader use of economic normatives, which open up scope for initiative and enterprise.
>
> ... We must sharply restrict the number of instructions, regulations and rules, which sometimes give wilful interpretations of Party and government decisions and fetter the autonomy of enterprises.[1]

At the Party Congress in early 1986, Gorbachev continued to espouse these same points. Additionally, he called for further centralization of decision-making, especially in the context of creating several super-ministries to better coordinate strategic planning across key sectors of the national economy. At the same time, paradoxically, Gorbachev indicated that greater reliance was to be placed upon loosening controls from above, as implied in the foregoing quotation. There was clearly some potential for confused signals. Gorbachev proposed, and the Party and government endorsed, a series of new initiatives. Control from the centre was to be relaxed. Enterprises would be given the freedom, indeed, would be encouraged, to arrange contracts and establish prices for some goods. Thus, the role of the State Committee on

Prices was to be trimmed. It would concentrate on determining prices in priority sectors. Where there was already adequate supply of goods or services, enterprises would be permitted to set prices within upper and lower ranges. It was expected that greater resource-use efficiency would result from this development, which was scheduled to take effect in the new Five Year Plan in 1991. Enterprises were to be judged by their profitability. If they recorded a loss, they were to be permitted to go bankrupt. This was a major change inasmuch as the state had always put huge amounts of financial capital into the accounts of non-profitable enterprises. By the end of the 1980s, all factories were to have switched to self-financing, in anticipation of a major price reform in 1991. Meanwhile, more foreign capital was to be enticed into the country on a joint venture basis, with majority control of course being vested in the Soviet counterpart. Indeed, there were a number of such initiatives. They included everything from companies facilitating technology transfers to establishing fast-food operations. All wanted to gain a foothold in the Soviet Union. Most have dramatically expanded their operations in the post-Soviet era.

One of the most successful, and controversial, Gorbachev initiatives was the fostering of co-operatives. By 1988, 150,000 people worked in the 14,000 cooperatives which had been created. While most were in the consumer goods and services sector, a few were engaged in manufacturing intermediate goods for Soviet industry itself. Their success was measured by the huge incomes many members were able to earn, in some cases ten to twelve times the average monthly wage of a white-collar worker. The Soviet income tax system was not geared up for such a phenomenon. In the first year, the levy was in the range of 2 to 3 per cent. By the third year, it increased to 10 per cent. But with personal incomes of several thousand roubles per month, a new class of wealthy Soviets was in the making. But their incomes would pale in comparison with those earned in the post-Soviet era.

It needs to be borne in mind that the Gorbachev initiatives, while pointing in some new directions, nonetheless followed on the heels of other major economic reforms. All failed to live up to expectations. Some, such as the *sovnarkhozy* experiment of the Khrushchev era, disappeared altogether. The Soviet economic system time and again proved enormously resilient when it came to absorbing, diluting, and fundamentally emasculating efforts to reform it. And in hindsight, the system did produce at various times significant rates of economic growth, at least until the 1980s.

The rate of growth of Soviet national income during the 1930s probably exceeded 5 per cent per annum, and rates of industrial growth 10 per cent. Following World War II, and the subsequent period of reconstruction, these annual rates were once again achieved. But from the 1950s on, there was a perceptible decline. This was the result of several factors. The basic economic structure was consciously diversified, the consumer goods sector garnered a larger share of investment funds, and from the 1960s, agriculture had come to claim a much larger share of investment capital. All of these developments, and others as well, served to dampen the return on investment, the rate of growth of industrial production and national income. The 1965 economic reform was a response to this decline in the rate of annual economic growth. But it, and other measures to promote efficiency and higher rates of growth since, did not alter these basic trends. As Fig. 3.4 indicates, the average annual rate of growth of

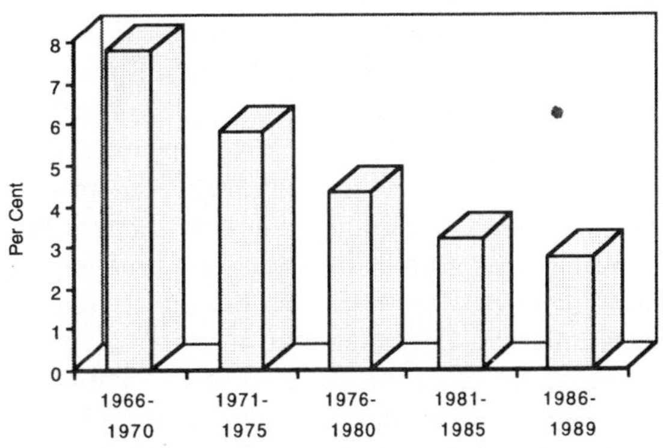

FIGURE 3.4 *Soviet era economic growth, 1976–1989*
(*Source*: Compiled from data in *Narodnoye Khozyaystvo SSSR v 1989g* (Moscow: Goskomstat, 1990), 9).

national income by five-year planning periods in the last few years of the Soviet era dropped substantially. For the 1966–1970 period, growth was nearly 8 per cent. Between 1986 and 1989, the average annual rate of growth of national income was only 2.7 per cent. And these were the official figures. Again, with the advantage of hindsight, it is now clear that the economic situation was actually much worse. By the early 1980s, the actual level of real growth of the Soviet economy was the subject of much debate. Some observers reckoned it was barely expanding one year to the next. A number of western economists argued that in fact the economy was experiencing negative rates of change. Whatever the actual state of affairs, there is little doubt that the Soviet economy was in considerable difficulty.

The problem was not simply attributable to the attitude of workers, but included such things as style of management and long-standing patterns of investment. For example, for years the preferred route to higher production levels was through investment in a new plant. Reconstruction, modernization, and maintenance of existing plants was a distinctly lower priority amongst managers. By the 1980s, past failure to re-equip and update existing enterprises caught up with planners and managers alike. Simply put, the standard management practice was one of extracting from existing facilities as much as possible without putting anything back. Of course, for the typical manager, shutting down a production line in order to modernize meant that overall production that year would drop, with the obvious negative impact on profitability and bonus income for all concerned. Given the continual emphasis on achieving targets set by central planners, such strategies were clearly impractical. Hence, short-term, rather than even medium-term, strategic management decision-making was encouraged.

After the reform of 1965, increased reliance on foreign technology was a common strategy in economic development. But here too the plans came upon hard times. Imported technology, especially from western countries, usually required payment in hard currency. From 1965 on, most of the Soviet Union's hard-currency earnings was derived from the sale abroad of energy resources, notably oil and natural gas. The decline in the price of oil on international markets after the early 1970s had an adverse effect on Soviet hard-currency earnings. As if this was not problem enough, hard-currency reserves, which had been so assiduously accumulated through the sale of such commodities as oil, gas, gold, and, whenever possible, manufactured goods, were eroded because of the failure of agriculture to perform according to plan. The Soviet Union was consistently a net importer of grain, in some years of very considerable quantity, from the early 1970s down to 1991. The inability of agriculture to meet planned output targets was an unanticipated drain on hard-currency reserves. One opportunity cost accruing from buying western grain was less western technology. It was small wonder that all Soviet leaders bemoaned the performance of the agricultural sector, regularly pointing to shortfalls despite the huge increase in the relative share of available investment capital being directed toward the countryside after 1965. In the final years of the Soviet system what helped to fill the void in production, especially but not exclusively in the consumer goods sector, was the existence of a shadow economy, or second economy.

The second economy

In the discussion of the Soviet economy thus far, attention has been focused on the formal sector, that is, the state sector, which was centrally planned. There was another dimension to the Soviet economic scene which was not planned, but which was nonetheless important. This was the informal sector, or second economy, which, while far from the absolute importance of the planned economy in terms of roubles of output, trade turnover, or whatever other measure might be employed, was nonetheless significant because it helped to bridge the gaps. In any economic system, there will be occasional scarcity of some goods or services. The Soviet Union was certainly no exception to this generalization. Indeed, as is implicit in some of the foregoing discussion, the very nature of central planning in a complex economy is likely to produce scarcity – either artificial or real. To a not inconsiderable extent, the second economy helped to offset shortages in both the producer and consumer goods sectors. The activities embraced by the second economy ran the gamut from those which were legal and sanctioned by the state, to those which were simply illegal and therefore subject to criminal prosecution when uncovered. In this brief discussion, only the first type of activity will be considered.

That central planning of the Soviet economy was a formidable task goes without saying. The system functioned, but it was in the very nature of the decision-making structure that periodic shortages occurred, and for the successful management of the economy these had to be overcome. Scarcity of both producer and consumer goods was thus a fact of life. Unlike those economic systems in which prices charged for goods and services reflect to some degree supply and demand, the official Soviet price system was never governed by the market, but by bureaucrats. As was noted, the price system was rigid, only infrequently revised and, for ideological reasons, did not always reflect scarcity. Soviet managers and Soviet consumers, however, were not without considerable talent in working within and around the system.

The penchant of Soviet enterprise managers to finesse as much as possible by way of inputs to the productive process out of central authorities has already been discussed, along with the consequences in terms of waste and inefficiency. Waste and inefficiency occur when inputs are used in excess of the necessary minima. An example is that of labour input, and this was one of the reasons for the typically lower level of labour productivity registered by the Soviets in most international comparisons. But as a managerial strategy, padding the labour supply made good sense, since demands for higher levels of output by central planners might be more easily accommodated if there was a 'reserve' labour pool available. *Perestroyka* was intended to change this ingrained practice. To acquire, in similar fashion, more raw materials, semi-processed goods and so on than was really required in the production of a particular commodity also provided a certain flexibility in this regard. Thus, despite repeated attempts to establish norms which really reflected the minimum input requirements for industrial production, the typical Soviet manufacturing process registered a higher level of resource consumption than western industrial states. The acquisition of excess supplies, however, had a purpose other than simply facilitating an adaptive response to increased output targets imposed by central planners. It played a part in the extensive bartering that went on between enterprises. Bottlenecks in materials supply were endemic to the Soviet system of central planning. These were at least partially overcome by means of 'unplanned' resource reallocations. Successful managers were those who were able to accumulate surpluses of something they did not require in order to exchange them for something they did. The intermediary in this process was the so-called *tolkach*, or pusher. This was the plant employee whose formal job designation was usually something other than what was done, namely expediting. It was his or her skill in finding other enterprises with which to enter into semi-official exchanges of mutually required commodities which helped to make the system function. Central authorities were clearly not oblivious to this type of activity. While it has historically operated outside the officially sanctioned inter-enterprise flows of material, the customary inability of the system to deliver what was required when needed resulted in such activities being tolerated. Indeed, with each move toward enabling enterprises to enter into formal contracts to accommodate such transfers, the state moved another step closer to institutionalizing formally what went on anyway. As a rule, unplanned transactions were strictly barter, with no financial payment made. The redistributive process was seldom so straightforward as to involve only two participants in the barter. The longer the chain, the more important the skill of the *tolkach*. Clearly, a resource in short supply relative to others available for barter will command a premium regardless of the official price.

It is in the consumer goods sector, including food supply, where there was the greatest variety of second-economy enterprise and initiative, however. There was also the greatest potential for personal reward with least risk. To be sure, enterprise played an important part in the successful management of a factory with the concomitant benefit of bonuses for meeting planned output targets. But bonuses have official limits and were relatively straightforward to monitor. We might illustrate how difficult it was to control personal income in some of the legally sanctioned markets which comprised the second economy outside the producer goods sector. The collective farmers market was one case in point.

Ever since the 1930s, peasants in the collective farm sector, and some other categories of rural inhabitants as well, were able to sell surplus products from their personal plot and privately owned livestock. The evolution of this form of free enterprise we will examine in some detail in a later chapter, but suffice it to say at this juncture that the flow of such foodstuffs to the urban market to be sold at prices which reflected supply and demand conditions and not the state price for the same commodity, had been an important supplement to the diet of the typical city family. To be sure, there was some monitoring of the prices charged, and theoretically, even in times of acute shortage, peasants were not supposed to charge in excess of three times the official state price, but this was frequently ignored. Thus, there was considerable scope for peasant entrepreneurs to exploit shortage. Many did, and the standard of living of some peasant families with access to urban markets was of a very high order indeed.

Within the state retail sector, the regular shortages were sometimes used to advantage by the sales personnel. It was a simple matter to advise friends or relatives of an incoming shipment of goods customarily difficult to obtain. Holding back a commodity for sale at the official price for someone who then either used it, or, more likely, sold it at whatever the market would bear and shared the proceeds, was commonplace. In such cases, the second economy functioned to allocate goods in short supply to those who were willing to pay a premium price. The opportunities for such manipulation by state employees were limited only by imagination and the police. The greater the risk, the higher the price, of course. But much of what commonly went on was borderline in terms of legality.

Rental housing, especially but not exclusively holiday accommodation, was another example of where private enterprise operated. Rental accommodation helped to fill an important niche in the Soviet system, but the line between legitimate rent and gouging was far from clear. Here again, the price charged for a scarce resource was determined very largely by what the market would bear. The absence of clear guidelines on rent to be charged and inadequate means of enforcement enabled the marketplace to exist. The sale of flowers by individuals in railroad or subway stations or in pedestrian underpasses was another common activity which was perfectly legal. Each city had at least one second-hand, or flea, market at which supplementary income generation was possible. In short, whether renting an apartment or house, purchasing food at the collective farmers market, hiring a repairman *nalevo*, or on the side, buying flowers on the street, there were innumerable instances of individual enterprise bridging the gaps which were associated with the central plan. Ignoring entirely all activities which were clearly illegal, the second economy performed an important service for the state. Thus, there was a pool of entrepreneurial talent already in place to take advantage of the new economy of the post-Soviet era.

Market reform in the post-Soviet era

The legacy of the Soviet central planning system left for the new post-Soviet states was dismal – vast bureaucracies, bloated labour forces in industry, insufficient consumer goods and services, inefficient agriculture, natural resources mis-management practices, and widespread environmental degradation. As noted earlier, in the final years of Soviet power the economy was in a state of decay. Industrial production across the Soviet Union fell by close to 20 per cent in the first months of 1991 alone. Light industry output, which was heavily geared to the consumer goods sector, shrank by about 40 per cent. The list of shortcomings in economic performance at the close of the Soviet era is long and distressing to recount. In 1917, the new Soviet leadership did not have any models of operating socialist systems upon which they could draw upon for guidance. In 1991, in contrast, there were many examples of market economies which the governments of post-Soviet states could examine in order to gain some insights into how best to chart a course out of the swamp of state ownership and central planning.

In addition to the models provided by western industrial economies of Europe and America, there was also the post-war experience of the fast growing southeast Asian states. While the market economies of these states might provide some lessons for post-Soviet states intending market reform, none of them had to deal with state ownership of all resources and an all-encompassing central planning system. Put simply, the concepts of private property and private profit were already an integral part of their social and political structures. Only in the Baltic states of Lithuania, Latvia and Estonia were there significant numbers of people with first-hand experience of the inter-war years of independence and non-socialist economies. For the post-Soviet states rather more relevant for the transition from a socialist economy to a market economy was the experience of Eastern European states. As part of the Soviet sphere of influence following World War II, their economies had been developed according to the Soviet model. Already in the 1980s most of them had started to reform their economies.

One economic reform approach aggressively promoted by a number of western advisors to new post-Soviet governments, and drawn from the economic transition experience of the former Soviet east bloc countries, was the so-called 'big bang' model. This had been followed in the first days of political independence in Poland, and later in the Czech Republic. In the 'big bang' approach, state subsidies, state control of production and prices, and state ownership of the means of production itself are jettisoned quickly. This approach to introducing private ownership, production and market-determined prices ostensibly has the appeal of bringing the benefits of a market economy to the people quickly, thereby enlisting their participation and political support for the reform process. But Poland, for example, is not necessarily an apt model for post-Soviet states. In the first place, Poland has the distinct advantage of a largely homogeneous population, both ethnically, and in terms of widespread adherence to the Catholic Church. This helps to unify the population during a time of rapid and profound change. In contrast, peoples of the new post-Soviet states are typically multi-ethnic, and religion tends to play a much less important part in their daily lives. Put simply, in the new political reality of post-Soviet states some people are more equal than others, thus there is more scope for dissension than consensus. Secondly, Polish agriculture has always been dominated by privately owned smallholdings. Their peasant owners were well acquainted with the relationship between supply, demand and profit making. Therefore, the prospect of food shortage during the transition phase was reduced. By way of contrast, there were some parts of the post-Soviet scene where the gross inefficiencies of socialized agriculture threatened food shortages. But even with these advantages, Poland's experience in transforming its socialized economy into a market-based one was by no means entirely positive. Initially, industrial production plummeted, potentially profitable businesses went bankrupt, unemployment grew, inflation was rampant, and most people suffered a decided drop in their standard of living. Poland has made progress since the first shock waves of reform to be sure, but the social and economic costs of the 'big bang' approach were palpable. The more gradual relaxation of state control and slower rate of privatization followed in, for example, Hungary and the Slovak Republic offered some alternative strategies, but nowhere in Eastern Europe was there a sure prescription for a painless, popularly supported transition from a command economy to a market economy. In short, there would be hardship meted out to the peoples of the new post-Soviet states whichever approach was adopted in restructuring their economies. And of no small concern in many post-Soviet states was the fear that should there be widespread social upheaval in consequence of market reform, democracy itself could well be at risk. The economic decline of the final years of Soviet power produced considerable hardship for ordinary people. While the prospect of market reform held out the promise of a better life, getting there with society still intact was the challenge. Thus, the economic reform process has begun, but is being pursued with varying degrees of commitment.

As Fig. 3.5 indicates, by 1994 there was definite progress in the movement toward a market economy amongst the post-Soviet states in terms of privatization of large and small firms, and the introduction of market prices. The data are arranged in such a manner as to highlight the Russian situation in relation to the western tier and southern tier states. Comparative data are included for Eastern Europe. Since 1991, Russia's transition to a market economy, as measured by the very general indicators listed in Fig. 3.5, has been reasonably respectable. The greater progress across most of Eastern Europe in market reform should be seen in the context of the earlier demise of Communist control there than in the Soviet Union. Before discussing the situation in Russia in some detail, the market reform process elsewhere in the post-Soviet scene will be reviewed briefly.

State	Privatization: Large Firms	Privatization: Small Firms	Market Prices
Russia	■■■	■■■	■■■
Western Tier			
Belarus'	■■	■■	■■
Ukraine	■	■■	■■
Estonia	■■■	■■■■	■■■
Latvia	■■	■■■	■■■
Lithuania	■■■	■■■■	■■■
Moldova	■■	■■	■■■
Southern Tier			
Georgia	■	■■	■■
Armenia	■	■■■	■■■
Azerbaijan	■	■	■■■
Kazakhstan	■■	■■	■■
Kyrgyzstan	■■■	■■■■	■■■
Tajikistan	■■	■■	■■■
Turmenistan	■	■	■■
Uzbekistan	■■	■■■	■■■
Eastern Europe			
Albania	■	■■■	■■■
Bulgaria	■■	■■	■■■
Croatia	■■■	■■■■	■■■
Czech Republic	■■■■	■■■■	■■■
Hungary	■■■	■■■■	■■■
Macedonia	■■	■■■■	■■■
Poland	■■■	■■■■	■■■
Romania	■■	■■■	■■■
Slovakia	■■■	■■■■	■■■
Slovenia	■■	■■■■	■■■

FIGURE 3.5 *Movement towards a market economy, post-Soviet and East European states, 1994*
(*Source*: Based on data in *The Economist*, 3–9 December 1994, 27)

Amongst the western tier states, Lithuania, Latvia and Estonia clearly led the pack in terms of privatization. In each country the government was quick to shake off the legacy of the Soviet era. Prices were freed from state control, enterprises of all types and in all sectors of the economy were privatized, separate currencies were introduced with central bank policies that would enable these currencies to be freely convertible in the shortest possible time, and issues of taxation were dealt with early on so as to provide as stable a basis as possible for the private sector to plan. In short, much happened quickly to offset the previous dependence on the rest of the former Soviet Union for raw materials and manufactured and consumer goods. In foreign trade, the ties with the east have been broken as most foreign trade is now with the west. In the energy sector, however, a critical dependence on Russia continues. It needs to be borne in mind when considering the fast pace of market reform and privatization in the Baltic states, as noted earlier, that each has an older generation with direct experience of a market economy from the inter-war years of independence. This cannot be said of the rest of the former Soviet Union. Save for a few exceptions, such as the territories along the Finnish border regained during World War II, the people have no collective memory of a market economy. Thus, the more rapid transition to a market economy in the Baltic region than elsewhere in the post-Soviet scene is understandable.

In Ukraine in the first years of independence, government policies to diminish the role of the state in favour of the private sector have been hesitant and sometimes contradictory. Priority was given to maintaining as much as possible of the socialist safety net in terms of subsidized foodstuffs, free health care, education and so on. Ukraine of all the western tier states has made least progress in the march to a market economy in the period covered by Fig. 3.5. Part of the hesitancy in Ukraine to push the reform process is tied to the fact that it was severely impacted by the dissolution of the Soviet system. Its very substantial industrial infrastructure was part of a system in which scale economies and regional industrial specialization were consciously fostered. In the post-Soviet period, sources of raw materials and semi-finished goods for its huge factories are located in as many as 14 independent states, each now possessing different economic development priorities. Unfortunately, as a major industrial region Ukraine lacks adequate energy resources to sustain domestic consumption. It has to cope with how to pay Russia for its energy supplies, the price for which is fast approaching world levels. Markets for Ukraine's products are also impacted as its traditional customers are located outside its borders. Many of its markets are in countries now experiencing serious economic problems. They have only limited ability to pay for imports. Under the former Soviet system, all business transactions were conducted in roubles. Trade deficits certainly occurred, but were invariably covered by subsidies from the Soviet government. Now settling trade debts is not only complicated by customers being in other countries, but by the replacement of the Soviet rouble by new national currencies. These generally do not have fixed exchange rates. Rampant inflation and bankruptcy in places with few laws and less enforcement regarding business contracts simply add to the confusion of trying to conduct foreign trade. Selling manufactured goods outside the post-Soviet realm is possible to be sure, but invariably they cannot compete with foreign output either in quality or in price.

Rapid privatization of heavy industry would have resulted in huge numbers of workers becoming unemployed. Many Ukrainian politicians advocated maintaining state ownership. Nonetheless, underemployment, if not unemployment, in Ukraine is commonplace, and constitutes a significant economic burden. Ukraine moved quickly to establish a separate currency, but unlike the Baltic states it did little else to foster a market economy. Its central bank, with the support of government, simply printed money in an effort to maintain the largely socialist economy. The outcome was predictable – the new currency quickly devalued. Severe inflation ensued and the economy faltered. Bartering goods began to gain favour over business conducted on the basis of a currency of rapidly diminishing worth.

Part of the limited progress in implementing reforms may well have to do with the rather complex nationality issues confronting Ukraine. Russians are predominant in the eastern, heavily industrialized region, Ukrainians dominate in the west. There are religious tensions between some Eastern Orthodox Russians and some Ukrainian Catholics. Russian-dominated Crimea, a region historically part of Russia until it was made part of Ukraine in 1954, remains a potential geopolitical 'hot spot'. The dispute over how to allocate the former Soviet Black Sea naval force and military bases in Crimea further complicates matters. In short, Ukraine had numerous reasons for putting a high priority on continuity in economic and social conditions and less on policies which might create social unrest.

By the mid-1990s, the pressure from outside sources of financial and development assistance, such as the International Monetary Fund and the World Bank, combined with the steadily deteriorating industrial economy to force Ukraine into adopting a much more pro-active position regarding privatization and economic reform. The emergence of former Communists in key positions in government is not unique to Ukraine, and therefore cannot be regarded as the principal reason for the comparatively slow progress. Indeed, not a few such people have become the most ardent proponents of a market economy. Leonid Kuchma, the recently elected President of Ukraine and a pro-market reformer, is a case in point.

Even before the collapse of the Soviet Union in 1991, the role of the state in the Moldovian economy was declining. Between 1988 and 1991, for example, the share of employment in the state sector dropped from nearly 80 per cent to 54 per cent as employees in the retail and trade sector in particular assumed ownership of enterprises. Since independence Moldova's overall progress in terms of the measures used in Fig. 3.5 is better than either Ukraine or Belarus', neither of which has had to deal with such serious internal political problems. Moldova's have to do with Ukrainian and Russian minorities who are resisting post-independence attempts by the government to promote a form of Moldovanization, that is, restrictions on citizenship, priority use of the Moldovan language and so forth. The presence in Moldova of a sizeable Russian military force is required in order to ensure peace. The economy of Belarus' was, and remains, closely integrated with that of Russia. Its government has been consistent in facilitating economic cooperation in practice, even if the rhetoric of its political leaders at times suggests that policies promoting greater economic independence from Russia would be pursued. Culturally and linguistically its population is inextricably bound up with Russia. Indeed, the first language of the majority of the Slavic Belorussian population is Russian, not the linguistically different Belorussian language. In the first days of independence, schools were established in which the language of instruction was only Belorussian. Initially popular, both interest in them, and attendance at them, has declined since.

Across the southern tier states, privatization of large and small firms and the adoption of market pricing is rather more variable than elsewhere in the post-Soviet scene. Amongst the Caucasian states, Azerbaijan has clearly lagged behind Armenia and Georgia. But in comparison with most other post-Soviet states, all three Caucasian states are laggards in terms of the criteria included in Fig. 3.5. However, these data do not cover all sectors of the economy obviously. As discussion in later chapters will indicate, there are significant levels of privatization in particular sectors of their economies. In the Middle Asian realm, Kyrgyzstan began the transition to a market-based economy quickly, as the data in Fig. 3.5 imply. The pro-market reform approach adopted by the Kyrgyzstan government has been endorsd by international development agencies. Nonetheless, the material standard of living of its population is now lower than it was during the Soviet era. Because of its relative location, lack of infrastructure and poorly developed manufacturing base, the prospects for rapid improvement in the economic well-being of its population are less than auspicious. But it does have mineral resources as yet untapped. Indeed, Kyrgyrzstan's fast track approach to market reform has proven to be attractive to foreign investment on the part of resource development consortia, and many are now actively developing a variety of mineral resources, including gold. Kyrgyzstan and Kazakhstan have followed fairly closely the economic reform model adopted in Russia. While according to Fig. 3.5, Kazakhstan had not made much progress by 1994 in market reform compared to other Middle Asian states, there is nonetheless a very pro-reform government in place. In this respect there is much more in common between the Kyrgyzstan and Kazakhstan than Fig. 3.5 suggests.

Tajikistan became embroiled in civil war soon after the collapse of the Soviet Union. It now maintains a measure of domestic order with Russian assistance, and the Russian army defends its borders against anti-government forces based in Afghanistan. Despite being beset within by social and political unrest and occasional outbreaks of civil war, Tajikistan has nonetheless made some progress in market reform (Fig. 3.5). Turkmenistan lags behind all the Middle Asian states in market reform, a fact seemingly of little concern to its government. As is the case throughout Middle Asia, its government is controlled by former establishment Communists who now wear another political hat. The social safety net which was so important a part of the Soviet-era 'contract' with the people still is much in evidence in Turkmenistan. Food prices are subsidized, education, medical care and other government services remain firmly in place, even if many of them are of low quality. Unlike some other states in the region, Turkmenistan's relatively small population can be sustained for the foreseeable future in the manner to which it had grown accustomed, because of the sizeable hard currency revenues from the export of natural gas, and substantial proven reserves upon which the government can rely for future income. Kazakhstan is in much the same position as Turkmenistan in terms of probable future revenue for the state coffers, because of substantial oil and natural gas reserves. The large Russian presence in Kazakhstan requires the government to demonstrate much greater sensitivity to public concerns and international relations than is demanded of the Turkmenistan government, for example, where there is also a Russian minority, albeit a small one.

Throughout Middle Asia, substantial state involvement in many sectors of the economy is likely to continue for a long while yet. All Middle Asian states will likely continue to share another characteristic which makes economic decision-making and management less complex than in many other parts of the post-Soviet scene. Their electorates, even in ethnically diverse Kazakhstan, have demonstrated repeatedly in the first years of independence that they are inclined to support strong governments, governments which in most Middle Asian states have operated in a manner more than a little reminiscent of the former Soviet top-down decision-making style. Indeed, the President of each Middle Asian state is a former Communist, presides over a Parliament dominated by former Communists, and manages by decree rather than by consensus. Post-Soviet referenda to extend terms of office of Presidents beyond the limits stipulated in their constitutions have been overwhelmingly supported by the masses. In states such as Turkmenistan there is even plentiful evidence of a 'cult of personality' surrounding the incumbent President, Saparmurad Niyazov. *Plus ca change*?

Data for the 11 Commonwealth of Independent States (CIS) for 1993 indicating the respective roles of the state and the private sector in retail trade are presented in Fig. 3.6. These data offer another perspective on the privatization process. As noted in the preceding chapter, the three Baltic Republics and Georgia refused to join the CIS when it was first formed in late 1991. Georgia was induced to become a member in 1994 as a result of overt political pressure from Russia. Georgia's retail trade pattern in 1993 was similar to those of Armenia and Azerbaijan, that is, the private sector dominated. As Fig. 3.6 indicates, Middle Asian and Caucasian states had the largest shares of retail trade in the private sector, a pattern somewhat in contrast to that suggested by the data in Fig. 3.5. Upon reflection this should not come as too great a surprise, however, since it was in the Caucasus and Middle Asia where the role of the personal sector in retail trade and consumer services was most highly developed during the Soviet era. The black market also flourished throughout these areas. The transition to privately-owned, and now legal small retail trade, was thus not a major step. Amongst those western tier states in the CIS, only Moldova had a majority of retail trade in the private sector (Fig. 3.6).

It should be noted that although privatization and market reform in Russia does not always rank first in terms of the information presented in Figs 3.5 and 3.6, in fact the Russian economic

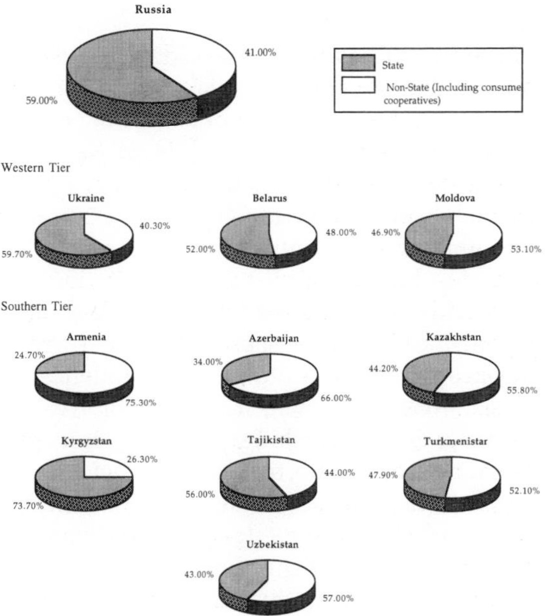

FIGURE 3.6 *Post-Soviet states retail trade turnover by sector, 1993*

(*Source: Ekonomika Sodruzhestva Nezavisimykh Gosudarst v 1993g* (Moscow: Statkomitet SNG, 1994), 20)

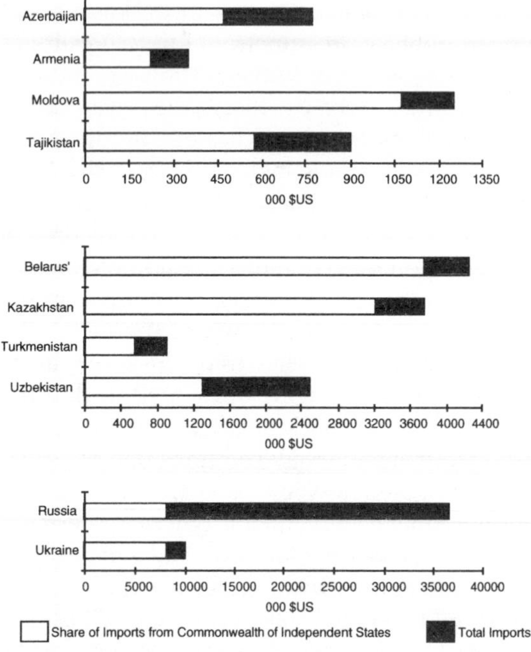

FIGURE 3.7 *Import structure of post-Soviet states, 1994* (*Source*: *Ekonomika Sodruzhestva Nezavisimykh Gosudarst v 1994g* (Moscow: Statkomitet SNG, 1995), 30)

reform model has been followed rather closely in a number of post-Soviet states. This may not always be emulation of best practice, but because of the continuing dominant role of Russia in their economies there is often strong economic argument for alignment with some Russian economic reform initiatives. The structure of imports for the member states of CIS in 1994 outlined in Fig. 3.7 is a good indicator of the inter-dependence of many post-Soviet states, and especially of their continuing dependence on Russia, a fact of life which again reinforces the need for some meshing of basic strategies regarding market reform. The inter-dependence of post-Soviet state economies still remains quite substantial despite the efforts of some states to extricate themselves from this legacy of the Soviet system. With this thumbnail sketch of some aspects of market reform across the post-Soviet scene in hand, the discussion will now turn to a more detailed examination of the process in Russia.

ECONOMIC REFORM IN RUSSIA

The Russian approach to market reform was based on the experiences of former Soviet bloc states in Eastern Europe. The difficulty in applying the lessons from this region, however, was that not one of those countries had to deal with such huge territory, heterogeneous population or, by 1991, such a seriously deficient domestic economy. Moreover, in eastern Europe there was, as in the case of the Baltic states, a proportion of the population with direct experience of a market economy from the pre-World War II years of non-Soviet domination. All told, Russia's nearly 150 million citizens in 1991 faced some exceedingly difficult years of economic reform. It was recognized that the transition to a market economy would consist of three phases – crisis, stabilization and restructuring. By the mid-1990s, there were beginning to be signs of stabilization. We will first describe some of the main features of the years of economic crisis.

The first task faced by Russia's largely reformist-minded government headed by Boris Yeltsin in the first few months of independence was that of privatization of the economy. By October of 1991, a basic programme had been put in place for both stabilizing the economy and laying the groundwork for the transfer of ownership of property and resources from the state to a new class of property owners. Yeltsin was granted widespread presidential powers at this time by the elected members of the central government in order to implement the reforms. By January 1992 a flurry of decrees had been issued. Minimum wage levels were established. Most wholesale and retail prices were freed from state control. Retail trading was opened to anyone who wished to become involved. Foreign investment restrictions were eased. Price increases for key resources remaining under government control, such as energy, were introduced. Within weeks prices for a wide range of goods and services rose dramatically. The cost of living soared, offset at first by recourse to savings, and after they were exhausted by cutting back on purchases and attendant hardship. Since Russia had assumed a substantial share of the former Soviet Union's nearly US$ 80 billion international debt, foreign creditors had a vested interest in the performance of the Russian economy. Progress in market reform in Russia was therefore important since repayment prospects were obviously related to how well Russia's economy performed. Thus, almost immediately the Russian government was under external pressure in terms of its fiscal policies, in which control over the operation of its central bank figured prominently. Government budget deficit was to be tightly monitored and each year reduced, subsidies to inefficient sectors of the economy still under state ownership and management were to be dropped, interest rates were to rise, and the rouble currency stabilized. All of this and more besides was demanded of Russia in order to ensure a moratorium on debt repayment and foreign financial assistance in the transition to market economy period. But what was expected by way of fundamental change was in many respects beyond the capability of the system for managing the economy to deliver very quickly. After all, the task at hand was huge. How was private ownership of factories and farms, land and resources, shops and homes, to be achieved? How would the existing codes of law be rewritten to deal with the new reality of private ownership, business contracts, and associated rights and responsibilities? And equally if not more important, how would decades of decision-making practice, ingrained values and related behaviour patterns be transformed? Privatization of state assets was begun almost immediately, and in the cities in particular the population wasted little time in taking advantage of the opportunities provided. They were encouraged to engage in now legally sanctioned private trade, if not driven in some cases to do so by the rapid decline in production in many sectors of the economy. This development not only dimmed employment prospects for the future, but since many enterprises and institutions soon had difficulty making salary and wage payments it had immediate financial consequences for hundreds of thousands of people.

Amongst the first components of the state-run economic system to be privatized were the retail trade and public catering enterprises. By late 1993, about one-third of all such establishments in Russia had been sold, primarily to previous employees and managers. However, the extent to which these facilities had been privatized varied from one region to another, as Fig. 3.8 illustrates. The tempo of privatization was very much influenced by the terms and conditions set forth as conditions of sale. And this was very much left to the discretion of local authorities who were the principal 'owners' of these facilities under the Soviet system. Two cities stand out as leaders in the process – St Petersburg with about 75 per cent of its retail trade and catering enterprises privatized by 1993, and Moscow with close to two-thirds. Kemerovo Oblast' in West Siberia was not too far behind with about 55 per cent (for location, see Fig. 9.12, p. 280). Needless to say, where authorities imposed fewest restrictions, the transition from state to private ownership was rapid, and the share of total employment in the non-state sector increased dramatically. In St Petersburg, for example, it increased more than eight-fold to 15.4 per cent between 1991 and 1992. In Moscow, the increase was even more substantial. Non-state employment accounted for 27 per cent of the total in 1992. Indeed, in few other parts of the country was the change so swift. Perhaps only in Nizhniy Novgorod was there a comparable change in the structure of employment. In this city of nearly 1.5 million people to the east of Moscow, government authorities had adopted a particularly aggressive approach with respect to turning city-owned assets over to private owners. However, the sheer size of the Moscow economy with its nearly nine million inhabitants meant that whatever was done there had enormous impact in absolute terms. Additionally, what was done in the capital city influenced decision-makers elsewhere. The pattern depicted in Fig. 3.8 actually understates the role of the private sector in retail and public catering. The legislation permitting individuals to engage in retail trade had resulted in literally hundreds of thousands of people taking to the streets to offer all manner of goods for sale at whatever price could be had. Street trading is difficult to monitor with any accuracy. Therefore,

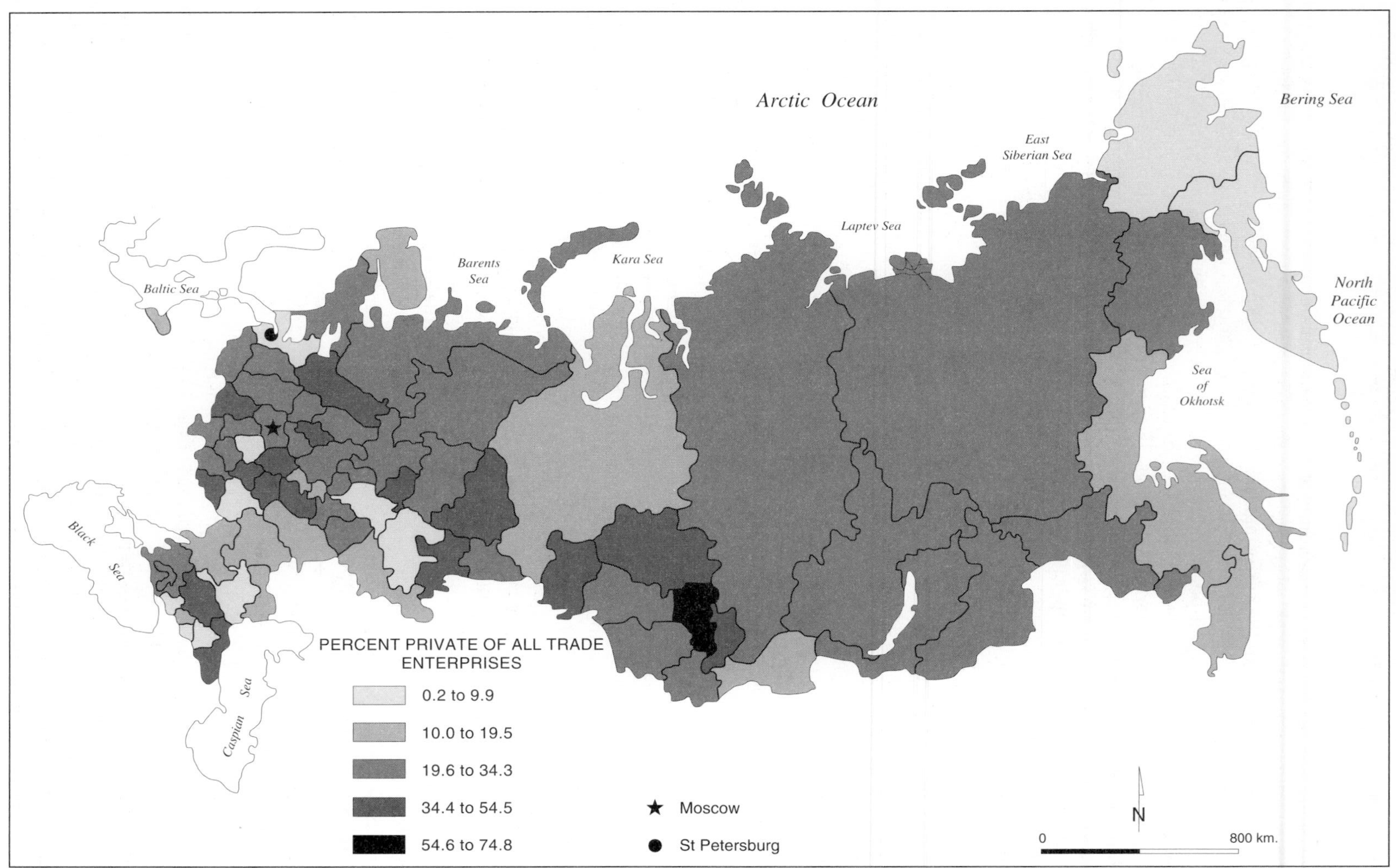

FIGURE 3.8 *Privatization of retail trade and public catering enterprises: Russia, September 1993*
(*Source*: T. Heleniak, *Economic Geography of the Russian Federation* (Washington: The World Bank, Country Department III, Europe and Central Asia Region, 1994), Vol.1, No. 22)

the number involved according to official data is certainly considerably less than the real total. It continues to be an important function in all Russian cities, and for many people is now the sole source of income.

A scheme for privatizing a substantial part of industry was adopted by the Russian government, but only after much debate over how best to achieve both equity and a measure of efficiency. The challenges were many. Simply turning over the means of production to the workers would be easiest, but was not acceptable. Those who happened to be employed in the few well managed and profitable enterprises or sectors would gain an unfair advantage over the majority who were not so fortunate. Offering such state assets for sale for money would simply give, it was feared, even more power to the former *nomenklatura* strata. Many of these people were fast becoming exceedingly rich because their connections provided opportunities to buy commodities or manufactured goods at low state prices, but sell at market prices, often for hard currency. Russia also wanted to avoid making available to foreign concerns state assets for what would be firesale prices. This would not only lessen potential future income for the state, it would diminish control by the government in restructuring the economy. Despite strong opposition from conservative elements in the central government, in the summer of 1992 Yeltsin introduced a privatization programme. It was to begin that October.

The programme was similar in many respects to that used in the Czech Republic. It involved giving to each of Russia's nearly 150 million citizens a privatization voucher valued at 10,000 roubles, ostensibly a per capita allocation equal to 35 per cent of the total value of the state assets (1972 prices) being converted to private ownership as of July 1992. At the time the value of the voucher was more than twice the average monthly income. The idea was that by providing each person with a voucher a new class of property owners would be created. Vouchers could be saved, sold or invested. Coupled with the decree that required all state enterprises employing more than 1000 people, or having a book value of 50 million roubles or more, to submit privatization plans to the federal government by November 1992, the voucher programme gave a substantial fillip to the privatization drive across Russia. Most enterprises opted for the privatization plan that gave employees the right to purchase 51 per cent of the stock of their enterprise at a fixed price. The balance of the stock could still be purchased by investors, but only in exchange for privatization vouchers, not cash. By allowing freedom to do with vouchers what one wished, millions were soon sold. As more and more vouchers were offered for sale, the price the market placed on them was soon less than the face value. Those businesses, banks and individuals accumulating large numbers of vouchers at discounted prices could use them to invest in enterprises the government decree had obliged to privatize, or they could hold them in anticipation of an increase in value. The voucher-based privatization programme was in force until July 1994. Thereafter, and for the first time in the market reform era in Russia, shares in enterprises could be bought for cash.

Despite the opposition to the voucher-based privatization scheme initiated by Yelstin as part of his reform programme, it did create the basis for a stock market. And the Presidential decree forcing certain of the state's enterprises to put in place a privatization plan by November 1992 did result in their conversion to some form of private ownership by the end of that year. As in the case of retail trade and public catering facilities, there was significant regional variation in the privatization of industry, something which Fig. 9.11, p. 278 clearly indicates. In large measure this was because the regions in Russia quickly took for themselves an increasingly autonomous role in deciding upon an economic development strategy. This was possible because of an implicit pact with Yeltsin.

In Yeltsin's on-going dispute with an increasingly obstructionist Russian Parliament, he was forced to look to the regions for sufficient political support to implement his reform agenda. Russia's democratically elected Parliament was formed in early 1990. Most members at that time belonged to the Communist Party. With the demise of the Party and then the Soviet Union, there emerged a wide array of new political parties and blocs in the Russian Parliament. Irrespective of party or bloc affiliation, however, following independence many members of Parliament became increasingly leery of the pace of economic reform, worried by the dismantling of the social safety net of the Soviet era, and very uneasy about the hardships the economic reform was imposing on the population, and perhaps more cynically, about the potential political consequences of social unrest for themselves. The democratically elected members of Parliament were, not surprisingly, also upset with the 'rule by decree' methods adopted by President Yeltsin in circumventing them in the implementation of the economic reform. This dispute culminated in the armed conflict of October 1993, described in the preceding chapter. In exchange for supporting President Yeltsin's reform initiatives, leaders of the ethnic and non-ethnic minority-based regions alike steadily co-opted economic decision-making responsibilities, and associated authority, previously the perogative of central government. Many leaders of the ethnic territories which had acquired Republic status within Russia already had garnered substantial autonomy in economic and other affairs (see Fig. 9.12, p. 280). But not all regional leaders acquiring more power were inclined to pursue with vigour market reform and privatization. Thus, the patterns of privatization depicted in Fig. 3.8, and Fig. 9.11, p. 278, reflect a variety of regional political and economic agendas. And, of course, across a country of such enormous extent, embracing such divergent regional geographies, differences in regional response to market reform were inevitable. Two models of regional economic management stand out, however. The first model or approach is that associated with Nizhniy Novgorod to the east of Moscow, the second with Ulyanovsk Oblast' also to the east of Moscow in the upper-Volga river region (Fig. 9.12).

In Nizhniy Novgorod radical economic reform was taken up with enthusiasm. Privatization was facilitated in every way possible, and laws to protect private property were introduced. Price controls and limits on profit taking were not imposed. The regional economy was opened to all forms of competition, that is, there were no restrictions on exports or imports. At the same time, much effort was expended enhancing employment prospects for the region's citizenry. And to ensure that marginal sectors of society were not impoverished completely in the reform process, a form of partial income support was provided by regional government. This provided an average per capita income sufficient to purchase goods and services worth at least twice the official subsistence level income for the region. This social safety net was also intended to prevent regional purchasing power capacity from being significantly reduced in the transition phase, thereby negatively impacting demand and related regional suppliers. All of these reform policies were endorsed by a coordinating council made up of representatives from both the executive and legislative branches of government in Nizhniy Novgorod. While this common political front on the part of the regional executive and legislative branches was an essential prerequisite for the strategy adopted, throughout the region there was in fact a high degree of public support for it.

In Ulyanovsk, there was a similar meshing of executive and legislative will, but in this region it was in support of a very conservative approach to economic reform. Here the main emphasis was on ensuring continuation of many vestiges of the former Soviet-era social contract, especially of controlled, and cheap, prices for basic foodstuffs. By adopting a self-sufficient, regional food policy, prices in the shops were held in check. But this could only be done by imposing a ban on food exports from the region. This could only be accomplished by local industries agreeing to supply agricultural enterprises with goods and services at prices low enough to ensure they could remain viable. A form of local sales tax provides regional government with sufficient income to maintain these subsidies. In the industrial sector of the economy, regional needs were also given the highest priority. This approach was possible only because Ulyanovsk Oblast' possesses a diversified agricultural-industrial economy. By enforcing a regional pricing policy, factories continued to be supplied, farmers were assured of necessary inputs, and consumers were guaranteed adequate supplies of low-cost basic foodstuffs. For the decision-makers the benefit was greater social stability, and continuing political support. The result was that food, and other basic consumer goods prices were significantly lower in Ulyanovsk Oblast' than in Nizhniy Novgorod. However, the region had always been a sizeable net importer of foodstuffs, and its leaders had to reform an economy that was completely dominated by a huge military industrial complex. Clearly, at the regional level different approaches to economic reform were much influenced by the nature of the economic structure. But the common denominator in both the Nizhniy Novgorod and Ulyanovsk approaches to regional economic management was the close working relationship between the executive and legislative branches of government. This permitted an adaptive and strategic response to the flood of Presidential decrees promoting economic reform. In many other places this did not hold true. Ad hocism prevailed. Many regions were as severely impacted by the lack of consensus, if not hostility, between executive and legislative branches, as was Moscow. As will be noted in ensuing chapters, the regions play a prominent role in the post-Soviet Russian scene.

FURTHER READING

Adams, W., Brock, J.W. 1993 *Adam Smith Goes to Moscow: a Dialogue on Radical Reform*. Princeton, NJ, Princeton University Press.

Aganbegyan, A. 1988 *The Challenge: Economics of Perestroika*. London, Hutchinson.

Anders, A., Layard, R. 1993 *Changing the Economic System in Russia*. London, Pinter; New York, St. Martin's Press.

Aslund, A. 1993 *Systemic Change and Stabilization in Russia*. London, Royal Institute of International Affairs.

Aslund, A. (ed) 1995 *Russian Economic Reform at Risk*. London, Pinter; New York, St. Martin's Press.

Aslund, A. 1995 *How Russia Became a Market Economy*. Washington, DC, Brookings Institution.

Bergson, A., Levine, H.S. (eds) 1983 *The Soviet Economy: Toward the year 2000*. London, George Allen and Unwin.

Buck, T., Cole, J. 1987 *Modern Soviet Economic Performance*. Oxford, Blackwell.

Buston, T.G. 1985 *Gorbachev. A Biography*. New York, Stein and Day.

Christensen, B.V. 1994 *The Russian Federation in Transition: External Developments*. Washington, DC, International Monetary Fund.

Churchyard, L.G. 1973 *The Soviet Intelligentsia*. London, Routledge and Kegan Paul.

Clague, C., Rausser, G.C. (eds) 1992 *The Emergence of Market Economies in Eastern Europe*. Cambridge, Blackwell.

Clarke, S. (ed) 1995 *Management and Industry in Russia. Formal and Informal Relations in the Period of Transition*. Aldershot, Edward Elgar.

Clarke, S., Fairbrother, P., Borisov, V. 1995 *The Workers' Movement in Russia*. Aldershot, Edward Elgar.

Earle, J.S., Rapaczynski, A. 1993 *The Privatization Process in Russia, Ukraine and the Baltic States*. New York, Central European University Press.

Gaidar, E.T. 1995 *Russian Reform. International Money*. Cambridge, MA, MIT Press.

Goldman, M.I. 1994 *Lost Opportunity: Why Economic Reforms in Russia Have Not Worked*. New York, W.W. Norton & Co.

Hanson, P. 1994 *Regions, Local Power and Economic Change in Russia*. London, Royal Institute of International Affairs.

Hardt, J.P., Kaufman, R.F. (eds) 1993 *The Former Soviet Union in Transition*. Armonk, New York, M.E. Sharpe.

Islam, S., Mandelbaum, M. (eds) 1993 *Making Markets: Economic Transformation in Eastern Europe and the Post-Soviet States*. New York, Council on Foreign Relations Press.

Klopov, E., von Otter, C. (eds) 1993 *The Russian Labour Market and the Beginnings of an Industrial Relations System*. London, Sage.

Kuznetsov, A. 1994 *Foreign Investment in Contemporary Russia: Managing Capital Entry*. New York, St. Martin's Press Inc.

Lazear, E.P. (ed) 1995 *Economic Transition in Eastern Europe and Russia: Realities of Reform*. Stanford, CA, Hoover Institution Press.

Le Houerou, P. 1995 *Investment Policy in Russia*. Washington, DC, World Bank.

Leitzel, J. 1995 *Russian Economic Reform*. London, Routledge.

Liu, G. 1994 *States and Markets: Comparing Japan and Russia*. Boulder, CO, Westview Press.

McKinnon, R.I. 1994 *Gradual Versus Rapid Liberalization in Socialist Economies: Financial Policies in China and Russia Compared*. San Francisco, CA, ICS Press.

McFaul, M., Perlmutter, T. (eds) 1995 *Privatization, Conversion, and Enterprise Reform in Russia*. Boulder, CO, Westview Press.

Menges, C.C. (ed) 1994 *Transitions from Communism in Russia and Eastern Europe: Analysis and Perspectives*. Washington, DC, George Washington University, Program on Transitions to Democracy; Lanham, MD, University Press of America.

Nelson, L.D., Kuzes, I.Y. 1994 *Property to the People: The Struggle for Radical Economic Reform in Russia*. Armonk, NY, M.E. Sharpe.

Nelson, L.D., Kuzes, I.Y. 1995 *Radical Reform in Yeltsin's Russia: Political, Economic, and Social Dimensions*. Armonk, NY, M.E. Sharpe.

Schipke, A., Taylor, A.M. (eds) 1994 *The Economics of Transformation: Theory and Practice in the New Market Economies*. Berlin, Springer-Verlag.

Shishkov, I.V., Yevstigneyev, V.R. 1994 *Economic Relations Within the CIS: How Relevant is the Experience of Cooperation and Integration in Western Europe?* London, Federal Trust. Translated by Vladimir R. Yevstigneyev.

Smith, A. 1993 *Russia and the World Economy: Problems of Integration*. London, Routledge.

Sutela, P. (ed) 1993 *The Russian Economy in Crisis and Transition*. Helsinki, Bank of Finland.

Wallich, C.I. (ed) 1994 *Russia and the Challenge of Fiscal Federalism*. Washington, DC, World Bank.

NOTE

1. Gorbachev, M. 1985 Communiqué on the Plenary Session of the Central Committee of the Communist Party of the Soviet Union. *Pravda* and *Izvestia*, 24 April 1985, in *The Current Digest of the Soviet Press*, 37(17): 3, 6.

4

POPULATION TRENDS IN THE POST-SOVIET SCENE

'My God, I'm in Russia – a refugee!'
(Line from a song of Slavic refugees from Tajikistan.
Nezavisimaya Gazeta, June 2, 1993, 3)

Economic and political uncertainties prevail in the post-Soviet scene. Family circumstances are often extremely difficult, and this is reflected in plummeting birth rates. Acute economic hardship, fast deteriorating public sector health care systems, and civil wars are just some of the reasons for rising death rates. Population growth in some post-Soviet states is now entirely dependent upon immigration, much of which is involuntary. The 1990s have witnessed hundreds of thousands of Russians fleeing areas of civil war, threatened war, anti-Russian popular sentiment and outright discrimination in former Soviet republics, the so-called 'near abroad'. The reception of these forced migrants in a homeland struggling to adapt to a new political and economic reality has often been less than enthusiastic. Scores of thousands of non-Russians have also taken refuge in Russia for much the same reasons. Their reception has often been hostile. As well, huge former Soviet military garrisons outside present day Russia have been disbanded. In consequence, hundreds of thousands of Russian military personnel have returned home to uncertain postings, but certain economic hardship. Millions more Russians continue to live in the 'near abroad', their possible future forced repatriation portending even greater economic and social dislocation. Thus, basic demographics and migration within and between post-Soviet states are changing in significant ways long-standing ethnic balances and geo-political relationships.

In this chapter, post-Soviet demographic trends, migration patterns and the distribution of minority groups in Russia, and of Russians in the 'near abroad', will be discussed. Before doing so, however, it is necessary to have some understanding of what went before. Thus, this chapter begins with a thumbnail sketch of demographic trends, demographic differences amongst nationalities, and the changing shares of national minorities of the total population during the Soviet period.

DEMOGRAPHIC TRENDS IN THE SOVIET ERA

The territorial expansion of the Russian Empire over the centuries was briefly described in Chapter 2. As noted, this brought under Russian hegemony a vast multi-national population. On the eve of World War I, birth and death rates were both very high, typical of a rural, and for the most part comparatively underdeveloped, economy. The rate of natural increase exceeded that of west European countries in the late nineteenth and early twentieth centuries owing to the deeply ingrained tradition of early marriage and large families amongst Russia's predominantly rural populace. The excess of births over deaths regularly exceeded 15 per 1000 population. Life for the vast majority of the Empire's subjects was hard, the bulk of the population remained illiterate, and life expectancies were low, especially, but not exclusively, in Muslim Central Asia.

Following the Russian revolution of 1917 and the onset of Soviet socialism, there ensued a flurry of policies and programmes intended to raise the quality of life for all citizens. Education and public health were accorded high priority. Attitudes toward population growth were initially somewhat ambivalent, although public policy did emphasize the importance of putting in place the basic structures and services for a healthy, educated young generation. Later on, government policies relating to population planning became decidedly pro-natal. Indeed, for several decades Soviet bloc and Catholic countries were aligned in opposition to United Nations' programmes promoting family planning through artificial means of birth control. Of course, the rationales for this political alignment were quite different – one was religious, the other ideological. Part of the reason for the Soviet opposition to such programmes was based on the premise that western capitalist countries sought to control population increase because of the fear that it would lead to growing unemployment, with all its attendant economic and political consequences, in developing countries. The Soviet response was that to link unemployment with overpopulation was fallacious. Unemployment, it was argued, was an inherent feature of the capitalist system. Under socialism all citizens are productively employed, by definition unemployment does not exist, and therefore population growth is to be encouraged. However, translating theory into reality in matters affecting human behaviour was rarely straightforward as Soviet demographic trends attest.

Any attempt to fully comprehend demographic trends, and therefore develop informed government policy, laboured under two inter-related handicaps during the Soviet era. Censuses were not conducted regularly, and demographic research early on fell into official disfavour. As a professional discipline, demography was abolished in the mid-1930s as part of the Stalin-led purge of all things that either did not cast appropriate light on, or did not conform to, government policies and programmes. Of course, compared to what happened to those unfortunate enough to fall into disfavour during Stalin's time, merely abolishing a profession and academic discipline was benign treatment indeed. It was not until the 1960s that demography was again legitimized by the state. What prompted this change in official stance was the realization, finally, that far more accurate information about the behaviour and values of the Soviet population was required for informed policy formulation on the part of political and government decision-makers. Miscalculations of population growth, flagging labour productivity, the unanticipated impact of urbanization on social values, these were just a few of the factors which helped to make research into the reasons for, and dimensions of, demographic change once again a legitimate exercise. The problem of inadequate published census data could not be so easily resolved, however.

During the pre-Soviet period only one national census was conducted. This took place in 1897. By the onset of World War I in 1914 there were approximately 160 million people living within the vast expanse of the Russian Empire. The next censuses occurred in 1920 and 1926. By the latter date, just over 147 million people lived within the new, and truncated, borders of the Soviet Union. Thus, the population base of the new Soviet state was substantially smaller than its Tsarist counterpart. Between 1897 and 1926 the combined impact of World War I, revolution, civil war and large scale emigration make accurate population counts impossible. Nonetheless, it is reckoned that the population loss they occasioned may have been as much as 25 million. Even allowing for sizeable error, the loss is colossal by any standard of measure, and clearly had an impact on basic demographic patterns. In fact, pre-World War I fertility and mortality rates were not attained until the mid-1920s. But this state of affairs was not to persist long, for the social and economic policies of the new Soviet government were soon to have an impact on demographic trends.

Shortly after the Soviets came into power, divorce proceedings were simplified and the legal distinction between legitimate and illegitimate births was erased. Ready access to abortion was facilitated. Experiments in communal living were not uncommon during the 1920s. Forging a new society in which women were emancipated meant change in traditional marital relationships. The demographic consequences of these pieces of social legislation, which in general constituted a challenge to the stability of the nuclear family, were greatest in the cities where the birth rate began to decline after the mid-1920s. This decline was inversely related to the increase in the rate of abortion. It should be emphasized, however, that in 1926 about three-quarters of the total population still lived in the countryside. Thus, traditional rural values continued to govern the lives of the majority of the population.

Economic development policies also had demographic repercussions. As the domestic economy slowly improved during the early 1920s, more and more people were drawn to the cities. Urban-industrial development was given a fillip with the introduction of Stalin's First Five Year Plan in 1928. This programme of forced industrialization was dependent upon rapid urban growth, which in turn required large scale rural–urban migration. To free up peasant labour for the factories, Stalin initiated the campaign to collectivize the peasantry, about which more will be said in a later chapter. We need only note here that peasant resistance to forced collectivization was fierce, agricultural production collapsed as a result, famine took hold in parts of Ukraine and the Lower Volga regions (the traditional breadbasket of the country), and food rationing was introduced in the cities. Within five years the programme of forced collectivization was virtually complete, but not before millions of lives had been lost through murder and starvation, and the pattern of natural increase of the general population once again severely disrupted.

While living conditions in the village were often extremely harsh, in the cities conditions of daily life and labour were also difficult. To be without a job in the early 1930s usually meant denial of a food ration. In most urban families, husbands and wives were both obliged to work in order to make ends meet. Raising children was in a real sense a luxury. Many women chose not to do so and there was widespread recourse to abortion which the state had earlier made freely available as part of its social development programme. Thus, during the 1930s urban birth rates were frequently lower than the death rates. Urban growth, therefore, was primarily a function of in-migration, just as it had been throughout the nineteenth century. The impact of these troubled times on demographic trends was soon evident.

The next census after the one in 1926 occurred in 1937. Before the census was taken, Soviet officials using vital statistics from the 1920s and extrapolating future total population scenarios assumed that the Soviet population would reach 170 million by 1937. However, the 1937 census recorded a population of only 162 million. The shortfall was unexpected, and in a planned economy undergoing rapid industrialization, a rather serious turn of events. The shortfall is attributable to several factors. Firstly, as a result of Soviet social policies, especially but not exclusively access to abortion, there was a precipitous decline in the crude rate of net natural increase, that is, the difference between unadjusted, crude rates of birth and death, which is usually expressed per 1000 population. This was especially evident in the cities, as already noted. Secondly, there was the huge loss of life associated with the forced collectivization of the peasantry, and the Stalin-initiated purges of Communist Party members and others who held, or were presumed to hold, dissident opinions. Small wonder the 1937 census was never published, the results left to languish on inaccessible archive shelves until the collapse of the Soviet system. The results of the next census taken in 1939 were only published during the Soviet era as residual columns in the 1959 census. Censuses were also taken in 1970, 1979 and 1989. While the state did little in terms of publishing the details of these censuses, it responded to the population 'loss' confirmed by the 1937 census by introducing a series of explicitly pro-natalist policies.

Beginning in 1936 abortion on demand was abolished. It was only to be available through official channels by meeting stringent medical criteria. Abortion otherwise was deemed to be a criminal offence. Divorce was also made more difficult to obtain. Mothers of large families were accorded public accolades. At the same time, the legal obligations of fathers toward illegitimate offspring were relaxed. In short, the state introduced a series of measures intended to strengthen the nuclear family and encourage larger families. Given the population loss which occurred during the 1930s there was reason enough for these policies. The catastrophic impact of World War II simply reinforced the continuing need for such measures.

After allowing for border changes, it is reckoned the losses inflicted upon the Soviet population during World War II were, at a minimum, between 25 and 30 million. Few families escaped the war untouched by disaster. The human costs are evident in the age–sex structure of the population for 1989 depicted in Fig. 4.1. Evident still was a sizeable imbalance of females in the older age groups. While female life expectancy customarily exceeds that of males, and therefore females normally predominate in the older age cohorts, the peculiar distortion in the Soviet age–sex structure reflects in large part the huge toll World War II exacted from the male population. A close inspection of Fig. 4.1 reveals that there are truncations of both males and females in particular age groups. An obvious one is the 45 to 49 age cohort born during World War II, but distinguishable as well are similar contractions for those born during World War I, in the ensuing years of revolution and civil war, and during the troubled 1930s. Beginning in the early 1960s there was a further contraction in the population base (note the 20 to 24 age cohort in 1989). Clearly, as the artificially reduced World War II age cohort entered the 1960s the number of births would have fallen independent of any change in the fertility rate. But there was a significant increase in the relative importance of the urban sector during the post-World War II period. In 1961 the share of the urban sector in the total population finally overtook the rural, something which happened in most western industrialized states decades earlier. Historically, urbanization changes fertility patterns; the per capita birth rate drops. As the urban share of the population increases relative to the rural, the impact of urban values on countryside customs is enhanced. By 1989, fully two-thirds of the Soviet

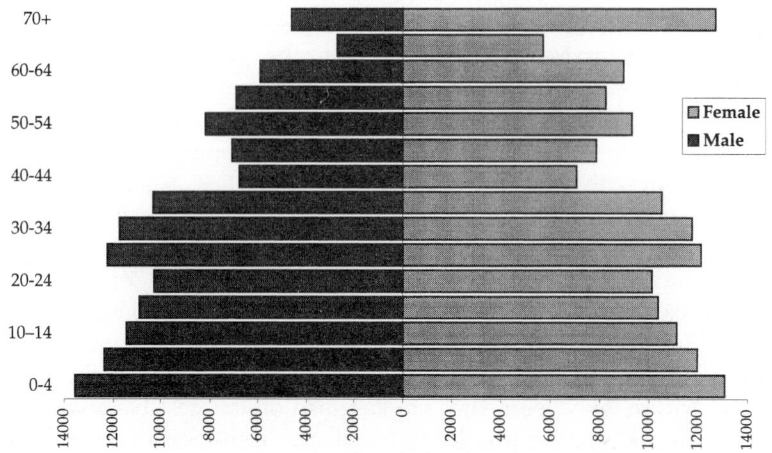

FIGURE 4.1 *Soviet era age–sex profile, 1989 (population in thousands)*
(*Source*: Compiled from data in *Narodnoye Khozyaystvo SSSR v 1989g* (Moscow: Goskomstat, 1990), 30)

population were classified as urban and the basic demographic situation was altered accordingly.

As Fig. 4.2 makes plain, throughout the twentieth century there was a general decline in both birth and death rates. With the advantage of hindsight, periods of crisis, either of domestic or external origin, are readily apparent from the data presented. However, a sharp decline in the birth rate which began in the late 1950s once again caught Soviet planners of the time off guard. For example, official estimates made in the latter half of the 1950s projected a total population of 250 million for 1970. In the early 1960s this projection was revised downward to 248 million. But even that adjustment proved far too optimistic, since the 1970 census recorded only 241.7 million people. The difference may seem merely academic, but it needs to be borne in mind that in a planned economy long-term investment decisions involve assumptions about future labour/capital ratios. Errors in projecting labour supply in a country which has virtually no net immigration obviously can have serious consequences.

Beginning in the 1960s, sociologists and demographers tried to account for the dramatic drop in the crude rate of net natural increase from 17.8 per 1000 population in 1960 to 8.9 in 1969. The fact that abortion was again made available upon demand in the mid-1950s after nearly two decades of tightly circumscribed legal access was of some importance in accounting for the decline, although this should not be overemphasized since recourse to illegal abortion as a means of contraception during the years of proscription was common. But on balance, high rates of abortion, a greater propensity to divorce, planned deferral of marriage amongst the rapidly expanding urban population, and a growing preference to trade off maternal/paternal benefits for a higher material standard of living combined during the 1960s with the arrival of the war-reduced age cohort to alter fundamentally previous patterns of fertility. The urban birth rate fell from 21.9 per 1000 in 1960 to 15.3 in 1968, its lowest point until the turmoil of the final year or two of the Soviet era. Amongst the rural population the birth rate declined from 27.8 per 1000 in 1960 to 18.7 in 1969. This was the lowest point in Soviet history.

As Fig. 4.2 indicates, the birth rate slowly edged upwards after bottoming out in 1968, and reached 20.0 per 1000 in 1986. On a disaggregated basis, both urban and rural rates also peaked at 18.1 and 23.6 respectively in 1986. During the 1970s the post-war baby-boom cohort began to enter the reproductive age. Thus, there were simply more prospective parents than in the early 1960s.

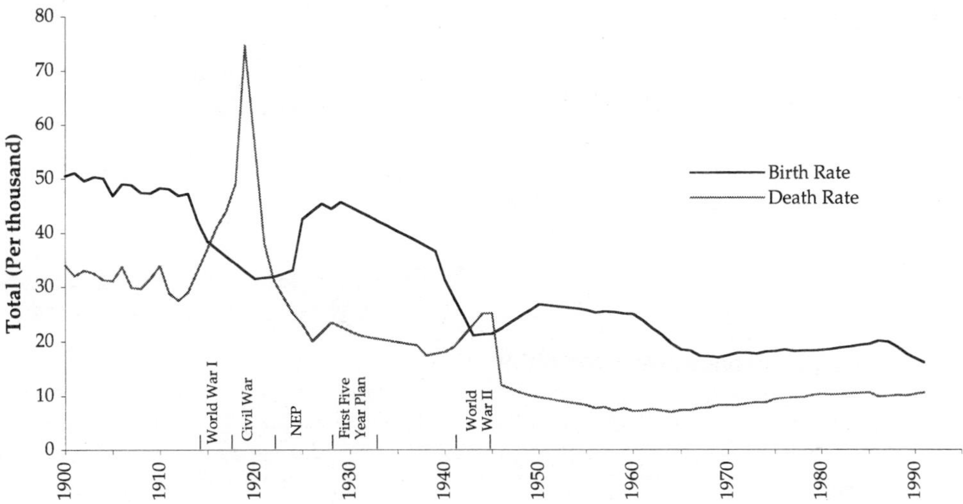

FIGURE 4.2 *Russian Empire and Soviet era birth and death rates, 1900–1990*
(*Sources: Based on J.H. Bater, The Soviet City. Ideal and Reality* (London: Edward Arnold, 1980), 70; *Narodnoye Khozyaystvo SSSR v 1989 g* (Moscow: Goskomstat, 1990), 38)

Female fertility seemed to have increased as well, at least to judge by the number of first births by females aged 15–19. State policy perhaps played some part in this development. For example, in the late 1960s the military draft age was lowered and the length of military service reduced. These changes in policy returned men to civilian status at age 20 rather than 22, as had been the case, and more of them started families earlier than they would have done otherwise. The government also augmented benefits paid to those having children. Family allowance payments were introduced early in the Stalin era, but later were reduced in value. In 1974, and again in 1981, payments were increased quite substantially. Maternity leave arrangements were also improved and, of particular importance for urban families, in 1981 eligibility for improved housing was linked to family formation and number of children. In 1982 the number of children born exceeded five million for the first time since the early 1960s. And the percentage of women with three or more children began to climb at this time as well. All told, at first glance it appeared that state policies intended to stimulate the birth rate were having some success. But upon closer inspection, it soon became apparent that it was in the non-Slavic parts of the Soviet Union where the birth rate edged up most significantly (Table 4.1). In the Slavic and Baltic realms, the trend to higher rates of birth was not sustained. And after 1986, birth rates steadily slipped for both the urban and rural categories of population right across the country.

As already noted, for most of the Soviet period the death rate declined. However, as Fig. 4.2 indicates, there are some obvious aberrations. These resulted from the devastations wrought by external factors such as world war, and internal policies such as the collectivization of agriculture. The death rate reached its nadir of 6.9 per 1000 in 1964. As the consequences of earlier catastrophes worked their way through the population structure, with the resultant gradual resumption of something approaching a normal age–sex composition, the death rate steadily edged upward, reaching a peak of 10.8 in 1984, and dropping slightly thereafter. The pattern of death rates during the 1980s was more variable than that of births, as Fig. 4.2 reveals.

The steady rise in the death rate during the 1970s and early 1980s was initially regarded as perfectly normal. After all, with the gradual restoration of a more balanced age–sex profile as the war- and collectivization-devastated age–sex cohorts in particular passed on, the death rate would inevitably increase. However, there soon emerged serious concern over the reasons for some deaths, a concern which resonates right down to the present day. Life expectancy of Soviet males, for example, fell from the mid-1960s until the early 1980s, a trend which then stood in stark contrast to the experience in all other industrialized countries. In the mid- to late 1960s, life expectancy at birth for males was at an all time high of about 66 years. In 1980 it was only 62. By the latter part of the 1980s it had risen to about 65 years. Simultaneously, infant mortality increased during the 1970s before dropping slightly in the 1980s. Statistical data for infant mortality and life expectancy were not published from the early 1970s to the mid-1980s. In the absence of sufficient data to permit a detailed analysis of actual trends, conventional wisdom has it that the decline in the life expectancy for males was related to health problems, especially those of a respiratory and alcohol-related kind. Infant mortality is a more contentious issue, and probably reflects serious deficiencies in health care delivery. But other reasons may play a part in explaining the trend. Improved accuracy of reporting infant mortality, especially in the Middle Asian republics, is amongst the most often cited.

The shifting balance amongst Soviet nationalities

By the late 1980s the economic conditions in the Soviet Union were clearly inauspicious, the need for change was apparent, and hence uncertainty was commonplace. Under such circumstances it is perhaps not surprising that overall population growth slowed. But the pattern was not everywhere the same, as the republic level demographic

Table 4.1 *Birth, death, and crude rate of net natural increase* per 1000 population, 1970, 1981–89. Average family size, 1979*

	1970			1981			1982			1983			1984			1986			1989			1979 average
Republic	**Birth**	**Death**	**CNNI***	**Birth**	**Death**	**CNNI***	**Birth**	**Death**	**CNNI***	**Birth**	**Death**	**CNNI***	**Birth**	**Death**	**CNNI***	**Birth**	**Death**	**CNNI***	**Birth**	**Death**	**CNNI***	**family size**
SLAVIC																						
Russia	14.6	8.7	5.9	16.0	10.9	5.1	16.6	10.7	5.9	17.6	11.1	6.5	16.9	11.6	5.3	17.2	10.4	6.8	14.6	10.7	3.9	3.3
Ukraine	15.2	8.9	6.3	14.6	11.3	3.3	14.8	11.3	3.5	16.0	11.5	4.5	15.6	12.0	3.6	15.5	11.1	4.4	13.3	11.6	1.7	3.3
Belarus'	16.2	7.6	8.6	16.3	9.6	6.7	16.3	9.6	6.7	17.5	10.0	7.5	17.0	10.5	6.5	17.1	9.7	7.4	15.0	11.3	4.9	3.3
BALTIC																						
Estonia	15.8	11.1	4.7	15.4	12.3	3.1	15.4	11.9	3.5	16.0	12.0	4.0	15.9	12.5	3.4	15.6	11.6	4.0	15.4	11.7	3.7	3.1
Latvia	14.5	11.2	3.3	14.0	12.6	1.4	14.6	12.2	2.4	15.9	12.5	3.4	15.7	12.9	2.8	15.9	11.9	4.0	14.5	12.1	2.4	3.1
Lithuania	17.6	8.9	8.7	15.1	10.3	4.8	15.2	10.0	5.2	16.3	10.3	6.0	16.2	10.9	5.3	16.5	9.9	6.6	15.1	10.3	4.8	3.3
MOLDOVA	19.4	7.4	12.0	20.5	10.3	10.2	20.6	10.2	10.4	22.5	10.9	11.6	21.9	11.1	10.8	22.7	9.7	13.0	18.9	9.2	9.7	3.4
CAUCASUS																						
Armenia	22.1	5.1	17.0	23.4	5.3	18.1	23.2	5.5	17.7	23.6	5.8	17.8	24.2	5.8	18.4	24.0	5.7	18.3	21.6	6.0	15.6	4.7
Azerbaijan	29.2	6.7	22.5	26.3	6.9	19.4	25.3	6.7	18.6	26.2	6.7	19.5	26.6	6.8	19.8	27.6	6.7	20.9	26.4	6.4	20.0	5.1
Georgia	19.2	7.3	11.9	18.2	8.6	9.6	17.9	8.4	9.5	18.0	8.4	9.6	18.5	8.8	9.7	18.7	8.8	9.9	16.7	8.6	8.1	4.0
MIDDLE ASIA																						
Kazakhstan	23.4	6.0	17.4	24.3	8.0	16.3	24.3	7.8	16.5	24.4	8.0	16.4	25.4	8.2	17.2	25.5	7.4	18.1	23.0	7.6	15.4	4.1
Kyrgyzstan	30.5	7.4	23.1	30.8	8.0	22.8	31.2	7.8	23.4	31.5	7.9	23.6	32.1	8.3	23.8	32.6	7.1	25.5	30.4	7.2	23.2	4.6
Tajikistan	34.8	6.4	28.4	38.3	7.8	30.5	38.2	7.7	30.5	38.7	7.6	31.1	39.8	7.4	32.4	42.0	6.8	35.2	38.7	6.5	32.2	5.7
Turkmenistan	35.2	6.6	28.6	34.3	8.5	25.8	34.7	8.0	26.7	35.6	8.5	27.1	35.2	8.2	27.0	36.9	8.4	28.5	35.0	7.7	27.3	5.5
Uzbekistan	33.6	5.5	28.1	34.9	7.2	27.7	35.0	7.4	27.6	35.3	7.5	27.8	36.2	7.4	28.8	37.8	7.0	30.8	33.3	6.3	27.0	5.5
USSR	17.4	8.2	9.2	18.5	10.2	8.3	18.9	10.1	8.8	20.1	10.3	9.8	19.6	10.8	8.8	20.0	9.8	10.2	17.6	10.0	7.6	3.5

*CNNI = crude rate of net natural increase

Sources: Narodnoye Khozyaystvo SSSR 1922–1982 (Moscow: Finansy i Statistika, 1982), 29; *Narodnoye Khozyaystvo SSSR v 1982g* (Moscow: Finansy i Statistika, 1983), 33; *Narodnoye Khozyaystvo SSSR v 1983g* (Moscow: Finansy i Statistika, 1984), 33; *Narodnoye Khozyaystvo SSSR v 1984g* (Moscow: Finansy i Statistika, 1985), 35; *Narodnoye Khozyaystvo SSSR v 1989g* (Moscow: Finansy i Statistika, 1990), 38; *Narodnoye Khozyaystvo SSSR za 70 Let* (Moscow: Finansy i Statistika, 1987), 406–407; David Lane, *Soviet Economy and Society* (Oxford: Blackwell, 1985), 118

data presented in Table 4.1 imply. These republic level data, however, do not accurately reflect the real differences in rates of natural increase amongst national minorities since the share of the titular group of the total republic population during the Soviet period was quite variable. Slavs traditionally comprised the majority of migrants into non-Slav republics and comprised substantial minorities in many of them. Indeed, in Kazakhstan and Kygyzstan, in-migration by Slavs and others had reduced the titular nationalities to minority positions. The demographic characteristics of the migrant Slavic population reflected its predominantly urban origin and destination. The net result was that its presence outside the home republics usually dampened the birth rate in the non-Slavic host republics. Thus, amongst rural Middle Asians, such as the Uzbeks and Tajiks for example, the average family size was much larger than it was for the republic population as a whole. Conversely, the 10 million or so Muslim people who at the close of the Soviet era were living in the Russian republic no doubt bolstered somewhat the per capita birth rate there. The impact of Slav migration into the Baltic republics can only be speculated upon, but as Estonians and Latvians were barely replacing themselves in the latter years of the Soviet era perhaps the presence of Slavs actually enhanced the crude rate of net natural increase. At any rate, the principal message conveyed by the rather incomplete data presented in Table 4.1 is that it was the Middle Asian and Azeri Muslim peoples who were increasing in number at the fastest rate.

The changing proportions of the major nationality groups outlined in Table 4.2 indicate why there was growing attention paid to questions of ethnic balance during the latter years of the Soviet era. Officially, Soviet propaganda stressed the equality of all peoples and supported the concept of *sliyaniye,* or the fusion of peoples of different ethnic backgrounds. In reality, relations between national minorities, and between national minorities and Slavs in general and Russians in particular, were not always ideal. Over the years there were unofficial reports of inter-nationality conflicts in Middle Asian cities, usually triggered, it would appear, by some local problem such as

Table 4.2 *Soviet-era population by major nationalities in the Soviet Union, 1959–2000*

	1959		1970		1979		1989		2000 (Projected)	
Nationality	Million	% of Total	Million	% of Total	Million	% of Total	Million	% of Total	Million	% of Total
USSR, total	208827	100.0	241720	100.0	262436	100.0	285743	100.0	300000	100.0
Russian	114114	54.6	129015	53.4	137397	52.4	145155	50.8	140000	46.7
Ukrainian	37253	17.8	40753	16.9	42347	16.1	44186	15.5		
Belorussian	7913	3.8	9052	3.7	9463	3.6	10036	3.5		
Estonian	989	0.5	1007	0.4	1020	0.4	1027	0.4		
Latvian	1400	0.7	1430	0.6	1439	0.5	1459	0.5		
Lithuanian	2326	1.1	2665	1.1	2851	1.1	3067	1.1		
Moldovan	2214	1.1	2698	1.1	2968	1.1	3352	1.2		
Armenian	2787	1.3	3559	1.5	4151	1.6	4623	1.6		
Azeri	2940	1.4	4380	1.8	5477	2.1	6770	2.4		
Georgian	2692	1.3	3245	1.3	3571	1.4	3981	1.4		
Kazakh	3622	1.7	5299	2.2	6556	2.5	8136	2.8		
Kyrgyz	969	0.5	1452	0.6	1906	0.7	2529	0.9		
Tajik	1397	0.7	2136	0.9	2898	1.1	4215	1.5		
Turkmen	1002	0.5	1525	0.6	2028	0.8	2729	1.0		
Uzbek	6015	2.9	9195	3.8	12456	4.7	16698	5.8		
Tatar	4968	2.4	5931	2.5	6317	2.4	6921	2.4		
Jews	2268	1.1	2151	0.9	1811	0.7	1378	0.5		

Sources: Murray Feshbach, 'The Soviet Union: Population Trends and Dilemmas,' *Population Bulletin,* Vol. 37, No. 3 (August 1982), Tables 6, 22; *Narodnoye Khozyaystvo SSSR v 1989g.* (Moscow: Finansy i Statistika, 1990), 30

food shortage. By the late 1980s, there were widely publicized large scale inter-ethnic riots in some cities. Ethnic stereotypes did exist, and for some people these shaped relationships with other nationalities. For instance, in Middle Asia indigenous peoples were sometimes referred to pejoratively in Russian as *zver'ye,* or animals. As the data in Table 4.2 indicate, the proportion of Slavs of the total population steadily declined in the post-World War II period, while that of Muslims increased. During the Soviet era, prying indigenous Middle Asians from the village and its traditions had not proven easy. The majority position of ethnic Russians was clearly obviously about to change in the 1990s. The factors giving rise to the demise of Russians as the majority are evident from even a cursory examination of the trends in the vital statistics presented in Table 4.1. From the perspective of Slavs in general and Russians in particular, there were few positive demographic trends in the late Soviet era. But there is little that is entirely positive about prevailing trends in vital statistics in most of the post-Soviet states either.

Demographic patterns in the post-Soviet 'near abroad'

As the comparative vital statistics presented in Fig. 4.3 indicate, Russia has much in common with the other Slavic states. In Ukraine and Belarus', birth rates are low, death rates are high, and the trend since independence is for the former to decline while the latter rise. Indeed, an excess of deaths over births in Ukraine first occurred in 1991, a year earlier than in Russia. Along with Estonia, Latvia and Lithuania, the three Slavic states share the dubious distinction of having in 1994 a negative rate of net natural increase. In other words, in all six states deaths exceed births. We will examine first the situation in Ukraine and Belarus'.

As the data in Fig. 4.3 indicate, in Ukraine the death rate in 1994 exceeded the birth rate by nearly five per thousand, and in Belarus' by nearly two. The trend in both countries since independence has been for the crude rate of net natural increase to be negative, and to increase. Thus, in Ukraine in 1991 it was only –0.8 per 1000 population; –2.0 in 1992; –3.5 in 1993; and –4.7 in

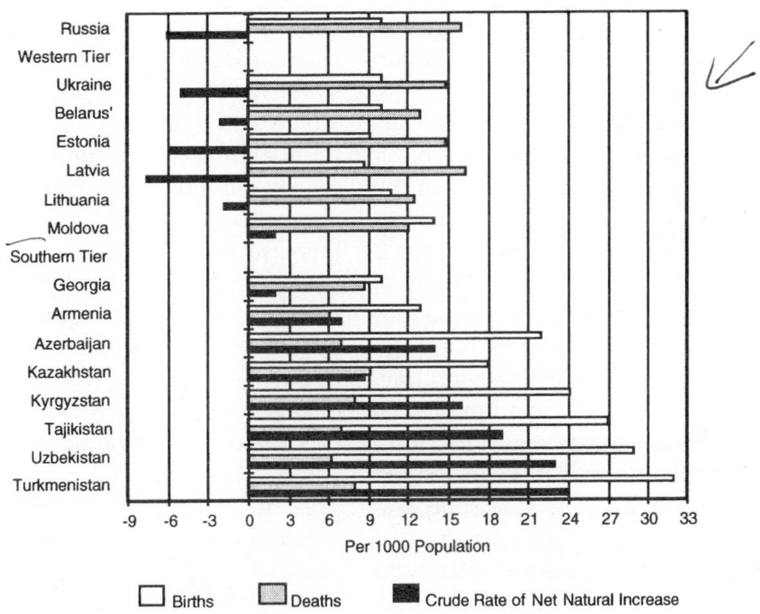

Figure 4.3 *Post-Soviet states birth, death and crude net natural increase rates, 1994*
(*Source: Ekonomika Sodruzhestva Nezavisimykh Gosudarst v 1994g* (Moscow: Statkomitet SNG, 1995), 16)

1994. In Belarus' the shift into negative rates of change occurred first in 1993 when it was –1.1. It reached –1.9 in 1994. The shift toward a market economy has been slow in these countries. Thus, on the economic front there are few reasons to have confidence in the future, and thus encouragement to raise a family. Moreover, since independence infant mortality has risen steadily in Ukraine and in Belarus'. For instance, in 1990 the rate per 1000 live births was 12.9 in Ukraine. In 1992, it was 14.0 per 1000; by 1994 it had reached 14.3. In Belarus' the rates for the same years were 11.9, 12.3 and 13.2, respectively. Compared to western countries where infant mortality rates are usually less than 10 per 1000 live births, the rates for Ukraine and Belarus' are certainly on the high side. Severe environmental degradation does little to assuage people's concern over raising a family either. In addition to the devastating consequences of the Chernobyl' nuclear power station meltdown of April 1986, industrial pollution throughout these states is a widespread, and serious, problem. From present day abnormally high levels of cancer, to the future well-being of seemingly healthy children now living in the vast areas of Ukraine and Belarus' impacted by Chernobyl' radioactive fallout, environmental health issues influence decisions regarding family formation. Not surprisingly, abortion levels remain high in Ukraine and Belarus'.

Of the three Slavic states, Ukraine had the largest share of its population in the 65 and older age group, 12 as compared to 10 per cent in the other two countries. While in most western industrialized countries this threshold captures reasonably well the portion of the population on pension, it considerably understates the share of the populace that is past the official working age in post-Soviet states (Table 4.3). The Soviet practice of having women retire at 55 and men at 60 has not changed. But basic demographic conditions have. In a situation of negative, or limited positive, population change (Fig. 4.3), comparatively early pension entitlement obviously results in a growing, dependent population relative to those still working. With nearly 22 per cent of its total population above working age, Ukraine exceeds both Russia (19.8) and Belarus' (20.5). Of course, being of pension age does not preclude having a job; indeed, under the present difficult economic conditions many pensioners have been obliged to supplement their income in order to survive.

While the data on life expectancy presented in Table 4.3 are for different years, these data, as with those for infant mortality cited above, are certainly consistent in portraying a generally poor, and deteriorating, quality of life. In Belarus' in 1993 life expectancies for males and females were 63.8 and 74.4 years respectively, down more than two years compared to the situation in 1989. In Ukraine, the most recent data available are for 1992, at which time life expectancies for males and females were 64 and 74. In 1989, the figures were 66.4 and 75.0 for men and women (Table 4.3). Given the trend in Belarus', and as will be discussed shortly, in Russia, there is little reason to suppose that this inauspicious decline has been reversed. With deaths exceeding births throughout the Slavic realm, only immigration can forestall continuation of absolute population decline.

If the Slavic realm presents a rather grim demographic picture, there is little in the Baltic states of Lithuania, Latvia and Estonia that offers much respite. They have long been characterized by very low rates of crude net natural increase (Table 4.1). As Fig. 4.3 indicates, all three Baltic states have an excess of deaths over births, and would have undergone an absolute decline in population even without the net out-migration which has taken place. Lithuania was notable amongst the Baltic states for its net natural increase in population in the immediate post-Soviet years. But by 1994 for the first time it too had an excess of deaths over births (46,486 deaths and only 42,832 births). While comparatively late in joining the ranks of countries with deaths exceeding births, already in 1992 Lithuania experienced an absolute reduction in total population, a trend which has continued down to the present. Between 1992 and 1993 this was the result of the out-migration of some 16,000 Russians and 7300 Ukrainians and Belorussians, amongst other peoples. In this period, people leaving Lithuania outnumbered immigrants by more than four to one.

Table 4.3 *Post-Soviet population age structures, 1993 and life expectancies, various years*

	Age structure (per cent)			Life expectancy at birth				Year
	Below working age*	Working age**	Above working age ***	In 1989		Various years		
				M	F	M	F	
Russia	23.8	56.4	19.8	64.4	74.4	57.3	71.1	1994
WESTERN TIER STATES								
Ukraine	22.5	55.9	21.6	66.4	75.0	64.0	74.0	1992
Belarus'	24.2	55.3	20.5	67.1	76.3	63.8	74.4	1993
Estonia	22.7	55.8	21.5	66.2	74.9	64.1	75.0	1992
Latvia	21.0	56.8	22.2	65.8	75.2	61.6	73.8	1993
Lithuania	23.7	56.5	19.8	67.1	76.4	64.9	76.0	1992
Moldova	29.5	55.1	15.4	65.5	72.1	63.9	71.9	1992
SOUTHERN TIER STATES								
Azerbaijan	35.1	53.1	11.8	66.4	73.5	66.3	74.5	1991
Armenia	31.7	54.9	13.4	64.6	67.4	68.7	75.5	1992
Georgia	26.1	55.8	18.1	68.0	75.4	68.7	76.1	1990
Kazakhstan	32.5	55.2	12.3	63.9	72.9	63.8	73.1	1990
Uzbekistan	42.9	49.0	8.1	65.8	71.6	66.1	72.4	1990
Turmenistan	42.2	49.1	8.7	61.8	68.2	62.3	69.3	1991
Kyrgyzstan	39.8	50.2	10.0	64.2	71.8	64.2	72.2	1992
Tajikistan	45.5	47.0	7.5	66.2	70.8	65.4	71.1	1992

*Age less than 16 years
**Age 16–55 for women; 16–60 for men
***Age over 55 for women; over 60 for men
Sources: Statistical Handbook 1994:States of the Former USSR (Washington: The World Bank, 1994), 14; T. Heleniak, 'Dramatic Population Trends in Countries of the FSU', *Transition.The Newsletter About Reforming Economies*, Vol. 6, Nos 9–10, September–October 1995, 4

Amongst all post-Soviet states, Latvia has the largest share of population in the above working age group, 22.2 per cent (Table 4.3). Estonia with 21.5 per cent was not far behind. Lithuania's share of total population in this category is a little lower still (19.8 per cent). With only small numbers of Estonians, Latvians and Lithuanians now living 'abroad', and restrictions on who may become citizens, there is little prospect for immediate solution to the basic question of how to ensure that the steadily growing population of elderly can be supported by the declining share of the total population of working age. In 1989, life expectancies for men and women taken together were 70.6, 70.4, and 71.8 for Estonia, Latvia and Lithuania respectively. As is so common elsewhere in the former Soviet Union (Table 4.3), life expectancies in the Baltic region are in decline. For men and women in Estonia in 1992 the figures were 64.1 and 75.0. In Lithuania they were 64.9 and 76.0 for the same year. In 1993, male and female life expectancies in Latvia were 61.6 and 73.8 respectively. The most precipitous drop was evidenced by males in Latvia. This may be related to the fact that Latvia has the largest ethnic Russian population of any Baltic state, and as will be discussed later, Russian males typically have the lowest life expectancy of any national minority irrespective of which part of the former Soviet Union they live in. In summary, there are very significant consequences of the deteriorating economic and social conditions everywhere in the Baltic states.

In Moldova, the basic demographic picture differs somewhat from that of the other western tier post-Soviet countries considered thus far. Unlike the Slavic and Baltic realms, natural population increase is occurring, but the crude rate of net natural increase is in decline. In 1990, the crude rate of net natural increase was 8.0 per 1000 population. In 1994, it had dropped to 2.4. Clearly, having children is not as high a priority in life in the post-Soviet period. Indeed, there is nothing very much to commend the Moldovan domestic economy. Political instability is also a serious issue for there exists the threat of civil war between the country's Moldovan nationalists and the Russians and Ukrainians living in the Trans-Dniester region who want to be independent of the new political reality. Thus, the fact that there are still more births than deaths in 1994 (Fig. 4.3) has more to do with the fact that Moldova is still predominantly rural (52 per cent) than it does to some inherently advantageous economic and political situation. Put simply, countryside customs, including having a larger family than in the city, still hold sway.

Throughout the southern reaches of the former Soviet territory, the basic demographic pattern stands in stark contrast to the western tier countries and Russia (Fig. 4.3). For many southern tier states, birth rates remain high, death rates are relatively low, natural population increase is substantial, and consequently the populations are, demographically speaking, quite young. In the Caucasus, Georgia has the lowest crude rate of net natural increase in this region, 2.1 per 1000 population, barely a quarter of the 1991 figure. In Armenia and Azerbaijan the rates in 1994 were certainly comparatively high, but in both countries there has been a decline since independence as well. As civil wars and economic chaos have prevailed for much of the post-Soviet period in all three countries, this trend is not surprising. Save for Armenia and Georgian, southern tier countries are Muslim. More than seven decades of Soviet promotion of atheism and restrictions on Muslim religious practice and training of mullahs was intended to lessen the role of traditional Muslim religious values and customs in daily life. The policy had some impact in the cities. But as the vast majority of Muslims remained firmly rooted in the countryside, continuity of values and customs rather than fundamental change prevailed. There is obviously a link between religion and demography in this part of the post-Soviet scene which is more important than the simple distinction between urban and rural shares of the total population. That urbanization will significantly dampen the birth rate, as has generally occurred in underdeveloped countries undergoing modernization elsewhere, is not entirely assured since throughout this southern tier region fundamentalist Muslim movements are a growing force in political and domestic life alike.

The data presented in Fig. 4.3 for Kazakhstan reveal a crude rate of net natural increase which is comparatively low for Middle Asia. In large measure this is a reflection of the fact that Kazakhs do not account for the majority of the population. Russians are almost as numerous, and together with other non-Middle Asian peoples their rates of natural increase dampen the overall average. Uzbekistan and Turkmenistan have crude rates of net natural increase which are still reasonably close to those prevailing in the late Soviet era. But even in these countries, the trend is for the birth rate to decline. Death rates have increased since independence, but most notably in Kazakhstan, Kyrgyzstan and Tajikistan.

Demographic trends in Russia

In 1989, Russians comprised only a slim majority of the Soviet population, 50.8 per cent, and based on prevailing trends were poised to become a minority. Even though control over critical centres of decision-making would not have diminished correspondingly, the prospects of minority status did not sit well with many Russians. In post-Soviet Russia, the situation is entirely different. Of the 148 million people living within the borders of the new state in 1995, nearly 82 per cent were Russian. Notwithstanding this numerical superiority, basic demographic trends continue to both trouble and challenge policy makers. In 1987 the TFR for the Russian republic was 2.2. The TFR, or total fertility rate, represents the average number of children that would be borne to each woman during her child bearing years if she had children at the same rate as women that age actually did in a particular year. A TFR of 2.1

is required to replace the present population. Thus, in 1987 the population was just barely replacing itself. Following the collapse of the Soviet Union the TFR plummeted. It dropped to 1.6 in 1992, and in 1993 was only 1.3. As the data presented in Fig. 4.3 indicate, there was a significant excess of deaths over births in 1994. In the short run, there is little that can be done to reverse the pattern.

In 1989 the birth rate in Russia was 14.6 per 1000 population. Two years later it had dropped to 12.1. By 1993, it had declined to only 9.4. Expressed differently, in 1993 only 1.4 million children were born. This was significantly below the average of 2.5 million children born each year in the Russian Republic during the less than auspicious 1980s, and was well below the lowest figure ever recorded during the entire post-war Soviet period – 1.8 million in 1968. Since the nadir in 1993, the birth rate has edged up marginally, although this was still not evident from the population pyramid for 1994 (Fig. 4.4). The increase is a function of the larger number of women in the prime child bearing age cohorts however, and not because conditions in Russia were judged more auspicious in terms of having children. In 1994, the birth rate was 9.6 per 1000. The impact on the age–sex profile of the population in Russia of this precipitous decline in the birth rate during these few years of political and economic restructuring is certainly apparent from even a cursory examination of Fig. 4.4. The contraction in the under five years age cohort was already very evident in 1992. But the collapse in the birth rate is not everywhere the same. It is much greater in the cities than in the countryside as might be expected. The regions outlined in Fig. 4.5 are the 11 economic regions and the enclave of Kaliningrad (for details, see Fig. 9.13, p. 282). As Fig. 4.5 indicates, across Russia there was considerable regional variation in the birth rate. It was above average in the far eastern and in the south European Russian regions. In the former area, the population is demographically younger.

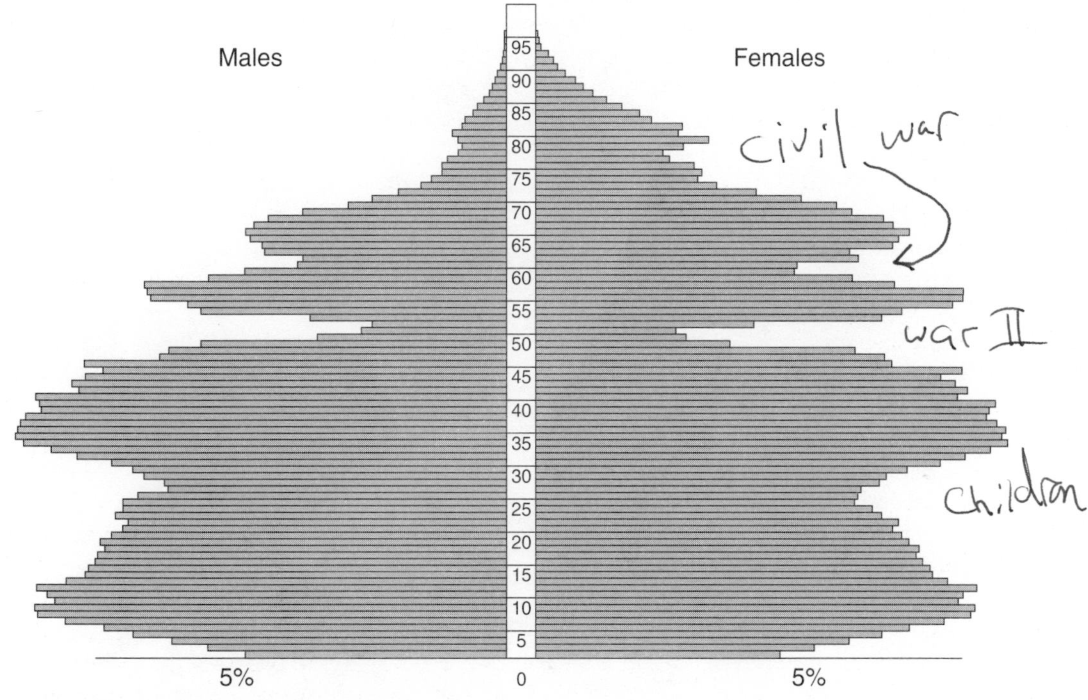

FIGURE 4.4 *Age–sex profile: Russia, 1994*
(*Source*: T. Heleniak, 'Dramatic Population Trends in Countries of the FSU', *Transition. The Newsletter about Reforming Economies*, Vol. 6, No. 9–10, 1995, 3)

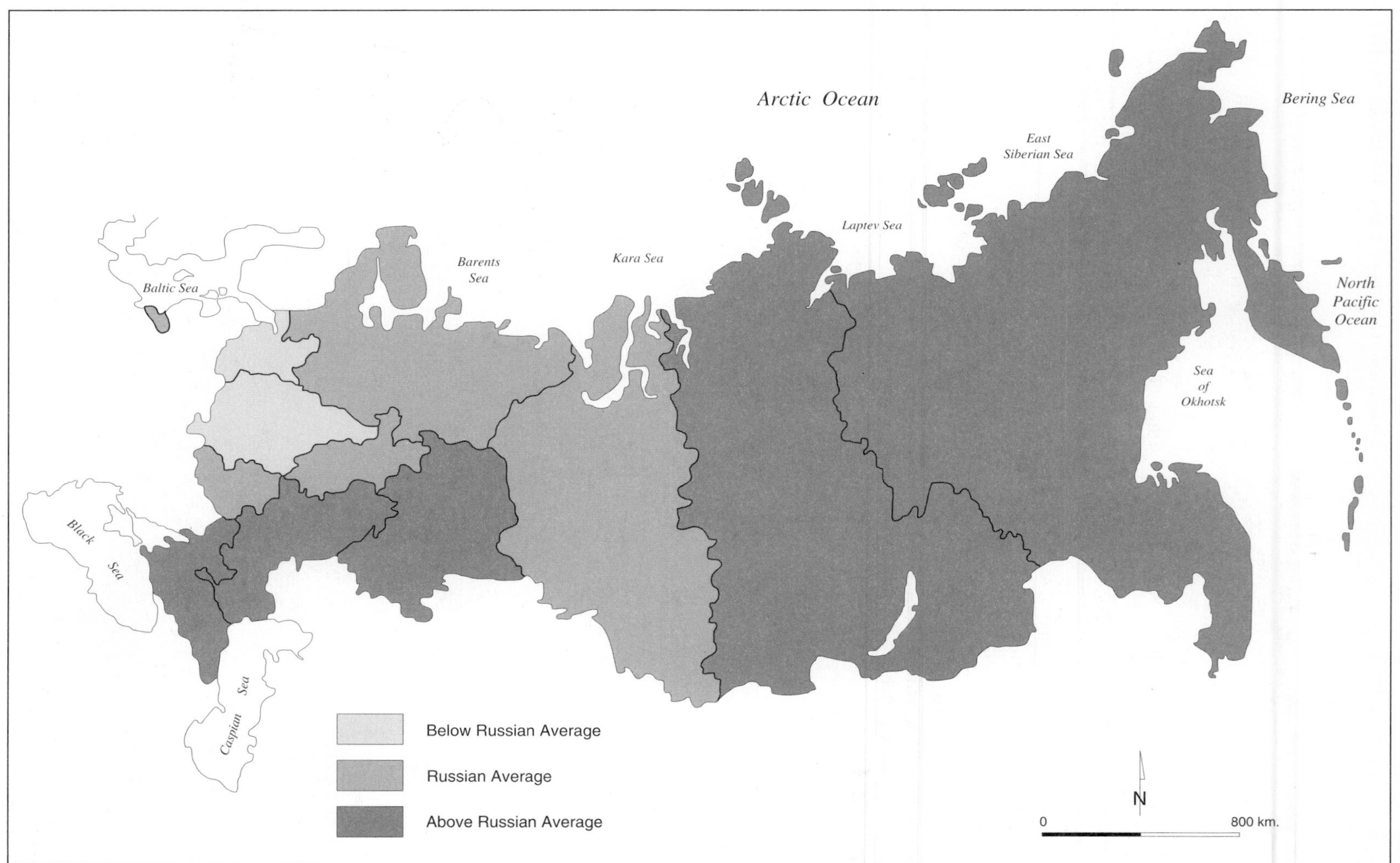

FIGURE 4.5 *Birth rates in Russia, 1993, by economic region*
(*Source*: *Prognoz Chislennosti Naseleniya do 2005 goda* (Moscow: Tsentr Ekonomicheskoy Konyunktury pri Pravitel'stve Rossiyskoy Federatsii, 1994), 71)

In the south European Russian region, the birth rate has been bolstered by an in-migration of Muslim refugees and other migrants from the Caucasus, a region of comparatively higher birth rates, as was noted earlier. In the predominantly Slavic Northwest and Central regions the birth rate is below average. Notwithstanding these regional differences, Russia in the 1990s has acquired the dubious distinction of having one of the lowest birth rates of any state in the world!

While the death rate in the Russian republic slowly edged upward during the final years of Soviet power, since 1991 it has surged. Deaths across all of Russia were 11.4 per 1000 population in 1991. By 1993, deaths reached 14.5 per 1000 and 15.6 in 1994. This represents a 36 per cent increase in just three years, a rate of change which was nearly double the percentage fall in the birth rate. Again these figures are just averages, and therefore mask some significant deviations at the regional and urban scales. Figure 4.6 indicates that in the core Slavic regions of European Russia the death rate exceeds the national average, obviously exacerbating the comparatively low birth rate over much of this region mentioned above. Within this core Slavic region, the impact of the general demographic situation is often all too apparent. In Moscow, for example, a demographically older population produced by decades of administrative control over in-migration, together with the upheaval of the present political and economic times resulted in a death rate in 1993 of 16.5 per 1000. With a birth rate in 1993 of only 7.0 per 1000 Muscovites, the crude rate of net natural increase was –9.5 per 1000! In 1985 the crude rate was 1.7. With continuing restrictions on in-migration, the consequences of this demographic situation are obvious. In January 1993, the total population living within Moscow's official boundaries was 8.881 million, 122,000 fewer people than two years earlier. What happened in the country's largest city was repeated at the national level.

In 1992, Russia registered an excess of deaths over births for the first time in its entire history. The difference that year was 219,800. In 1993, the situation was even worse. The total excess of deaths over births in 1993 was 753,300. Of this 'deficit', 557,800 was attributable to an excess of deaths over births in Russia's cities. Significant regional variations in population change occurred in consequence of different rates of birth and death. For example, between 1990 and 1991 deaths exceeded births in 30 of the country's 89 political administrative regions. Almost all of the regions across central and northwestern European Russia were so characterized. In the south and east, on the other hand, there were positive crude rates of net natural increase. These were typically in the low, single digits, except in those political administrative regions in which Muslim peoples predominated. Rates there were typically in the high teens. Across the country, the crude rate of net natural increase for 1991 was a mere 0.7 per 1000 population.

Despite substantial immigration of Russians and a growing stream of non-Russian refugees from the 'near abroad' since the demise of the Soviet system, this has not offset basic demographic trends. For example, in 1992 Russia's net migration balance was close to 140,000. Yet between 1991 and 1992 the total population in 44 of Russia's 89 political administrative regions either dropped, or remained unchanged. Such population growth as there was occurred principally in the North Caucasus, Povolzhskiy and Urals regions, once again mostly in areas with sizeable non-Slavic populations and where there were positive rates of crude net natural increase. In the few areas which recorded quite substantial population growth rates, for example the North Caucasus, it was often because in-migration of non-Russians from the 'near abroad' bolstered the prevailing relatively high crude rates of net natural increase. The North Caucasus region was the primary destination of tens of thousands of Azeris, Armenians and other minority groups fleeing the civil wars which had flared up in parts of Azerbaijan, Armenia and Georgia. Negative population change in 1991–1992 intensified in those regions where it occurred the preceding year, that is, throughout central and northwestern European Russia. Rural depopulation across much of European Russia had been the prevailing pattern for several decades, and in consequence older women dominate the age–sex pyramid,

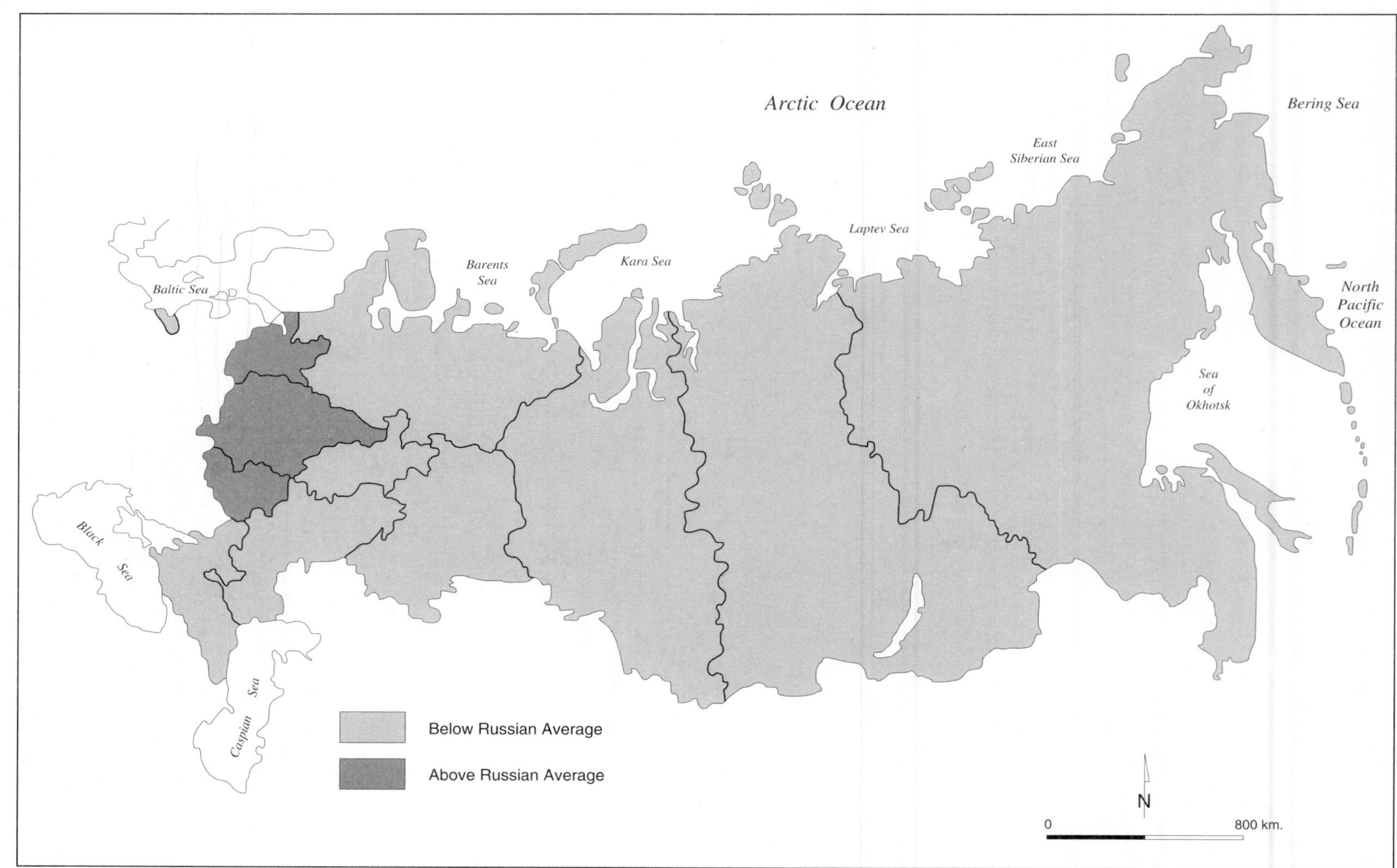

FIGURE 4.6 *Death rates in Russia, 1993, by economic growth*
(*Source*: *Prognoz Chislennosti Naseleniya do 2005 goda* (Moscow: Tsentr Ekonomicheskoy Konyunktury pri Pravitel'stve Rossiyskoy Federatsii, 1994), 66)

family formation is limited, the birth rate is low and the death rate was already relatively high. Current trends, therefore, simply exacerbate a long-standing problematic demographic picture. In general terms, ethnic Russian regions across the country evidenced little if any population growth, and in all too many instances experienced an absolute decline.

Population forecasts for the year 2005 based on the trends in vital statistics up to 1992 offer little solace (Fig. 4.7). The best case scenario is that the total population will drop from the 148.3 recorded for 1992 to 138.7. Using a more conservative set of assumptions produces a projected total population of only 131.5 million. In either case a situation in which the population will be reduced by roughly 10 to 17 million, by about 7 to 11 per cent, in just over a decade does not auger well for the social and economic well-being of the state. Because of the historic difference between urban and rural birth rates, the population loss would be most severe in the urban sector. Best and worst case projections would see the urban population decline by nearly 9 and 14 million, respectively. And of course, not all regions of Russia would be affected to the same degree. For example, in the Slavic Northwest region, the best and worst case forecasts give absolute population reductions of about 13 or 17 per cent, respectively. The reasons for the variation in projections are related to assumptions regarding trends in vital statistics and migration. For example, the best case scenario assumes, amongst other things, that the birth rate will begin to increase after the mid-1990s. Because of this assumption alone, there is good reason to opt for the more pessimistic scenario.

The principal means of birth control in Russia remains abortion, just as it was in the Soviet era. In 1991, some 3.5 million were performed in state run medical institutions. In addition, there were non-registered abortions in private medical facilities and abortions conducted 'illegally'. By 1994, the number of registered abortions had risen to more than four million. Including registered, non-registered and 'illegal' abortions, the total was rather more than three times the number of children born. The escalation in the number of abortions clearly dampened the birth rate, but it also helped to boost the death rate. Russia already has a maternal death rate many times higher than occurs in other developed countries

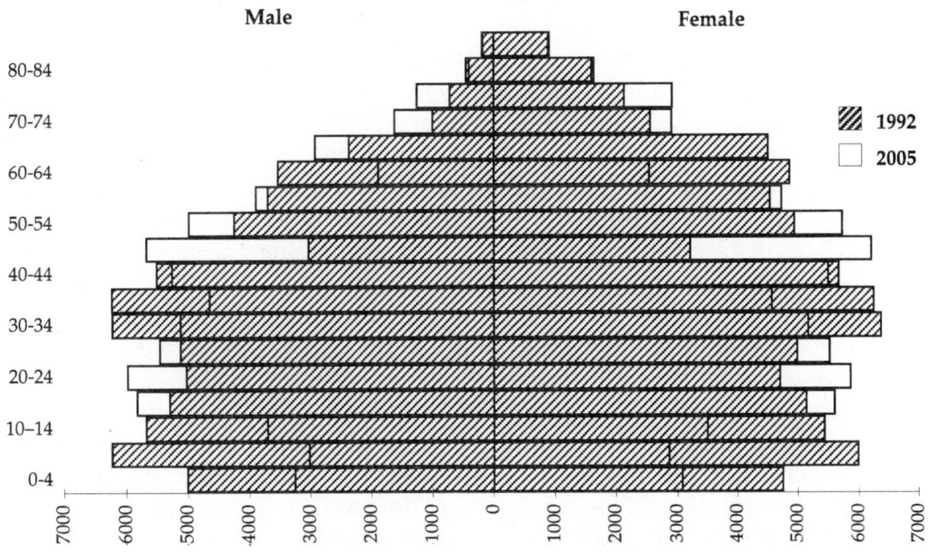

FIGURE 4.7 *Age–sex profile: Russia, 1992 and 2005 (population in thousands)*
(*Source: Prognoz Chislennosti Naseleniya do 2005 goda* (Moscow: Tsentr Ekonomicheskoy Konyunktury pri Pravitel'stve Rossiyskoy Federatsii, 1994), 52)

(about fifteen-fold). The reason for this, in part, lies in the exceptional dependence on abortion as a means of birth control. Most maternal deaths are deemed to be preventable if proper medical attention and treatment were provided. But under the present poor economic conditions this is not the case, and consequently about a quarter of all maternal deaths occurs during the abortion process.

Beginning in the mid-1990s, there will be slightly more women in the 20–29 age cohort than there were in the first half of the decade (see Fig. 4.4). Under normal conditions, a corresponding increase in the birth rate would be expected, thereby giving some small respite to an otherwise rather dismal demographic prospect. And as noted earlier, there has been an increase in the birth rate from 1993 to 1994. However, the possibility of even higher rates of abortion reducing the anticipated absolute increase in the number of children born has become a major concern for the Russian government. Thus, as occurred almost six decades earlier during the Stalin era when there was a similar preoccupation with enhancing the rate of population growth, a decision was taken in 1994 to restrict ready access to free abortion. Notwithstanding the opening of high quality, private sector fee for service medical facilities since the collapse of the Soviet system, the vast majority of abortions were still conducted in state medical institutions as part of the universal health care benefit package. After the 1994 decree, only abortions deemed necessary for strictly defined medical reasons are to be permitted 'free of charge' in state medical institutions. All others will be on a fee for service basis. The anticipated cost for an abortion at the time the decree was issued was between 40,000 to 80,000 roubles, a not insignificant sum for a great many people. Of course, this fee would be in addition to the gratuities customarily given to the underpaid staff in hospitals and clinics in exchange for ostensibly 'free' medical services. Whether this measure will have the desired consequence of increasing the birth rate is a moot point. More certain is that, as also occurred in the past, more women will resort to cheaper, illegal abortions with inevitably more abortion related maternal deaths. The present population dilemma was deemed serious enough to warrant interventionist state policies to boost the birth rate. However, the precipitous fall in births is exacerbated because deaths are rising at a similar rate.

The reasons for the escalating death rate are numerous and complex, and seemingly spare few socio-economic groups in Russian society. They include deteriorating public health standards and resultant increases in infectious diseases, alcohol related illnesses, respiratory illnesses, industrial accidents, and in many regions extremely serious environmental degradation which is directly linked to health problems and higher than average death rates. There is obviously a kind of scissors effect whereby a falling birth rate and rising death rate combine to produce a negative population change. This situation is extremely difficult to rectify in the short run. Indeed, for some categories of death the current state of affairs is likely to worsen before it improves. Infant mortality is a case in point.

Infant mortality in Russia remains disproportionately high for an ostensibly developed country, and is increasing each year. In 1989, infant mortality in the Russian republic stood at 17.8 per 1000 live births. It was 18.0 in Russia in 1992 and 18.7 in 1994. The reasons for rather gloomy short-term prospects for an improvement are partly statistical, partly related to the continuing deterioration in health care and the worsening health of prospective mothers. From the perspective of statistical procedure, until 1993 Russia used the long-standing Soviet system for classifying infant mortality. Only infants weighing more than 1000 grams, or reaching 35 centimetres in body length, were counted as live births. Smaller babies were classified as miscarriages, and therefore did not figure in the infant mortality rate. The equally long-standing World Health Organization criterion of 500 grams was finally adopted by Russia at the beginning of 1993. It is estimated that if this change in statistical procedure is instituted everywhere and followed as a criterion it will increase the infant mortality rate by 15 to 20 per cent. The deterioration in the health of expectant mothers, in part attributable to the poor economic circumstances, poor quality health care infrastructure, inadequate supplies of even simple medicines, the poor quality of available obstetric services, and the often abysmal state of the environment, are judged to be the major contributory reasons for the increase in infant mortality since 1991. Clearly, none of these factors is amenable to immediate improvement.

The drop in life expectancy of Soviet males registered during the 1970s was a singular event amongst industrialized states, as noted earlier. The situation improved somewhat during the final years of Soviet power, partly as a result of a crackdown on the production and sale of alcohol. In the Russian republic in 1980–1981, for example, the average life expectancy of males at birth was 61.5 years; for females, it was 73.1. A peak was reached in 1986–1987 when newborn males were expected to live 64.9 years, still a little below the all time Soviet era high of 66 years which prevailed two decades earlier. Females were expected to live 74.6 years in 1986–1987. For both sexes, average life expectancy had risen from 67.6 to 70.1 years from 1980–1981 to 1986–1987. While the improvement was certainly welcome, life expectancies in 1986–1987, the high point in the decade, were still substantially below those typical of western industrialized countries. For example, for all of Europe in 1992, the figures were 71 and 78 years for males and females respectively. In North America, the statistics were 72 and 79 years.

By 1991 the average life expectancy of those born in Russia had slipped from the mid-1980s peak of 70.1 years to 69.0 (63.5 years for males, 74.3 for females). A mere three years later, preliminary estimates indicated that male life expectancy had dropped more than six years to 57.3. Females born in 1994 could expect to live only 71.1 years, a drop of more than three years. Of course, these figures are just averages. In some parts of the country the situation is much worse. For example, in the Tuva Republic in Siberia male life expectancy in 1993 was a mere 52.3 years, and 64.3 for females. As a frontier region in a harsh climatic region a lower than average life expectancy might be anticipated. But the same reasons would seem to have little relevance for a region such as Leningrad Oblast′ which adjoins the city of St Petersburg. There males lived to only 55.7 years in 1993. And in the countryside they could expect to live only 54.9 years, as opposed to 56.1 in the Oblast′ urban settlements. Obviously, there are some serious problems in both regions. Across Russia male life expectancy in 1993 was greatest in the Republic of Dagestan in the north Caucasus. There males could expect to live to a ripe old age of 65.7 years. In North America and Europe, male life expectancy was in excess of 70 years at this time, and is increasing not declining. While as yet there are no conclusive data to explain the plummeting life expectancy rate, industrial accidents, including poisoning, have soared since the onset of the new political economy, and account for about half of all deaths amongst the working age population. Alcohol related illnesses and respiratory ailments also take a frightful toll of human lives across Russia. Alcohol production continues to be important because of the tax revenue it generates. Since independence imported brands are widely available and growing in market share. In short, alcohol consumption is again at very high level on a per capita basis. As noted already, infant mortality rates have risen. With the decision to bring Russian statistical procedures for counting infant deaths in line with World Health Organization standards, there will be further negative consequences for life expectancy rates. There is little to indicate improvement in the short run to be sure.

One of the demographic consequences of the differential life expectancy between Russian males and females is that the historic imbalance between the sexes is being exaggerated. In 1989, females comprised 53.3 per cent of the population of the Russian republic. They accounted for 54.1 per cent in 1994, and all indications are that their share will continue to increase. Another demographic consequence of present day trends in vital statistics is that the precipitous drop in births will accelerate the aging of the total population, something which a comparison of the age–sex data for 1992 and 2005 presented in Fig. 4.7 makes quite clear. Yet compared to other western industrialized countries, Russia in 1992 did not have an especially adverse ratio of 15 years and under, and 65 and over, age cohorts. The proportions of these age cohorts were 23 and 10 per cent respectively of the total population. For all European countries, for example, the comparable figures were 20 and 14 per cent. The dependent populations in both situations represented about the same share of total populations. However, in the post-Soviet scene 65 as a threshold for the pension age population is not very useful, as noted earlier. The proportion of Russia's population in the above working age group, that is, 55 for women and 60 for men, was 19.8 per cent in 1993. But in more than one-third of the 89 political administrative regions across the country in 1994, pensioners comprised between 20 and 27 per cent of the population. This simply means that a smaller share of the population in Russia is called upon to support the dependent young and pensioners than in western industrialized states, and that within the country there are significant regional differences. Current demographic trends would result in little alteration in the share of pensioners of the total population according to both the best and worst case forecasts for the year 2005. Indeed, because of the collapse in the birth rate, the share of the working age citizens will increase from 56.5 per cent in 1992 to either 64.1 or 65.7 per cent in 2005. While in one sense this is good news, the underlying reasons for this to occur remain a serious challenge for policy makers in the Russian government. It is small wonder that in the Russian press the word 'crisis' was frequently used to describe the present and portended situations. For some observers, a possible partial solution lies in repatriating more of the Russian population living in the 'near abroad'.

Contemporary migration between Russia and post-Soviet states

Migration during most of the Soviet era was controlled, at least to the degree that it was possible to do so. In a planned economy it was important to know how many more people had to be housed and serviced in a particular place from one year to the next. Thus, a vast bureaucratic system for monitoring internal migration was put in place in the early 1930s. Toward the close of the Soviet era, internal migration became more and more difficult to control and direct. Migration patterns increasingly reflected myriad personal decisions rather than state decrees. Indeed, in the first half of the 1980s there already was a net negative migration balance in all republics save for Russia, Estonia, Lithuania and Latvia. The latter three held out the not inconsiderable attraction of having the highest material standards of living in the Soviet Union. Thus, the Baltic states were the most popular destinations in terms of the ratio of in-migrants to total population. Still, Russia's net in-migration from 1979 to 1984, for example, was about three-quarters of a million people, more than six-fold the combined total for the three Baltic states. Because of its resource base and sheer size, Russia obviously offered the greatest number and range of jobs. Of course, not all those moving to the Russian republic were Russians, but a substantial share was. Many had departed the Middle Asian republics where life was already becoming uncomfortable because of overt anti-Russian sentiment. For

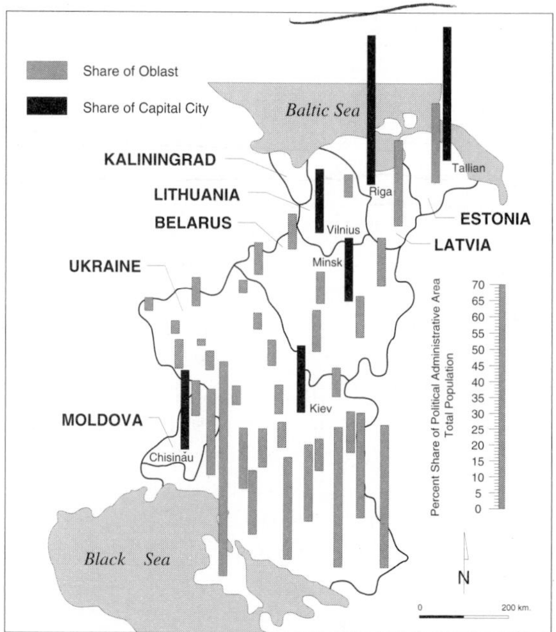

Figure 4.8 *Russians in western tier Soviet republics, 1989* (*Source*: Based on C. Harris, 'The New Russian Minorities: A Statistical Overview', *Post-Soviet Geography*, Vol. 34, No. 1, 1993, 13)

example, during the 1979 to 1984 period, there was a net out-migration of nearly one-half million people from the five Middle Asian republics, the bulk of whom were Russian. From the Caucasian republics there was a substantial net out-migration as well , though in this case the proportion who were Russian was undoubtedly much smaller simply because fewer lived there in the first place (see Fig. 4.9). At any rate, a net out-migration of Russians of whatever scale from the dominantly Muslim southern tier republics was scarcely in accord with the vested political interests of the state. Notwithstanding the exodus during the 1980s, according to the census of 1989 there were still 25.3 million Russians living outside the Russian republic. At this date Russians living in non-Russian republics still retained the advantage of speaking the official language of the state, were assured access to Russian language schools for their children, and in general were accorded all the benefits of the dominant ethnic group.

The relative shares of Russians of sub-regional populations in the fourteen non-Soviet republics in 1989 are depicted in Figs 4.8 and 4.9. In absolute terms the largest number, about 11.4 million, were living in the Ukrainian republic. Russians were certainly the predominant nationality in the eastern regions and in the Crimea. A migration of Russians to the coal fields and industrial centres

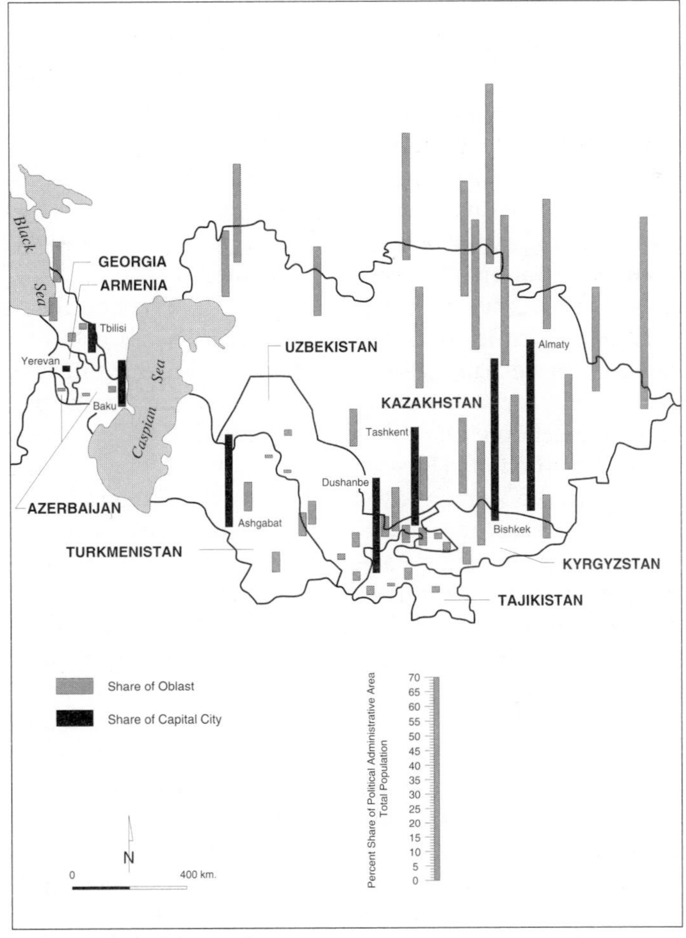

FIGURE 4.9 *Russians in southern tier Soviet republics, 1989*
(*Source*: Based on C. Harris, 'The New Russian Minorities: A Statistical Overview', *Post-Soviet Geography*, Vol. 34, No. 1, 1993, 14)

in the Donets region of Ukraine began in the late nineteenth century, and intensified during the early years of the Soviet industrialization drive. Throughout this region and the Crimea, Russian language and culture was the norm in the towns and cities, something reinforced by the customary minuscule use, or in some cases outright absence, of instruction in Ukrainian in the local school systems. Although there was not a single Ukrainian language school in Russia to serve the needs of the more than four million Ukrainians living there, there were 5000 Russian language schools in Ukraine, the majority of which were located in the aforementioned heavily Russianized regions. More than half of all children in Ukraine were enrolled in Russian language schools. About 1.4 million Russians lived in Belarus', a small fraction only of the number in Ukraine. The share of Russians in sub-regional populations declined with distance from the Russian border as Fig. 4.8 indicates. But even along the Russian border, no region was as relatively heavily settled by Russians as was the case in the eastern Ukraine and Crimea. Still, in comparison with the eastern Ukraine, for example, the penetration of Russian language into the school system was at least as extensive. In the late 1980s about three-quarters of all school textbooks used in Belarus' were in Russian, and the same share of the state's children were in Russian language schools. Of course, at the post-secondary level, instruction in Russian was the norm everywhere in the former Soviet Union. It is evident from Fig. 4.8 that the close to two million Russians residing in the Baltic states were more heavily concentrated in parts of Latvia and Estonia than in Lithuania.

Russians were clearly a dominant presence in many regions of the Middle Asian realm, a situation in stark contrast to the Caucasus (Fig. 4.9). Throughout the southern tier republics, and most especially in Kazakhstan, Russians dominated the populations in many regions. Indeed, the more than six million who resided in Kazakhstan outnumbered the titular population. As elsewhere in the Soviet Union in 1989, the Russians outside the Russian republic were mostly urban-industrial, rather than rural, and in consequence they customarily enjoyed a higher standard of living than the indigenous peoples. The titular nationalities throughout the southern tier region were typically more rural than the averages for their republics.

The demise of the Soviet Union and the rise of independent post-Soviet states created havoc in the lives of many people of various nationalities who, for whatever reasons, lived outside their 'home' republics. While Slavs in general and Russians in particular, comprised the largest number of those who suddenly found themselves living 'abroad', Soviet policies over the decades had variously encouraged, or forced, relocation of all national minorities. Thus, the map of nationalities on the eve of the system's collapse was exceedingly complex. The reasons to depart locales where families sometimes have lived for generations are now often quite compelling. From civil war, to harassment over civil rights, to job discrimination, to change in official status of the Russian language, must be added the collapse of domestic economies and the arrival of widespread unemployment.

As Fig. 4.10 indicates, from 1991 to 1994 there was each year a considerable population movement into Russia from the other post-Soviet states, save for Ukraine and Belarus'. But even in these states there has been a reversal in net migration balance, since 1993 in Ukraine and 1994 in the case of Belarus'. For the period from 1989 to 1994, only Ukraine and Belarus' have a positive net migration balance with Russia. But because of the scale of the migration to Russia in 1994 especially, the numbers are not large. For Belarus' it was a mere 3000 and for Ukraine 21,800. It is worth reiterating that notwithstanding substantial in-migration, total population in both Ukraine and Russia declined in 1992. The large number of Russians remaining in Ukraine and Belarus', as already noted, tended to be well ensconced in a culturally, if not politically, compatible environment. The proportions of children enrolled in Russian language schools in these states, about half and three-quarters respectively, has not changed. Indeed, just over one-half of the Belarus' population claim Russian as the mother tongue. The basic pattern of net out-migration from most post-Soviet states to Russia intensified in 1992 and again in 1994 (Fig. 4.10). While there have been variations in the flow into Russia, in no case was the net migration balance smaller in 1994 than in 1991. Within Russia, the destination of migrants is geographically concentrated. Most have gone to European Russia, the Urals and West Siberian economic regions, and to the north Caucasus region, especially to Krasnodar Kray, Stavropl' Kray, and Rostov Oblast' (see Fig. 9.12, p. 280).

Ever since the Baltic states were forcibly brought under Soviet control during World War II, they have consistently had positive net migration balances. Their traditionally higher standard of living was an attraction to those who could exercise some choice over where they lived and worked, as noted already. But it was Soviet policy to move Russians especially into the region in order to secure more effective political and economic control and mute any independence aspirations. The consequences were profound, most particularly in Estonia and Latvia. For example, from 1934 to 1989 the proportion of ethnic Estonians in the total republic population declined from nearly nine-tenths to just over three-fifths. The situation in Latvia was similar. In the mid-1930s, just over three-quarters of the population was ethnic Latvian. By 1989 there was a bare majority of only 52 per cent. As most migrants were destined

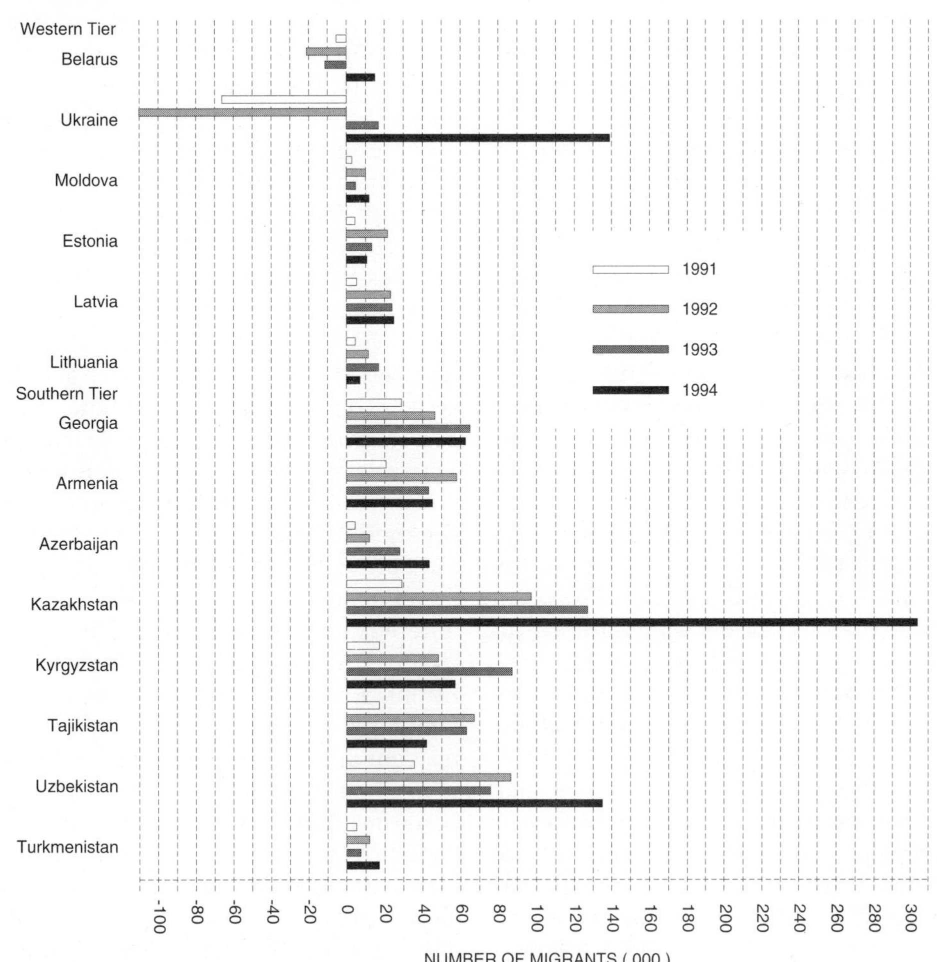

FIGURE 4.10 *Net migration balance: Russia with post-Soviet states, 1991–1994*
(*Source*: Based on data in T. Heleniak, 'Economic Transition and Demographic Change in Russia, 1989–1995', *Post-Soviet Geography*, Vol. 36, No. 7, 1995, 454)

for the cities of the Baltic states, the impact there was often overwhelming. For instance, in Estonia more than 90 per cent of all immigrants were urbanites. In some cities, such as Narva, less than a tenth of the population was Estonian on the eve of independence.

Since independence in September 1991, across all three Baltic states the migration balance has shifted from net inflow to net outflow. And the majority of those leaving are Russian. In Lithuania, for instance, Russians comprised nearly 60 per cent of the 27,324 emigrants in 1992. And this is from the Baltic state which has adopted the most liberal stand with respect to the future status of resident non-ethnic Lithuanians. For example, citizenship is open to anyone who is able to establish that they were permanently resident in the state before 1989, and whose income is from a legitimate source. Of course, Lithuania could afford a more generous definition for citizenship inasmuch as non-ethnic Lithuanians in 1994 comprised less than 20 per cent of the total population. Nonetheless, amongst Russians the possibility of acquiring Lithuanian citizenship was not pursued very vigorously since, among other things, it was not clear if by so doing the right to return to Russia at some time in the future might be put at risk. In 1992 there was still some migration into Lithuania, including amongst other nationalities, Russians. But the total number of immigrants that year was only 6206.

As a result of the net out-migration from each of the post-Soviet Baltic states, the share of the titular populations is slowly edging upward. In Latvia, where the numerical position of the titular nationality was the most precarious, Latvians increased their share of the total population from 52.5 to 53.5 per cent between 1992 and 1993 alone. Of perhaps greater geo-political significance in the changing migration balance has been the final withdrawal of the huge complement of former Soviet military personnel from the Baltic states. It should be noted that movements of military personnel are not included in the statistics presented in Fig. 4.10. During the first couple of years of independence the continuing presence of huge garrisons of troops had been used by the Russian government as leverage to extract concessions from the Latvian and Estonian governments in particular since in both states the absolute number of civilian ethnic Russians was large, and policies interpreted as prejudicial to the civil rights of Russians living there had been enacted. Of course, the fact that in Russia there were insufficient housing and employment opportunities for all military personnel returning from former Soviet territory and Eastern Europe figured prominently in the protracted negotiations over troop withdrawal as well.

Most of the Russian government's concern over civil rights has focused on who is eligible for citizenship in Latvia and Estonia, and what exclusion means to those not qualifying even if they have lived there for several decades. As ownership of property and participation in the privatization of former state enterprises is restricted to citizens, many people have been denied opportunities to participate in the new economies of Latvia and Estonia. Becoming a citizen is not straightforward. In the new Estonian state, for example, citizenship is restricted to those who have lived there for 16 years and who can demonstrate proficiency in the Estonian language. As the vast majority of Russians living in the Baltic republics on the eve of the collapse of the Soviet system could not speak the language of the indigenous peoples, this represents a major problem. The reason for this preoccupation with linguistic matters in Latvia and Estonia, in contrast to the situation in Lithuania, is understandable given the differential impact of migration into the Baltic republics over the years, and the consequent different degrees of Russianization of the local culture which ensued. For example, in Latvia and Estonia Russians comprised about a third of the total republic populations in 1989 compared to 9 per cent in Lithuania. The Russian language was more frequently heard on the streets, in educational institutions, and in government and industry offices than was Latvian or Estonian. Ethnic Russians really had little incentive to learn the local language. It is therefore not surprising that the Estonian and Latvian governments have introduced such stringent language requirements as part of the list of criteria determining just who is eligible for citizenship. Indeed, in 1989 there already had been amendments to each of the Baltic republic constitutions enshrining the local language as the state language. At the time this was largely a symbolic act. However, even if Russians meet residency requirements and are able to speak, for example, Latvian quite fluently, taking out Latvian citizenship naturally entailed swearing allegiance to the new state, something which might prejudice prospects of returning 'home'. In late 1994, the Russian government again announced steps to guarantee Russians still living in the 'near abroad' access to Russia by granting them the right of citizenship in Russia, something first recommended in 1990 in response to the wave of secessionist declarations in the non-Russian Soviet republics. Despite the important role language plays in the politics of the new Baltic states, the share of Russian language schools in each of them has not changed since independence. The shares are 10, 21 and 6 per cent respectively for Estonia, Latvia and Lithuania.

About 560,000 Russians lived in Moldova in 1989. In relative terms they were concentrated in the capital city of Chisinau, and especially in the region east of the Dniester river. This latter region has historic ties with the Russian and Soviet empires, has a majority population comprising Ukrainians and Russians, in contrast to the situation throughout the rest of Moldova, and has declared itself to be independent of the new state. Caught between Romanian speaking nationalists on the one side and Ukrainian settled areas on the other, out-migration of Russians since 1992 has been sizeable. The presence of the Russian military in the trans-Dneister region has helped to maintain order.

Amongst the southern tier states, the stimulus for Russians and others to leave varies considerably. In Georgia, Armenia and Azerbaijan, for example, nationalist movements have been instrumental in reducing significantly the shares of Russian language schools. In Georgia they had gone from 25 to 12 per cent, in Armenia from 7 to 2, and in Azerbaijan from 20 to 14 by 1994. In Middle Asia, Russian language schools remain more or less intact despite legislation giving priority to the local languages. It is therefore only a matter of time before access to such schools is reduced. Russian remains everywhere in the region the principal language used in higher education. But that said, the use of the Cyrillic alphabet in the languages of a number of the new states in the southern tier has been dropped (Azerbaijan, Uzbekistan) or will be soon (Kyrgyzstan and Turkmenistan), to be replaced by the Latin alphabet. This will help to create an additional barrier between Russian which uses the Cyrillic alphabet and the other languages. Interestingly, it was precisely because the use of the Cyrillic alphabet would help to bridge the languages of the region with Russian that the Latin alphabet was dropped after a period of experimentation in the early Soviet period.

Civil war, of course, figures prominently as a stimulus for out-migration from the Caucasus and Tajikistan to Russia (Fig. 4.10). In both regions Russia has a sizeable military presence intended to secure borders, to act as a peace keeper and, as in Moldova, to ensure a measure of protection for Russians still living there. Notwithstanding a large Russian military presence in war-torn Tajikistan, close to 80 per cent of the country's 380,000 Russian speaking inhabitants had departed by 1994. In Kazakhstan the more than six million Russians have long dominated the titular nationality, and despite promulgation of legislation promoting the rights of Kazakh peoples and language, it is not yet so difficult there as elsewhere in the region. Still a substantial out-migration began in 1992. About 200,000 Russians left in 1993, an out-migration offset in part by the arrival of 160,000 Russians, mostly it would appear coming from other Middle Asian states. Overall, Kazakhstan had a negative net migration balance with all other states each year since independence. Moreover, according to one opinion poll, fully one-third of the Russian population still resident in Kazakhstan wishes to depart. The prospect of perhaps two million migrants from this source alone is certainly unsettling for the Russian authorities.

Close to three million Russians lived in Uzbekistan, Kyrgyzstan, and Turkmenistan in 1989, principally, though not only, in the urban centres. In Kyrgyzstan for example, where the indigenous peoples were traditionally migratory herders, a sizeable number of Russian migrants had taken up agriculture. From the first two states there has already been substantial out-migration, and as the domestic economies continue to deteriorate and overtly discriminatory legislation is enforced, more Russians are likely to join the outflow. About one-sixth of Kyrgyzstan's Russian-speaking population had left by the end of 1993. The reasons are economic as well as political. A large proportion of resident urban Russians was employed in defence industry enterprises. The disintegration of the Soviet Union brought this sector to a virtual standstill and resulted in widespread unemployment. There have been some efforts by the government to allay the concerns of remaining Russians by seeking funding from Russia to convert former defence industry enterprises in which Russians are predominant amongst the workforce into joint ventures. However, little actual improvement in the domestic economy has occurred, and along with continual harassment or worse of non-native peoples, has served to sustain the out-migration from Kyrgyzstan. Some certainly comprised a share of 160,000 Russian migrants who arrived in Kazakhstan in 1993. The promotion of indigenous tongues as the official languages in each of the new Middle Asian states has unnerved many Russians, who during the Soviet era typically held good jobs and led comparatively privileged lives. But the reality is that only a tiny fraction, 2–3 per cent at best, can speak the local language. That the number of Russians living in Turkmenistan increased during the 1991–1994 period owes as much to Turkmenistan serving as an interim place to live for those departing more hostile Middle Asian locales, as it does to efforts of the government to provide some assurances that Russians living there will be well treated under the new order. For example, the Turkmenistan government passed legislation in 1994 enabling any person who wishes to do so to hold dual citizenship, the first Middle Asian state to do so. Despite this and other overtures, it is anticipated that as indigenization of the Turkmen culture and economy gains momentum, the Russian presence will contract. The steady replacement of the Russian language by Turkmen in government, business and education will ensure this occurs.

Of the 25.3 million Russians living outside the Russian republic in 1989, substantially over a million had returned 'home' by 1994. They were mostly, but certainly not only, from the southern tier countries. Areas caught up in civil wars, that is, the Caucasus and Tajikistan, have witnessed the wholesale departure of the Russian populations. By 1994 there were well over two million refugees and forced migrants of all nationalities in Russia. The precise number is not known because far from all people in these categories are legally registered. Estimates of the potential scale of the out-migration from the countries of the 'near abroad' vary. But a conservative, middle range, figure based on opinion surveys of sample populations of Russians living in the 'near abroad' suggest that as many as two to three million can be expected to leave for Russia over the next few years. Should there be further outbreaks of civil war, or rapid deterioration of domestic economies, the number would be higher. To some degree this in-migration will be offset by emigration, which since 1991 has picked up tempo. Some emigrants from Russia are destined for locations in the 'near abroad'. The large scale exodus of Ukrainians is a case in point (see Fig. 4.10). A sizeable number of people are leaving the post-Soviet scene altogether.

In general, emigration from the former Soviet Union tends to comprise mostly well educated people who are employable elsewhere. The potential outflow from all parts of the post-Soviet scene is of no small concern, especially in Europe which has already difficulty in keeping unemployment within acceptable bounds. In 1991 alone, just over one million people departed, principally but by no means exclusively, for Europe and North America. Some surveys have suggested that as many as seven million more people can be expected to emigrate over the next decade. Of course, the desire to emigrate can only be realized if a host country can be found. From Russia the outflow has also been substantial. Since independence in 1991, each year has witnessed an average net out-migration to destinations outside the former Soviet Union of just over 100,000 people. Most have gone to Germany. The next most common destinations were Israel and the United States.

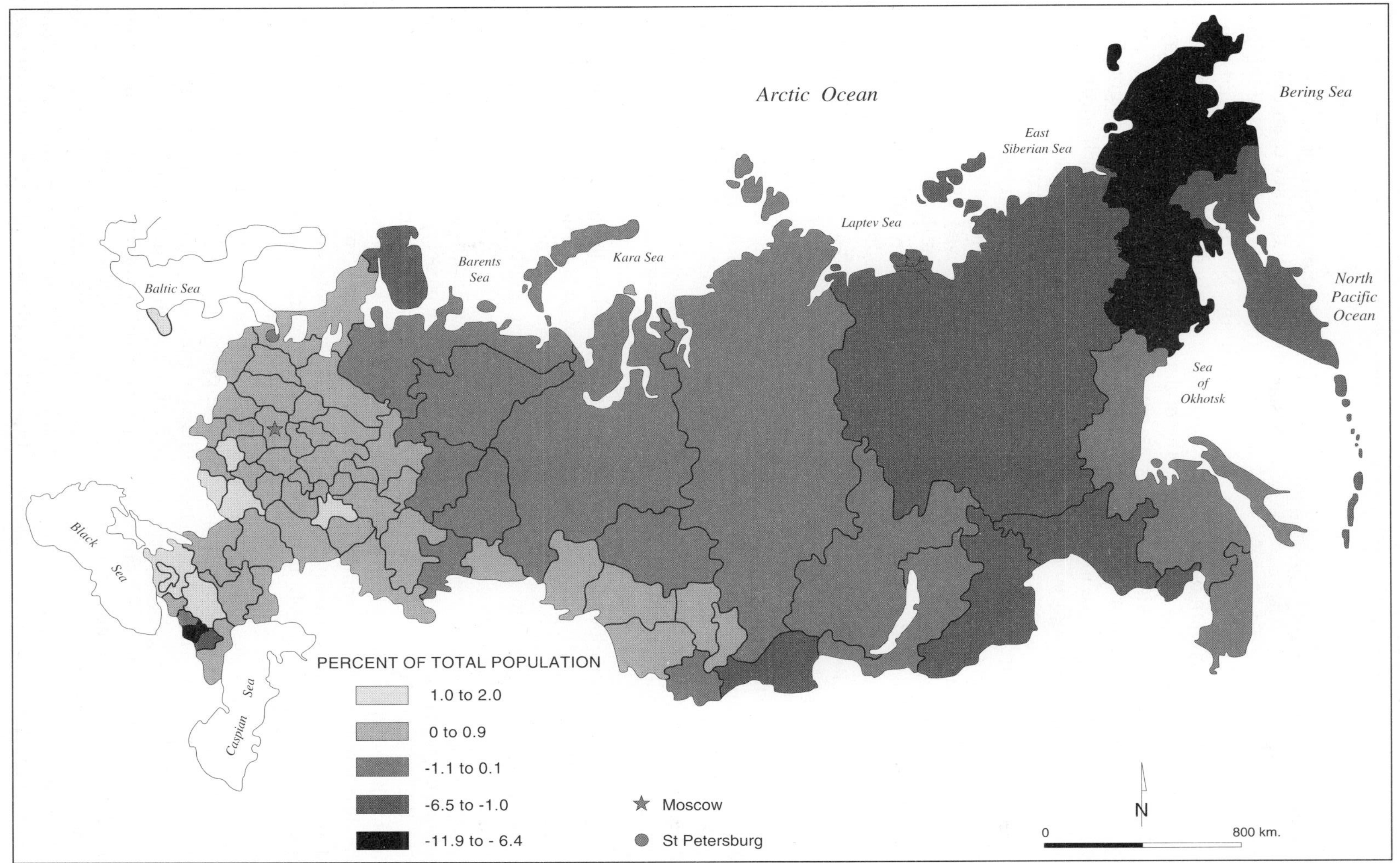

FIGURE 4.11 *Net migration balance: Russia, 1992*
(*Source*: T. Heleniak, *Economic Geography of the Russian Federation* (Washington: The World Bank, Country Department III, Europe and Central Asia Region, 1994), Vol. 1, No. 6)

Migration and Minorities within Russia

The general pattern of net migration balances in Russia in 1992 is the opposite of that which prevailed during most of the Soviet era. Put simply, then migration was from west to east, and from all regions into the north. Wage incentives in the post-World War II era had figured prominently as the reason for this pattern. As Fig. 4.11 shows, it was the western part of Russia which stands out in terms of migrants as a percentage of regional total population in 1992, whereas throughout the northern and eastern regions significant out-migration occurred. Not only are Russian citizens able to move with less bureaucratic interference now than in the past, employment in the predominantly resource-based industries of the eastern and northern regions is increasingly problematic. Many former state sector enterprises were heavily subsidized and no longer are. For many enterprises still operating, simply paying salaries and wage bills on a regular basis is a major challenge. Meanwhile, privatization has brought higher prices in a traditionally high cost of living environment. When operating enterprises cannot even pay workers regularly, when wages in comparable sectors of the European Russian economy are rising quickly, when the long-standing wage incentives paid to draw workers to the climatically harsh eastern and northern regions in the first place are increasingly irrelevant in the new economy, there is little incentive to stay if opportunities exist elsewhere. Indeed, the Russian federal government has now introduced a resettlement programme intended to facilitate relocation of families from northern locations to southern ones where costs are lower. Thus, as Fig. 4.11 indicates, there are few regions east of the Ural mountains holding their own in terms of net migration. Many more in the western part of the country are beneficiaries.

Within Russia are more than 100 different nationalities, 31 of whom have their own autonomous regions. These regions, of course, are residual features of the former Soviet nationalities policy. In terms of the share of the titular nationality of the population in each of the autonomous regions, in only eight did they comprise a majority. These are shown in Fig. 4.12. Half this group is located in the North Caucasus. And only six contained more than half of the total number of the titular population across the country as well as having the titular group a majority in the autonomous region. These are highlighted in the data presented in Table 4.4, which are drawn from the census of 1989, the most recent source available. Since 1989, the pattern of internal migration has been such that the relative position of a number of national minorities within their own territory has improved. This is because out-migration of Slavs from the national minority regions in the eastern part of the country has been characteristic, whereas net in-migration has occurred in some regions in the European Russia, especially those which are Muslim (see Fig. 4.11). Thus, for the first time in decades the ethnic balance in some of these territorial-administrative regions is slowly shifting in favour of the titular nationalities. Given the political movements feeding on secessionist aspirations in a number of these national minority regions such as the republics of Tatarstan and Bashkortostan, the position of Russians living there is different now than in the Soviet era when they were clearly the dominant ethnic group.

In many of these ethnic territorial-administrative regions the level of economic and social development is below average for Russia as a whole, a source of not inconsiderable grievance. Indeed, in national minority group regions which are resource rich, the regional disparity issue figures prominently in regional government's demands for a larger share of the tax revenue generated through resource development. This issue is not exclusively a national minority group concern. Indeed, there is considerable sympathy and support for regional governments across Russia in their efforts to exercise a greater degree of fiscal autonomy within the Russian Federation. The executive arm of central government under Boris Yeltsin has acceded to pressure for greater regional economic control in exchange for political support for his policies and programmes. Thus, nationality issues at the regional scale have

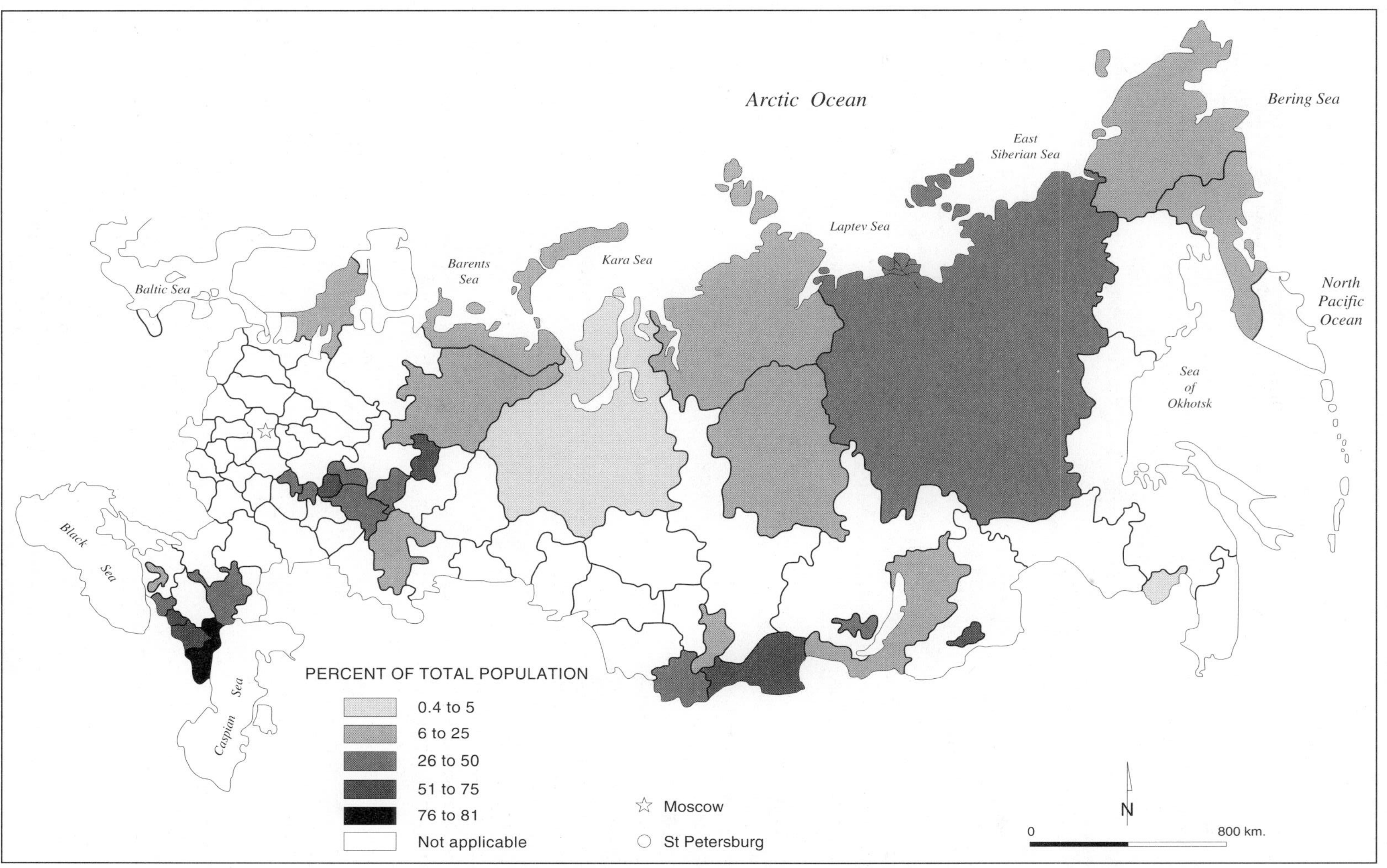

FIGURE 4.12 *Share titular population in autonomous regions of Russia, 1989*
(*Source*: T. Heleniak, Economic Geography of the Russian Federation (Washington: The World Bank, Country Department III, Europe and Central Asia Region, 1994), Vol. 1, No. 8)

Table 4.4 *National minority groups in Russia by territorial unit, 1989*

	Total population	Russians	Titular nationality	% Russian	% Titular nationality	Status in 1989*
RUSSIAN FEDERATION	147022	119866	119866	81.5	81.5	
Northern Region	6124	5017	–	81.9	–	
Karelia Republic	790	582	79	73.6	10.0	ASSR
Komi Republic	1251	722	292	57.7	23.3	ASSR
Nenets AOkr	54	35	6	65.8	11.9	AOkr
Northwestern Region	8241	7461	–	90.5	–	
Central Region	28544	26631	–	93.3	–	
Volgo-Vyatsk Region	8464	6353	–	75.1	–	
Mariy El Republic	749	356	324	47.5	43.3	ASSR
Mordvinia Republic	964	586	313	60.8	32.0	ASSR
Chuvashia Republic	1338	357	907	26.7	67.8	ASSR
Central Chernozem Region	7733	7363	–	95.2	–	
Povolzhskiy Region	16397	12052	–	73.5	–	
Kalmykia Republic	323	122	146	37.7	45.4	ASSR
Tatarstan Republic	3642	1575	1765	43.3	48.5	ASSR
North-Caucasus Region	16629	11234	–	67.6	–	
Adygea Republic	432	294	95	68.0	22.1	AO
Dagestan Republic	1802	166	1445	9.2	80.2	ASSR
Kabardino-Balkaria Republic	754	241	434	31.9	57.6	ASSR
Karachay-Cherkessia Republic	415	176	170	42.4	40.9	AO
North Ossetia Republic	632	189	335	29.9	53.0	ASSR
Chechnya and Ingushetia Republics	1270	294	898	23.1	70.7	ASSR
Urals Region	20239	14769	–	73.0	–	
Bashkortostan Republic	3943	1548	864	39.3	21.9	ASSR
Udmurtia Republic	1606	945	497	58.9	30.9	ASSR
Komi-Permyat AOkr	159	57	95	36.1	60.2	AOkr
West Siberian Region	15013	12749	–	84.9	–	
Altay Republic	191	115	59	60.4	31.0	AO
Khanty-Mansi AOkr	1282	850	18	66.3	1.4	AOkr
Yamalo-Nenets AOkr	495	293	21	59.2	0.4	AOkr
East Siberian Region	9153	7652	–	83.6	–	
Buryatia Republic	1038	726	250	69.9	24.0	ASSR
Tuva Republic	309	99	198	32.0	64.3	ASSR
Khakassia Republic	567	450	63	79.5	11.1	AO
Taymyr AOkr	56	37	7	67.1	10.7	AOkr
Evenki AOkr	25	17	3	67.5	14.0	AOkr
Ust-Orda Buryat AOkr	136	77	49	56.5	36.3	AOkr
Aga AOkr	77	31	42	40.8	54.9	
Far Eastern Region	7950	6347	–	79.8	–	
Sakha Republic	1094	550	365	50.3	33.4	ASSR
Birobijan AO	214	178	9	83.2	4.2	AO
Koryak AOkr	40	25	7	62.0	16.5	AOkr
Chukotka AOkr	164	108	12	66.1	7.3	AOkr

* ASSR: Autonomous Soviet Socialist Republics; AOkr: Autonomous Okrugs; AO: Autonomous Oblasts
Source: T. Heleniak, *Economic Geography of the Russian Federation.* (Washington: The World Bank, Country Department III, Europe and Central Asia Region, 1994), 1,2

become inextricably bound up with politics at the national level. There is a growing resentment to the transformation wrought by market reform and the opportunities it has created for what some Russians perceive to be a form of extortion on the part of a number of ethnic territorial-administrative units. It is sometimes contended that those national minorities whose territory happens to possess resources developed during the Soviet era, often by non-indigenous labour as well as capital, are holding the rest of the country to ransom.

Combined with the demise of the privileged status of Russians and the Russian language and culture in the post-Soviet scene, there is the potential for political and social unrest between minorities within the Russian state and the Russian majority. In Russia there are calls for a resurrection of the old order, for stability and for respect for the majority population. Russia is not immune to the ethnic conflicts which have enveloped parts of the 'near abroad'. The flashpoints are often in the cities. Ironically, it was the Soviet city which was to epitomize a higher order of economic, social and political existence than characterized the western world. The Soviet urban legacy and its transformation in the post-Soviet era is the subject of the next chapter.

Further reading

Akiner, S. 1983 *Islamic Peoples of the Soviet Union*, 2nd edn. London, Kegan Paul.

Allworth, E. (ed) 1971 *Soviet Nationality Problems*. New York, Columbia University Press.

Allworth, E. (ed) 1980 *Ethnic Russia in the USSR*. New York, Pergamon.

Arditis, S. (ed) 1994 *The Politics of East–West Migration*. Basingstoke, Macmillan Press.

Azrael, J.R. (ed) 1978 *Soviet Nationality Policies and Practices*. New York, Praeger.

Besemeres, J.F. 1980 *Socialist Population Policies*. White Plains, NY, M.E. Sharpe.

Chinn, J. 1977 *Manipulating Soviet Population Resources*. London, Macmillan.

Defosses, H. (ed) 1981 *Soviet Population Policy: Conflicts and Constraints*. New York, Pergamon Press.

Grandstaff, P.J. 1987 *Interregional Migration in the USSR: Economic Aspects 1959–1970*. Durham, Duke University Press.

Hamilton, K.A. (ed) 1994 *Migration and the New Europe*. Washington, The Center for Strategic and International Studies.

Katz, Z. (ed) 1975 *Handbook of Major Soviet Nationalities*. London, Macmillan.

Karklins, R. 1986 *Ethnic Relations in the USSR: The Perspective from Below*. London, George Allen and Unwin.

Kolstoe, P., Edemsky, A. 1995 *Russians in the Former Soviet Republics*. London, Hurst & Company.

Kozlov, V. 1988 *The Peoples of the Soviet Union*. London, Hutchinson.

Lewis, R.A., Rowland, R.H. 1979 *Population Redistribution in the USSR: Its Impact on Society 1897–1977*. New York, Praeger.

Lorimer, P. 1946 *The Population of the Soviet Union*. Geneva, S.D.N.

Lutz, W., Sherbov, S., Volkov, A. (eds) 1993 *Demographic Trends and Patterns in the Soviet Union Before 1991*. London, New York and Laxenburg, Routledge and IIASA.

Ryan, M., Prentice, R. 1987 *Social Trends in the Soviet Union from 1950*. London, Macmillan.

Ryvkina, R.V., Turovskiy, R. 1993 *The Refugee Crisis in Russia*. North York, Ontario, York Lanes Press. Edited and with an introduction by Robert J. Brym.

Shkolnikov, V.D. 1994 *Scientific Bodies in Motion: The Domestic and International Consequences of the Current and Emerging Brain Drain from the Former Soviet Union*. Santa Monica, Rand.

Shlapentokh, V., Sendich, M. and Payin, E. (eds) 1994 *The New Russian Diaspora. Russian Minorities in the Former Soviet Republics*. Armonk, NY, M.E. Sharpe.

Wixman, R. 1984 *The Peoples of the USSR: An Ethnographic Handbook*. Armonk, NY, M.E. Sharpe.

5

PRIVATIZATION OF THE SOVIET CITY

'One of the compelling ironies of urban existence is the great gap between the promise of modernity, the vision of abundance and equality, and its fulfilment by the cities which are its agents'
(H. J. Dyos, *Urbanity and Suburbanity*, 1973, 5)

The Russian revolution of 1917 imposed a new ideological blueprint for a society moulded over the centuries by the values and precepts of an absolute autocracy. Urban-industrialization became one of the central features of the Soviet development process, and cities were soon being built in accordance with socialist precepts. All resources having been nationalized, urban land uses were to be planned, not determined by the marketplace. In short, the Soviet socialist city was to become both an agent and an example of directed social and economic change. It differed in many ways from cities in Western Europe and North America, but nowhere so palpably as in the central city where there was no equivalent of a business district and where land was not used intensively. The dissolution of the Soviet Union has resulted in new political structures and economic reform. But what do these changes portend for cities bearing the indelible impress of decades of Soviet socialist planning?

In this chapter several issues will be examined in order to answer this question. However, in order to assess the degree and direction of change since 1991 it is necessary to have in hand some sense of the urbanization process during the Soviet era as well as some understanding of the form and social fabric of the Soviet socialist city. Thus, the urban growth process during the Soviet era will be briefly described. The ideals which informed the planned development of the socialist city will then be contrasted with the reality of life within it. With this overview of the legacy of the Soviet era in hand, the discussion will turn to the post-Soviet scene. The urban development process in general and the privatization of housing in particular in the new post-Soviet states will be examined first. How urban management issues are being addressed will then be considered in some detail. This discussion will conclude with an assessment of the impact of market reform and privatization on the geography of the city in Russia.

SOVIET-ERA URBAN DEVELOPMENT

In a Soviet type economy it was theoretically possible to manipulate population movements so as to conform with state objectives. But theory and practice were not always congruent as the discussion of migration trends in the preceding chapter makes plain. Nonetheless, for more than four decades the notion of an optimal size city was widely accepted as an essential element in urban policy, though what was to be the optimal number of people was far from consistent. Early concepts articulated in the 1920s reckoned that a population of 50–60,000 was ideal since this was a large enough number to make the provision of the necessary goods and services economic, yet small enough to permit a sense of community and a communal, socialist ethos to be fostered and maintained. At the time about half the urban population resided in cities of less than 50,000, and about three-quarters of the country's total population were rural. However, the Stalin-initiated industrialization drive brought about a seemingly unending circular and cumulative process of urban growth. Unable to hold urban growth in check, ideas about what constituted the optimum size began to change. By the mid-1950s the most frequently cited figures ranged between 150,000 and 200,000. This was both a pragmatic accommodation with the reality of the urban growth process, and a reflection of the changing notions about the urban economies. Although the urban population only overtook the rural in 1961, more and more of them were living in larger and larger cities. By the mid-1960s the optimum size had been bumped into the 200,000 to 300,000 range. Despite the steady inflation of what constituted the optimal size of a Soviet city, the pace of urban growth outstripped it. Perhaps the clearest indication of the failure of state planning to control the growth of cities during the Soviet era is the escalation in the number of them with more than one million inhabitants. In 1926, there were only two: Moscow and St Petersburg. By 1959 Kiev had joined them. A decade later seven more had been added to the roster. By 1991, more than two dozen cities were in the million or more category, fully one-fifth of the urban population lived in them, and more than two-thirds of the total population were urbanites. But long before then reality had forced reassessment of the theory behind the concept of an optimal size for the Soviet city.

Such concentration of the urban population was originally perceived to be a negative phenomenon, an inherent feature of capitalist societies where urban growth was both spontaneous and uncontrolled. The reason for the failure of a flood of decrees to limit city size, the failure of a vast bureaucracy to monitor and direct migration, and the failure of the system of residence permits (*propiski*) which gave the legal right to live in a particular city and was prerequisite to receiving state housing, had to do with basic economics. Decision-makers in the various ministries and enterprises knew that there were real external economies to be gained by locating production and other facilities in cities. This pushed up the demand for labour thereby fostering more in-migration. The urban growth process was circular and cumulative. And if location in a major city was absolutely impossible, then locating as close as possible to it was the next best thing. Thus, outlying settlements were often the recipients of much new, and unplanned, investment. While controls over population movement inhibited to some degree entry to the major cities, satellite communities experienced explosive growth. A case in point is the Moscow region. Between the late 1950s and the early 1980s, the population of Moscow itself increased from five to eight million, or by more than 57 per cent. But the population of urban settlements in the Moscow region grew five-fold over the same period. Despite many directives to locate production facilities in smaller, less well developed centres, by the end of the Soviet era close to one half of total investment in industrial production was concentrated in the roughly five dozen cities having one half million or more residents. This was not only in flagrant violation of a host of government policies and decrees, it spawned huge agglomerations. Literally millions of people who were denied the necessary residence permit to live in a major city commuted to work each day from smaller centres outside its borders. By the late 1980s the vast majority of the Soviet urban population lived in an agglomeration, not infrequently in a small dormitory settlement with deficient housing, consumer and cultural services. The definition of an agglomeration varied, but was customarily regarded as a network of urban and rural settlements linked to a city of at least 250,000 inhabitants by a transportation system that permitted journeys from dependent settlement to the core city within a two hour travel time. The concentration of population in the central city was rarely less than one half of the total metropolitan population of the urban agglomeration, a rather high share compared to western countries. This was largely owing to the low level of private automobile ownership and to the absolute preponderance of apartment living as opposed to detached, individually owned housing so characteristic of North America for instance.

Ironically, the Soviet approach to town planning was typically confined to a jurisdiction which stopped at the city border. The Moscow region was one of the few exceptions. The principal objectives of planning in the Moscow region were the containment of urban sprawl and the protection of the forest park belt as a much needed accessible recreation zone for the bulk of Muscovites who were dependent upon public transport. But it was precisely in the intended green belt where the aforementioned satellite city growth was fastest. Thus, even in the best planned city region in the country containment policies were not working. In less well planned areas, plan and reality were likely to be even further apart. Beginning in the 1970s, the agglomerations as spatial entities began to figure in national urban development planning strategies. But in very few places indeed was planning actually carried out at such a regional scale. Given the economic importance of urban agglomerations, the poor transportation infrastructure throughout the huge territorial expanse of the Soviet Union, and the state's mostly antiquated electronic communication system, small and remote places were further removed from the mainstream of national economic development than were similarly situated communities in North America.

The Central Urban Planning Institute located in Moscow was responsible for creating the conceptual basis and planning framework for addressing medium and long term settlement development needs across the country. As this institute was the country's principal planning bureaucracy, its staff was certainly aware of the prevailing urban growth trends and their social and economic consequences. But it was not empowered to implement remedial measures, only to recommend policy. To address the reality of urban agglomerations in the Soviet Union, the Central Urban Planning Institute in the early 1970s introduced with considerable fanfare a plan to create an interconnected system of settlements on a national scale – the so-called General Scheme for the System of Settlement. This General Scheme was to provide the spatial frame of reference for the development of the economy between 1976 and 1990. As Fig. 5.1 implies, the General Scheme was intended to integrate rural and urban settlements. Comprising a network of hierarchically organized settlement systems, 60 large, 169 medium, and 323 small, when complete it would have embraced more than 90 per cent of the total Soviet population. The General Scheme was also intended to help realize the long-standing goal of Soviet socialism to eradicate the differences between town and countryside. While it demonstrated that the concept of planning for agglomerations had certainly taken root, it did not result in the creation of juridically-based planning regions throughout the country, something which was a necessary but essentially unfulfilled objective during the Soviet era. Meanwhile, throughout the country the large city grew faster than was intended, spawning unplanned development in its shadow, and leaving more remote places to get on as best they could with limited investment and resources right down to the end.

Planning the Soviet socialist city

The creation of a new urban form figured prominently in the task of inculcating the values of a proletarian culture. Individualism and privatism in all their manifestations were to be supplanted by the proletarian principle of collectivism. In the first few years of revolutionary fervour this led to much experimentation, especially in art and architectural design. In terms of actual city construction, however, the chaos of Civil War and reconstruction of the national economy meant that little was accomplished outside the municipalization of existing real estate and the reassignment of housing space. Indeed, the massive urban–rural migration, which the chaos of the time occasioned, significantly eased the long-standing housing crisis. The benefits, however, were both relative and temporary.

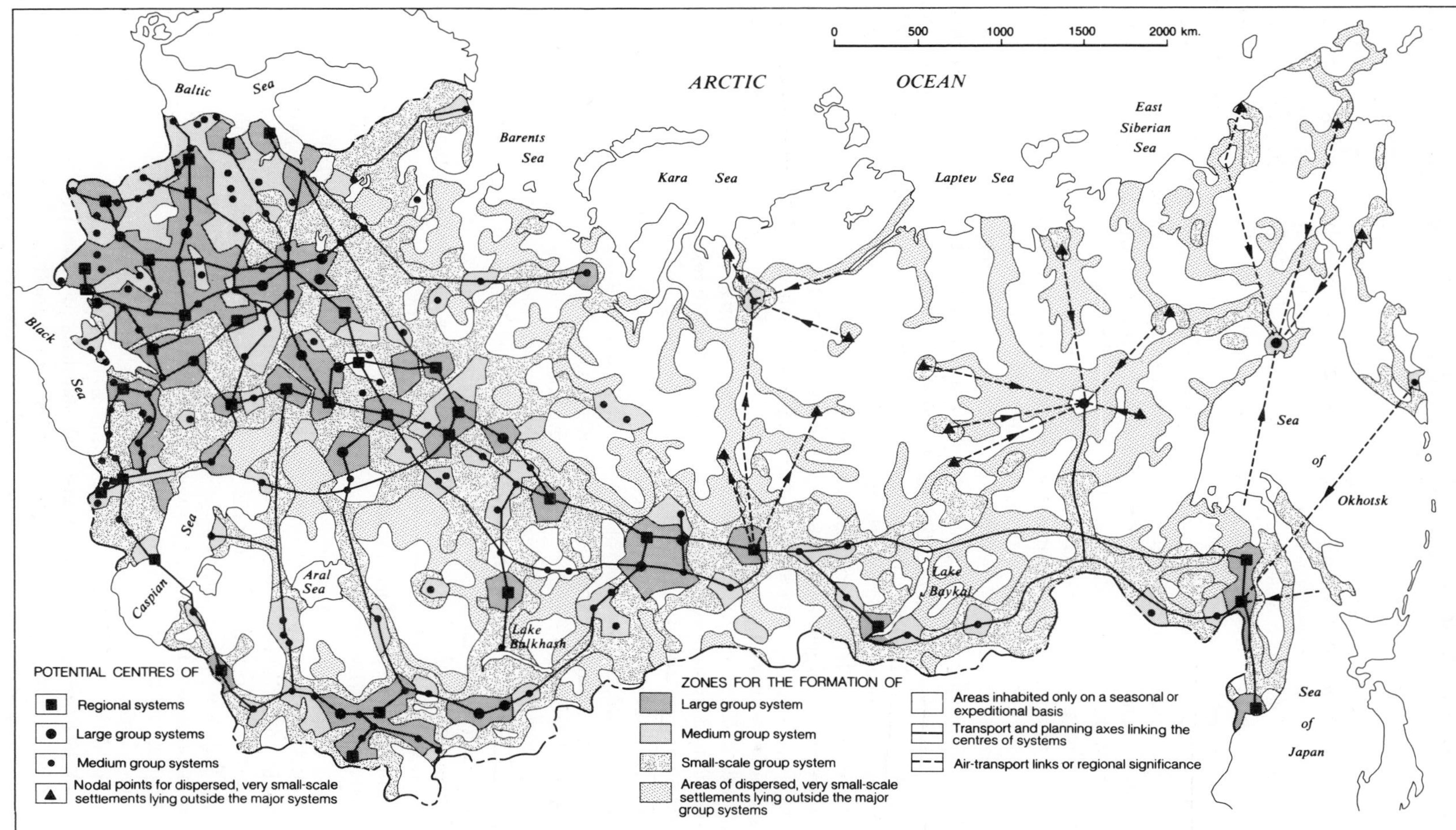

FIGURE 5.1 *General settlement scheme*
(*Source*: A. Brown *et al.* (eds), *The Cambridge Encyclopedia of Russia and the Soviet Union* (Cambridge: CUP, 1982), 348–9)

The fundamental contradiction between city and countryside, which Marxists claimed was produced under capitalism, somehow was to be resolved in a socialist society. Much was said about the role of communal living in the new order, some enthusiasts even arguing that it would eventually replace the nuclear family. These were just two of the many potential social changes introduced by the revolution which had some important implications for urban design. But there were no readily available models to accommodate the profound changes set in motion. While not much was actually built during the early 1920s, debate over what form the new Soviet socialist city should take was intense. There soon emerged two principal and opposing schools of thought, generally labelled the urbanist and de-urbanist schools.

The ideas of Soviet architect L. Sabsovich are typical of the urbanist approach to the socialist city. He proposed a system of largely self-contained urban centres in which multi-storeyed collective living facilities would de-emphasize distinctions between the various strata of society. The nuclear family, it was presumed, would eventually give way to an entirely communal way of life. Housing therefore would serve a dormitory purpose only. Filling the void would be a host of state services ranging from day care to communal food provisioning. A population of around 50,000 per urban centre was assumed to be the ideal size. Within Sabsovich's socialist city major design features would include: strict land use zoning, proximity of home and workplace and thus a reliance on pedestrian movement, a non-commercial city centre, and ample green space for recreation.

The de-urbanists argued that a scheme such as Sabsovich's simply replaced large cities with smaller ones. They did not result in a distinctly socialist urban environment. De-urbanists such as M. Okhitovich and M. Ginsberg, for instance, proposed an essentially townless, socialist society in which the population would be dispersed over all the habitable parts of the state. Settlement would occur in ribbon-like developments, individual dwellings being located in natural surroundings, but within easy access to communal centres for dining, recreation and so forth. While some measure of privacy would be provided in terms of living quarters, the lifestyle itself would be communal. All centres of employment and consumer services were to be located in these same ribbon developments, but in such a manner as to minimize travel time to them. Spatial mobility was predicated on universal use of the automobile, a marked contrast to the urbanists' emphasis on pedestrian journeys. Given the radical nature of the town planning proposals put forward by both the de-urbanist and urbanist schools, it is scarcely surprising that few schemes ever got past the drawing board. However, one idea did make the transition from drawing board to bricks and mortar – N. Miluitin's linear city.

Miluitin's adaptation of the concept first developed by the Spanish architect Soria y Mata in the 1920s was used in the planning of Stalingrad (now Volgograd) on the Volga river, and in part in the design of the new iron and steel centre in the southern Urals, Magnitogorsk. As Fig. 5.2 indicates, the linear city incorporated about a half dozen strictly segregated zones. A green buffer served to separate the industrial and transport zones from the residential zone, with due attention to prevailing wind direction. The parallel development of industry and housing was intended to facilitate short, pedestrian journeys to work. Consumer services were to be distributed throughout the city rather than concentrated in the centre in order to ensure equal accessibility for all. The city size was again to be restricted to about 50–60,000.

Whatever the merits of the architectural and town-planning schemes produced during the 'cultural revolution' of the 1920s, most were denied a place in the real world. As in literature and art, this visionary, utopian and, not uncommonly, quite impractical experimentation had little impact of the ordinary Soviet citizen. Indeed, the vast majority of the population still endured a life of grinding poverty in the village. With the advent of the Stalin-era forced urban-industrialization, the long-term development priorities of the state were laid down quite clearly. They did not include a fundamental restructuring of the settlement system or of the cities in it. The debate over the form of the Soviet socialist city was concluded in 1931 when it was decreed that

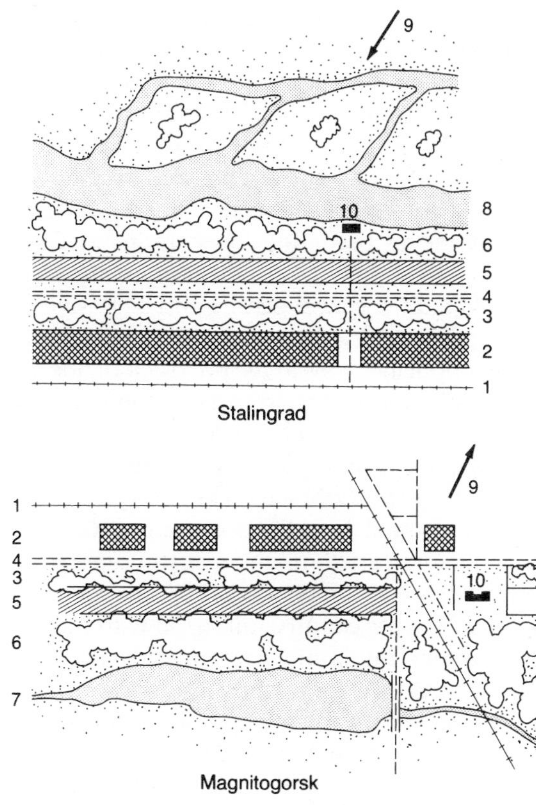

1, railway; *2*, industrial zone; *3*, green zone; *4*, thoroughfare; *5*, residential zone; *6*, park; *7*, Ural River; *8*, Volga River; *9*, prevailing wind; *10*, House of Soviets.

FIGURE 5.2 *Miluitin's scheme for a socialist city*
(*Source*: Based on F. Parkins, *City Planning in Soviet Russia* (Chicago: University of Chicago Press, 1953), 22)

all Soviet cities must be socialist by virtue of their being part of the Union of Soviet Socialist Republics. While this pragmatic policy statement, coming as it did with the full authority of the Communist Party, served to stifle further public discussion, and especially the oft-cited criticism that the existing urban system being capitalist in origin and in form should be eradicated, it did not mean that the lively debate of the 1920s was without any consequence. In fact many of the basic ideas of the urbanist and de-urbanist schools were incorporated, albeit in suitably modified form, in the Plan for the Reconstruction of Moscow adopted in 1935. The Party limited the terms of reference for the international competition to reconstruct the Soviet capital to a moderate reworking of the existing urban form, and thus ruled out the possibility of a new capital being built from scratch. The principles upon which the plan was based are important because they became the guidelines for town planning throughout the country during the Soviet era. Among the more important were: limited city size; an ideological role for the city centre; state control and allocation housing; spatial equality in the distribution of consumer and cultural services; and a limited journey to work. While the optimal city size at this time was still reckoned to be 50–60,000 the population of Moscow already exceeded three million. Public space in general, and central city space in particular, were assigned an ideological or symbolic role. The central city had to accommodate massive, orchestrated public ceremonies, as well as serve everyday needs, a rather challenging urban design problem. State control of housing was intended to ensure egalitarian allocation at a nominal rent.

Over the years the general principles for urban development were translated into myriad norms for planners to follow. For example, one of the most important norms pertained to housing entitlement. The sanitary minimum per capita housing allotment was set at nine square metres of living space in 1922 and was unchanged throughout the Soviet era. Living space included only the principal rooms in an apartment. Excluded was the floor space taken up by kitchen, bathroom, hallway, and storage. The journey to work time was set at 40 minutes. Many norms were established on a per 1000 population basis. For example, there were to be 35 cinema seats per 1000 urban inhabitants. The adoption of these and literally hundreds of other norms for the development of the Soviet socialist city was intended to ensure an equitable allocation of resources. This would help to lay the foundation for the creation of a new and higher form of society, one in which collectivism would supplant privatism, one in which the traditional cultural values and religions would be obliterated, one in which the primacy of the family would eventually disappear. But plan and reality were seldom close in the Soviet urban development process. The approach did not allow for differences in need according to demographic, ethnic or geographical differences across the country. And it presumed equality in treatment of the citizenry. However, contrary to original principles, Stalin began in the 1930s to entrench a system of privileges, benefits and salary differentials intended to ensure participation in the industrialization drive, and subsequently the war effort and post-war reconstruction. Put simply, in the Stalin era private values were converted into public values. The middle class – the *meshchanstvo* of the Russian imperial era – was legitimized, and hence so was privatism. Thus, even if the state had been able to meet the norms it set for its citizens, there would have been many who would have been disadvantaged. As it turned out, many of the norms remained goals rather than minimum standards during all of the Soviet era. The reason for this situation had much to do with political power and decision-making authority.

In theory, the city Soviet, or municipal government, and its bureaucracy including an architecture and planning department, was responsible for all that happens within city boundaries. However, in practice the structure of decision-making was such that municipal government did not figure very prominently in terms of its political influence, in terms of its claim to financial resources, or in terms of its ability to determine the course of actual development. In the Soviet model of centralized sectoral, or ministerial, planning, the theoretical rights of the city Soviet were often subordinate to the interests of particular ministries. Thus, the chief architect, to whom the architecture-planning department was subordinate, was responsible for ensuring that physical planning norms were met and that all development conformed to the general plan for the city, was often in an untenable position. Important ministries frequently side-stepped city planning regulations. This resulted in new, unplanned enterprises being built and unanticipated expansion in existing production facilities, all of which demanded more labour. Urban growth regularly exceeded what the planners had anticipated, therefore confounding honest efforts to meet the various consumer, cultural and housing norms, to say nothing of the basic urban infrastructure such as transportation, water and sewage systems. There were numerous attempts by central government to strengthen the role of the city in controlling its own affairs from the late 1950s right down to the end of the Soviet era. But decree after decree proved ineffectual. Those who lived in cities were destined to endure the consequences of urban officials never becoming 'masters in their own house' as each new piece of legislation intended. Undoubtedly the most glaring and pervasive failure of the system to meet a basic norm for the urban population was in the provision of housing.

It required more than seven decades before the average per capita allocation of living space reached nine square metres. When the norm was adopted in 1922 the average was six square metres. As the data in Fig. 5.3 indicate, there was considerable variation amongst the republics in 1989. Obviously, millions of people in urban places across the country still coped with less than the sanitary minimum. Urbanites in the Baltic region were the most liberally endowed, fully in keeping with their reputation for having the highest material standard of living in the Soviet Union. The republics of Middle Asia were obviously the worst off. Given the family size differences between indigenous peoples and immigrant Slavs for example, it is clear that these average figures understate the degree of overcrowding experienced by the former. The fact that it had taken so long simply to get the average for the country to nine square metres of living space per capita is a reflection of the limited investment in housing during the Stalin period. Indeed, just before Stalin's death in 1953 the per capita average was less than four square metres. A not uncommon situation in many Soviet cities at this time was not an apartment per family, but a family in each room of an apartment. Nine square metres of living space per person was beyond the ken of the average person. But for the privileged members of society, not for them the cramped and shabby quarters of the average citizen. Indeed, in censor-approved, Socialist-realist literature of the period, the hero contributing in his own special way to the development of Stalin's socialist state could be portrayed as a dedicated, career-minded manager aspiring to own a house, and perhaps a

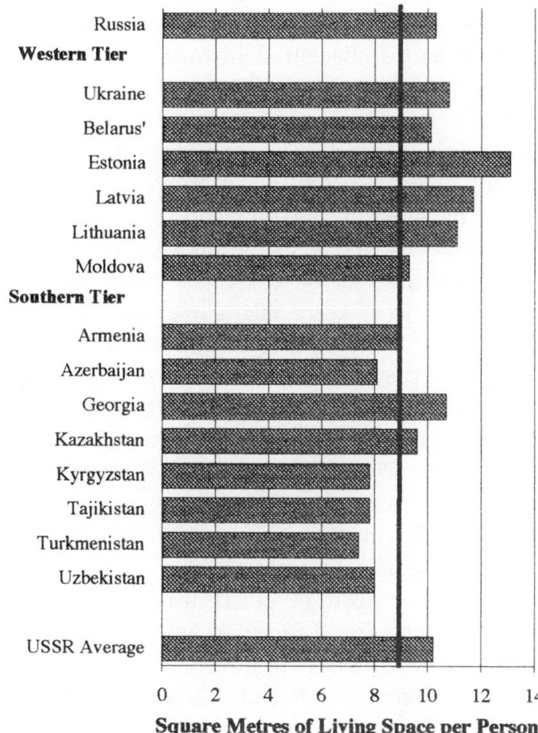

FIGURE 5.3 *Per capita urban living space, Soviet republics, 1989*
(*Source*: Compiled from data in *Narodnoye Khozyaystvo SSSR v 1989g* (Moscow: Goskomstat, 1990), 166)

dacha (or summer cottage) as well. He could even be depicted as driving his own car! Meanwhile, in the real world most people lived poorly, travelled six days a week on an overcrowded tram, returning home to scarcely less crowded conditions.

In the years following Stalin's death in 1953 doing something about the abysmal state of housing of the fast growing urban population was finally given priority by government. Construction technology for apartment building was transformed as a result of the application of factory floor organization principles. Standardized, prefabricated housing units became commonplace in the industry, and by the late 1960s high-rise apartment buildings were beginning to change the skyline of cities across the country, as the average number of floors in them steadily increased. In 1963 the typical apartment building had five storeys. A decade later it had ten. Between 1960 and 1975 fully two-thirds of the urban population were assigned improved, if not new, housing. This was a significant achievement after decades of deprivation, but as the foregoing discussion indicated it fell far short of what was needed to meet the sanitary minimum per capita allocation of living space established decades earlier.

The standardization of housing construction was extended to the design of residential areas in cities, and this is the reason for the uniformity, if not drabness, which still characterizes much of the urban landscape across the former Soviet Union. The typical approach was to plan for a three-level hierarchy in housing development, in which the *mikrorayon* was the key organizational unit. Each *mikrorayon* usually accommodated 8000–12,000 people, and had a radius of approximately 300–400 metres. Within each *mikrorayon* were several smaller housing units variously called *kvartali*, super blocks, or living complexes, each having a population of 1000–1500, and a radius of perhaps 50–100 metres. The highest order element in the hierarchy was the residential complex. Comprising several *mikrorayoni*, the residential complex customarily would have a total population of about 50,000–60,000. The idea was to have standardized apartment blocks occupied by a representative mix of the city's socio-economic and ethnic groups, ample green space, perimeter thoroughfares with public transportation, day care, educational and health services. All day-to-day requirements were to be satisfied by facilities located within a short pedestrian journey. Higher order goods and services were to be strategically located with the *mikrorayon* and residential complex. Only infrequent journeys to the central city were expected, and then more for the purpose of mass culture than personal consumption.

By the late 1980s roughly one-half of the Soviet urban population lived in a *mikrorayon*, in apartment complexes that were more standardized than differentiated. The *mikrorayon* was intended to provide the physical environment appropriate to the task of engendering a sense of neighbourliness, of collective responsibility; in short, a communal ethos. There is scant evidence that this occurred very often. Indeed, alienation rather than a sense of collectivism was a more frequently reported psychological trait of residents or *mikro-rayoni.* This was manifested in the usual ways – withdrawal from community endeavours, indifference to the maintenance of public space, vandalism and hooliganism amongst the young. Given the massive rehousing programme that began in the late 1950s which disrupted countless individual social networks, it is not surprising that alienation was widespread amongst residents of the vast housing developments that arose on barren suburban tracts.

While the state directed the urban development process, it did not own all the housing. The state, however defined, owned about 70 per cent of the total housing stock in the final years of the Soviet period. Municipal government, in theory the principal arena for decision-making in matters related to housing, actually controlled only about one-half of the state housing supply. The balance was owned, operated and maintained by ministries, departments and enterprises, whose self-interest in attracting labour was usually best served by retaining control over a housing supply. Despite years of promoting the concept of municipal jurisdiction in housing matters, central authorities were unable to persuade ministries, departments and enterprises to turn housing over to municipal government. Since housing was used by all 'owners' as one means of rewarding the privileged members of Soviet society, there was a strong disincentive to forego control over such an important way of dispensing benefit. Elites had always done much better than the average figures for per capita allocation of living space cited above might suggest. Such benefits have provided the already privileged with huge windfall gains in the post-Soviet era as will be described later.

The general thrust of Soviet housing policy was to create residential areas in which social class and ethnic segregation would not take place. To some degree, the policy was successful, but never to the extent claimed. This was partly owing to the long-standing practice of treating elites differently in the allocation of state housing, both in terms of the absolute amount and its location. It was also partly owing to the fact that there were two other forms of housing ownership – private and cooperative. The construction of private and cooperative housing had a checkered history. A substantial number of cooperatives were created during the early Soviet period when the state's housing construction efforts were decidedly limited. An obvious need was met in an ideologically more acceptable manner than by means of construction of privately owned housing. But official sanction was withdrawn in the late 1930s when all existing cooperatives were taken over by the state. They were, so to speak, nationalized and the former cooperative member became a tenant, albeit one paying the characteristic nominal rent. With the official recognition of a major housing crisis in the post-Stalin era, came a more liberal attitude toward cooperative housing construction. In 1962 cooperative apartment ownership was again legal. The state assisted by providing loans to cover up to 60 per cent of construction cost. In some cities a flurry of activity occurred. But those who benefited were not the lower classes, rather it was the middle class and elites who had both the necessary financial resources and the connections to secure site, building materials and the construction crews. Thus, co-operative apartments became very much a Soviet middle class housing form. Outwardly the state-owned and cooperative apartment block were not easily distinguished. However, conventional wisdom had it that there were visible signs that the two types of housing accommodated different strata of society. In the state-owned building lights went on earlier in the morning as the inhabitants prepared to leave for work. In cooperatives on the other hand lights came on later in the morning, a reflection of the different circumstances of their presumably more professional than working social-class composition. On the eve of the collapse of the Soviet system, cooperatives accounted for close to 10 per cent of all urban housing floor space.

Privately owned housing always existed, but restrictions had long been in place to control its construction in terms of floor space and location. Ostensibly its construction was prohibited in cities of more than 100,000 population. While there were examples of well built, albeit modest scale, private homes, characteristically they were poorly maintained, were frequently not integrated into the municipal services infrastructure, such as sewer and water networks, and were invariably occupied by workers and not professionals. Within cities private housing tended to be more prevalent on the periphery than in the centre. The private housing share of the total urban housing stock was steadily eroded, and was about 20 per cent at the end of the Soviet period. But there were significant regional variations. Throughout the Caucasus and Middle Asia, for instance, it accounted for as much as 40 per cent. The land occupied by such housing belonged to the state, of course. As a general rule, the shares of private and cooperative housing varied inversely according to city size. The larger the urban centre, the smaller the share of privately owned housing and the larger the share of cooperatives. In Moscow, for example, privately owned housing accounted for only 0.03 per cent of total housing floor space in 1990.

Residential segregation occurred on both ethnic and socio-economic bases in Soviet cities, despite proclamations to the contrary. In Middle Asia urban centres were dominated by immigrant Slavs. But the indigenous population was not entirely submerged since in most cities there were many who continued to occupy houses in the traditional quarter. The situation in Samarkand was typical. The built environment in this city of nearly 400,000 people in Uzbekistan mirrored the cultural values of distinct eras as Fig. 5.4 indicates. The traditional quarter was more or less the exclusive territory of Uzbeks and reflects the centuries old tradition of social custom and architecture. The late nineteenth century colonial outpost of the Russian government comprised part of the central city. With its broad thoroughfares liberally endowed with shade trees and laid out with some attention to the classical notions of town planning of the eighteenth century, it still stands in stark contrast to the hotch-potch of streets and lanes in the traditional quarter from which it was once separated. Of course, not all indigenous Middle Asians lived in the old quarters. The standardized state apartment blocks which surround the colonial and traditional quarters were home to many, but these newer areas were dominated by migrant Slav and other nationalities from elsewhere in the Soviet Union.

Figures 5.5 and 5.6 illustrate segregation along ethnic and socio-economic lines in Kazan'. Located at the confluence of the Volga and Kama rivers in European Russia, this city of more than a million people was the capital of the Tatar Autonomous Soviet Socialist Republic. As in the case of Samarkand, the indigenous population comprised about one-third of the total. Ethnic-dominated neighbourhoods were as much a part of the social geography of Kazan' as Samarkand's (Fig. 5.5). The peripheral dominance of workers is apparent from an examination of Fig. 5.6, as is the central city orientation of the professional class. Typically, Soviet central city districts offered a higher quality of living in terms of access to cultural amenities, social and consumers services and public transportation, and were therefore desirable residential locations. As a result of the prestige of the central city and state policies dispersing many economic and administrative services throughout the city, the core area retained a substantial residential population, something which differentiated it from the central business district of western cities.

During the debate over the future Soviet socialist city in the 1920s the notion that there would be no distinguishable centre was embraced by those proposing a radical departure from past, and indeed contemporary, trends. For them the only acceptable design solution was to create an entirely new urban environment consistent with the ideals of the society being forged. The 1935 Moscow Plan, however, did not endorse such a radical departure from the reality of the existing urban form. Nonetheless, the conviction that human behaviour could be modified by the built environment, and therefore by the design process, remained firmly entrenched. There would be a city centre but it was to be the nucleus of urban social and political life, not of commerce. By means of unified and uniform architectural ensembles, thoroughfares and squares, the city centre was to cater to massive public demonstrations. The design problem was one of striking a reasonable balance between occasional public functions and the day-to-day purposes these

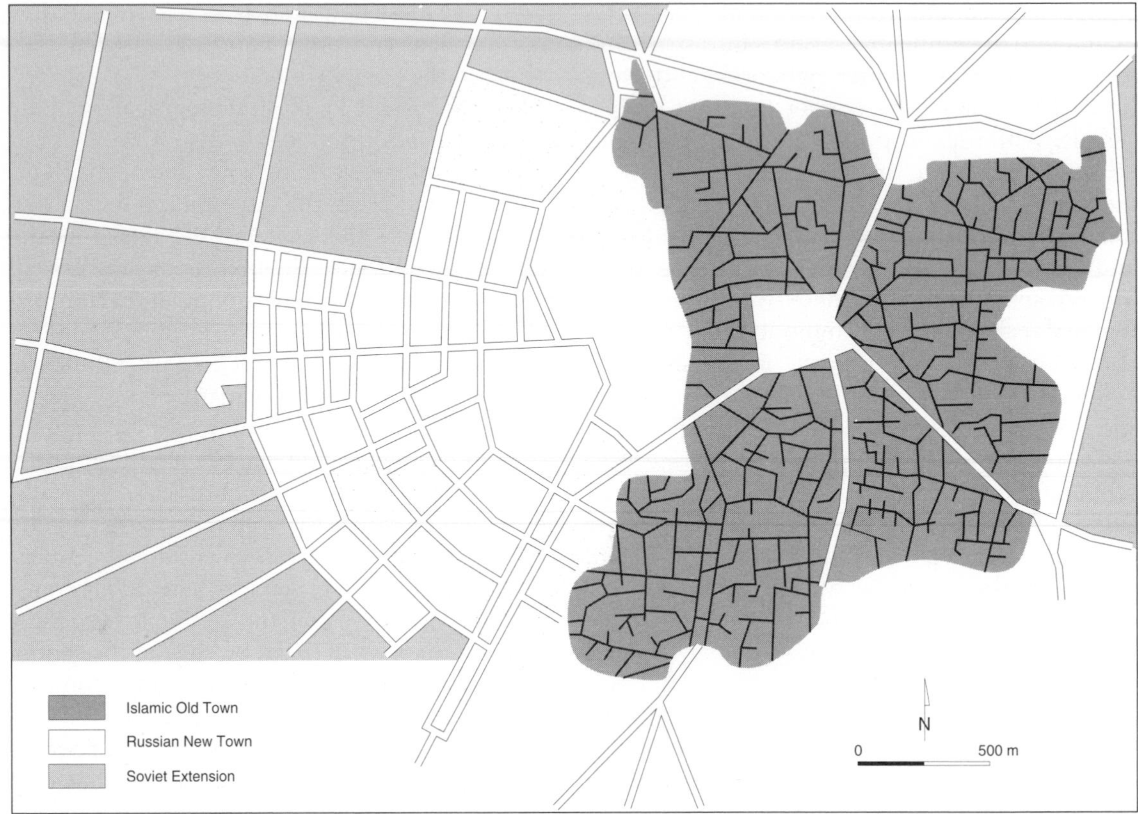

Figure 5.4 *Samarkand: three eras*
(*Source*: Based on E. Giese, 'Transformation of Islamic Cities in Soviet Middle Asia into Socialist Cities', in French, R.A. and Hamilton, F.E.I. (eds), *The Socialist City* (New York: John Wiley, 1979), 158)

same thoroughfares, squares and buildings had to serve. Thus, unified and uniform architectural ensembles and thoroughfares were imposed on central cities throughout the Soviet Union in order to accommodate periodic mass-orchestrated public demonstrations. Such orchestrated public events were deemed to be of great importance in inculcating values consistent with the state ideology. In further contrast with the west, the Soviet central city would retain a substantial residential population. This objective was easily realized since the state controlled the construction and allocation of housing. Notwithstanding the intentional decentralization of a wide array of consumer services in order to enhance its role as the locus of political and cultural life, the Soviet central city remained well provisioned in comparison with peripheral, suburban areas. It is therefore scarcely surprising that in the 'deficit culture' created by the failure of the state to meet planning norms, the comparatively well-off central city was frequently a preferred residential location amongst Soviet elites, something which the distribution of broad socio-economic groups in Kazan' implies (Fig. 5.6). Obviously the relationship between centrality and land value figured prominently in shaping land use in central cities in market economies. During the Soviet era, of course, all land ownership was vested in the state, and urban land use was determined by planners, not the market. Still, central city land was recognized as intrinsically more valuable than land on the periphery, though it was never utilized efficiently even by prevailing Soviet standards.

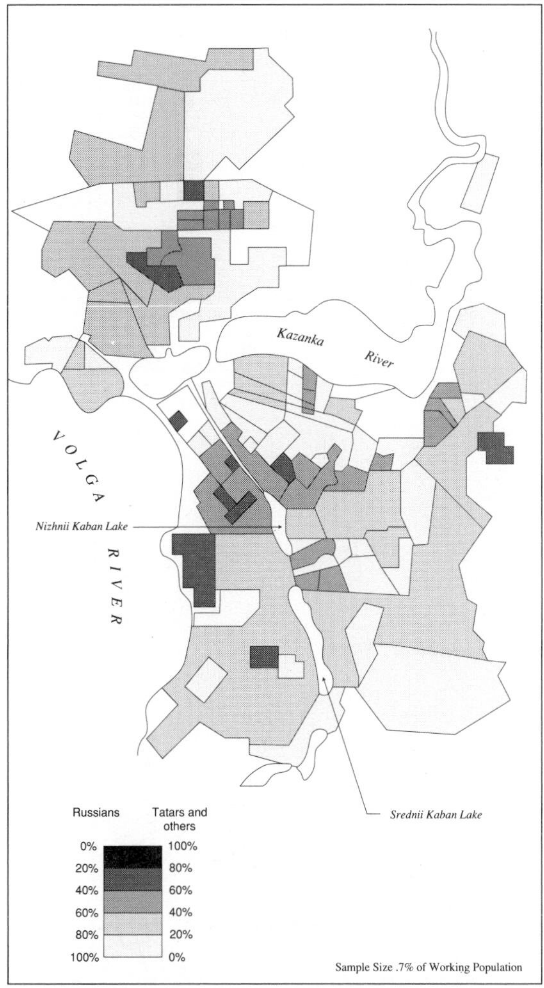

FIGURE 5.5 *Kazan: ethnic groups*
(*Source*: After V.O. Rukavishnikov, 'Ethnosocial Aspects of Population Distribution in Cities in Tataria', *Soviet Sociology*, Vol. 8, No. 2, 1978, 71)

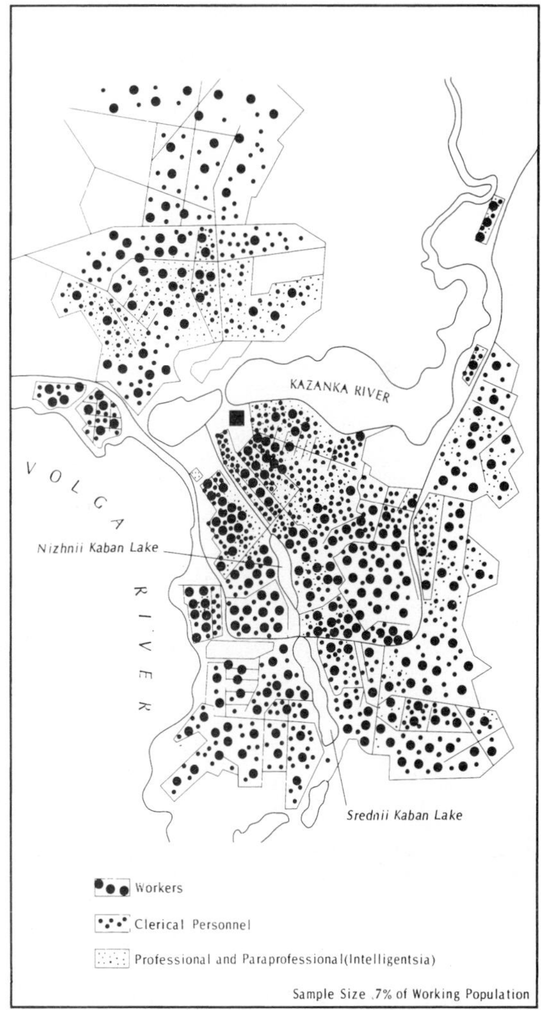

FIGURE 5.6 *Kazan: socio-economic groups*
(*Source*: After V.O. Rukavishnikov, 'Ethnosocial Aspects of Population Distribution in Cities in Tataria', *Soviet Sociology*, Vol. 8, No. 2, 1978, 70)

The dissolution of the Soviet Union has resulted in new political structures and economic reform. But what do these changes portend for cities shaped in part by decades of Soviet socialist planning? What, for example, has been the impact of market reform on the economy of the central city? What are the consequences of the development of a housing market for different social class groups in the new political economy? Is there any evidence yet of land use change? Before examining these and other issues in some detail, the urban growth process and privatization of housing across the post-Soviet scene will be described in general terms.

POST-SOVIET URBAN DEVELOPMENT IN THE 'NEAR ABROAD'

Throughout the Soviet period there was a steady increase in the urban share of the total population

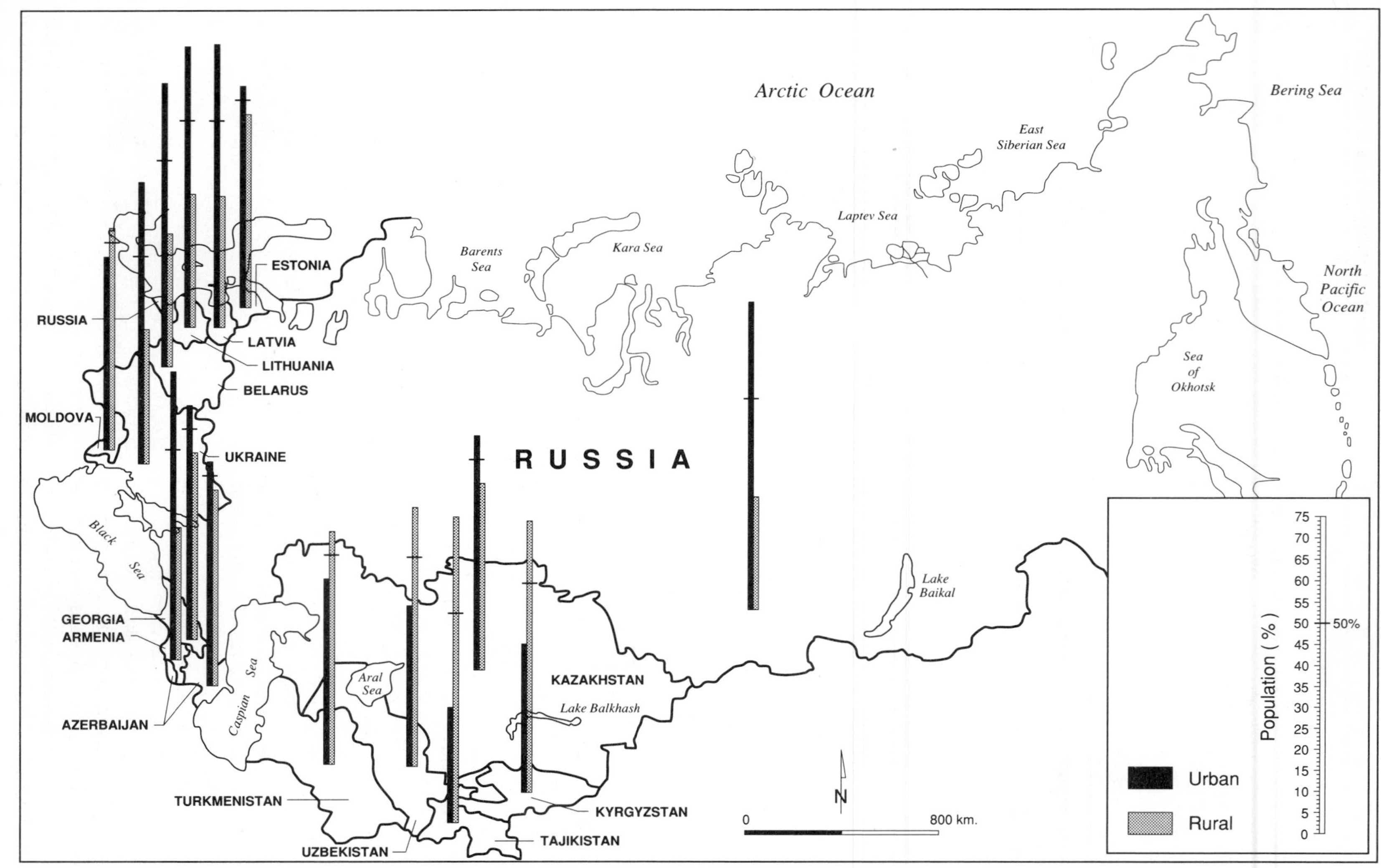

FIGURE 5.7 *Urban/rural population in post-Soviet, 1995*

(*Sources*: Compiled from data in *Demograficheskiy Yezhegodnik 1994* (Moscow: Statkomitet, 1995), 11; *Demographic Yearbook, 1994 (Vil'nius: Lithuanian Department of Statistics, 1995), 11–12; Macroeconomic Indicators of Latvia* (Riga: Central Statistical Bureau of Latvia, 1995), 6)

commensurate with the basic economic development objectives of the state. The only exceptions to this general trend occurred during periods of civil and world war (see Fig. 4.2, p. 93). On the eve of the system's demise, urbanites accounted for two thirds of the total population. The 1995 urban and rural proportions of total population by state are presented in Fig. 5.7. The western tier countries stand out clearly as the most urbanized, with Russia having pride of place in terms of the largest urban share (74 per cent). Moldova is the single exception by virtue of the importance there of the rural sector. The balance between the urban and rural shares of total population shifts quite markedly along the southern reaches of the post-Soviet scene. Indeed, in all but one of the southern tier states the rural population, not the urban, is largest. The exception is Kazakhstan, where the majority of the urban population is in any event Russian not Kazakh.

In the short period from 1990 to 1993, economic dislocation, political upheaval, civil war, forced migration, and some fundamental changes in demographic trends have interrupted seven decades of steady increase, save for periods of civil or world war, in the relative importance of the urban population. Notwithstanding an absolute increase in urban population in ten of the fifteen post-Soviet states during this period, the urban share in the majority of them declined anyway. As the data in Fig. 5.8 reveal, in 13 post-Soviet states the urban sector declined in relative importance between 1990 and 1993. In five states, the urban population actually shrank in absolute terms. Between 1993 and 1995 the absolute decline in urban population persisted, and came to characterize even more post-Soviet states. At this latter date, Russia, Ukraine, Moldova, the Baltic states of Estonia, Latvia, and Lithuania, Georgia, Kazakhstan, Kyrgyzstan and Tajikistan all had fewer urban inhabitants than in 1993. Clearly, the changes during the past few years have had a pronounced impact on the post-Soviet urban scene.

Between 1990 and 1993 the urban population increased in absolute in terms in most southern tier states, despite the sizeable exodus of the predominantly urban Russians from Tajikistan and Kyrgyzstan in particular. But natural increase in

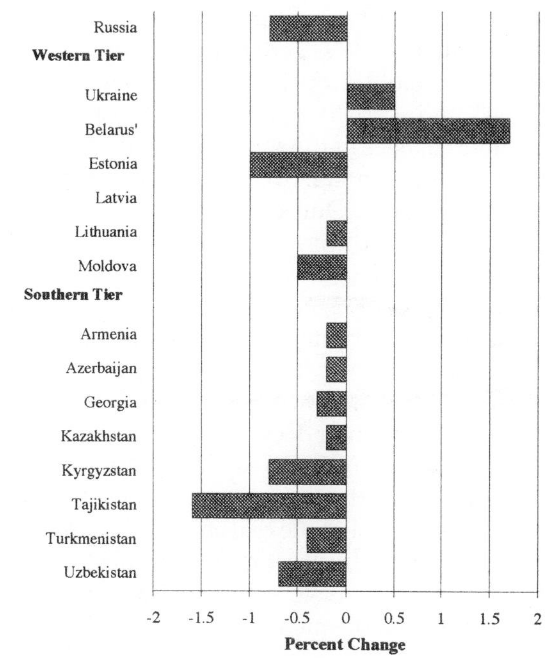

FIGURE 5.8 *Urban population change in post-Soviet states, 1990–1993*

(*Sources*: Compiled from data in *Narodnoye Khozyaystvo SSSR v 1989 g* (Moscow: Goskomstat, 1990), 19–24; *Statistical Handbook 1994. States of the Former USSR* (Washington: The World Bank, Studies of Economies in Transition, 14, 1994), 14)

the countryside was substantially higher. The largest relative decline was in Tajikistan where the urban share of the total population dropped from 32.0 to 30.4 per cent (Fig. 5.8). The relative changes were also substantial in Kyrgyzstan and Uzbekistan where the urban population declined from 38.0 to 37.2 and from 40.6 to 39.3 per cent respectively. In the Caucasus, the urban share also decreased in each state, primarily owing to a substantial increase in the number of rural inhabitants. In Georgia, there were actually 25,000 fewer urban inhabitants in 1993 than in 1990. This singular occurrence was not unique to Georgia. In Moldova, Estonia, Lithuania and Russia the urban shares dropped in both relative and absolute importance also. In Latvia, the absolute number of urban dwellers declined by 1993, but the relative share was unchanged (Fig. 5.8). Amongst western tier countries, only Ukraine and Belarus' maintained the Soviet era pattern of steady growth in the urban sector at the expense of the rural in the 1990–1993 period. In both states there was an absolute drop in the rural population: 209,000 and 139,000, respectively. Most of the negative changes in the share of the urban population depicted in Fig. 5.8 are just a fraction of a percentage point. Outside of Russia, the largest absolute drop during the 1990–1993 period occurred in Latvia, where the net loss in the urban sector was 71,000. The absolute change in Russia was of an altogether different order of magnitude, however. There the net reduction in the urban population was 1,231,000! The reasons for this quite singular decline will be examined in a later section. Suffice it to say for the time being, that urban growth across the post-Soviet scene has faltered, and that the reasons for the negative changes in the urban sector in many post-Soviet states are not amenable to short-term correction, even if that was deemed a desirable objective by government.

In many cases the demographic shift which has given rise to a significant excess of deaths over births in the city cannot be offset through in-migration, either because controls over population movement put in place during the early Soviet period have not been rescinded, or because political/economic conditions are so bleak there is little incentive to relocate from countryside to city as was so common in the past. All post-Soviet states must cope with the legacy of serious housing shortage, and this constitutes a major impediment to migration to the city where this is possible, and is one reason for the continuation of the restrictions over in-migration where the system of residence permits has been maintained. It was noted earlier that throughout Middle Asia in the late Soviet era, living space per capita was much below the national average (Fig. 5.3). The exodus from the region of several hundred thousand Russians and other non-indigenous national minorities has not dramatically eased the problem. There is still a sizeable gap between the demand for decent housing and its supply. In those areas impacted by civil war, such as Tajikistan, the situation is acute. In Dushanbe, the capital, refugees entering the city have greatly outnumbered the exodus of, amongst others, Russians. And in conditions of actual, or potential, civil war, which have prevailed in parts of the Caucasus and Moldova since independence, construction of new housing obviously does not figure amongst the priorities for government or the nascent private sector.

Even in comparatively well provisioned areas spared the upheaval of civil war, serious housing problems still exist. A case in point is Latvia. Despite the relatively high per capita allocation of living space in its cities (see Fig. 5.3), about 6 per cent of the total housing stock in Latvia is classified as unfit for habitation. And as is commonplace elsewhere in the post-Soviet scene, the indigenous population is disproportionately represented amongst the poorly housed. During the Soviet period, incoming Russian families were often given priority in housing. Thus, a larger share of Latvians occupy sub-standard accommodation in older apartment buildings than do Russians. For example, in Riga, 29 per cent of the Latvian population live in apartments built before 1918 compared to only 18 per cent of non-Latvians. A substantially larger share of Latvians than Russians live in communal apartments. Apartment buildings constructed during the past three decades or so, that is, buildings with a relatively high level of basic services, are in the main occupied by non-Latvian immigrants from the former Soviet Union. Thus, given the slowing down, if not complete cessation, of new apartment construction by the state, there remain serious shortages of decent accommodation for present inhabitants, let alone additional residents. In the Baltic states there is an additional complication. Since they were not brought under Soviet rule until World War II, there are many former owners of properties who are now entitled to reclaim them. In the case of Latvia, in the first two years of independence more than 14,000 claims for recovery of housing illegally 'nationalized' by an 'occupying power' were registered with government authorities. About 1500 of these property claims, mostly in the cities and towns, were settled with properties being returned to former owners or heirs.

The faltering growth of cities in most post-Soviet states theoretically permits some small advantage in terms of urban development planning. Clearly, where accommodating rapid growth is now less pressing, some opportunities for strategic planning of future development are presented. But in many, if not all, of the new states simply trying to cope with the problems of the day is an all-consuming task. Nonetheless, in a few places there are some positive signs. For example, the jurisdiction of urban government during the Soviet era typically did not extend past the official city border. Thus, urban-centred regions were not part of the planning tradition, notwithstanding the fact that they were long ago recognized as a high priority. There were a few notable exceptions, Moscow and St Petersburg included among them. But these were situations in which city government was especially powerful politically, and therefore could extend its influence to the oblast' or regional, level. Since the breakup of the Soviet Union, some governments have proceeded to revamp the traditional approach to planning. In Latvia, for instance, foreign technical assistance in urban and regional planning has been enlisted to put in place a system to manage development in the Riga urban region. The new Riga-centred planning region will extend far outside its 'official' city borders. It will, for example, include Jurmala, a coastal resort complex some 30 kilometres west of Riga. Jurmala is only one of many satellite urban centres in the region whose development for the first time will be linked in an integrated manner to that of the primate city.

In the Soviet period, urban government was financed in the main through transfer of funds from higher levels of government. Within cities municipal government was not 'master of its own house', as was discussed earlier. In addition to not being unable to influence the location decisions of ministries and enterprises, and not controlling all of the state-owned housing within its borders, in a number of cities parts of basic municipal service systems such as water, sewer and public transportation were built and operated by ministries and enterprises. City consumer, cultural, and even health care facilities were regularly outside the authority of urban government for the same reason as well. As privatization expands, as the state sector is eroded in importance throughout the post-Soviet scene, what used to be viewed by ministries and enterprises as assets in attracting and holding labour are now often perceived to be costly liabilities. Shifting responsibility for the ownership and maintenance of housing to urban government is a case in point.

The data in Table 5.1 indicate that privatization of state sector housing was quite variable in its impact by 1992. In Russia, while only about 8 per cent of the urban housing stock had been converted from some type of state ownership to private, this was still more than in either Ukraine or Belarus'. But since 1992 there has been a significant increase in the private sector housing share in Russia, as will be discussed shortly. Although the information for the Baltic states is incomplete, if it is assumed that the data for Lithuania are representative of what has occurred across the region, then obviously there was not much change in the level of private ownership during the first couple of years of independence (Table 5.1). Thereafter, the change was significant. Where an increase is also evident was in Georgia, Uzbekistan and Kazakhstan. Even during the Soviet era, the Caucasus in general and Georgia in particular had highly developed systems of non-state activity. These ranged from legal activities such as personal auxiliary farming and professional services offered by medical doctors, to illegal black market trading. More than a third of Georgia's urban housing stock was privately owned even before the government's decision to proceed with a major state housing privatization initiative. Nonetheless, in the short time from March 1992 to early 1993, an additional 28 per cent of the state urban housing stock was converted to private ownership. By simply transferring ownership free of charge, save for minimal transfer and registration fees, and without restrictions as to when the privatized housing might be sold, Georgian authorities ensured rapid transfer to the private sector. Elsewhere restrictions have been placed on what may be done with newly privatized housing. In Turkmenistan, for example, privatized housing may not be sold for ten years, a policy intended to block the disposal at profit of apartments by people intending to depart the

Table 5.1 *Privatization of urban housing stock in post-Soviet states, 1990–94*

State	Per cent of total floor space				
	1990	**1991**	**1992**	**1993**	**1994**
Russia	0.2	0.3	8.1	24.0	32.0
WESTERN TIER					
Ukraine	0.6	1.7	–	–	–
Belarus'	0.4	2.4	4.4	12.5	13.8
Estonia	–	–	–	–	–
Latvia	–	–	–	–	–
Lithuania	9.2	10.3	–	82.0	93.0
Moldova	1.0	–	–	–	–
SOUTHERN TIER					
Georgia	–	–	>50.0	>50.0	>50.0
Armenia	1.8	2.2	–	–	–
Azerbaijan	0.4	0.9	4.4	–	–
Kazakhstan	1.3	6.1	32.0	42.4	–
Kyrgyzstan	1.8	1.5	19.4	–	–
Tajikistan	2.7	4.2	–	–	–
Turkmenistan	0.8	1.7	32.0	33.0	–
Uzbekistan	2.2	2.4	25.2	90.5	–

Sources: *Strany-Chleny SNG, Statisticheskiy Yezhegodnik.* (Moscow: Finstatinform, 1993); *Rossiyskiy Statisticheskiy Yezhegodnik, 1994* (Moscow: Goskomstat, 1995), 231; *Lithuania's Statistical Yearbook, 1992* (Vilnius; Lithuanian Department of Statistics, 1993), 97; *Key Economic Indicators in Lithuania, 1993–1994* (Washington: The World Bank, unpublished report, 1994), 3; *Lietuvos Ekonomikos Apzvalga, 1995 m. geguze* (Vilnius: Statistikos Departamentas Prie Lietuvos Respublikos Vyriausybes, 1995), 88, 89; *Respublika Belarus' v Tsifrakh, 1994* (Minsk: Ministerstvo Statistiki i Analiza Respubliki Belarus', 1995), 205; *Statisticheskiy Byulleten', No. 1, 1995* (Almaty: Gosudarstvennyy Komitet Respubliki Kazakhstan po Statistike i Analizu, 1995), 79; *Narodnoye Khozyaystvo Respubliki Uzbekistan v 1993 g* (Tashkent: Gosudarstvennyy Komitet po Prognozirovaniyu i Statistike Respubliki Uzbekistan, 1994), 167, 169; *Turkmenistan v 1993 godu. Statisticheskiy Yezhegodnik, Tom 1* (Ashgabat: Goskomstat Turkmenistana, 1994), 88, 89.

state at the earliest possible opportunity. In Uzbekistan and Kazakhstan, where there has also been a quite substantial shift in ownership of housing from the state to personal ownership since 1990, the approach to privatization is much less restrictive. For example, in Tashkent, the capital of Uzbekistan, 160,000 apartments were privatized in 1992 alone. One-half of them were transferred free of charge, the balance at a nominal cost. The government of Kazakhstan has placed a high priority on the privatization of all state assets in its attempt to modernize and restructure the economy. Privatization of housing in Kazakhstan has certainly been substantial (Table 5.1). While a large share of housing has been turned over free of charge, the procedures from one city to another in Kazakhstan are far from consistent. In those places where apartments are to be sold to tenants for cash or exchanged for tenant's privatization vouchers, the privatization process has been understandably slow. Nonetheless, by 1995 the privatization of housing in many places across the post-Soviet scene was significant.

Everywhere in the post-Soviet scene, the question of who will pay for apartment building maintenance looms large in the minds of tenants and owners alike. Since much of the housing stock is in need of major capital repair this is not an inconsequential matter. It is sometimes assumed that by remaining a tenant, rather than assuming ownership of an apartment, the maintenance costs will continue to be paid by the building owner. What is tending to happen is that a growing proportion of the cost of maintenance is being transferred to apartment occupants, irrespective of their status, through increased rent or special fees. Not surprisingly, the privatization of housing has been most expeditious where costs of assuming ownership have been minimal, where restrictions on what can be done with the property are few, and where information as to maintenance responsibilities and costs have been clarified.

Market reform and urban development in Russia

The urban share of population of Russia reached its zenith of 74 per cent in 1989. It remained at that level for the next couple of years and then began to decline. The absolute drop of 1,231,000 urban residents between 1990 and 1993 is certainly indicative of the turmoil produced by the dissolution of the Soviet system. The reasons for the contraction in the number of people living in Russia's cities are varied, but most important is the simple fact that the urban death rate exceeded the birth rate by a wide margin. Traditionally, a 'natural' demographic deficit in a particular city would be offset by the continuing sizeable in-migration from the countryside. However, a diminished propensity to leave the village for possibly an even more uncertain life in the city, and the less than accommodating attitude toward potential rural migrants on the part of city authorities in many parts of Russia, have reduced the volume of rural–urban migration in the recent past.

It was observed in the preceding chapter that there was net out-migration from most of the northern and eastern regions in Russia in 1992 (see Fig. 4.11, p. 115). In many remote, resource-based settlements across these vast territories it is no longer possible to sustain production in the new economy. In contrast to the Soviet era when regional development policies promoting the northern and eastern regions gave rise to large subsidies for industry and substantial wage incentives for workers, the Russian government now has a northern resettlement programme which pays families to move out of the region. In many places subsidies are being scrapped, industries are folding, and jobs paying high salaries are fast becoming just a memory. The government has come to the rather obvious conclusion that it is far cheaper to provide the necessary services for redundant or displaced workers in a southern location than it is to support northern settlements now without an economic base. Thus, out-migration has been sizeable. While many migrants are relocating to cities elsewhere in Russia, some are returning to the village. Still others are leaving Russia altogether for their national homelands in the 'near abroad'. And, of course, the latter two responses to present uncertainties are not restricted to residents of remote, high cost of living settlements. Of lesser importance in accounting for the absolute decline of the urban population in Russia between 1990 and 1993 is the reclassification of some small, urban-type settlements as rural. This is typically the result of a change in the employment structure consequent upon market reform in general and the demise of previously subsidized, inefficient industries in particular. Urban-type settlements are currently defined as having a minimum of 2–3000 inhabitants in which 85 per cent of the labour force is engaged in non-agricultural sectors. There were just over 2100 such places in Russia in 1993.

The vast majority of Russia's 89 political-administrative regions were more than 65 per cent urban in 1993 (Fig. 5.9 and Fig. 9.12, p. 280). Only a handful did not have a majority of their population in the urban sector. These were mostly non-Russian ethnic areas located along the northern flank of the Greater Caucasus mountains, and in Siberia. Underlying the general pattern depicted in Fig. 5.9 is a system of urban places heavily skewed towards the large city. More than a quarter of all urbanites lived in Russia's 13 cities of one million or more, and close to 40 per cent in cities of 500,000 or more (Table 5.2). Despite the Soviet-era policies to contain the growth of large cities, they have steadily garnered a greater share of the population. As yet there is no urban settlement development strategy for all of Russia comparable to the General Scheme for the System of Settlement in the Soviet Union put forward in the 1970s. There have been a number of suggestions

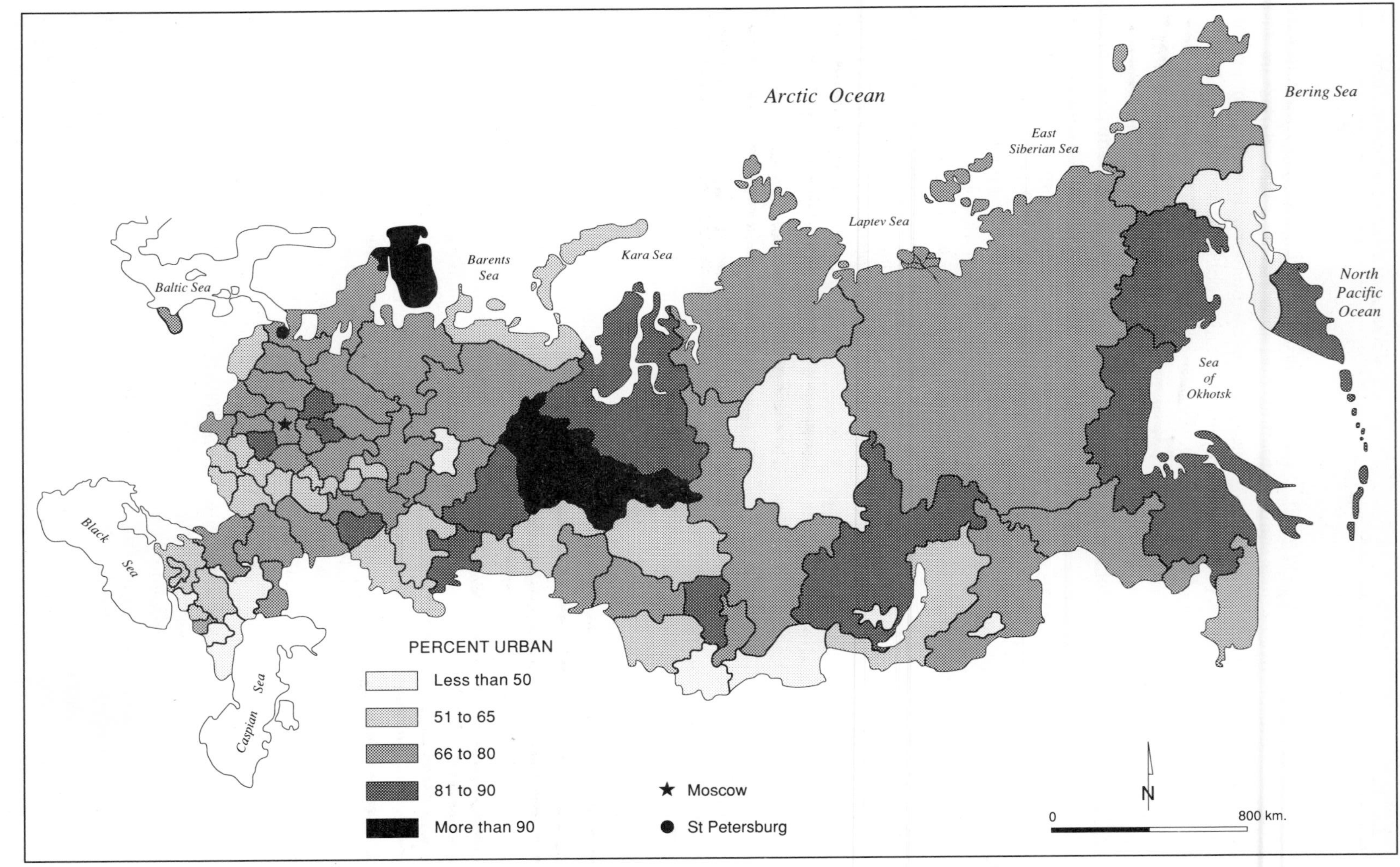

FIGURE 5.9 *Urban population: Russia, 1993*
(*Source*: T. Heleniak, *Economic Geography of the Russian Federation* (Washington: The World Bank, Country Department III, Europe and Central Asia Region, 1994), Vol. 1, No. 5)

Table 5.2 *Urban population in Russia by city size group, 1979–1994*

<table>
<tr><th>City Size Group</th><th colspan="2">1979</th><th colspan="2">1992</th><th colspan="2">1994</th></tr>
<tr><th></th><th>Millions</th><th>% of Total</th><th>Millions</th><th>% of Total</th><th>Millions</th><th>% of Total</th></tr>
<tr><td>Less than 3000</td><td>1.0</td><td>1.1</td><td>1.1</td><td>1.0</td><td>1.0</td><td>0.9</td></tr>
<tr><td>3000–4999</td><td>2.3</td><td>2.4</td><td>2.1</td><td>1.9</td><td>1.9</td><td>1.8</td></tr>
<tr><td>5000–9999</td><td>5.2</td><td>5.5</td><td>5.5</td><td>5.4</td><td>5.4</td><td>5.0</td></tr>
<tr><td>10,000–19,999</td><td>6.9</td><td>7.3</td><td>7.9</td><td>7.2</td><td>7.7</td><td>7.2</td></tr>
<tr><td>20,000–49,999</td><td>11.8</td><td>12.4</td><td>12.5</td><td>11.4</td><td>12.3</td><td>11.4</td></tr>
<tr><td>50,000–99,999</td><td>9.3</td><td>9.8</td><td>11.5</td><td>10.5</td><td>11.6</td><td>10.8</td></tr>
<tr><td>100,000–249,000</td><td>13.0</td><td>13.7</td><td>13.4</td><td>12.3</td><td rowspan="2">29.2</td><td rowspan="2">27.2</td></tr>
<tr><td>250,000–499,999</td><td>13.8</td><td>14.5</td><td>15.7</td><td>14.4</td></tr>
<tr><td>500,000–999,999</td><td>12.7</td><td>13.4</td><td>13.3</td><td>12.2</td><td>13.7</td><td>12.7</td></tr>
<tr><td>1,000,000 and over</td><td>18.9</td><td>19.9</td><td>26.2</td><td>24.0</td><td>24.7</td><td>23.0</td></tr>
<tr><td>Russia, total</td><td>94.9</td><td>100.0</td><td>109.2</td><td>100.0</td><td>107.5</td><td>100.0</td></tr>
</table>

Sources: *Narodnoye Khozyaystvo Rossiyskoy Federatsii*, 1992, (Moscow: Goskomstat, 1992), 9; *Rossiyskiy Statisticheskiy Yezhegodnik 1994* (Moscow: Goskomstat, 1995), 25

as to how potentially strategic, but little developed, regions in Russia could be put to better use, however. For instance, there has been some publicity accorded to a very preliminary scheme for resettling several hundred thousand former military personnel in new urban settlements in the still lightly populated corridor between Moscow and St Petersburg. While such schemes may well have some merit, long-term strategic settlement planning is simply not a priority for a government trying to cope with contemporary economic, social and political problems. Meanwhile, the market reform and privatization initiatives are having a significant impact on the urban scene across Russia.

Amongst the first components of the state-run economic system to be privatized were the retail trade and public catering enterprises. By late 1993 about one-third of all such establishments in Russia had been sold, primarily to previous employees and managers. However, the extent to which these predominantly urban facilities had been privatized varied from one region to another (see Fig. 3.8, p. 84). Two cities stood out as leaders – St Petersburg with about 75 per cent of its retail trade and catering enterprises privatized, Moscow with close to two-thirds. The tempo of privatization was very much influenced by the terms and conditions set forth as conditions of sale. Needless to say, where authorities imposed fewest restrictions, the transition from state to private ownership was rapid, and the share of total employment in the non-state sector increased dramatically. In St Petersburg, for example, it increased more than eight-fold to 15.4 per cent between 1991 and 1992. In Moscow the increase was even more substantial. Non-state employment accounted for 27 per cent of the total in 1992. Indeed, in few other parts of the country was the change so swift. Only in Nizhniy Novgorod was there a comparable change in the structure of employment. In this city of nearly 1.5 million people to the east of Moscow, government authorities had adopted a particularly aggressive position with respect to turning city-owned assets over to private owners, as noted in an earlier chapter. However, what was done in the capital city influenced decision-makers elsewhere. And in Moscow there have been decidedly reformist City Councils and Mayors since the advent of the era of the new economy. Privatization of Moscow's municipal economy illustrates well what is happening throughout urban Russia.

In 1992, almost 27 per cent of Moscow's 4.363 million strong labour force was employed in the non-government sector, up from barely 22 per cent a year earlier. What is interesting is that according to official data the total active labour force in 1992 was more than one-half million members smaller than the year before. It is estimated that about 288,000 of this apparent loss in labour force numbers are accounted for by the non-productive branches of the urban economy, that is, the service sector. The official number of unemployed in the city was only 23,500 in 1992, and therefore this group does not account for more than a fraction of the reduction in the number of employed. For many people no longer officially counted amongst the gainfully employed, informal and unregistered street trading provided an income, and one often considerably larger than that which can be earned through working for a government sector enterprise or institution. Something of the tempo of the privatization process is conveyed by the data presented in Fig. 5.10. Clearly, the increase in employment in the non-state sectors between 1993 and 1994 was significant. In the latter year, the state accounted for less than one-half of the total employment for the first time. By 1995, the proportion of the labour force in the non-government sector was fast approaching two-thirds of the total. The labour force was also larger in absolute terms in 1995 than in 1992. In part this was because more

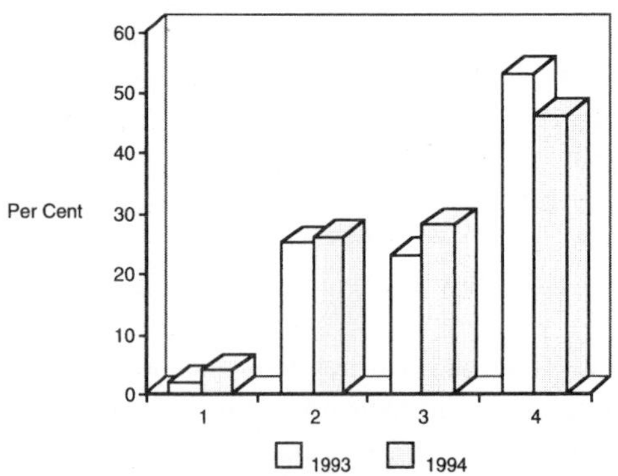

FIGURE 5.10 *Employment in state and private sectors: Moscow, 1993–1994*
(*Source*: Compiled from data in *Moskva v Tsifrakh '94* (Moscow: Moskovskoy Gorodskoy Komitet Gosudarstvennoy Statistiki, 1995), 9)

people in the private sector were caught up in the statisticians' surveys, but also because difficult economic conditions had forced many pensioners back into the workforce in order to bolster their incomes.

The extent of privatization of municipal enterprises by sector and by administrative region (see Fig. 5.13) of the city in mid-1993 is detailed in Table 5.3. Close to one-half of the 10,166 enterprises sold by the city was in the retail trade sector. Public catering facilities and consumer services were the next most important sectors, and taken together these three sectors accounted for more than 98 per cent of all privatized municipal establishments. For the city's transportation and construction industries, the respective shares privatized were 28 and 25 per cent. However, in terms of the value of privatized municipal assets, the 159 construction enterprises ranked first amongst the six sectors listed in Table 5.3. They were valued at 4390 million roubles. The more than 4700 privatized retail trade establishments had a combined value of 3934 million roubles. By late June 1993, municipal assets with a total value of 10,392 million roubles had been privatized, and close to one-quarter million Muscovites had become at least part owners of a business. By this date nearly a million other people were employed in the non-government sector. From the perspective of Moscow as a whole, the central city stands out in terms of privatization. That it should figure so prominently in both the privatization of existing enterprises and in the creation of new ones is scarcely surprising. The economic reform has not lessened in any way its inherent locational advantages. As elsewhere, central city accessibility, infrastructure, and the circular and cumulative impact of an expanding complement of service sector functions simply increase the competition for, and cost of, central city space. The share of the private sector has increased steadily since 1991. In 1995, more than 60 per cent of the approximately 15,200 retail trade and consumers services establishments were in the private sector.

In Moscow there have been opportunities to invest and participate in business, and not solely in small scale, consumer service-oriented

Table 5.3 *Number of privatized municipal enterprises, Moscow, June 21, 1993*

Region	Sector					
	Trade	Catering	Consumer services	Goods transport	Industry	Construction
Central	929	342	744	–	–	–
North	460	74	270	–	–	–
Northeast	395	154	229	–	25	–
East	509	103	289	–	–	–
Southeast	398	115	263	–	–	–
South	474	94	300	–	–	–
Southwest	408	94	252	–	–	–
West	352	68	252	–	–	7
Northwest	210	38	126	–	–	–
Zelenograd	44	6	33	–	–	–
Not specified	570	586	684	97	20	152
Total	4749	1674	3442	97	45	159

Source: Operativnye dannye o khode privatizatsii v Moskve na 21.-6-93, svedeniye (Moscow: Meriya, Pravitel'stvo Moskvy, Otdel Informatsii, 1993), unpublished data base

enterprises. For example, managers of the huge ZIL automotive plant put on the auction block more than one million shares of its stock in the spring of 1993 and attracted buyers from across Russia. Many research institutes have now developed marketing strategies and created business plans for their wares. Consulting firms abound, as do joint ventures with foreign companies. Often what is marketed is the ability to do business in Russia, that is, to provide appropriate introductions to the right people in the right places. Given Moscow's role as the national capital this often means nothing more than providing an opportunity to gain the ear and possibility the endorsement of key bureaucrats, of whom there remain literally tens of thousands. While there have been many opportunities for Muscovites to participate in the privatization process, it has still not made much inroad into the state-owned, as opposed to municipally owned, sector, especially the military-industrial complex.

Along with privatized retail shops scores of thousands of Muscovites participate in informal retail trade on the streets either out of choice or necessity. Indeed, there is more than a little truth to the quip that Russia's capital has become one vast market since street trading was legalized in the spring of 1992. Attempts to organize and control street trading, especially in central Moscow, by establishing large scale 'market' venues such as that at the Luzhniki sports complex have had limited success. For probably the majority of street vendors such informal business is little more than a necessary supplement to other sources of income. However, for many Muscovites street trading is now the only source of income, and not necessarily a modest one. For not a few people such petty trading is now big business since it provides considerable scope for criminal activities ranging from extortion to disposal of stolen goods. How Muscovites earn a living in the new economy is certainly an important dimension of the privatization process, but as important is the question of how privatization is affecting the fundamental issue of shelter.

Legislation to permit the privatization of state-owned housing in Russia was introduced in July 1991. The law permitted occupants of apartments owned by ministries, enterprises and local soviets, that is, local governments, to acquire personal ownership of 27 square metres of useful space per family member free of charge. Useful space is the total floor space of an apartment as distinct from living space which includes only principal rooms. The legislation suggested that allocations above this threshold be paid for, but a fair degree of discretion was permitted in this regard. As a result, some local administrations with resolutely reformist governments permitted sitting tenants to privatize their apartments free of charge irrespective of the amount of useful space per family member. This occurred in the city of Moscow, and in Amursk, Smolensk, Saratov and Tula Oblasts. Another ten or so political-administrative authorities established thresholds of about 40 square metres of useful space per family member, that is, approximately 50 per cent higher than the federal legislation recommended. However, the vast majority of Russia's 89 political-administrative regional authorities followed fairly closely the recommended guidelines.

Privatization of apartments was in fact quite slow in the months immediately after the legislation was introduced, even where restrictions were minimal or non-existent. In part this was due to some initial administrative difficulties in putting procedures in place, but more generally there was hesitancy amongst occupants of state-owned apartments owing to uncertainty concerning the obligations privatization would bring. The central concerns were maintenance and tax liabilities. Thus, across Russia only 122,000 apartments out of a total stock of nearly 34 million were transferred from the state to private ownership from July to December 1991. The initial impact of privatization in the housing sector was obviously minuscule.

The pattern of privatization of apartments up to September 1993 is presented in Fig. 5.11. By this date about 21 per cent of Russia's apartments had been privatized. Political-administrative areas in the North Caucasus, Central European Russia and south central Siberian regions stand out with more than 25 per cent. But the Adygea and Karachay-Cherkessiya Republics and Stavropol' Kray in the North Caucasus, and the Altay Republic in West Siberia were clearly the leaders nationally, as all had in excess of 40 per cent of their apartments in private hands by autumn 1993 (see Fig. 9.12, p. 280). At the other end of the privatization continuum were Ul'yanovsk Oblast' and the Tatarstan Republic in the Povolzhskiy region, and Sakha Republic in the East Siberian region. Tatarstan Republic had pride of place amongst the laggards with a mere 0.7 per cent of apartments in private ownership. Not surprisingly, it was the jurisdictions with fewest restrictions that led in the privatization process. Moscow, where the transfer was free of charge and without restrictions, ranked first in absolute number of apartments transferred to sitting tenants. By late 1993 more than 30 per cent of the city's residents owned their apartments, and by April 1995 nearly 56 per cent of the city's apartment floor space had been privatized. Privatization of housing in Moscow has proceeded more quickly than in any other large city. Given the demonstration effect of market reforms in Moscow throughout Russia, it is again worth examining what has happened there in more detail.

Moscow's housing stock comprises nearly 3.2 million apartments, and a relative handful of *kottedzhi*, the Russian version of single family homes, the vast majority of which are of post-Soviet vintage. Thus, in Moscow as elsewhere in urban Russia, home is invariably an apartment. Nine-tenths of all apartments are in buildings constructed since 1956; most of them built quickly and often badly. A shade more than one-fifth of the housing stock consists of single-room apartments, about 46 per cent have two rooms, and 28 per cent three rooms. Apartments having four or more rooms account for only 3 per cent of the total, or about 96,000 units. A good many of the apartments with four or more rooms are part of the pre-revolutionary housing inherited by the Soviet government, and are located mostly in buildings in central Moscow. Some Soviet-era apartment buildings constructed for members of the former Soviet elite also comprise part of this small, but important segment of the housing stock. Moscow is no different than any other city in Russia; it has a serious housing shortage. In early 1994 some 740,000 Muscovites, or more than 8 per cent of the total population, were on the list for improved housing. Of this number, 123,000 had been waiting for an apartment for ten years or more. Nearly 250,000 people were living in dormitories, about 78,000 of whom were children. Some 130,000 families still lived in communal apartments, typically in one room and sharing kitchen, bathroom and toilet facilities. Perhaps worst off were the 21,600 families occupying apartments in nearly 2200 buildings officially

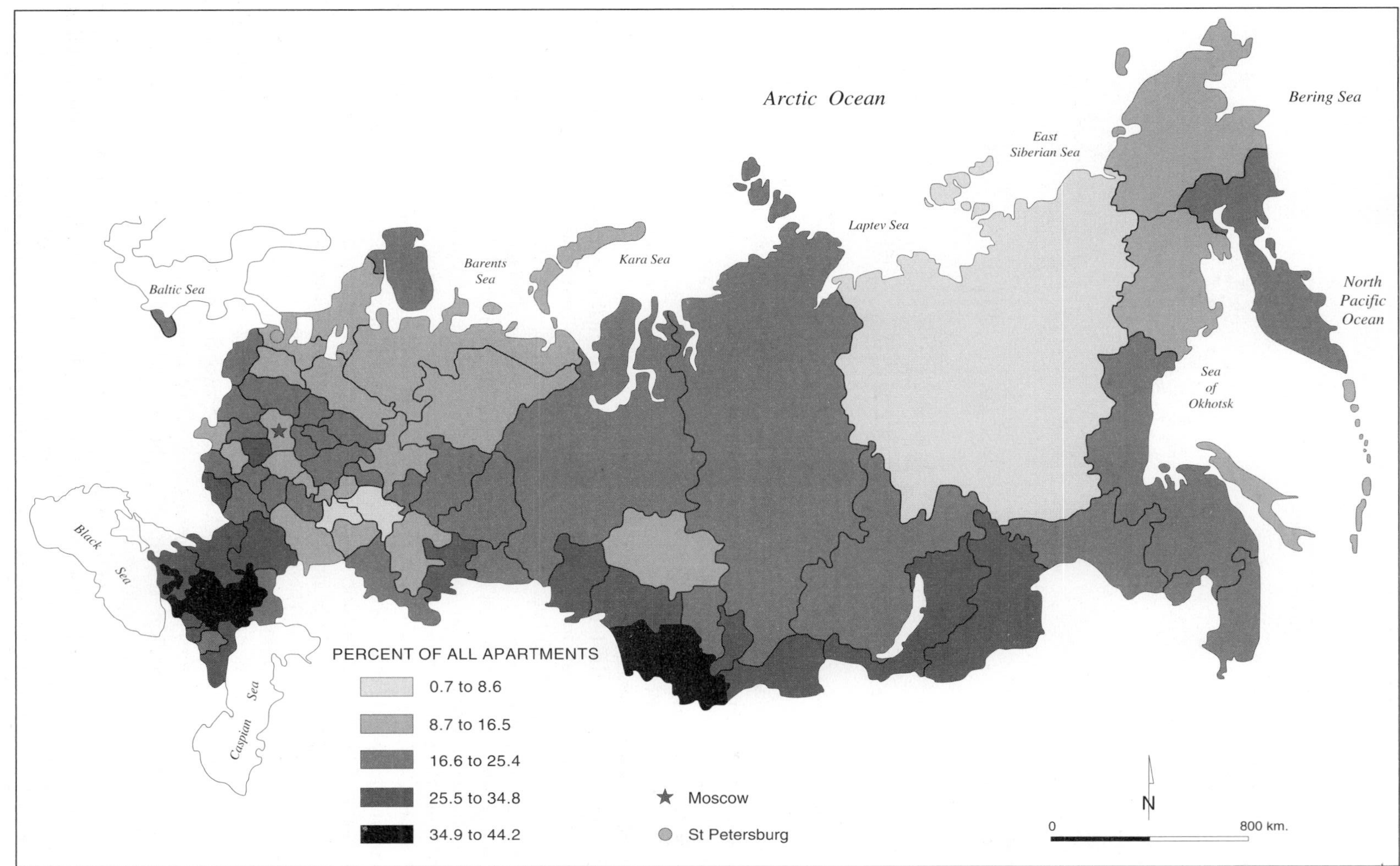

FIGURE 5.11 *Privatization of apartments: Russia, 1993*
(*Source*: T. Heleniak, *Economic Geography of the Russian Federation* (Washington: The World Bank, Country Department III, Europe and Central Asia Region, 1994), Vol. 1, No. 20)

Table 5.4 *Moscow housing space by Prefecture, January 1, 1992*

Prefecture	Useful space (thousand sq. m)	Living space (thousand (sq. m)	Per capita useful space (sq. m)	Per capita living space (sq. m)	Share living space of useful space per capita (%)
Central	17052.7	10867.8	25.5	16.2	63.5
North	17638.5	11298.1	18.2	11.7	64.3
Northwest	10813.3	6869.1	17.8	11.3	63.5
Northeast	19292.0	12179.5	17.3	10.9	63.0
South	23766.0	15011.4	17.6	11.1	63.1
Southwest	18149.4	11650.0	19.0	12.2	64.2
Southeast	14726.9	9393.7	17.4	11.1	63.4
West	17318.3	10866.5	17.8	11.1	62.4
East	21384.7	13767.2	17.7	11.4	64.4
Zelenograd	3025.8	1889.5	17.7	11.1	62.7
Moscow	163167.6	103792.8	18.4	11.7	63.6

Source: *Administrativnye Okruga Goroda Moskvy v 1991 godu* (Moscow: Goskomstat RF, 1992), 58

designated as unfit for habitation. Thus, there is considerable variation in apartments in terms of number of rooms, total floor space, quality of space, age of building and neighbourhood amenity. Not surprisingly, within Moscow there are widely recognized differences in social class standing amongst city districts.

Table 5.4 provides some data on the distribution of housing space for each of the ten Prefectures, or political-administrative areas, in Moscow (see Fig. 5.13). In the Central Prefecture each resident has, in absolute terms, more housing space than anywhere else. Only in the Southwest Prefecture do residents have average per capita housing allocations higher than the average for the city. In all other Prefectures, the citizens have less. Clearly, if egalitarian principles had been more consistently observed in housing allocation practice during the Soviet era, there would be much less variation amongst the Prefectures. The singular position of residents of the Central Prefecture in terms of average per capita housing allocation must be seen in the context of its social class structure. Historically, the western part of the city was a preferred residential location in comparison to the eastern regions, and a residence in the centre was preferred over one on the periphery. The ability of the Soviet state to dictate what was built and where, and to allocate housing as a perquisite helped to accentuate the social class status of the central part of the capital. The 1971 General Plan for the development of Moscow also played an important part in this process. It designated the central city core region a cultural and prime residential area as opposed to the earlier town planning emphasis on locating government and administrative functions there. As unskilled and skilled workers in significant numbers were relocated from the central city to new housing on the periphery, more housing in the centre was made available for elites, and thus their relative importance as a social class in the city core increased accordingly.

A 1993 survey of a sample population from two of the ten municipal districts comprising the Central Prefecture revealed considerable differences in housing conditions (Fig. 5.12). About 10 per cent of the respondents lived in one-room apartments, 35 per cent in two-room units and 23 per cent in three rooms or more. In this latter category were certainly to be found representatives of social class elites. That said, the most telling statistic was that nearly one-third of all respondents lived in communal apartments! For the residents of the Central Prefecture to have, on average, the highest per capita housing allocation clearly means that elites there are extremely well housed. While many Central Prefecture residents still live in single rooms in large apartments decades ago converted to communal usage, many such apartments are occupied by single families of the former Soviet elite. Indeed, some apartment buildings in the Central Prefecture constructed during the Soviet era were specifically ear-marked for this component of Soviet society and are similarly spacious. Put simply, it is entirely possible that the Central Prefecture has both some of the worst and the best of housing in Moscow. While there were always districts commonly associated with elites or the lower class, the market reform and privatization of housing have the potential to further differentiate this facet of the urban landscape.

Although privatization of municipally-owned apartments in Moscow dates from 1989, it was only in early 1992 that the process began to gather real momentum. Legislation introduced at the beginning of 1992 enabled tenants in apartments owned by agencies of the federal government to participate in the privatization process. In the spring of 1992, people who had a cooperatively-owned apartment were encouraged to privatize even though they already 'owned' their residence. This was to ensure that no questions would arise in future regarding their status as owners. As remarked earlier, cooperatively-owned apartments were 're-legalized' in the Soviet Union in 1962, after having been nationalized by the state in the late 1930s. By the time of the collapse of the Soviet system, about a tenth of Moscow's housing stock was cooperatively owned. About 72 per cent was owned by the city (roughly three times the average across Russia), and the balance was owned by other higher level government ministries and departments. Only a fraction of one per cent of the housing was privately owned, as noted earlier as well. To the end of 1991, less than 4000 apartments had been privatized throughout Moscow; a year later the total exceeded 365,000 units. By early 1995 the majority of Muscovites had legal title to their apartments. Quite conceivably the quality of apartment occupied had some relationship to the propensity to privatize. Conventional wisdom has it that those who had done well in the Soviet era in terms of housing were amongst the first to take advantage of the privatization legislation. For example, renting out spacious, privately-owned apartments in central Moscow for substantial sums in US dollars is not uncommon.

Privatization of housing has given rise to a vibrant and highly speculative market for apartments. Many citizens who now have legal title to their home have become 'wealthy', on paper at least, by virtue of the seemingly insatiable demand for apartments in the capital, and the escalation in prices for them. Figure 5.12 provides some insights into the spatial variation in prices for one, two and three-room apartments in 1993. Prices quoted in Moscow are typically in US dollars per square metre of useful space, not in roubles. Useful space includes all floor space in an apartment, that is, principal rooms, corridors, bathroom, kitchen and storage. From Fig. 5.12 it is apparent that there is the customary price gradient from the centre of city to the periphery. Most of the Central Prefecture fell within the highest cost area for apartments, that is, US$ 900–1100 per square metre of useful space. Indeed, prices there have reached levels comparable to those fetched in the major West European capitals. The West and Southwest Prefectures shared the most extensive tract of land falling in the next highest apartment price category. The least desirable apartment location was in the heavily industrialized and ecologically problematic Southeast Prefecture, where on average apartments were typically fetching less than US$ 250 per square metre of useful space. The price for an apartment is related to the metro system, for easy pedestrian access to a metro station brings a premium. The price asked per square metre for apartments in close proximity to metro stations can be quite a bit higher than the average cost for the region in which they are located, as the data presented in Fig. 5.12 indicate.

For those individuals who have privatized their apartments at effectively no cost, the incentive to sell and relocate to a less expensive location either in Moscow, or elsewhere, is certainly substantial. There are many people seeking to improve their housing and are prepared to spend significant amounts of money. Clearly, such real estate transactions belong to the world of the newly rich or to firms seeking a central city office location. While it is frequently foreign companies or joint ventures which are involved in setting up offices, by no means are Russian companies simply onlookers. For those who are still obliged to live in

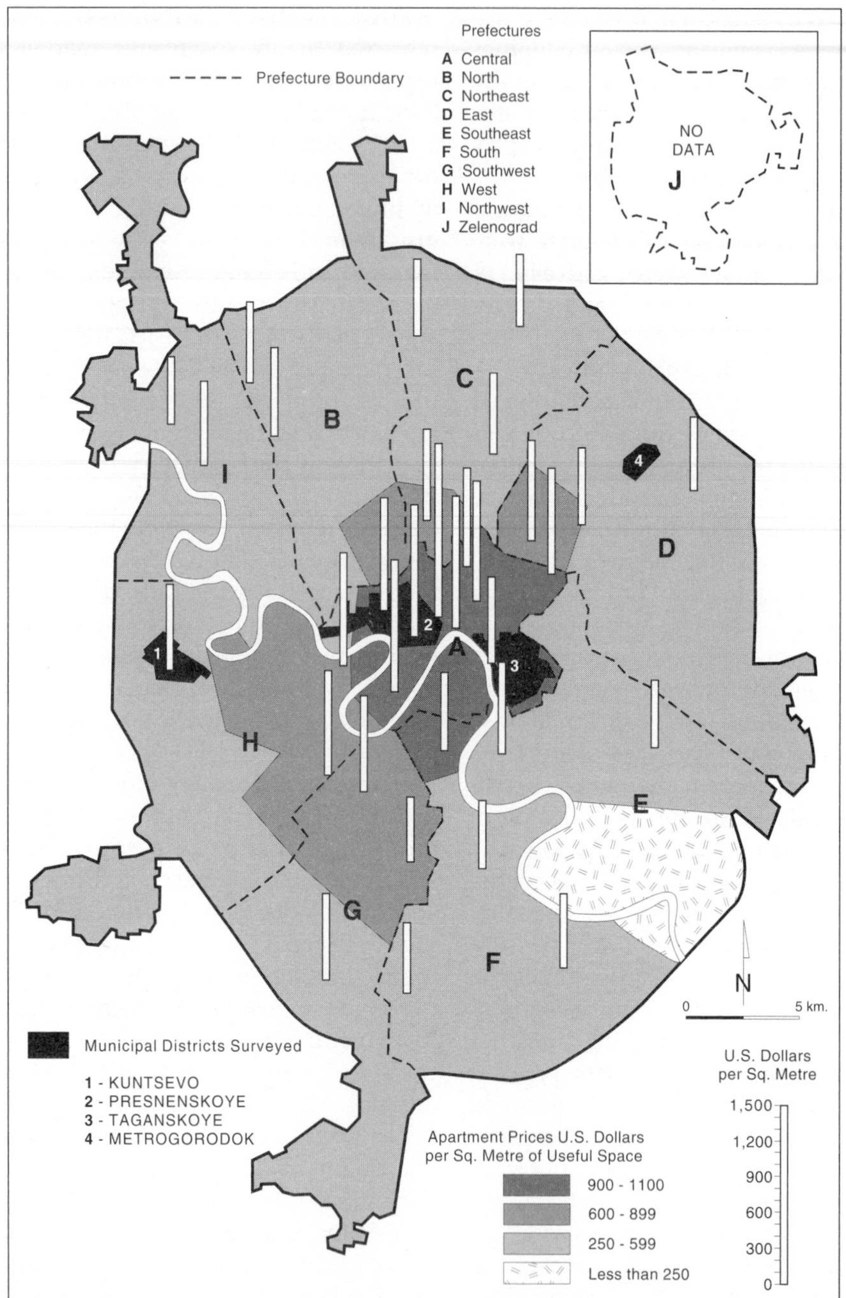

FIGURE 5.12 *Apartment prices: Moscow, 1993*
(*Sources*: Based on *Moskovskiye Novosti*, June 6, No. 23(669), 1993, 5a; J.H. Bater, A.A. Degtyarev, V.A. Amelin, 'Politics in Moscow: Local Issues, Areas and Governance', *Political Geography*, Vol. 14, No. 8, 1995, 674)

a single room in a communal apartment, or who remain on the waiting list for improved accommodation, their reactions to seeing advertisements for individual apartments with between 100 and 200 square metres of useful space for which the asking price occasionally exceeds

US$ 250,000, and regularly exceeds US$ 100,000, can only be imagined. The consequences of the development of a housing market are most evident in the Central Prefecture, both in terms of average apartment prices and in terms of the erosion of the stock of housing.

The conversion of housing space into commercial uses, especially but by no means exclusively in the Central Prefecture, has greatly exacerbated the housing shortage. However, the pressure to convert housing to commercial use in the Central Prefecture is not restricted to the substantial flats of the elites which might be rented or sold. There is growing concern over the conversion of communal apartments, since the occupants are frequently elderly, marginalized members of contemporary Russian society whose economic situation is decidedly bleak. Instances of manipulative real estate 'firms' facilitating the privatization and sale of single rooms in communal apartments in order to eventually gain control of the whole apartment unit which then can be sold at substantial profit are not uncommon. Across Moscow about 2000 buildings had been sold by 1993; half of them located in the Central Prefecture. In that year alone, a million square metres of floor space were taken out of their various uses in the Central Prefecture for redevelopment. By 1997 at least six million square metres of floor space are to be refurbished or built anew. According to agreements between local government and firms engaged in building reconstruction, 50 per cent of the refurbished floor space is supposed to be returned for allocation as housing. However, by mid-1994 less than 10 per cent of the one million square metres under redevelopment in the Central Prefecture in 1993 had been reassigned to the Prefecture housing stock. What has happened to the rest is an intriguing question. Some buildings no doubt were still under reconstruction. Equally likely is that some floor space intended for housing has gone into commercial use where large profit can be made. In any event, many thousands of people 'temporarily' relocated as a result of building redevelopment in the Central Prefecture have no way of returning to the city centre should they wish to do so. Legislation introduced by the city government in 1993 is supposed to control the conversion of housing space into commercial or other non-housing usage. More recent legislation is intended to protect citizens displaced by central city redevelopment by guaranteeing the right to return. But throughout Moscow such conversion and forced relocation occurs despite restrictions, and this is obviously facilitated by privatization.

Housing allocations differ according to occupation, as perhaps might be expected from what has been said already. The 1993 survey of a random sample of residents in the Central Prefecture referred to above also included representative samples of residents drawn from a municipal district with a distinct lower class image, Metrogorodok in the Eastern Prefecture, and from one widely associated with elites, Kuntsevo in the Western Prefecture (Fig. 5.12). Metrogorodok was first developed in the 1930s to house construction workers for the Moscow metro system. Kuntsevo was originally an area of *dacha*, or summer cottage, settlement on the outskirts of Moscow before being enveloped by urban expansion. Parts of Kuntsevo still retain something of a bucolic atmosphere. Metrogorodok, where the skyline is dominated by one of Moscow's huge heat and power stations, is decidedly industrial in character. The responses from the sample populations in the four municipal districts surveyed have been aggregated and are presented in Table 5.5. In order to highlight housing differences by occupational group, only the proportion of respondents occupying a separate apartment, the proportion in this category with three or more rooms, and the proportion living in a communal apartment, that is, in an apartment in which it is necessary to share with other, unrelated people, kitchen, bathroom and toilet facilities, are listed.

As the data in Table 5.5 make clear, there are certainly differences in housing situations across the occupational groups used in the survey. It is notable that the skilled worker and apprentice occupational group would appear to enjoy rather better housing than any other listed, at least insofar as these data connote quality of accommodation. That it should is consistent with some precepts of the former Soviet socialist system. This group has the smallest share in

Table 5.5 *Housing by occupational group in four municipal districts: Moscow, October 1993 (per cent)*

Occupational group	Separate apartment (%)	Apartment of 3 or more rooms (%)	Communal apartment (%)	Privatized apartment (%)
Business owner	84.3	43.1	15.7	29.4
Member of profession (i.e. doctor, teacher, etc.)	75.9	37.0	24.1	21.8
Artistic professional	89.3	39.3	10.7	53.6
Government employee	87.1	31.9	12.9	15.2
Non-state office worker	83.3	30.6	16.7	19.7
Trade service sector Employee	83.7	37.2	16.3	26.1
Skilled worker/apprentice	92.8	44.1	7.2	33.3
Blue collar worker (*rabochiy*)	85.0	35.0	15.0	14.2
Student	91.6	35.4	8.4	18.8
Homemaker	83.3	37.5	16.7	22.5
Working pensioner	87.7	29.2	12.3	45.5
Non-working pensioner	87.5	26.1	12.5	42.9

Source: Questionnaire survey, 1993
*Total sample, 993
**Taganskoye, Presnenskoye, Metrogorodok, Kuntsevo

communal apartments, and the largest occupying apartments with three or more rooms. Since the economic reform has created a market for housing, it is interesting that those describing themselves as business owners rank second in terms of the share (43 per cent) occupying what in Moscow is quite substantial accommodation, that is, a separate apartment with three or more rooms. Yet it is equally notable that a substantial share of those fully engaged in the new economy still live communally. Of course, what constitutes a business is subjective. Clearly, being a member of the health care or teaching professions does not bring much evidence of benefit in terms of housing, Indeed, this group had the lowest share occupying a separate apartment, and therefore the largest share living in a communal one. Artistic professionals along with skilled workers seem not to have fared too badly in housing allocation under the Soviet system. There is no reason to assume that the broad differences in accommodation amongst occupational groups in Moscow would not be found in other Russian cities.

Within the city of Moscow land cannot be bought and sold, only leased. Only in mid-1995 was a federal law passed that prohibited the sale of land in Russia's large cities. But before such legislation was introduced, in Moscow, St Petersburg and a number of other major cities, local authorities took steps unilaterally to impose such restrictions. Leasehold agreements for periods of 5–49 years are typical. While the highly competitive bidding for leaseholds often has generated far more income for city coffers than anticipated, the high cost of leaseholds for developers of course is simply passed on to end users. Within municipal government there is concern that if the land conversion process is not carefully managed, ecologically significant open space will be eroded. Many parts of Moscow, but notably the heavily industrialized southeast, are already suffering from environmental degradation. But as Soviet cities historically were developed without much regard to effective utilization of land, as will be discussed in the ensuing chapter, there is certainly scope for considerable infilling. And some areas are eminently suited to housing.

There are at least twelve of Moscow's municipal districts in which land has been leased to private sector developers for construction of *kottedzhi.*, the post-Soviet Russian equivalent of individual houses. As Fig. 5.13 indicates, these municipal districts are typically on the fringe of the built-up area. Close to 1500 hectares were designated for this purpose as early as 1991. It has been estimated that land within the city's borders which could be

FIGURE 5.13 *Areas for cottage development in Moscow, 1993*
(*Sources*: J.H. Bater, 'Housing Developments in Moscow in the 1990s', *Post-Soviet Geography*, Vol. 35, No. 6, 1994, 320)

turned over to private house construction is limited to around 7000 hectares. The leasehold cost of the land for *kottedzhi* is certainly not a critical factor in the decision to buy or lease one of these new homes. Such 'cottages' offer 200–350, or even more, square metres of useful space, frequently including a garage, and sometimes a sauna. Western in terms of scale, they are obviously geared to newly rich Russians, deep-pocketed 'ex pats', foreign firms or governments. Selling prices of US$ 250,000 or higher are not uncommon. Such accommodation obviously is far beyond what ordinary Muscovites might ever aspire to occupy.

Private sector housing construction owes much to a Presidential decree issued in early 1992 which outlines a scheme for development of individual homes in the Moscow region. Because of the perceived limited supply of land for private house construction within the capital itself, Yeltsin's Presidential decree focuses on Moscow Oblast'. It proposes that 40,000 hectares of land within a 50 kilometre radius of the capital be designated for individual house construction during the 1990s. By the middle of the next century perhaps as much as one-quarter of a million hectares would be involved in the project. Envisioned are new homes for 1.2 million families, two-thirds of them Muscovites. The concept itself is not original, although reliance on the private sector to build homes is obviously post-Soviet. The scheme was worked out jointly by architectural/town planning agencies of the city and oblast'. Proposed limitations on types of homes constructed, per capita average amounts of useful space within them, plot sizes, and the relative proportions of low, average and high income families who will occupy these homes are more than a little reminiscent of Soviet social engineering. High income families are ostensibly limited to a 5 per cent share of the housing space, with 35 and 60 per cent to be allocated to low and middle income families respectively. While perhaps admirable social planning objectives, it would seem that *kottedzhi* advertised for sale at present are beyond the financial reach of all but the very wealthy. As was so often the case in the Soviet era, plan and reality are far from congruent.

While grand in concept, the progress in plan implementation to date has been relatively modest. Approximately 4000 hectares were to be designated for development in the first year, 3000 of which were to be allocated to Muscovites for house construction. The balance was to be available to citizens of Moscow Oblast'. Actual land allocations agreed to by the various levels of government involved – city, oblast', region – are much more limited. Figure 5.14 depicts the locations and areas set aside under the terms of this 1992 Presidential decree by early 1993. Just over 1400 hectares are involved in this phase. Over a ten-year period, 140,000 *kottedzhi* are to be built. The agreements between the newly privatized, or newly constituted land development and construction organizations, and the government agencies controlling the sale of land for housing, are interesting. For example, in the project outlined in Fig. 5.14, the developers acquire land with limited road access and no infrastructure and are obliged to turn over to the city and oblast' housing authorities, free of charge, 10 per cent of the *kottedzhi* built, and 40 per cent of them at cost. The balance can be sold at the prevailing market price.

Aside from the need to determine appropriate levels of compensation for the owners of the lands in question, mostly state and collective farms, there are major infrastructural needs such as road, water, sewage and power networks that must be installed and paid for. Since all levels of government are looking for ways to save not spend, these are not inconsequential issues. There is no shortage of private sector development companies, since forming such a business is not especially difficult. The same cannot be said for ways of raising the necessary capital and facilitating the sale of *kottedzhi* to individuals. Notwithstanding a recent Presidential decree advocating the creation of some mechanism for providing financial assistance to those wanting to purchase a home, to date there is still no mortgage market. Buying homes usually requires a 30 per cent up front cash payment, 50 per cent during construction, and the balance upon assuming occupancy. Proximity to Moscow and accessibility to road and commuter rail line play a predictable part in determining prices for existing houses. As expected, prices for what appear to be comparable housing according to real estate listings decline the more distant the location in Moscow Oblast' from the capital. For a detached house boasting ample floorspace, with basement, garage and readily accessible to Moscow by car, prices in the US$ 200,000 and higher range are quite usual.

Of course, such prestige homes account for only a very small share of the privately owned, detached housing stock outside the city of

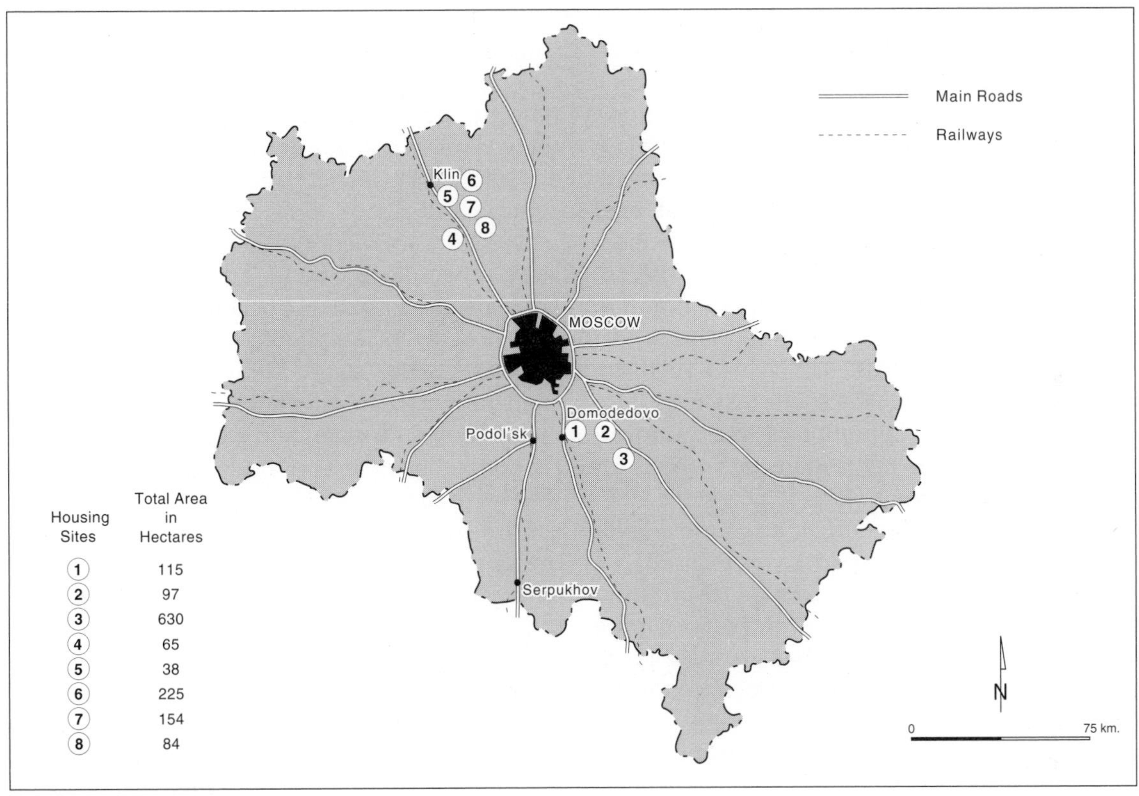

FIGURE 5.14 *Areas for cottage development in Moscow Oblast', 1993*
(*Source*: J.H. Bater, 'Housing Developments in Moscow in the 1990s', *Post-Soviet Geography*, Vol. 35, No. 6, 1994, 323)

Moscow's borders. Most housing of this type is modest, if not primitive, and priced accordingly. Much of it is for summer time use only – either the traditional and simple cottage, or *dacha*, or more modest still, summer garden structure – and has minimal services. Some 2.5 million Muscovites have such accommodation at present. As a result of recent agreements between oblast' and city authorities, the former made available to Muscovites an additional 17,000 hectares for such use in 1991 and 1992 alone. Over the preceding four decades there had been no more than 50,000 hectares in total freed up by oblast' authorities for the same purpose. Land allocated for this type of seasonal accommodation obviously does not require the services necessary for year-round homes. The transfer of land for the construction of *kottedzhi*, on the other hand, raises numerous financial and indeed social issues, and as already noted has not progressed as fast as Yeltsin's Presidential decree of 1992 proposed. While the income generated by the sale of land for prestige housing is certainly attractive to the vendor, whether government or privatized organizations such as former collective and state farms, the social consequences of such developments in Moscow Oblast' remain of some concern.

The privatization of the existing housing stock is an important part of the market reform process. Whether an intended outcome or not, the cleavages in the social class structure are deepening in consequence of the economic advantage bestowed upon all those who had benefited from the Soviet system of housing perquisites. The already advantaged are able to capitalize on the value of their housing at effectively no cost. Privatization of housing obviously only affords economic advantage if there is a market for apartments which might be offered for sale or rent. The market reform has created a new class of citizen, one whose business acumen, established connections with the remaining and still powerful bureaucracy, and/or criminality, have generated huge personal incomes, and for whom material possessions, including improved housing, are more important than money itself. Combined with a growing number of western businesses and 'ex pat' staff prepared to pay the going rate for commercial or domestic space, there is no immediate prospect of diminution in demand. Moreover, in addition to the disruption that privatization of municipal housing construction trusts has created, there are institutional, legal and, in some instances, attitudinal barriers to significant and speedy adjustments in supply. The net result is that targets for housing construction are largely fiction. In 1992, for example, barely one-fifth of the planned addition to Moscow's housing stock was realized.

While the foregoing discussion of privatization of housing has focused on Moscow because more has happened faster there than in most other cities, it nonetheless has much relevance for the rest of urban Russia. In cities across the country the same processes are at work. Apartments are being privatized. Those people who are benefiting economically from the market reform usually put improved housing as a high priority in their personal lives. A market is developing in which apartments are being bought and sold. Prices are rising rapidly because of the imbalance between the supply and demand for housing, though Moscow will no doubt continue to have pride of place amongst Russian cities in terms of the cost of accommodation simply because of its size and role as primate city. In Russia's small cities, land may now be bought and sold, not just leased as is the case in Moscow. In such cities, the newly wealthy are sometimes building substantial houses for themselves. But irrespective of whether land is owned or leased, private house construction occurs in or around most of the larger cities, and many smaller ones as well.

The development of a market is not just for housing obviously. With bidding for leaseholds on land in Moscow pushing prices to levels much in excess of what is obtained in many western capitals, control over land clearly has the potential to bring substantial profit. The rapid escalation in the price of apartments has even prompted some reassessment as to whether or not facilitating tenant privatization of enterprise-owned housing is now such a good strategy. Previously the cost of maintenance in relation to rent paid encouraged all 'owners' of housing to relinquish control, and hence obligation for upkeep. Now real estate is seen less as a liability and more as an asset, especially as a hedge against inflation.

Clearly, real estate is the chief resource of any city, Moscow included. As it increases in value in the marketplace, so too does the potential municipal property tax revenue. That land tax is a major source of revenue for cities in developed economies is not lost on those managing cities in Russia. In the case of Moscow, by extending city control over more territory in Moscow Oblast', the opportunities for trading off housing Muscovites there on less valuable land, while facilitating commercial development, especially in high value central city areas, are enhanced. The official city position is that land acquisitions in Moscow Oblast' are not speculative ventures. Far less ambiguous in terms of motivation is the current struggle for control over property between the city of Moscow and the federal government.

Within Moscow's borders are a vast array of enterprises traditionally controlled by the state as opposed to municipal government. A substantial share of state enterprises in post-Soviet Russia has now been privatized, but in Moscow the role of state enterprises in the urban economy is still significant. Privatization vouchers issued to all citizens in 1992 play an important part in the transition from state control to some form of private ownership of former state enterprises. The issue in Moscow is that the city's very influential Mayor believed that not only are the owners of privatization vouchers invested in these 'new' enterprises not realizing any real return, but that the 'price' paid does not reflect the real value of the enterprise. Moscow therefore stopped registering joint-stock companies brought into existence under federal government legislation permitting voucher-based privatization. In early spring 1994, the city government endorsed a move initiated by the Mayor to create a network of so-called financial-industrial groups. This new structure would embrace banks, factories, trading houses and holding companies. To participate in it requires that firms take on a partner – the city of Moscow. What Moscow will contribute financially to the private sector enterprise participating is, in essence, the 'value' of the land upon which the enterprise sits but does not own. In many instances this is perhaps the enterprise's most valuable asset. As noted already, land in Moscow cannot be bought and sold, only leased according to terms and conditions established by the municipal bureaucracy responsible for land development. The incentive for firms to participate is obvious – to do business of any kind in Moscow requires cooperation of city administrative structures. And many enterprises seek to do business with Moscow itself. This ploy by the city to retain some measure of control over the economic activities of former state enterprises, and new private sector activities as well, is being stiffly resisted by the federal government's State Property Committee. It had the mandate to bring the first stage of voucher-based privatization to a conclusion in July 1994, and then to proceed to enhance the opportunities for privatization based on investment capital. The arrogation of a potentially significant decision-making role in private sector organizations on the part the city of Moscow is simply another manifestation of the politics of property development. The rationale offered by city officials is that this protects Muscovites by ensuring that the value of a basic resource, land, is not squandered by market-place machinations. The criticism of the move is obvious – it perverts the course of privatization by establishing a post-Soviet version of the Soviet era style of managing the economy. This battle for control over property is not confined to Moscow. While it takes on a different guise in those cities in Russia where land can be bought and sold, given the value of what is at stake, no doubt the war between competing interests will continue.

FURTHER READING

Andrusz, G.D. 1984 *Housing and Urban Development in the USSR*. London, Macmillan.

Bater, J.H. 1980 *The Soviet City: Ideal and Reality*. London, Edward Arnold.

Brumfield, W.C., Ruble, B.A. (eds) 1993 *Russian Housing in the Modern Age: Design and Social History*. New York, Woodrow Wilson Center Press and Cambridge University Press.

Chase, W.J. 1987 *Workers, Society, and the Soviet State: Labour and Life in Moscow, 1918–1929*. Urbana, University of Illinois Press.

Colton, T.J. 1996 *Moscow. Governing the Socialist Metropolis*. Cambridge, MA, Harvard University Press.

Friedgut, T.H., Hahn, J.W. (eds) 1994 *Local Power and Post-Soviet Politics*. Armonk, NY, M.E. Sharpe.

French, R. A. 1995 *Plans, Pragmatism & People. The Legacy of Soviet Planning for Today's Cities*. London, UCL Press.

French, R.A., Hamilton, F.E.I. (eds) 1974 *The Socialist City*. New York, Praeger.

Harris, C.D. 1970 *Cities of the Soviet Union. Studies of their Functions, Size, Density and Growth*. Chicago, Rand McNally.

Hoffman, D.L. 1994 *Peasant Metropolis: Social Identities in Moscow, 1929–1941*. Ithaca, New York and London, Cornell University Press.

Kotkin, S. 1994 *Magnetic Mountain. Stalinism as Civilization*. Abingdon, Oxfordshire, Carfax Publishing Company.

Lane, C. 1981 *The Rites of Rulers: Ritual in Industrial Society – The Soviet Case*. Cambridge, Cambridge University Press.

Lappo, G.M., Petrov, N.V. and Adams, J. 1992 *Urban Geography in the Soviet Union and the United States*. Lanham, Rowman and Littlefield.

Lewis, C.W., Sternheimer, S. 1979 *Soviet Urban Management: With Comparisons to the United States*. New York, Praeger.

Miliutin, N.A. 1974 *Sotsgorod – the Problem of Building Socialist Cities*. Cambridge, MA, Massachusetts Institute of Technology Press. Translated by P. Sprague.

Morton, H.W., Stuart, R.C. (eds) 1984 *The Contemporary Soviet City*. Armonk, NY, M.E. Sharpe.

Parkins, F. 1953 *City Planning in Soviet Russia*. Chicago, University of Chicago Press.

Ruble, B.A. 1990 *Leningrad: Shaping a Soviet City*. Berkeley, University of California Press.

Ruble, B.A. 1995 *Money Sings: The Changing Politics of Urban Space in Post-Soviet Yaroslavl*. Washington, DC, Woodrow Wilson Center Press; Cambridge and New York, Cambridge University Press.

Ross, C. 1987 *Local Government in the Soviet Union: Problems of Implementation and Control*. London, Croom Helm.

Savas, E.S. and Kaiser, J.A. 1985 *Moscow's City Government*. New York, Praeger.

Turner, B., Hegedus, J., Tosics, I. (eds) 1992 *The Reform of Housing in Eastern Europe and the Soviet Union*. New York, Routledge

6

NATURAL RESOURCES MANAGEMENT IN THE SOVIET AND POST-SOVIET ERAS

'Let us not, however, flatter ourselves overmuch on account of our human victories over nature. For each such victory nature takes its revenge on us'
(F. Engels, *Dialectics of Nature*, 1974, 180)

Landscapes the world over bear mute testimony to the values and attitudes of the people who have occupied, and wrested a living from the land. The relationship between society and the natural environment, and the impact of the former on the latter, is influenced both by the stage of economic development attained by the society, and its fundamental attitudes or values. In this latter respect, one of the distinguishing features of the Soviet system was that ownership of all natural resources was vested in the state. Private ownership of land and natural resources was equated with profligate development for personal benefit, and environmental degradation in the pursuit of profit. State ownership was presumed to prevent misuse, entrench rational development practices, and thus benefit society as a whole. From countryside to city, the manner in which natural resources were managed under the Soviet system left an indelible imprint on the landscape. And rarely was it a desirable one.

In the new economy of the post-Soviet era, one of the most contentious issues confronting the new states is who should have proprietary rights to land and natural resources. Exactly how market reform and privatization will alter the landscape across the post-Soviet scene is not altogether clear. But that change will occur is certain. In this chapter, the current status of land ownership and resources management across the post-Soviet scene will be reviewed. But before doing so, it is necessary to put contemporary developments into context. Thus, the evolution of Soviet natural resources management principles, and some examples of the legacy of resource management practice, will be described. In ensuing chapters, the development and management of a number of the more important natural resources will be examined in greater detail.

FIRST PRINCIPLES OF SOVIET RESOURCES MANAGEMENT

The revolution of 1917 witnessed the demise of the imperial era with its deeply conservative system of beliefs, and the advent of a Bolshevik party ostensibly guided by the principles of Marxism. How this political transformation affected the structure of the political-administrative system, the nature of decision-making in managing the economy, and the country's population and urban environment has been examined in the preceding chapters. Marxist theory also spelled change for the management of natural resources. Once nationalized, natural resources were to be allocated free of charge to agents of government whose responsibility it was to develop them according to the dictates of the central plan. The need to demonstrate socialist society's ability to conquer, to subjugate, the physical environment was also a central theme in the new Soviet ideology. Over the years, the Soviet approach to resources management was altered, though usually too little and certainly too late to have much positive, lasting impact.

According to the Marxist scheme of things, under conditions of scientific socialism the door was open to a future of enhanced material prosperity, if not abundance. Under the new order it was presumed, for example, that the capacity of the soil to feed an ever larger Soviet population was virtually unlimited. Gone was the Malthusian doctrine which envisioned the world's limited agricultural land base incapable of producing enough food to meet the basic needs of an ever growing population. Such pessimistic scenarios were true only if land remained privately owned, and therefore exploited by the class of capitalists who controlled its use. Under socialism, science and technology would be brought to bear on behalf of society at large, not just one privileged part of it. Under socialism, there was no teleological order of things in which men and women could be subjugated by the natural environment. Socialist society was not perceived to be in harmony with nature, was not perceived to be in some ordered relationship with nature. Socialist society was instead viewed as being in conflict with nature. The natural environment was often described by Marxists as being in a state of disorder, or chaos, and it was the task of socialist science and technology to conquer nature, to transform it. For progress to occur, the inherent contradiction between the state of the natural environment as it is, and as it ought to be, had to be resolved. Thus, for much of the Soviet era the notion of conquering nature played a central part in the official ideology. Soviet scientists often took literally the clarion call for action espoused by the political propagandist.

The private ownership of land and natural resources was seen by the Marxist as the principal reason for their exploitation and waste. In contrast, nationalization of all resources and their planned development was equated with wise husbandry and conservation. Under socialism, resource management conflicts and environmental pollution simply would not exist. Much of the early legislation reflected this somewhat benign attitude. We need only illustrate this point by way of a few examples.

The decree nationalizing land and water resources was issued in January 1918, and followed a decree of November 1917 which transferred the ownership of large agricultural estates from individuals to the state. By June 1918, a decree had been issued which gave to local authorities the responsibility of managing the nation's forest resources. At about the same time, the management of wildlife was assumed by the state insofar as the control over hunting and firearms was vested in local authorities. As time passed, the scope of state control broadened. Legislation prohibiting the discharge of polluted waters by industrial and commercial enterprises was adopted in 1923. Up until Lenin's death in 1924, at least 15 decrees pertaining to the management and conservation of the state's natural resources were enacted. Many were attributable to the personal interest and initiative of Lenin himself. But Lenin was only one of many theorists convinced that state ownership of natural resources, combined with proper and legal allocation of authority for their management, was sufficient to remove the possibility of misuse and consequent environmental degradation. The legal apparatus put in place by the state at this time was assisted by educational programmes and voluntary organizations. A central role in this latter regard was assumed by the All-Russian Society for the Protection of Nature founded in 1924. While the concept of conservation or protection of nature was implicit, if not explicit, in much of this early legislation, the purpose of conservation was not simply to set something aside never to be developed by the state. In fact, most legislation of a conservationist hue was intended to 'save' resources so they would be available for economic development at a later date. Perhaps the single most important exception was the creation of a system of nature preserves, or *zapovedniki*, which occurred in 1919.

While environmental degradation was presumed incompatible with socialism, it ought to be remembered that the principal objective of the new state was to develop the domestic economy. With the death of Lenin, and the assumption of control by Stalin, industrialization became the single most important goal. The role of science was to facilitate this endeavour, and thus the role of technology figured prominently in the planned development of the natural resource base. A more rational management of resources might have been possible had there not been such attachment to the concept of 'conquest of nature'. Similarly, the concept of value in Soviet economic theory also proved to be prejudicial to rational resource management.

In nineteenth century Marxist thought, labour was regarded as the basic element in determining the value of commodities. The enhancement of the price of a commodity as it passed from producer to consumer under capitalism did not reflect real value, but simply the manipulation of supply and demand and the exploitation by those who controlled the means of production of those who did not. Put simply, to the Marxist a commodity was without value until labour was applied to it. The value of a manufactured good was therefore equal to the sum of the labour required to make each of its component parts. Natural resources, on the other hand, were without value until developed by human labour. It followed, therefore, that in the Soviet Union all land, water, forest, mineral and other natural resources were allocated to users free of charge. The consequences of this Soviet interpretation of Marx's labour theory of value were unforeseen. Perhaps not surprisingly, the treatment of natural resources as free goods was inimical to their rational utilization and conservation.

Technology and Soviet schemes for the transformation of nature

With the advent of the Five Year Plan and Stalin's industrialization drive the application of socialist science and technology to the resolution of the nation's problems assumed an importance previously unmatched. And so too did the role of ideology. Typical of the attitude which soon took hold was that pertaining to water resources management. Even during Lenin's time large-scale hydro-engineering projects were promoted enthusiastically. Thus, during the 1920s a number of dams were approved for construction on the rivers of European Russia. These projects were an integral part of the GOELRO Plan, or the State Commission for the Electrification of Russia, one of the earliest of the Soviet regional development schemes. While the Plan included thermal as well as hydroelectric stations, it was the latter which captured the attention of the propagandist. Where else could one find such clear testimony to the ability of socialist science to conquer nature? The first large-scale, for the time, project to be completed was at Volkhov on the Volkhov river in northwest European Russia near St Petersburg (then named, Leningrad, see Fig. 8.1, p. 220). It was commissioned in 1926 amidst considerable fanfare. In this regard, the propagandist was greatly aided by one of Lenin's more famous dictums: 'Communism is Soviet power plus the electrification of the whole country'. Under Stalin such sloganeering was raised to new heights, and so was the assault on nature.

Stalin supported a broad range of initiatives which were customarily publicized in terms of a struggle between socialist science and technology and the vagaries of nature. Large-scale water resource development projects continued to figure prominently in this scenario. Typical of some of the worst features of the Stalin era was the construction of the White Sea–Baltic Canal. For the time it was a huge undertaking, yet it was completed in less than two years thanks to the widespread use of forced labour, and little regard for human life. The development of the hydroelectric power potential of the Volga river was accorded high priority, and construction of a number of power stations was begun during the 1930s. Conceived as a multi-purpose river basin development scheme, it was closely linked to urban-industrialization in the Volga region. As time passed and technology improved, the scale of each power station increased accordingly. Even after Stalin's death in 1953 individual power stations were regularly publicized in terms of conquest of nature by Socialist 'man'. The considerable contribution of women labourers to these development projects was still accorded little attention While the achievements of technology were customarily lauded by public figures and the media, basic science had an obvious role to play in the creation of socialist society as well. But where basic science became the handmaiden of state ideology there were bound to be problems. Genetic science in particular and biology in general offer a case in point.

The name of T.D. Lysenko is inextricably linked to the early years of experimentation in Soviet agriculture. At a time when the Soviet state had embarked on a vast programme of industrialization there was little money, and even less will, to invest in agriculture. Thus, schemes to improve production which involved limited financial outlay were compelling. If they had overtones of a new science created under the conditions of building socialism they were even more so. Thus, the 'peoples' scientist' Trofim Lysenko was able to gain Stalin's personal support for his projects, most of which were of dubious scientific value. One example was 'yarovization', or the pre-heating of seeds. This sped up the process of germination, thereby permitting crops to be grown in regions not normally suited to them. This was only one of the presumed advantages of this widely applied experiment. Lysenko also contended such a process would alter the basic genetic structure of the plants involved. The seeds of such plants, he claimed, would 'adapt' to the new, more harsh, natural environment. Lysenko continued to hold an influential position even after Stalin's death. He died in 1976 at a time when a more realistic assessment of his contribution to Soviet science could be made without fear of reprisal. Lysenko is generally credited with blocking real advances in Soviet genetic science for the better part of three decades. Something of the flavour of Lysenko's deficient science is also reflected in the Stalin Plan for the Transformation of Nature adopted in 1947.

During the 1930s the role of the geographical environment in the development of society became a contentious topic of debate. Stalin eventually dictated the official view of the Party in 1938 by declaring that the economic factor, not the geographic environment, was the dominant influence in the process of societal development. He allowed that although in some circumstances the geographical environment could accelerate or retard societal development, under the conditions of socialism this would be impossible. The development of a socialist society would be quick, and easily outstrip any changes which might occur in the geographical environment. What is more, the advancement of science under socialism would permit the acceleration of the process, indeed, would allow the process of societal change to occur at a fundamentally different level than was possible under capitalism. An example of the thinking of the time was the idea that climatic conditions over large areas could be altered through human intervention. This notion gave rise to one part of the Stalin Plan for the Transformation of Nature, namely the creation of a network of shelterbelts. Put simply, science and technology were assumed to be capable not just of bringing order to the environment, as in the case of hydraulic engineering, but of actually changing the natural order.

The shelter belt component of the Stalin Plan is outlined in Fig. 6.1. It, like other ameliorative programmes introduced in 1947, sought to stabilize agricultural production in the critically important south Ukraine and Kuban regions of the North Caucasus. Here the variability of rainfall during the growing season made planning agricultural output problematic in the extreme. Highly variable yields were related in large measure to the prevalence of the *sukhovey*, or very dry airmass movements from the northern part of Middle Asia during July and August. Originating in the region north of the Aral Sea, these air masses tracked west, frequently devastating crops in their path. In the era of Lysenkoism it was thought possible to mitigate this prevailing climatic regime by introducing into a wooded steppe, steppe and semi-desert environment, four huge shelterbelts. The planting of the four major shelterbelts outlined in Fig. 6.1 was complemented by an equally vast programme of planting windbreaks around individual fields. Such schemes of planting vegetation in agricultural regions had a long history. Indeed, shelterbelts were widely used in both the United States and Canada during the 1930s. But in the latter places, protection of individual fields so as to enhance moisture retention, and therefore enhance yields, was the rationale. In the Soviet Union this reasonable objective was carried a significant step further. There science and technology in the employ of the state were given the task of ameliorating the climate of a vast region of steppe and semi-desert. Clearly, objective scientific opinion did not support such an undertaking. With the death of Stalin in 1953 the scheme was quietly shelved. The landscape, however, had been permanently altered. A careful perusal of vegetative maps of the region in later decades will reveal relict features of this programme, but only in the zone of the wooded steppe. Whether the existence of these relict features has been of any real economic benefit to agriculture in the region is a moot point. Some studies of the effects of large-scale shelterbelts in other countries suggest that not only do shelterbelts raise moisture consumption during the summer, they prevent a more widespread distribution of snow cover, and hence reduce moisture availability in the regions affected. Needless to say, the climatic regime in this region of the former Soviet Union was not altered by the Stalin Plan.

During the Stalin period, and for that matter in the ensuing Khrushchev era as well, the attitude towards the environment reflected in official proclamations did not diverge very much from the theme of conquering nature. If Khrushchev was perhaps more sanguine in his view as to what science and technology could achieve in reworking the natural environment, many of the schemes closely associated with his name were nonetheless technologically dependent. For example, his promotion of the so-called Virgin Land Schemes entailed the cultivation of a vast area of grassland in Kazakhstan and West Siberia. Huge state farms were established, monoculture of grain introduced, and vast amounts of agricultural machinery were dragooned from elsewhere in the country to run the farms. Over 40 million hectares were put under the plough in a few short years. In a region previously the domain of the pastoral nomad, the land was cropped annually. Some of the consequences of this scheme will be examined in the next chapter, but the point to be made here is that the ideology of a technological imperative prevailed. Perhaps in this era of Soviet achievements in space, especially the highly successful Sputnik space craft and the first ever human space flight in it, such a firm belief in technology was to be expected. But in the former Soviet Union, as elsewhere in the world during the final decades of the twentieth century, the assumption that the transformation of the environment could be achieved without cost slowly changed in the face of mounting evidence of environmental degradation.

From the middle of the 1960s on, there seems to have been much more explicit concern with protecting the environment, at least to judge from some Soviet legislative initiatives and policy decisions. This suggests that the Stalin-era preoccupation with the conquest of nature, with its purposeful transformation, had been tempered somewhat. However, it would be incorrect to suggest that the traditional belief in the role of science and technology in achieving the state's objectives had been jettisoned entirely.

From Transformation of Nature to Conservation of Nature?

One of the central problems confronting the Soviet leadership right down to the end was the dismal state of agricultural production. Beginning in the early 1970s, the Soviet Union shifted from being a net exporter to a net importer of grain. After the 1965 economic reform, which amongst other things sought to stimulate agricultural production, this sector commanded in excess of

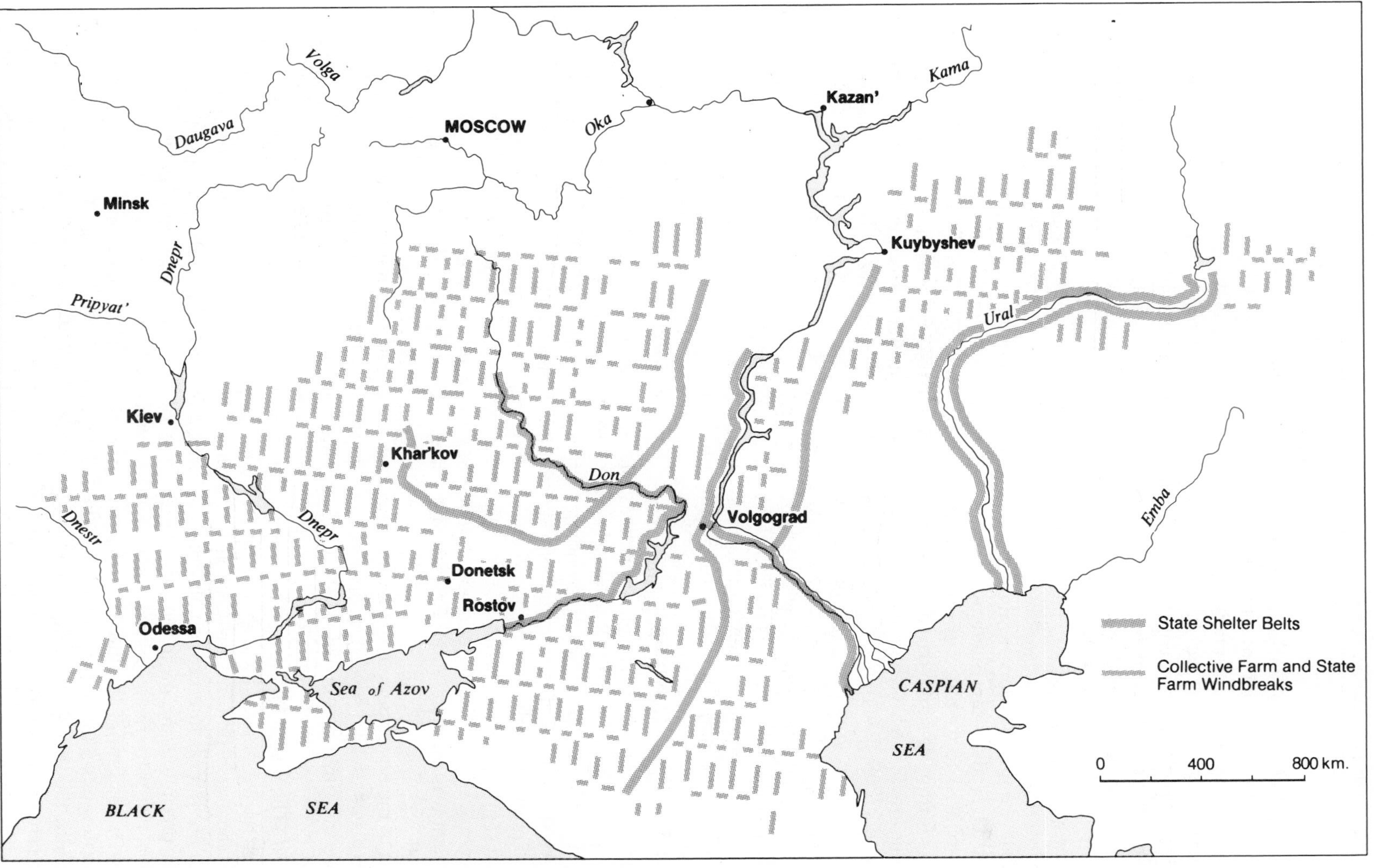

Figure 6.1 *Stalin plan for the transformation of nature*
(*Source*: A.E. Burke, 'Influence of Man upon Nature-the Russian View: A Case Study', in Thomas, W.L. (ed.), Man's Role in *Changing the Face of the Earth* (Chicago: University of Chicago Press, 1956), 1043)

one-third of available investment capital annually. Yet despite the huge annual investment, grain production still did not meet planned targets. The problem was many-sided, as we shall see in the next chapter, but part of it was environmental. Put simply, where there was adequate heat to grow crops, there was generally inadequate moisture. This broad relationship is described by Fig. 6.2. The zone of greatest agricultural potential, that of chernozem soil, is characterized by considerable variability of precipitation during the growing season. Over the years there were many schemes proposed to fix this vagary of nature. They ranged from Lysenko-based agronomy, which being unscientific did not work, to vast projects in which the flow of major rivers would be reversed. Such grandiose schemes were obviously perfectly compatible with the Stalin-era preoccupation with transforming nature. As it happened, river diversion projects remained on the drawing board owing to their cost, not for want of enthusiastic support. In the post-Stalin era the basic problems did not disappear, but technological advances made some of the river diversification projects more realistic alternatives and the apparent benefits seemed to outweigh the costs, at least insofar as Soviet economics was concerned. Aside from improving domestic agricultural production, the case for river diversion gained support from some other quarters in the final years of the Soviet era. Prominent amongst them was the state of the Caspian and Aral Seas.

For most of the Soviet period, the level of the Caspian Sea had steadily declined. In total, it dropped nearly three metres before stabilizing in the late 1970s and then rising once more in the late 1980s. The reason for the decline appears to have been related principally to changes in the long-term climatic regime in the Caspian Sea watershed. But human intervention also played a role in the precipitous decline in the level of the Caspian Sea. The Volga river, which accounts for more than 70 per cent of the inflow, had been extensively developed for water power since the Stalin era. By the 1960s the Volga river was basically a series of very large reservoirs. Urban industrial development at the numerous large-scale hydroelectric power stations dramatically increased water consumption. In the lower Volga region, substantial volumes of water are also diverted for large-scale irrigation projects.

Human intervention into natural ecological and hydraulic systems negatively impacted the flow of the Amu and Syr Dar'ya rivers in Middle Asia. The areas adjacent to the Amu and Syr Dar'ya offered considerable potential for irrigation. Indeed, the region has always relied on irrigated agriculture for much of its food production. However, in the post-war period especially the region's agricultural potential was put more fully into the service of the state. Huge areas were brought under irrigation for the purpose of cotton monoculture. During the period from 1965 to 1991, the total area under irrigation in the former USSR doubled, and most of the expansion occurred in this region. As these two rivers are the principal sources of water for the vastly expanded territory under irrigation, the resultant drawdown of the Aral Sea was even more dramatic than that of the Caspian from the 1920s to the 1970s. In the final decades of the Soviet era, the level of the Aral Sea dropped by more than 10 metres. The rate at which the Aral is shrinking is now slower since there is virtually no more water in the Amu and Syr Dar'ya rivers which can be diverted, but it has not stopped because of continuing evaporation from what is left of the Aral Sea. By early next century, the Aral Sea will have largely disappeared. Already there are profound consequences, as discussion of the environmental and public health impacts of the reduction in its surface area in Chapter 10 will make plain. Thus, there was great pressure during the Soviet era to do something to ameliorate the low water level of these two major inland seas. There was also pressure to increase, or at least stabilize, grain production through expanding the area under irrigation. And in Middle Asia, continued expansion of the irrigated cotton fields was a high priority, and linked to this objective was that of providing employment for the fast growing rural population in the region. All of these problems, and more besides, gave a real fillip to proposals to divert the waters of several north-flowing rivers to the south. After all, was not the northerly

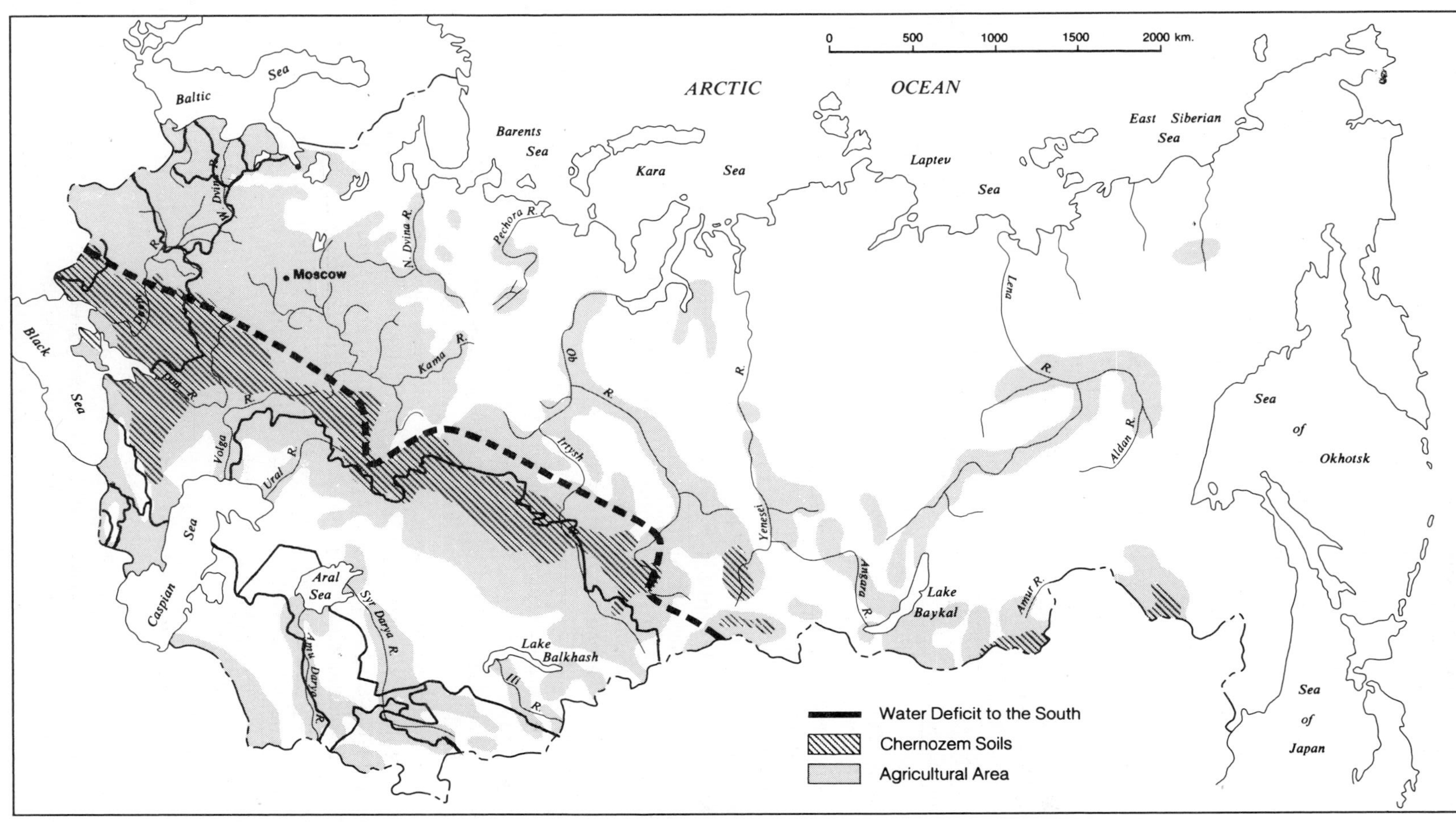

FIGURE 6.2 *The effective moisture divide*
(*Source*: Adapted from J.R. Tarrant, 'The Significance of Variability in Soviet Cereal Production', *Transactions, Institute of British Geographers, New Series*, Vol. 9, No. 4, 1984, 392)

flow of such an important resource testimony to the need to rework the natural environment? Engineering surveys were initiated on the assumption that such massive river reversals were in the national interest.

While in parts of northern European Russia some comparatively minor projects were initiated, they were minor only in the sense that in terms of the volume of water diverted and the cost involved they paled in comparison with the proposed diversion of Ob′ and Irtysh rivers. As is portrayed in diagrammatic fashion in Fig. 6.3, the so-called SIBARAL scheme entailed taking water from the upper reaches of the Ob′ river to the Irtysh, and then reversing the flow of the Tobol′ river at its confluence with the Irtysh. A combination of gravity and pumping stations would see nearly 30 cubic kilometres of water move from the north to the arid south (see also Fig. 6.2). The cost was estimated to be in the billions of roubles, but the presumed benefits were greater. Not only would the area under irrigation be greatly expanded, bringing with it an increased and more predictable level of agricultural production, but the problem of the declining levels of the Aral Sea, and possibly the Caspian Sea as well, would be addressed. Up until 1985 the project seemed to have the full backing of Party and government. Middle Asian political figures and technical personnel alike were enthusiastically in support. And as noted, water for irrigation was seen in the region as the key to resolving both the pressure for land and the need to more fully employ the rapidly growing rural labour force. Many Soviet scientists and engineers were keen to see the implementation of the century's most massive example of the transformation of nature. Already a huge amount of time and money had been committed to the scheme. Notwithstanding the environmental benefits to be gained from stabilizing the levels of the major inland seas, and in the process forestalling a change in climatic regime which portended greater aridity, there were costs. And these began to be articulated with increased urgency from the early 1970s on.

The principal concern of those opposing the project centred on the inadequate attention accorded to environmental impact assessment, and the apparent absence of a thorough comparative cost evaluation of alternative solutions to the problem. In the northern reaches of European Russia there was much local opposition to possible loss of significant cultural and archaeological artefacts due to flooding. Other people were worried about the impact on the *tayga*, the principal vegetation and the mainstay of numerous local settlements. Climatologists outside the Soviet Union, as well as some inside, expressed grave concern over the potential climatic change for the northern hemisphere. A diversion of fresh water of the scale intended, the argument went, would affect the extent of the ice cap, albedo, and upper air circulation patterns. The fresh water comprises a relatively thin layer above the Arctic Ocean's salt water. As it freezes at a higher temperature than salt water, any substantial reduction in fresh water in the Arctic Ocean would result in a shrinkage of the ice cover. Still others, and probably more persuasively, focused their criticism on the fact that a huge expenditure was required to implement the scheme, and that it could be at least two decades before any return on this investment would occur. And all of this expenditure to bring water to a region in which existing irrigation projects were notoriously wasteful of water.

As late as 1984 it appeared that the protagonists had the upper hand in this increasingly contentious debate. The SIBARAL project was included in early discussions of the draft guidelines of the 12th Five Year Plan (1986–1990), which were published in 1984. Throughout 1984 and early 1985 the project appeared to have the full support of the leadership. But the criticism continued, and gained momentum with the arrival of Mikhail Gorbachev in March of 1985. By November of that year, the 12th Five Year Plan was released for discussion. It was ratified by the Party and government in early 1986. The Plan did not include any reference to the SIBARAL project. In August of 1986, the Central Committee of the Communist Party and the Council of Ministers adopted a resolution to discontinue design and preparatory work not just on the diversion of the Ob′ and Irtysh waters (SIBARAL), but on the diversion of part of the flow of northern European Russian rivers into the Volga system as well. Further study

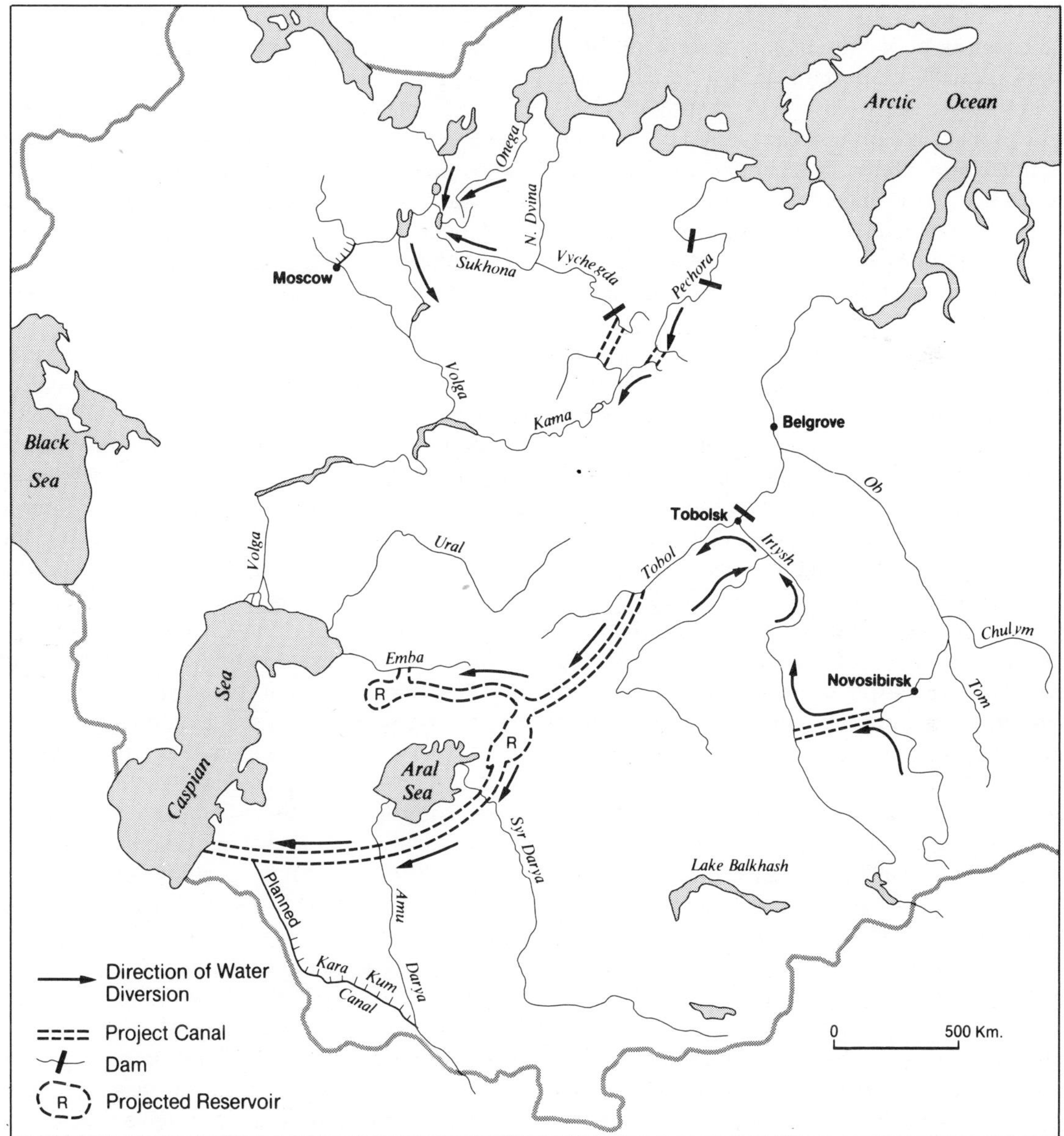

Figure 6.3 *The SIBARAL project*
(*Source*: L. Symons *et al.*, *The Soviet Union, A Systematic Geography* (Totowa, NJ: Barnes and Noble, 1983), 85)

of the ecological impacts of such schemes, and of their benefits and costs, was ordered. Meanwhile the various ministries and agencies involved in water management were instructed to find other means to conserve water and to develop alternative strategies for the intensification of agricultural production.

There is no doubt that the huge cost of the SIBARAL project was the principal reason for the decision to discontinue work, and to drop it and other similar schemes from the 12th Five Year Plan. But a case also can be made for the growing importance of an environmental or conservationist perspective. The Gorbachev emphasis on *glasnost'*, or openness, clearly played a major part in facilitating public discussion of some of these issues. Equally important in legitimizing the environmental perspective, was the growing catalogue of instances of mis-management of natural resources attributable to technological intervention. While by no means was this list restricted to the management of water resources, certainly some of the worst examples of environmental degradation were related to their mis-management.

In the early 1970s it was determined that construction of a huge dike separating the Caspian Sea from its appendage, the Kara-Bogaz-Gol lagoon, would help to stabilize the level of the Sea by preventing excess water entering the lagoon and evaporating. However, by the time the dike was completed in 1980 the decline in the level of the Caspian Sea had stabilized; indeed, it had already increased slightly. The dike was also supposed to reduce slowly the area of the Kara-Bogaz-Gol lagoon. This was intended to benefit the chemical industry which had developed on its shore by permitting easier exploitation of its vast deposits of salts and other chemicals. These had been created through the evaporation over the centuries of the salty Caspian waters. Most forecasts estimated that the Kara-Bogaz-Gol would continue to exist as a lagoon for a couple of decades. In fact, by November of 1983 it had all but dried up, threatening the very future of the chemical industry in the process. Moreover, the creation of a vast salt bed posed a major environmental hazard for the extensive tracts of irrigated land in Middle Asia since wind-blown salts were now a distinct possibility. The solution required construction of a system of aqueducts to once again replenish the Kara-Bogaz-Gol with Caspian waters. By the mid-1990s, the rising level of the Caspian Sea necessitated diversion of water back into the Kara-Bogaz-Gol in order to minimize shoreline damage and flooding of coastal settlements. Similar miscalculations had an equally serious impact on the level of Lake Sevan in Armenia. There water diverted to generate hydroelectricity resulted in a far greater draw-down than anticipated. Given the historic importance of Lake Sevan in Armenian folk culture, to say nothing of the economic consequences, this unexpected environmental impact was widely criticized. To the ordinary Armenian citizen, aesthetics of landscape clearly had a value, a value given no weight in the original plans.

The Ministry of Power and Electrification figured rather prominently in many of these controversies. Throughout the Soviet Union, the Ministry's management of reservoir levels had long been severely criticized as having deleterious consequences for other vested interests in river basin management. Economies achieved were often decidedly short-run. For instance, in many of the low-gradient river systems of European Russia, plans to construct dikes or berms around reservoirs to minimize the area affected were shelved in the interest of cost-saving. In consequence losses of agricultural land were in excess of what were really necessary. In Siberia, 'savings' were achieved by not clear-cutting the forest cover from the intended zone of inundation of the Angara and Yenisey hydroelectric power projects. Not only was the economic loss in terms of unharvested timber significant, but prospects for a reservoir-based fishing industry were severely compromised. Even the generation of electricity was affected since the intake pipes to the turbines were frequently clogged with wood, thus reducing the volume of water intended to pass through the plant. Calculations were wide of the mark in respect of several other operational parameters as well. At the Bukhtarma hydro plant on the upper Irtysh river, the reservoir never did accumulate the planned volume of water. The available reservoir waters typically were released during the peak demand for power periods – autumn and winter. In the spring and summer the reservoir was replenished, but downstream there was often a shortage of water for agriculture as a result. In the case of the Bukhtarma hydro plant, this situation existed for more than two decades, and resulted in substantial losses in agricultural production in the region. Such opportunity costs were never adequately incorporated into the benefit–cost assessments of large-scale hydroelectric power projects. There were clearly conflicts of interest between the various groups of ministerial decision-makers. In the late Soviet period, the so-called 'departmentalist' mentality was increasingly the target of serious criticism, not just from the Gorbachev-era economic reformers, but from environmentalists as well.

The main thrust of much of the criticism of resources management from the environmental point of view focused on the absence of a holistic view of the impacts of technological intervention in natural systems. As in any industrialized state, there were numerous instances of technology developed for its own sake at the expense of other interest groups. In St Petersburg, flooding was a problem ever since the city was founded in 1703, because the city was built on the delta of the Neva river. When heavy autumn rains swell the outflow at the same time that on-shore winds drive water up the Gulf of Finland toward the city, the risk of flooding is very high. The technological solution was to construct a barrier dike across the Gulf to 'protect' the city. Local scientists contended that this posed an environmental hazard and not a solution to the problem of flooding. Their concerns were ignored. The dike was built, and natural water exchange and currents were altered. The considerable effluent carried by the Neva was impounded and resulted in pollution of both water and air. By the mid-1990s, the dike remains in a limbo, awaiting authorities to determine whether the structure should be modified or demolished. Scientific opinion from a different perspective than the protagonists of technological intervention into natural systems was usually ignored. Additionally, the economic criteria employed in decision-making was frequently flawed. For example, in irrigation schemes the 'irrigators' were customarily remunerated on the basis of the volume of water they processed. The more diverted to irrigation, the higher their wages. Since water was a free good for most of the Soviet era, there were no incentives to conserve; indeed, the system encouraged waste. A not dissimilar situation prevailed in the Khrushchev-initiated Virgin Lands scheme. There the area planted was expanded annually in response to the central plan. The dependence on a high degree of mechanization was related to the huge scale of the individual state farms, frequently in excess of 40,000 hectares of cultivated land. Against the advice of some agronomists the land was cropped annually and not allowed a period of fallow. Both water and wind erosion took hundreds of thousands of hectares of valuable land out of production as a result.

For most of the Soviet era, public or scientific criticism of high-priority state projects was not common, and when it did occur it was not especially effective. In the final years, *glasnost'* opened the door to public scrutiny of all issues, including those concerning negative environmental impact. Past reliance on well-intended legislation was questioned, and new approaches to resources management recommended. However, it would be wrong to contend that the Gorbachev-era tolerance of public comment on environmental and resource management matters signalled a complete break with previous practices. After all, the environmental disaster by which all others in the Soviet period pale in comparison, the nuclear meltdown at Chernobyl', was not reported immediately in the Soviet media. To be sure, it soon had headline attention, but the control over the media by central authorities remained a fact of Soviet life right to the end. Still, there is some basis for contending that the long-standing preoccupation with transforming nature was changing, albeit slowly. Natural resources were slowly being recognized as something to be managed wisely. And central to their wise management was the need to introduce prices into the world of state-ownership of all resources.

The Soviet approach to valuation of resources

Resource allocation procedures became the subject of intensive debate during the 1960s for some of the same reasons which prompted similar discussions elsewhere – a growing appreciation of the scarcity of natural resources and the recognition that greater economy in their extraction and consumption was long overdue. For most of the Soviet period, natural resources were allocated to state enterprises free of charge based upon some notion of presumed requirements or prescribed norms. The entirely rational behaviour of a natural resources extractive industry enterprise manager charged with the task of meeting an ever increasing production quota is, by way of illustration, as follows. Having successfully put a mine into operation, the manager is assigned a larger production target each year. Initially these targets are met, indeed, exceeded, thereby bringing manager and worker alike a bonus payment. The increased production is achieved by hiring more labour and using more capital equipment. At some point, however, it becomes difficult to maintain output, let alone increase it, since the deposit is steadily being depleted. Cost of production increases. Profitability of the operation declines in consequence. So long as the manager is able to successfully argue the need to move the operation to a new site, for which there is no charge and at which the marginal and average costs of production are lower, then there is no incentive to extract from the first deposit the maximum which is both technically and financially possible to achieve. In most mining operations, the enterprise would have to pay less than half the geological prospecting costs involved in finding another site. Thus, the request to move operations approved, the enterprise relocates. A rational decision for the management and workers; indeed, a rational decision for the ministry concerned since its success was dependent upon the performance of the various enterprises under its jurisdiction. Better to have more enterprises able to meet a demand for increased production imposed by Gosplan, and be profitable, than not. However, from the perspective of the national economy as a whole such decision-making behaviour was sub-optimal. Put simply, it was the antithesis of rational resource management, since natural resources development was not pursued with much regard to resource scarcity and value. The all too typical outcome was profligate waste and increasingly serious environmental degradation. Not surprisingly, in virtually every measure of consumption of resources per unit of production, the Soviet Union compared very poorly with western countries. Thus, a growing chorus of voices began to argue for a reassessment of the ideological obstacles to pricing natural resources. A major stimulus for change was the 1965 economic reform, and its emphasis on profit as the key measure of economic performance. It was now widely acknowledged that the desired efficiency in resource use would never occur if they continued to be allocated free of charge. The process of trying to introduce charges for natural resources thus began in the 1960s, and continued down to the final days of the Soviet Union. There was considerable ingenuity employed in attempting to invoke some semblance of a market economy factor in resource pricing during the final years of the Soviet era. There were also some small achievements. A few examples will suffice.

Land use within cities was irrational. In virtually every assessment of Soviet town planning in the final decades of the Soviet period the question of how better to allocate urban land uses arose. The high proportion of one-storeyed structures was often cited as an example of wasteful use of urban land. Even in cities with more than a million inhabitants, such buildings often comprised more than half of city land area. Not all such structures were housing. Indeed, industrial and institutional use of land was often criticized for taking up more land than the norms prescribed. But since land was free, all enterprises were encouraged to acquire as much as possible, by whatever means, simply as hedge against future need. It was contended that extensive land use produced abnormally high infrastructure and service costs, thereby adding to the overall expense of maintaining the urban system. Even in the city centre, land use was often extensive rather than intensive. The skyline of central cities began to change in the mid-1960s, but the arrival of the Soviet version of the skyscraper, which rarely exceeds 30 storey, did not solve what was widely regarded as inefficient use of a valuable resource. The standard argument once again was that in the absence of any charge for urban land there was no incentive to use it economically. A number of possible solutions to this problem of inefficient central city land use were proposed, but one of the most interesting of them entailed the use of surrogate measures of a market economy to place a value on the economic, engineering, social and environmental dimensions of the urban scene.

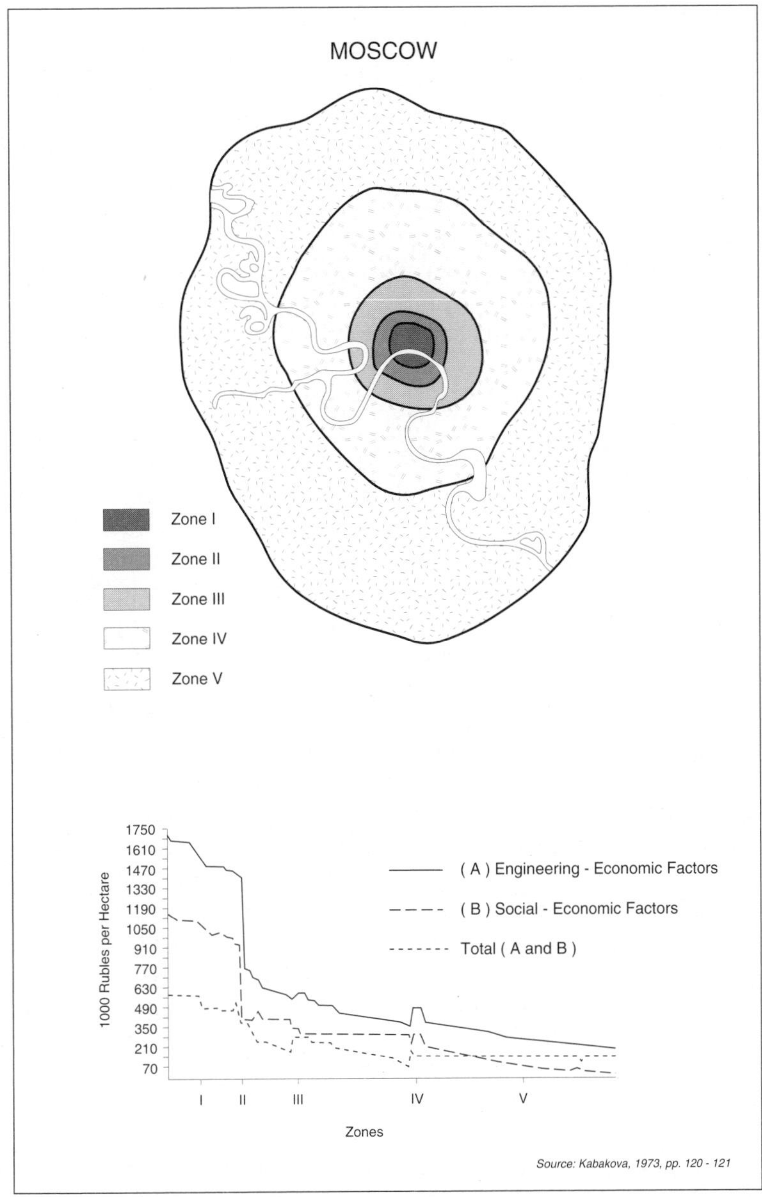

FIGURE 6.4 *Land value surface: Moscow in the 1970s*
(*Source*: Based on S.I. Kabakova, *Gradostroitel'naya Otsenka Territory Gorodov* (Moscow: Stroyizdat, 1973), 125)

What appears in Fig. 6.4 was akin to a land-value surface for the city of Moscow for the early 1970s. Similar analyses were undertaken to determine the spatial variations in the value of land in a number of other cities. The objective of the exercise, which it should be noted entailed a considerable amount of subjective evaluation, was to introduce charges for urban land which would reflect the inherently higher value of central city locations. The issue of how to actually charge realistic differential rents for urban land remained unresolved to the end of the Soviet era, however.

In the post-war period the loss of agricultural land to urban development had been considerable.

While the legislation which governed land use and conversion included norms for the amount of land required for particular purposes, it did not establish prices since land remained a free good to the end of the Soviet period. Not only had this situation encouraged rapid conversion of high quality agricultural land to non-agricultural uses, it biased evaluations of the benefits and costs of large-scale projects such as hydroelectric power stations. More than six million hectares of high quality agricultural land were permanently lost to reservoirs during the Soviet era. In the absence of a price for land, and therefore some ability to determine its discounted value in the future, it is small wonder that the calculation of the cost of a kilowatt-hour of electricity from hydropower stations had been criticized as unrealistically low. The prices paid by the state for agricultural commodities such as grain eventually came to be spatially differentiated. Thus, in more marginal agricultural regions, the price paid for commodities was higher than for the same product in more bountiful agricultural areas. While not quite the equivalent of differential rent, it was nonetheless a vast improvement over the flat rates of payment. The price paid by the state, however, never did reflect prevailing conditions of supply and demand.

Beginning in the late 1960s, most of the legislation pertaining to the management of natural resources was revised. Included were new Principles for the use of land (1968), water (1970), minerals (1975), forests (1977), air quality (1980) and the animal world (1980). Land remained a free good, notwithstanding growing criticism of the consequences. In the case of water, it too continued to be allocated free of charge to most users. In certain cases token charges could be levied. For example, in a special joint resolution of the Party and government in 1979, a charge for water used by industry was introduced. This special measure permitted a levy of relatively modest proportion to be exacted for consumption up to the maximum stipulated by the norm for the production process involved. Above-norm usage was supposed to bring a steep surcharge into play. Set at 400 per cent of the within-norm rate, it was intended to force water conservation. Of course, to charge for water requires some mechanism for monitoring actual rates of consumption, something not universally available then, or now for that matter. For example, domestic consumption was not metered and was paid for through an imputed fixed charge incorporated into rents for state-owned apartments. Not surprisingly, water in the typical Soviet urban household was not used moderately. As in the case of industry, domestic consumption was greatly in excess of what was used in comparable situations in western industrial states. While a charge for irrigation waters could be levied, this was rarely done. Thus, in general for agriculture, which accounted for about half of the total consumption, water remained in practice a free good. In the case of mineral resources, there were some measures to encourage full utilization, but they were indirect and largely ineffective. Aside from a modest levy to cover a proportion of geological exploration costs, in the 1980s resource 'sites' were still being allocated to users free of charge. Clearly some sites were advantaged because of favourable geological conditions for resource development. In such cases, higher taxes were imposed by the state on the value of the output each year. But such turnover taxes were in any event not very high. Whether the deliberately lower rates which were imposed on more marginal operations actually encouraged enterprises to stay in business longer remains a moot point. Certainly the pressure on enterprise decision-makers was to gain access to sites more geologically favourable, better located, or ideally both. In the 1977 legislation on the use of forest resources there were direct charges stipulated for the right to harvest a stand of timber. While there was debate as to the extent to which then prevailing prices for natural resources actually reflected their real value, they did represent a gradual shift in policy, and hence ideology, from the position adopted in the early Soviet period. With the advent of the Gorbachev-era economic reforms, scarcity and the market place were accorded unprecedented importance in pricing of natural resources. The process of reforming Soviet resource management practice had started, but too late to have any lasting impact. The collapse of the whole system ushered in a new era of resource ownership and management.

Land ownership in the post-Soviet era

The transition from a socialist to a market economy in each post-Soviet state has included much debate, often heated, over the issue of resource ownership. Who should own land, and what can be done with it figures prominently in the debate. Reformists everywhere contend that a fundamental first step in the transition process involves establishing in law the right of individuals and juridical bodies to own land, and to sell it in a free market. In many post-Soviet states, there is still a long way to go in the formation of a land market. In this section, the status of the land privatization process across the post-Soviet scene will be reviewed. The situation in Russia then will be discussed in greater detail.

Amongst the western tier states, Lithuania, Latvia and Estonia moved quickly to create the legal framework for the private ownership of land. Of course, in these states there were many people who still remembered the inter-war years of independence and non-socialist economic conditions. There was therefore a strong grassroots support for the replacement of rights lost with the forced incorporation of these countries into the Soviet Union during World War II. Laws were passed on the privatization of land in 1991, and since then additional legislation has been enacted to refine the process. One of the first challenges in each of the Baltic states was to make legal provision for the return of land and property confiscated at the time of the Soviet occupation. Indeed, privatization really began with the restitution of ownership rights. In Latvia, the law on land reform adopted in 1991 initially forbade land to be owned by juridical entities, for instance legally incorporated institutions whether domestic or foreign, or by foreign citizens. An amendment to the law in 1995 now permits both to own land for business purposes. Lithuania has restricted land ownership to citizens only, excluding both domestic and foreign legal entities and foreign citizens. However, there have been a number of amendments proposed to ease the restrictions. While all had been rejected up to 1995, in the competition between the Baltic states for foreign investment there will no doubt continue to be efforts to enable foreign and domestic companies to buy land related to business needs. Long-term leases are also an option. Legislation to effect such an arrangement was put forward in 1994 in Lithuania, but had not been approved a year later. In Estonia there are similar obstacles and opportunities. Restitution of land and property to former owners, or their heirs, was one of the first priorities. But sorting out such matters as legal ownership entitlements five decades after confiscation is fraught with difficulties, and as elsewhere, there is much at stake in terms of economic development. Legislation dealing with land and real estate ownership was not passed until 1993. According to it, both land and real estate necessary for business may be bought by foreign enterprises, provided the transaction is approved by local municipal authorities. Since local authorities are likely to be more sympathetic to job creation initiatives than those in central government, permission of municipal officials is unlikely to be a significant barrier. Without laws both permitting and protecting the right to own land on the part of legal entities, especially foreign ones, investment by western companies will be more difficult to attract. All of these countries are keen to join the European Union. Restrictions on foreigners in terms of their buying land does little to promote the case for inclusion. Thus, Lithuania will come under growing pressure to adjust its laws to be more compatible with those of Latvia and Estonia.

In Moldova, laws pertaining to the ownership and disposition of land were enacted in 1992. This law recognized the right of individuals to own real estate, including land. The state was also prepared to enter into leasehold arrangements for periods up to 99 years. But while the law on land ownership was progressive in spirit, the right to dispose of privately owned land was restricted since land could not be bought or sold until the year 2001. Clearly, such caveats do little to accommodate the interests of market reformers. But such cautious steps in the privatization of land process were commonplace in many other states as well. In Belarus' even before independence a land lease law was enacted. This 1990 legislation was intended to lay the legal framework for lifetime inheritable leases of land for private farming. But the leases permitted could not be sold or mortgaged, nor could land be bought. This rather tentative first step was followed by a law on land ownership in 1993. This law allows ownership of small plots of land for such purposes as housing, *dacha,* or a cottage, orchards, gardens and the like. For small scale farmers, the law permits leasing of up to 50 hectares. More recent legislation permits the mortgaging of land lease contracts. A new code to govern the disposition of land was also passed in 1990 in Ukraine. The code regulated the formation of private farming, and dealt only with lifetime inheritable land allocation issues. Subsequent legislation passed in 1991 and 1992 reflected the new political and economic reality of independence. Private ownership of land was recognized in the legislation, along with state and collective ownership. The role of the state remains pre-eminent. In agriculture, land could be privatized in the creation of a peasant farm, but as will be discussed in Chapter 7 Ukraine has made little progress in diversifying the traditional organization of agriculture. Peasant farms were still only a very minor element in the rural scene in 1995. According to the legislation, even this land cannot be bought or sold until 1998. As for ordinary citizens, legal entities, and foreigners, buying and selling of land in a free market is still some distance down the road.

Southern tier states are quite mixed in their legislative response to the opportunities presented by the new economic reality. In the Caucasian states of Armenia and Georgia, privatization legislation dealing with land ownership was quickly adopted. In Georgia, for example, the legal framework was in place by early 1991, and by mid-year some 50 per cent of agricultural land was privately owned. A moratorium on transfer of ownership, on selling land, was imposed for two years. In the cities, there was a significant share of the housing stock which was privately owned already in the Soviet era. Extending right of ownership to the land was a logical step. Armenia adopted privatization legislation at the same time, but moved even faster than Georgia to create peasant farms. By late 1991, the bulk of the agricultural land had been privatized. In Azerbaijan the process was a good deal slower. A land code to facilitate the privatization of agricultural land was only adopted in 1994. The delay was related to the politically sensitive nature of the principle of private land ownership in the countryside. Agricultural land cannot be owned by foreigners. In the cities privatization of real estate is proceeding apace, but the issue of foreign ownership remained unresolved in 1995.

In the Middle Asian states of Kazakhstan, Kyrgyzstan, Tajikistan, Uzbekistan and Turkmenistan there are clear differences in the attitude to, and hence progress towards, privatization. The first two are proactive and have made substantial progress, the latter two are moving only hesitantly. In Tajikistan the internal turmoil makes implementation of any government policy somewhat problematic. In Kazakhstan laws facilitating the privatization of land were introduced early, and formally ensconced in the draft constitution which appeared in 1995. Private agricultural plots and kitchen gardens were declared to be private property. But the constitution goes further, including the right for enterprises of all descriptions to own the land necessary for their operation. The government retains control over water and natural resources, something which is common everywhere in the post-Soviet scene. The private ownership of land being guaranteed as part of a state constitution is certainly a major step amongst most post-Soviet states. Kyrgyzstan property law stipulates that all land, and natural resources of all description, belong to the state. Leasehold arrangements are possible in the countryside and are the basis for the formation of peasant farms. Leaseholds, while inheritable, cannot be sold. A similar policy applies in Tajikistan, a result of the constitution which was adopted in 1994 which declares that the state is the owner of all land and natural resources. Turkmenistan adopted land reform legislation in late 1993.

Agricultural land may be leased or inherited, but only in limited quantity, the ceiling being 50 hectares. Once allocated, land cannot be subdivided. The minimal lease period is ten years, a period deemed long enough to promote good management. Historically, the land was vested in kinship groups and extended families. There not being any history of private land ownership amongst Turkmens there is no grass-roots pressure to introduce it. This applies throughout most of the region in fact. In Uzbekistan as well the land is state owned, although leaseholds are permitted in the rural sector and are perceived to be a step toward privatization at some point in the future. A government decree of 1993 permits the leasing of buildings and land by foreign investors for periods up to 100 years.

Land ownership and resource development in Russia

In Russia, some important steps toward facilitating property rights for individuals had been taken even before independence. Two laws on land reform were adopted in late 1990, the intent of which was to end the state's monopoly on land ownership. Subsequently, a series of Presidential decrees and legislation newly adopted or amended served to further entrench the principle of land as a commodity. The new constitution specifies the right of land ownership. The civil code confirms ownership of private property as a right. But it is the land code which provides the mechanism for land to be bought or sold. A land code first proposed in 1991 underscored the rights of individuals by specifying precisely what can be bought and sold. Foreign ownership is not permitted, unless part of a joint venture arrangement. The question of opening up a market for land was a very contentious political issue from the outset of the economic reform, and hence the proposed federal land code was not approved and the whole matter remained in a state of legal limbo. This left the disposition of land in the hands of local government officials who interpreted existing laws and Presidential decrees to suit their own purposes. As in the Soviet era, principle and practice were not always the same. Indeed, this applies across the post-Soviet scene for wherever there are powerful vested interests opposed to a particular law, there are usually a host of ways in which the interpretation of the law can be misconstrued, or its implementation derailed. In Moscow and St Petersburg, for instance, the federal civil code stipulating that land could become private property was simply rejected by the Mayors. They declared the code invalid within their jurisdictions, and then sought special Presidential exemptions to sustain their position, which in this case were forthcoming. The result was that land could not be bought or sold in these cities, only leased. Just outside the city border there quickly developed a very active market for land. In smaller cities, the buying and selling of land did occur, though the rights of the land owners were often not especially certain. By 1995, many important matters pertaining to land ownership remained in a kind of legal twilight zone. For example, the rights of owners of apartments in small cities to the land on which their building sits is not clear. Developers of housing or commercial properties to be put on the market do not always have clear tenure of ownership themselves. The purchasers of such properties are even in a less satisfactory situation with respect to the right of ownership of the land involved. The list goes on.

The question of land ownership has not been resolved to the satisfaction of all simply because the very concept is still rejected by many people on principle, while many others who stand to benefit from the creation of a land market wish to control what happens rather than let the market decide. The whole issue of private ownership of land came quickly to the fore in the drafting of a land code in the State Duma in 1994. The task was turned over to a standing committee dealing with agricultural issues. Dominated by Duma members sympathetic to the maintenance of the status quo in the countryside, a draft land code finally gained the necessary support of the Duma in July 1995. The draft document was then forwarded to the Federation Council for consideration. The draft document greatly restricts the existing rights of peasants to acquire land for a private, peasant farm. In the draft land code, land in large Russian cities cannot be bought or sold, only obtained on long-term lease. The bill must receive the endorsement of the Federation Council before becoming law. But President Yeltsin can veto it if it does pass. The problems associated with the new Russian land code simply mirror the deeply held opinions that land should not become yet another commodity.

The legal ramifications of de-nationalizing natural resource development not surprisingly has been the subject of intense debate in Russia. Anti-reform groups contend that the state must retain a monopoly position in resource development otherwise the land and its resources will be ravaged by foreign firms bent upon taking advantage of Russia's present day needs. The principal law dealing with subsurface mineral resources was adopted in 1992, and established very clearly the role of the state as the owner of these resources irrespective of who owned the surface land area in question. Russian companies and joint ventures with foreign firms, however, are entitled to lease exploration and production rights for limited periods of time from the government. For an exploration license, the period was five years; for actual production the leasehold was for 20 years. In most instances, licenses were issued by the government on the basis of competitive tenders. Royalty and taxation payments were to be paid according to the prevailing rates. But in Russia the whole question of ownership by government of resource rights has been complicated by the existence of Republics within the federation, many of which claim ownership of natural resources, some of which have negotiated special arrangements with Moscow to have a substantial say in how they will be developed. The leverage such Republics have is that in Russia the political authority of the centre is waning, and there are many minority group regions which are more than ready to assume greater control over their own economies, and especially over those natural resources which are highly valued by the west. Oil and natural gas, rare metals, diamonds – the list is long, and so too is the list of interested foreign firms willing to deal with Russia's regions even in the murky legal and taxation environment which surround transactions with regional governments. In 1994, a new version of the law dealing with underground resources was passed by the State Duma. This law recognizes the joint responsibilities of both the federal government and the regions in issuing licenses for resource development. To the extent that this law clarifies how business will be done, and especially what the taxation regimes will be, then the door will be opened a little wider to much desired foreign investment. In the oil and gas sector, for example, the number of federal, regional and local taxes in some regions is close to 40. Tax avoidance is fostered in such an environment.

Across the post-Soviet scene the concerns are broadly the same, although not all countries have gone as far as Russia in market reform and breaking down the old government monopolistic structures. In all countries the ownership of natural resources is vested in the state and development is controlled either by state companies or through government ministries or departments. As in Russia, foreign companies or joint ventures are obliged to bid for licenses, increasingly within the context of competitive tenders. However, in most post-Soviet states special licenses negotiated with the appropriate government agency are still possible to obtain. Indeed, a growing trend amongst those countries which had decentralized decision-making in the resource industries, established quasi-private companies to manage development, and had promoted privatization and market reform, is for government to attempt to put in place single authorities to supervise and control resource exploitation in order to secure greater control over income generation and revenue for the state. In Russia in late 1995, for example, there was discussion of the need to create some type of overarching government organizational structure for the oil industry, an industry dominated by some 30 or more privatized production companies. But where government intervention is rising, there are likely also to be inefficiencies and problems of a different kind. A case in point is the attempt to replicate the urban land market.

It is now possible to buy land and obtain legal title to it as private property in Moscow Oblast', but its sale within the city has been prohibited as a result of special legislation adopted by the City of Moscow. There is long-term leasing, however. In order to establish a rational basis for lease rates and various forms of taxation, there has been continual refinement of the non-market procedures to establish the differential value of land throughout the city. This began in the 1970s as noted earlier. A more recent version of a differential system of land values is that adopted by the

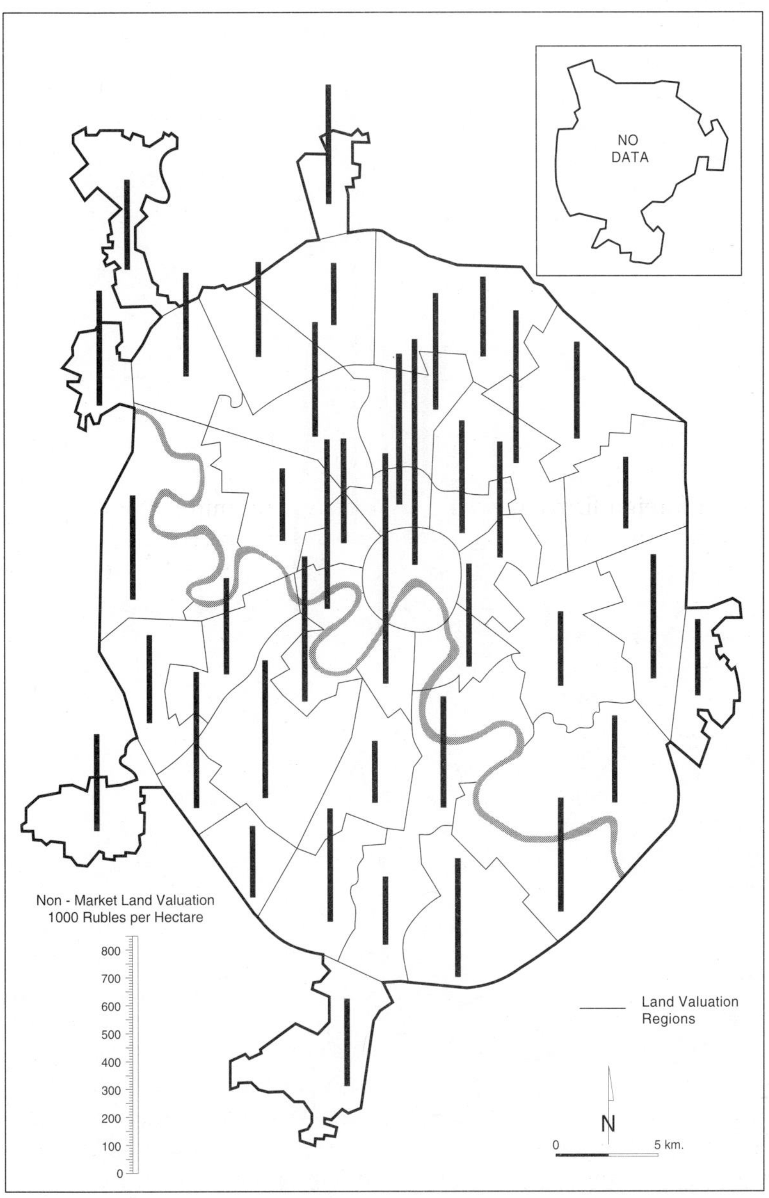

FIGURE 6.5 *Land value surface: Moscow, 1992*
(*Source*: J.H. Bater, V.N. Amelin, A.A. Degtyarev, 'Moscow in the 1990s: Market Reform and the Central City', *Post-Soviet Geography*, Vol. 35, No. 5, 1994, 257)

Moscow City Council in late 1992. The 40 zones within Moscow are depicted in Fig. 6.5. It is apparent that the city centre, which is demarcated in this scheme by the *Sadovoye Kol'tso*, or Garden Ring, was assigned one of the higher valuations. But at 810,000 roubles per hectare, it was not the highest assessment. Pride of place went to Zone 6, immediately to the south of this

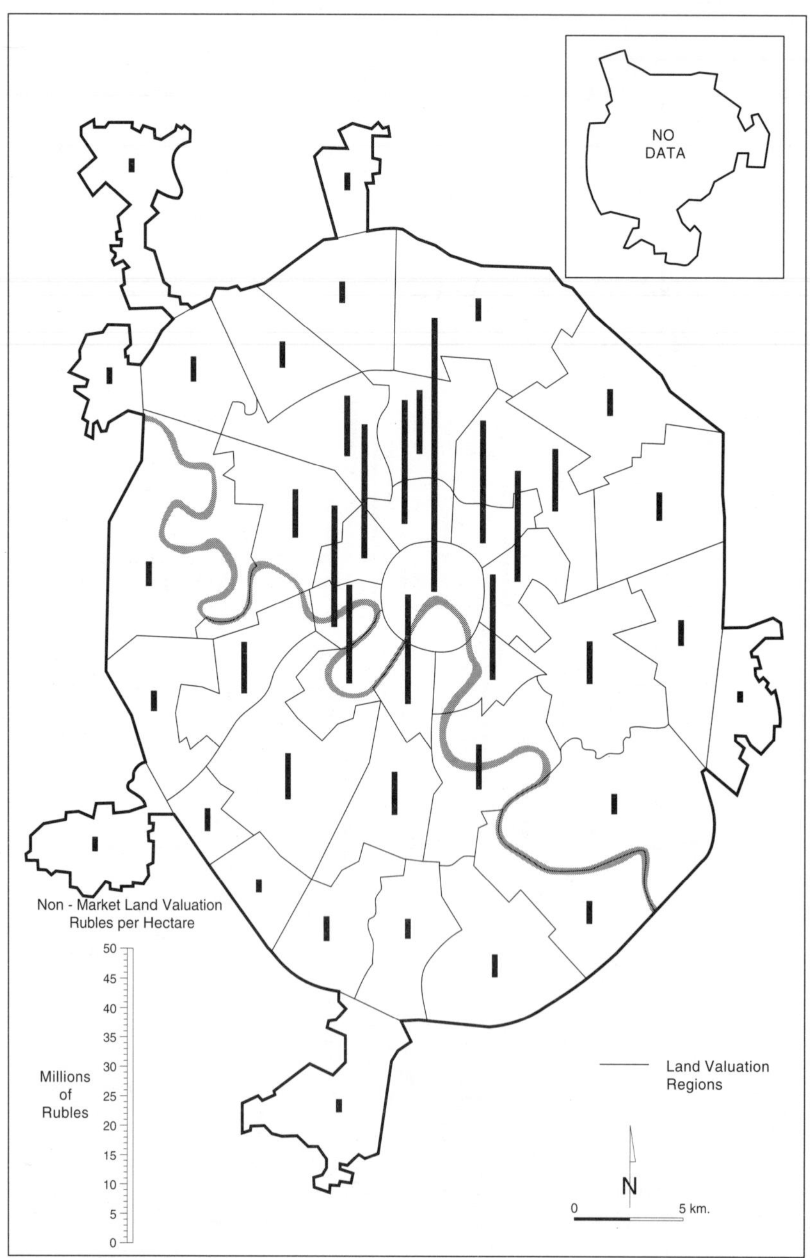

FIGURE 6.6 *Land value surface: Moscow, 1994*
(*Source*: J.H. Bater, V.N. Amelin, A.A. Degtyarev, 'Moscow in the 1990s: Market Reform and the Central City', Post-Soviet Geography, Vol. 35, No. 5, 1994, 258)

central zone, where land was 826,000 roubles per hectare (or about US$ 413 at the then prevailing exchange rate). Although there was certainly a core–periphery gradient in the pattern of land valuations portrayed in Fig. 6.5, there were clear anomalies. These may well have been quite appropriate, but there is no substitute for the marketplace in establishing land values. There is also the not inconsiderable problem of adjusting for the current value of the rouble in a period of high inflation. Auctions of municipally-owned buildings in 1992 and 1993 in Moscow invariably produced far greater returns to the city than anticipated by the authorities. While this was undoubtedly a positive outcome for municipal coffers, it did not imply a full understanding of current market conditions. Government at all levels is not unaware of the gap between official land valuations and market prices, nor of the importance of property taxes in western cities as a source of municipal revenue. In January 1994, the schedule depicted in Fig. 6.5 was again revised. In the new schedule, depicted in Fig. 6.6, there is much clearer recognition of central city land values, most notably in the Central Prefecture. Valuation of land in the core region, the Garden Ring, was increased to 46.6 million roubles per hectare, or roughly $US 23,300 at prevailing exchange rates. Land there was now twice the value of the region with the next highest assessment. This was certainly a justifiable increase. But even this change cannot keep pace with the market. The asking prices in US dollars per square metre for apartments in Moscow in mid-1993 are portrayed in Fig. 5.12, p. 146. Comparison with the values per hectare assigned in Fig. 6.6 using surrogate measures underscore the difficulties of any taxation system to stay abreast of market changes. In market economies the cost of land has some bearing on the land-use decision-making process, both in government and the private sector. As yet, this cannot be adequately factored into the decision-making process in Moscow.

The market reform process, privatization of property and the right of citizens and legal entities to own real estate and land has started throughout the post-Soviet scene. Resource development is increasingly being brought into the world economy, with mixed consequences for the countries and peoples concerned. How these forces for change are impacting the fundamentally important business of feeding the populations of these new states is the subject of the following chapter.

FURTHER READING

DeBardeleben, J. 1985 *The Environment and Marxism–Leninism: The Soviet and East German Experience*. Boulder, CO, Westview Press.

DeBardeleben, J., Hannigan, J. 1985 *Environmental Security and Quality after Communism: Eastern Europe and the Soviet Successor States*. Boulder, CO, Westview Press.

Graham, L.R. 1972 *Science and Philosophy in the Soviet Union*. New York, Random House.

Graham, L.R. 1990 *Science and the Soviet Social Order*. Cambridge, MA, Harvard University Press.

Graham, L.R. 1993 *The Ghost of the Executed Engineer: Technology and the Fall of the Soviet Union*. Cambridge, MA, Harvard University Press.

Jancar, B. 1987 *Environmental Management in the Soviet Union and Yugoslavia*. Durham, Duke University Press.

Joravsky, D. 1961 *Soviet Marxism and Natural Science, 1917–1932*. New York, Columbia University Press.

Pryde, P.R. 1995 *Environmental Resources and Constraints in the Former Soviet Republics*. Boulder, CO, Westview Press.

Rozengurt, M. 1991 *The Soviet Water Crisis: Exposing an Environmental Disaster*. Boulder, CO, Westview Press.

Turnbull, M. 1991 *Soviet Environmental Policies and Practices: The Most Critical Investment*. Brookfield, VT, Dartmouth.

Vari, A., Tamas, P. 1993 *Environment and Democratic Transition: Policy and Politics in Central and Eastern Europe*. Dordrecht, Kluwer Academic Publishers.

Weiner, D.R. 1988 *Models of Nature-Ecology, Conservation and Cultural Revolution in Soviet Russia*. Bloomington, IN, Indiana University Press.

Ziegler, C.E. 1987 *Environmental Policy in the USSR*. London, Frances Pinter.

7

MANAGING THE LAND: FROM PEASANT TO PROPRIETOR

'the commune is the cradle of the social organism'
(A. De Gurowski, *Russia As It Is*, 1854,192)

There are few countries in the world in which the agricultural sector is not affected by government policies and programmes. From the provision of an infrastructure to facilitate shipment of foodstuffs from countryside to city, to price support programmes to ensure farmers a reasonable return on investment of labour and money, to marketing boards to assist in coordinating the sale of production, to quotas and duties on imported foodstuffs, governments the world over endeavour to assist and protect the agricultural sector. But relatively few countries followed the Soviet model of state ownership of the land, forced collectivization of the peasantry, and state planning of agricultural output. Those that were obliged to adopt the system because they were part of the post-World War II Soviet sphere of influence have been fast to shed it when opportunity arose. Other countries which emulated the system have been forced to make changes when inadequate levels of output threatened to choke the supply of food to the cities. It is conventional wisdom that farmers, and by extension the states they live in, do better when they own their land and work for profit.

The history of agriculture in the Russian Empire is not one of a large class of family farmers working their land independently, of course. At the time of the Russian Revolution, three-quarters of the approximately 160 million total population were rural. Most of the rural population lived in villages, conducted their daily affairs in harmony with age-old customs, and cultivated their separate strips of land in accordance with the collective decisions of the rural commune, or *mir.* The revolution and subsequent socialization of agriculture changed in a fundamental manner the relationship of the peasant to the land, and to the crops sown and harvested. But collectivism itself was deeply engrained in the cultural traditions of most rural Slavic peoples. Cultural tradition notwithstanding, the forced collectivization of the peasantry during the Stalin era did not achieve the state's objective of creating a productive agriculture. The state-run farming sector was simply inefficient. The most productive components of the Soviet socialized agricultural system were the tiny private plots and few livestock which individual rural households could legally cultivate and own. Cultivated assiduously and nurtured with great care by tens of millions of peasants during their free time, yields from these small islands of private endeavour exceeded many times over those in the vast socialized agricultural sector in which peasants were daily obliged to labour long hours.

Privatization of the agricultural sector was widely perceived to be a panacea for the flagging production which characterized the final decades of the Soviet period. Indeed, during the Gorbachev era some first, hesitant steps were taken to facilitate the formation of a Soviet version of the family farm as part of the strategy to bolster agricultural output. The collapse of the Soviet system seemingly would have removed all obstacles to widespread transformation of socialist agriculture into a more productive system characterized by the private farm geared to the marketplace. A few post-Soviet states have made substantial progress in the privatization of agriculture, but most have not. Private ownership of land, upon which fundamental reform of agriculture is predicated, remains a major obstacle to economic reform. How, or indeed even why, land should be privatized is still an emotionally charged political issue in many post-Soviet states. This is especially so in Russia, Ukraine and Belarus'. Notwithstanding some success in privatizing the rural economy in some states, nowhere has aggregate agricultural output increased. In Russia, for example, imported foodstuffs are steadily garnering a larger share of the domestic market. Farming, like industry, has been severely impacted everywhere by the disintegration of the Soviet economic system. Thus, the transition from a socialist agricultural system to one in which individual farmers respond to the opportunities and challenges of a market economy is far from complete. In this chapter, we will examine the process, and progress, of post-Soviet agrarian economic reform. Before doing so however, it is necessary to have some appreciation of the physical resource basis for agriculture across the post-Soviet scene, and some understanding of the nature of the agricultural system imposed during the Soviet period.

THE PHYSICAL BASIS FOR AGRICULTURE

From the standpoint of growing crops, there is much that is disadvantageous about the physical geography of the post-Soviet territories. An ideal balance between heat, moisture and light is an all too rare occurrence. In Chapter 1 the basic features of the physiography, climate and zones of natural vegetation were outlined. It was noted that the combination of northerliness and continentality means severe temperatures during the winter months are commonplace. Indeed, in Russia for example, permafrost conditions affect most of the country. At the other extreme, much of Middle Asia suffers from desert heat. While the harsh conditions imposed by too much cold or too much heat, too much or too little moisture, do not rule out agriculture, crop options and types of husbandry are clearly restricted. The largely latitudinal pattern of the vegetative zones is very much a reflection of the impact of climate. The wedge-like strip of productive wooded steppe and steppe which diminishes in north–south extent from Ukraine and the Kuban through to Eastern Siberia is one example of the influence of climate (see Fig. 1.5, p. 17). The basic pattern of 'accumulated heat', that is, the number of degree-days above the 10 degrees Celsius threshold, which tends to be the temperature at which plant growth can be sustained, is perhaps even more striking in this regard (Fig. 7.1). The desert and semi-desert region of Middle Asia clearly stand out on this map as the principal storehouse of accumulated heat. Tapping this potential climatic resource comprised part of the rationale for the scheme to divert the north flowing Ob' and Irtysh waters to the south, as was noted in the preceding chapter. Unfortunately, for most agriculturists the distribution of effective moisture depicted in Fig. 7.2 (see also Fig. 1.4, p. 16) is roughly the obverse of the distribution of accumulated heat (Fig. 7.1). The area of greatest agricultural potential, namely the belt of rich chernozem soils, of which Ukraine has a significant share, also lies within the all-too-extensive zone of potential moisture deficiency (see Fig. 6.2, p. 161). Put simply, altogether too much of the land in most post-Soviet states is decidedly marginal in terms of agricultural potential.

As Fig. 7.3 reveals, the largest agricultural region is roughly coincident with the tundra and northern tayga and is characterized by reindeer

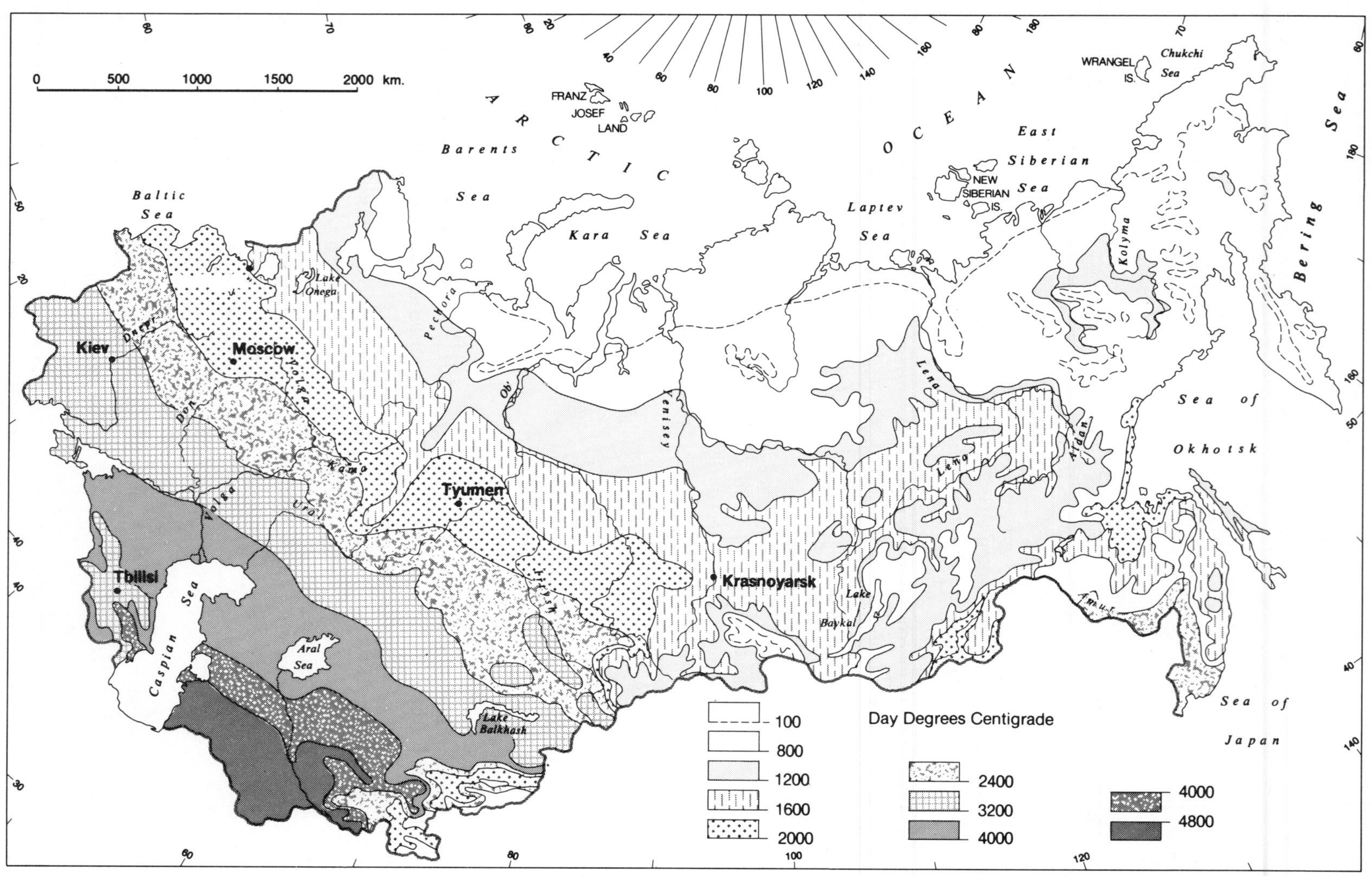

FIGURE 7.1 *Distribution of accumulated temperature*
(*Source*: L. Symons *et al.*, *The Soviet Union. A Systematic Geography* (Totowa, NJ: Barnes and Noble, 1983), 40)

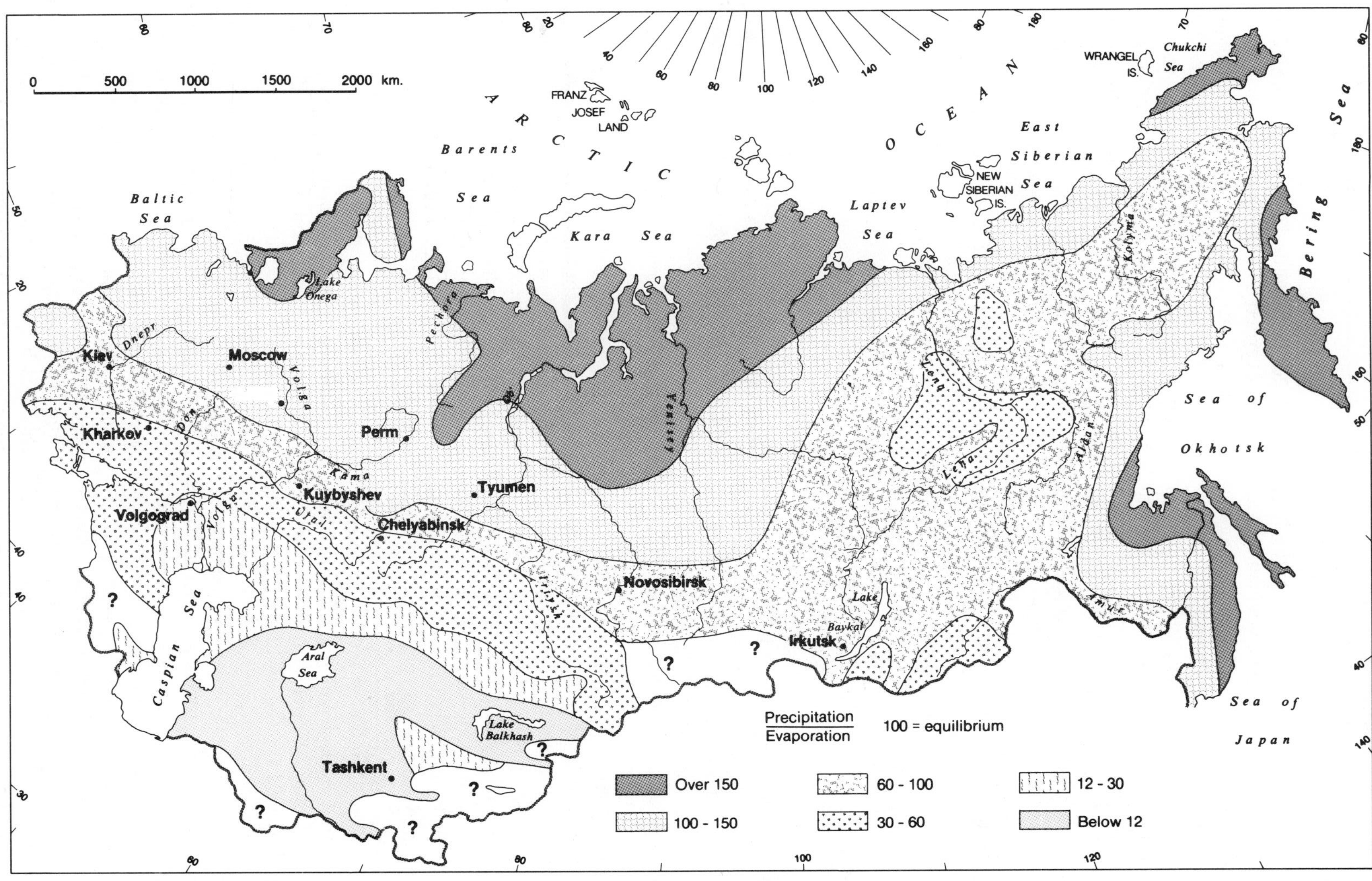

FIGURE 7.2 *Distribution of effective moisture*
(*Source*: L. Symons *et al.*, *The Soviet Union. A Systematic Geography* (Totowa, NJ: Barnes and Noble, 1983), 40)

husbandry. Where settlement exists, and where local soil conditions allow, some vegetable production occurs. But more often than not, vegetable production is only possible in the artificial environment of the greenhouse. While this is of obvious local importance in terms of both employment and food production, the region as a whole is still a net importer of all food products. To the south is a zone of tayga in which reindeer husbandry is complemented by more extensive pockets of cultivation and livestock rearing. Agriculture here is usually pursued in conjunction with other economic activities and is highly dependent upon inorganic fertilizer supplements to the typically heavily podzolized soils of the forest and marsh. Dairying occurs in the tayga, but mostly in association with major northern settlements such as Murmansk. If these areas are combined with the vast track of desert and semi-desert pastoralism, it is apparent that for over much of the post-Soviet scene agriculture is a distinctly extensive as opposed to intensive activity. In order to ensure a balanced diet for the resident population of these huge territories food must be imported. Elsewhere dairying, livestock raising, grain production and market gardening are found in differing combinations as determined by the natural environment, local tradition, and the demands of urban settlement. An 'agricultural wedge', broadly defined by a line joining St Petersburg, Odessa and Irkutsk, encompassed the bulk of agricultural production, and not surprisingly most of the population as well, of the former Soviet Union and Russian Empire before it. As Fig. 7.3 indicates, there are relatively few specialized agricultural regions. Most are found in the more temperate realms near the Black and Caspian Seas, in the sub-tropical enclaves of the Caucasus, and in the oases and irrigated tracts of Middle Asia. The intensity of agricultural activity is obviously higher in these few regions where the physical basis for agriculture is propitious. It is also a function of accessibility to urban markets. Around most cities there has long been more specialized, intensive, market gardening agriculture geared to the needs of urbanites. But excluding dairying and market gardening under greenhouse conditions, the range of agricultural activities to a large extent is still governed by climate and physical geography. To be sure, soil drainage and irrigation, chemical fertilizers, genetic science and mechanization permit more efficient use of the agricultural land base, but it is still impractical to grow corn in the tayga or cotton in the steppe. Clearly, the nature of the physical environment imposes some constraints on agriculture. But during the Soviet era, the institutional environment which enveloped those whose daily life and labour were tied to the land was probably a greater impediment to enhancing agricultural production than was the natural environment. The following section discusses the evolution and efficacy of the Soviet-era institutional arrangements imposed on the agricultural sector.

THE ORGANIZATION OF AGRICULTURAL PRODUCTION IN THE SOVIET ERA

At the end of the Soviet period about one-third of the total population was classified as rural. According to Soviet definitional procedures, this was the share of the total population which lived in places *not* qualifying as urban-type settlements or cities. The total number of people actually involved in some facet of the agricultural sector exceeded 60 million, that is, slightly more than one-fifth of the total population. By any standard of comparison this was an inordinately large share of the total for a modern industrial state. It is worth reiterating that until 1961 the rural population was more numerous than the urban. An additional 11 per cent, some 33 million people, lived in the countryside, but were not directly dependent upon agriculture for their livelihood. Many of these people were part of the urban workforce, commuting daily from their homes in a rural settlement to the city out of necessity, or occasionally choice. The rest comprised workers in factories located in the countryside and a wide range of technical specialists (part of the *intelligentsia*, not peasant, social class). From school teachers, to scientific personnel, to government bureaucrats, the support structure for the

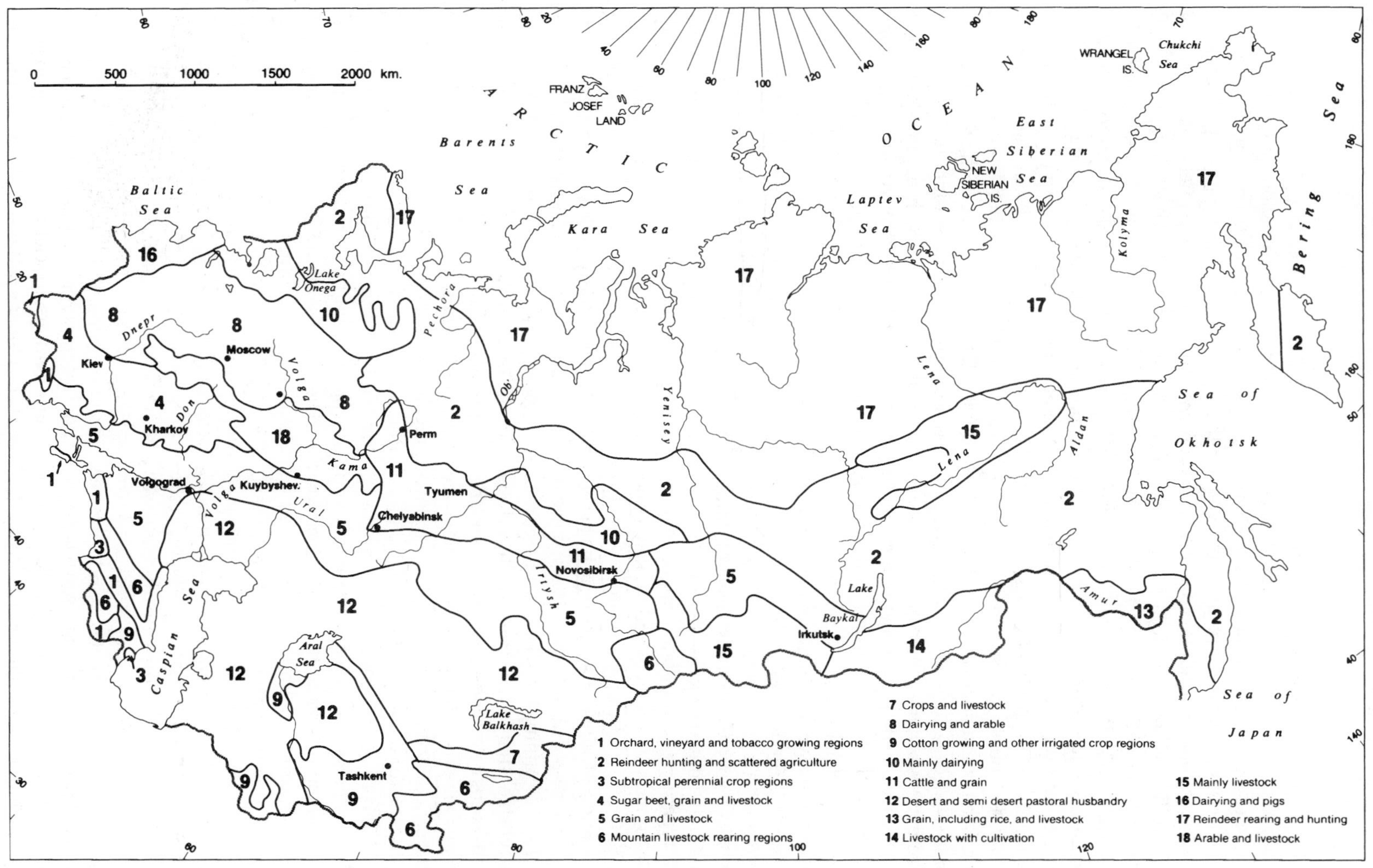

FIGURE 7.3 *Agricultural regions*
(*Source*: L. Symons *et al.*, *The Soviet Union. A Systematic Geography* (Totowa, NJ: Barnes and Noble, 1983), 120)

millions of people directly engaged in agriculture had grown rapidly in the post-war period. This development brought considerable change to the rural social class structure, not least of which being a host of other, more subtle, manifestations of social position than the official categories of social class strata (see Table 2.3, p. 44). For example, *kolkhozniki* who were first to hold an internal passport clearly had status in the Soviet scheme of things. Education and positions in management brought a wider array of perks and privileges as well. Of course, beneath the trappings of a modernizing, industrializing society, lay the traditional social class structure in which peasant values determined status. Thus, both in the city and the countryside there were many nuances in determining social distance. In the countryside, however, the combination of modernity and traditional cultural values made for a much more complex social system. Attempts to reorder rural society to meet the requirements of the Soviet socialist model met with strong resistance, and thus with only limited success.

To the typical peasant of late imperial Russia, legal title to land was often beyond his ken. Functionally illiterate, and deeply imbued with the values of the communal lifestyle which the *mir* offered, he went about his tasks with a sense of duty to the community as a whole, and secure in the knowledge that what was produced on his various strips of land could be used to feed his family. The Stolypin Agrarian Reform and estate agriculture, modern and mechanized, were slowly changing the rural scene in late imperial Russia to be sure, but the resurgence of the communal ethos at the time of the Russian Revolution and the redistribution of land was a measure of how important centuries-old traditions remained in the countryside. In central European Russia, peasants who had opted out of the commune in accordance with the Stolypin Reform often were forcibly returned. For many peasants living in the city, an opportunity to return to the land was compelling. No doubt the severe hardships of urban life during the revolution and ensuing years of Civil War were further incentive to leave. Thus, the number of peasant households was larger following the revolution of 1917 than before it (25 million as opposed to 21 million). Many of the western regions traditionally characterized by individual farming units were 'lost' to the Soviet Union through independence movements and border adjustments at the time of the Treaty of Versailles. These included the Baltic States, and parts of the west Ukraine and Belarus'. As a result of this process, the role of the commune in agriculture was intensified in the immediate post-revolutionary years. But it was not the same everywhere. From the European Russian core to the peripheral regions of the new Soviet state the influence of the *mir* declined, just as it did prior to the revolution.

While the revolution brought a redistribution of land, the Civil War brought the domestic economy to the brink of ruin and gave rise to forced requisition of foodstuffs from an increasingly beleaguered peasantry. Subsistence agriculture prevailed since those who could produce a surplus were disinclined to exchange agricultural commodities for a currency that was worthless. Lenin took the major step of permitting the peasantry a measure of free enterprise with the introduction of the New Economic Policy in 1921. This had several consequences. It stabilized the domestic economy. Since forced requisition was ended and replaced by taxes in kind or in money, the peasantry was encouraged to produce food for the market. While the government promoted the replacement of the commune by collectives and state farms, the lure of profit had an even greater impact on the communal system through the expansion of private farming. The stated owned the land, but capitalistic forms of agriculture developed rapidly during the 1920s nonetheless.

The success of the New Economic Policy in restoring agricultural production to pre-World War I levels, thereby ensuring an adequate food supply for the rapidly expanding urban populace, was double-edged. The capitalistic tendencies unleashed by the New Economic Policy were facilitating the rise of a class of wealthy peasant farmers, or *kulaki*, in the countryside, which at the time was home to four-fifths of the total population. The growing commercial significance of the *kulaki* underscored the apprehension of some leaders over the political reliability of the peasant population as a whole. Soviet authorities had formalized the earlier Russian practice of separating the peasantry into three categories, poor peasants (*bednyaki*), middle peasants (*serednyaki*), and rich peasants (*kulaki*). In endeavouring to exercise political control over the massive rural population, the government strategy was based on fostering tensions between the members of the poor and middle peasantry and the *kulaki*. Yet the official position over how to deal with the peasantry was subject to considerable debate. After Lenin's death in 1924, rival Party strategists, Bukharin and Stalin, came to represent the two predominant, but conflicting, views. Bukharin argued that the nationalization of land, the tax system, and agrarian reform initiated by the revolution, were leading to a genuine convergence of classes in the countryside, and consequent diminution of *kulak* authority. Many took Stalin's analysis as correct, however. He contended that the rural scene was becoming more, not less, stratified as the *kulak* element steadily acquired more rented land and hired labour. The fear that a politically unreliable peasantry might thwart the drive to industrialize the country by withholding food from the cities loomed ever larger. The Stalinist view prevailed. A programme of collectivization was adopted as part of the First Five Year Plan (1928–1932) introduced by Stalin.

In some regions, the authority of the commune as a social and economic institution had been challenged by the nascent forms of capitalistic agriculture which flowered under the New Economic Policy. Thus tensions already existed, and were consciously exacerbated where possible by Soviet authorities. The initial plan to collectivize 20 per cent of the cultivated area made use of the poor peasants (*bednyaki*) in the assault on the *kulaki*. Formation of collective farms entailed the redistribution of wealth, and in some villages this objective found popular support. Resistance to collectivization was common, however. Where the art of persuasion failed, armed force eventually brought compliance. In 1930, the tempo of the collectivization drive was escalated, the total elimination of the *kulak* element now being a central goal of the leadership. By 1935, the programme had transformed the Soviet countryside. The costs were staggering. Those villagers unfortunate enough to be labelled *kulaki*, and often the label had little to do with wealth but rather popularity, were dispossessed, relocated or shot. As crop production plummeted in the chaos of the time, famine took hold, notably but not exclusively in Ukraine. Population loss cannot be determined with any real precision, but minimally amounted to four million. Livestock holdings were decimated as peasants slaughtered their animals rather than turn them over to the state. Two-fifths of cattle and horses and two-thirds of sheep, goat and pig holdings were lost. The age-old link with the land was severed. What was ushered in were new institutional arrangements for the management of agriculture – the collective farm or *kolkhoz*, the machine and tractor station (MTS), and the state farm or *sovkhoz*.

To the casual observer, the new collective farm, or *kolkhoz*, may not have appeared much different from the age-old commune, or *mir*. Peasants lived in their own homes in the villages of before. Most peasants still worked land they did not legally own. The ownership of land in each *kolkhoz* was vested in the state, but it was now formally leased in perpetuity to the *kolkhoz* membership and cultivated collectively rather than in separate strips allocated to individual households. The mix of crops and their rotation were now determined by higher authority, but the collective's well-being was still determined by the weather. Peasants still could not move freely about the country. Travel was governed by the *kolkhoz* management, whereas in a not-so-distant era the elders of the commune determined who could leave, and for how long. But there were also significant departures from past customs. Peasants were now obliged to work the *kolkhoz* land for a specified, and over time a larger, number of days a year. Each *kolkhoz* had to deliver to the state specific quantities of agricultural product according to the dictates of the plan. Payment for peasant labour rendered was unpredictable since it was based on the assumption that there would be a surplus after compulsory deliveries to the state. Until the mid-1960s, the return on peasant labour was based on the concept of the work-day (*trudoden'*). Each task on the *kolkhoz* was measured in a work-day equivalent, which varied according to the skill required. After the production quota set by the state had been delivered, the remainder of the harvest could be sold to the state at a fixed price or sold, legally, at the prevailing market price in the so-called free collective farmers markets in the cities, presuming the surplus commodities were marketable. In any event, the cash earned from the sale of the above quota production, and/or the surplus in kind, was shared amongst the *kolkhoz* peasants according to the total individual labour contribution as measured by labour-day equivalents. The steady escalation in the required number of days to be worked for the *kolkhoz* was a reflection of the fact that the payment for labour was completely unpredictable. On the all too frequent occasions when there was no surplus left after the state quota was met, there was simply no return for the time spent labouring in the *kolkhoz* fields. The willingness to work hard for the collective farm was thus compromised, and yields fell in consequence. As one of the most important institutions in the agricultural sector, the *kolkhoz* was clearly deficient. Most peasants preferred to toil instead on their private plots or take care of their privately owned livestock. As we shall describe in detail shortly, it was the personal farming activity legally sanctioned by the Collective Farmers Charter of 1935 which sustained the peasant family for decades. Notwithstanding government restrictions on the personal sector from time to time, and the obligatory tax imposed on presumed income from it, this sector was critically important in total food production. Thus, the state extracted food and income from the countryside by imposing quotas for each *kolkhoz* to meet, by taxing the personal sector, and until 1958 by extracting payment in kind for the use of machinery provided by the Machine and Tractor Stations (MTS). After that date the MTS was abandoned and the equipment sold to the collective farms, whether it was wanted or not.

As time passed, the position of the collective farm peasants, or *kolkhozniki*, steadily deteriorated. And in turn, so did the diet of the fast growing urban population. As described in the preceding chapter, the prevailing ideology of the conquest of nature resulted in schemes to expand the arable without investing in essential infrastructure, such as an adequate all-weather road system to enable harvested crops to be delivered to the railhead. To this day, much of the countryside comes to a standstill during the periods of *bezdorozh'ye*, or roadlessness, associated with the spring thaw and autumn rains. Given the dismal prospects for the *kolkhoz* operation, especially in the more marginal regions for agriculture, two options were frequently pursued. One was to amalgamate in order to achieve some possible benefit from scale economies. Thus, from 1928 to 1950 the total number of *kolkhozy* was halved. The number of households and arable land in each roughly doubled. But the larger the operational unit, the greater the difficulty to coordinate activities, for as Fig. 7.4 illustrates, the typical collective farm comprised several separate settlements. By the end of the Soviet era, *kolkhozy* numbered about 26,000, down from more than 240,000 in the late 1930s. On average, each *kolkhoz* had several hundred households under its jurisdiction, and about 6500 hectares of land to administer, of which 3600 hectares were cultivated. Members of many collective farms chose a second option to improve their conditions of daily life and labour – changing the institutional arrangement from *kolkhoz* to *sovkhoz*, or state farm.

The *sovkhoz* offered one distinct advantage to the impoverished *kolkhozniki*. By changing status they became employees of the state. To be sure, they were still raising crops, but no longer were they dependent upon the vagaries of nature to produce a surplus over quota in order to obtain payment for their labour. State farm employees were just that – salaried workers who were paid irrespective of the level of production. Some 23,000 collective farms changed status, many of them amalgamating in the process. Prior to the collectivization drive initiated by Stalin, there were perhaps 1000 state farm operations. For the most part they had been set up on former estates and were run as model agricultural enterprises. From the outset they tended to be more specialized, and better equipped, than the typical *kolkhoz*. They always owned their own machinery, and were therefore not dependent on the MTS

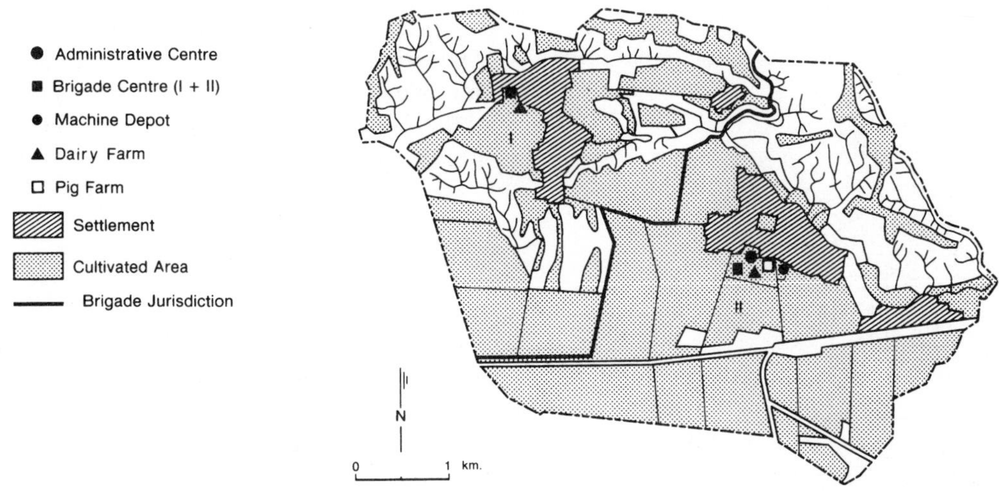

FIGURE 7.4 *Collective farm*
(*Source*: Based on *Dierke Weltatlas* (Braunschweig: George Westermann Verlag, 1978), 128)

during its existence. During the early 1930s, some 3000 *sovkhozy* were created in the open steppe to produce grain. This development was in response to the collapse of grain production from the *kolkhozy* and peasant sector during the chaos and famine of 1931–1932. Under Khrushchev, the state farm was given another fillip since it was the principal institutional organization associated with the Virgin Lands Scheme. As a result, the cultivated areas under *sovkhoz* jurisdiction tripled to more than 22 million hectares between 1953 and 1956. At the end of the Soviet period, the total number of *sovkhozy* was nearly 23,000. Inasmuch as the labour force was governed by norms relating workers to area cultivated, or to numbers of livestock, there was not as large a dependent population as on the collective farm. But because of its typically large size, the average labour force was substantially greater, and unlike the *kolkhoz* it was predominantly male, not female. Each *sovkhoz* administered on average about 16,000 hectares of land, nearly 5000 of which were under cultivation. Many *sovkhozy* in the steppe region specializing in grain production were huge by any standard of measure. *Sovkhoz Gigant*, portrayed in Fig. 7.5, was aptly named for it occupied 48,000 hectares of steppe near Rostov-on-Don. It is apparent from Fig. 7.5 that *Sovkhoz Gigant* had a decentralized administrative structure, a feature in fact common to all the larger *sovkhozy*. By way of comparison, the *kolkhoz* illustrated in Fig. 7.4 covered barely 2000 hectares. The typical *kolkhoz* settlement was the ancient linear village, houses gable-end to the road, and stretching sometimes for several kilometres. Such villages still number in the many tens of thousands.

Thus, Soviet agricultural production was vested in three types of institutional arrangement, the *kolkhoz*, the *sovkhoz* and the personal sector. The few individual farms missed in the collectivization drive owing to their remote location eventually disappeared, but on the eve of the World War II they still accounted for about one-tenth of the total area under cultivation. The distribution of cultivated land according to institutional type is presented in Table 7.1. The principal trends were a steady reduction in the share of arable under *kolkhoz* jurisdiction owing to conversion to *sovkhoz* status, an increase in the share of *sovkhoz* land owing to conversion, and expansion of the arable in frontier regions, and a steady decline in the importance of the personal sector commensurate with the absolute decline in rural population and number of households. On a more general level, the total area under the plough fluctuated quite substantially. It reached

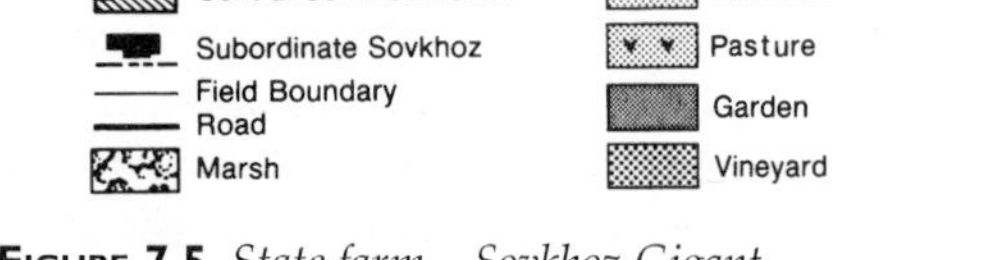

FIGURE 7.5 *State farm – Sovkhoz Gigant*
(*Source*: Based on *Dierke Weltatlas* (Braunschweig: George Westermann Verlag, 1978), 128)

a peak during Khushchev's last years as leader owing principally to the expansion of the arable under the Virgin Lands Scheme. Marginal land was taken out of production immediately following his demise as leader. And some land previously used for grazing, but ploughed up during the period of expanding the arable at all costs, was returned to pasture. The cultivated area peaked again in the mid-1970s, not quite reaching the apogee of 1963, however. Poor harvests, owing mainly to recurrent drought conditions, resulted in a steady contraction during the last two decades of the Soviet era. The area under the plough clearly has some bearing on the volume of agricultural commodities produced. But yields per unit area cultivated are obviously the critical factor, and in this respect the attitude of those who till the land is of no small importance. Before examining the question of incentive in the Soviet agricultural system, it would be appropriate to describe briefly production trends.

SOVIET-ERA AGRICULTURE PRODUCTION TRENDS

Centralization of decision-making in agricultural production is problematic at best. Historically, the controls from the centre were excessive. Up until the mid-1950s, detailed planning from Moscow governed the crop mix, and even the time of sowing and harvesting for each agricultural operation. Such detailed control from the centre was finally acknowledged to be pointless. The demise of the MTS along with many other adjustments to the management of agricultural inputs and output signalled the steady shift toward a more rational management system. By the end of the Soviet period, targets were set nationally for the Five Year Plan by Gosplan, but at the farm level discretion was given both to *kolkhoz* and *sovkhoz* management teams by the Ministry of Agriculture in determining how their individual operations could best contribute to achieving these targets. A host of other organizations had been created to enhance agricultural production. Prominent among them were Agro-Industrial Production Associations and Intercollective Farm Cooperatives. The former were associations of farming operations and support services such as factories processing agricultural commodities, distribution and research enterprises. Their purpose was to expedite the production and delivery of higher quality agricultural commodities. The Cooperatives concentrated more on improving infrastructure and the management of the land resource itself. However, despite countless government and Communist Party decrees exhorting peasants to produce more, the poor correlation between planned output and actual delivery of agricultural commodities to the state in final years of Soviet power did not improve.

Table 7.1 *Soviet-era sown area – 1940–1989 (million hectares)*

	1940	1960	1963	1970	1975	1980	1985	1987	1989
*Sovkhozi**	13.3	73.2	97.8	100.9	112.9	115.2	112.6	112.4	–
Kolkhozi	117.7	123.0	114.0	99.1	98.2	95.9	92.0	92.2	–
Personal sector	19.6	6.7	6.7	6.7	6.6	6.2	5.7	5.7	–
Total	150.6	203.0	218.5	206.7	217.7	217.3	210.3	210.3	209.8

*Includes other state agricultural enterprises
Sources: Narodnoye Khozyaystvo SSSR v 1960g (1963g, 1970g, 1975g, 1980g, 1985g) (Moscow: Finansy i Statistika, 1961, 1964, 1971, 1976, 1981, 1986); *Narodnoye Khozyaystvo SSSR za 70 Let* (Moscow: Finansy i Statistika, 1987), 225; *Narodnoye Khozyaystvo SSSR v 1989g* (Moscow: Finansy i Statistika, 1990), 375

As already noted, the Russian Empire, and until the 1960s the Soviet Union, were consistently net exporters of grain, even if segments of the domestic market had to do without. A more enlightened approach was adopted in the early 1960s. Thereafter, the Soviet Union emerged as a major market for surplus grain produced in other parts of the world. The shift from net exporter to net importer of grain was related to two issues. The first was the realization by the Soviet leadership that consciously depriving the population during peace time was no longer politically tenable. Food riots had wracked Poland in the late 1950s. There were reports of similar, but smaller scale, disturbances in Soviet cities. Put simply, tampering with the food supply was especially fraught with risk. The second was related to a perceived need to do more to improve the diet of the average citizen. In the early 1950s, the typical Soviet meal was heavy in carbohydrates, light in protein, and generally deficient in both quantity and selection of vegetables and fruits. Basic dietary improvement was closely linked to increasing grain production, especially coarse grain for livestock fodder. Steady improvements in the supply of higher quality foodstuffs did occur in subsequent years. By the late 1970s the diet of the ordinary citizen was higher in protein, lower in carbohydrates than decades before, and it included a reasonable variety of vegetables and fruits, both domestic and imported. The disruptions in food supply which characterize the final two or three years of the Soviet period therefore were especially upsetting to, indeed deeply resented by, the public at large who now were obliged to join ever longer queues to purchase what little was available. Near empty shelves in the state food shops were a clear indication that the whole Soviet system was in a state of crisis.

Protein intake was inextricably linked to the size of livestock holdings, and expansion of the latter was dependent upon fodder supply. When fodder supplies were inadequate to sustain animals over the winter, distress slaughtering in late autumn or early winter months was the inevitable outcome. For years distress slaughtering had frustrated efforts to quickly build up livestock holdings. This situation was exacerbated by the devastation of livestock holdings at the time of collectivization. The impact of the collectivization drive is readily apparent from Fig. 7.6. For many peasant households, slaughtering their animals was preferable to turning them over to the state. Even by 1950, the total numbers of cattle and pigs barely equalled the 1929 levels, and the number of sheep remained substantially below. Horses, the principal draught animal during the 1920s, suffered a similar decline, but owing to mechanization of agriculture were destined never to regain their former importance. While there had been a modest recovery by 1950, the population was still worse off than in 1929. Put simply, there were about 50 million more mouths to feed in 1950 than in 1929. Small wonder that recipes of an earlier, more bountiful era, were rendered useless. If in the early 1950s an egg had become a luxury in many urban households, what was a fruit pie with meringue topping? Indeed, the content of Soviet cookbooks eventually began to reflect the new, more austere, reality. In later years when the programmes to improve the diet had begun to have a perceptible impact, cookbooks published

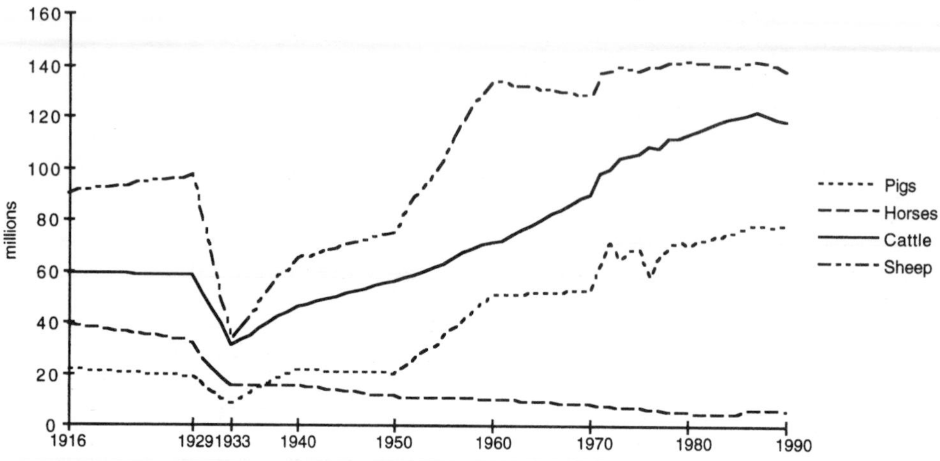

FIGURE 7.6 *Soviet livestock holdings*
(*Source*: Compiled from data in various issues of *Narodnoye Khozyaystvo SSSR v 1964 g ... 1989g* (Moscow: Goskomstat, 1965 ... 1990))

in the 1920s and early 1930s became something of a collector's item for in them were recipes no longer easily obtained.

The substantial increase in livestock holdings between the 1950s and 1980s outlined in Fig. 7.6, was translated into large gains in the production of beef, veal, lamb, and pork. A vast expansion of coarse grain production underlies this change. While silage corn was introduced during the Khrushchev years as a major fodder crop, especially in south Ukraine and the Kuban region of the north Caucasus, it did not displace the traditional Russian and Soviet emphasis on oats and barley. These two crops accounted for about three-quarters of coarse grain production. Corn made up perhaps one-sixth in a good year. In terms of maintaining, and enhancing, livestock holdings, a substantial share of the wheat harvested in the Soviet Union of necessity had to be fed to animals instead of humans. In times of poor harvest, the Soviet Union bought grain in order to sustain livestock holdings and meat production. As well, meat imports steadily increased from the 1960s. During the 1970s they averaged about 300 million tons per year. They were even larger during the 1980s, a trend which simply reflected the worsening performance of Soviet agriculture.

The pattern of Soviet grain production since the early 1960s is presented in Fig. 7.7. During the 1960s there was a steady, if irregular, increase in production, and save for 1963 and 1965 it was a decade of grain exports. This general trend in grain production was a reflection of policies introduced the preceding decade. Overall, agricultural output was 40 per cent higher in the second half of the 1950s compared to the first half. In bringing about this change, the Virgin Lands Scheme figured prominently. But as has been noted earlier, expansion of the cultivated area into marginal lands previously the domain of the pastoralist was a calculated gamble. Constant cropping raised the stakes. The disastrous harvest of 1963 simply confirmed informed observer's opinion of the risk involved. Still, recovery was rapid as average or higher amounts of precipitation during subsequent growing seasons produced reasonable yields and the traditional export of grain was resumed, albeit at a relatively small scale. Yields were higher owing to better weather conditions, but they also owed something to a much greater investment in fertilizer, machinery and, as will be discussed shortly, the wages paid to farm workers. What the 1970s and 1980s brought was quite different. Harsh weather, including too much heat, moisture or cold at critical times, produced a dependence on world grain markets. What the data presented in Fig. 7.7 reveal is, first of all, an overall stagnation in agricultural production

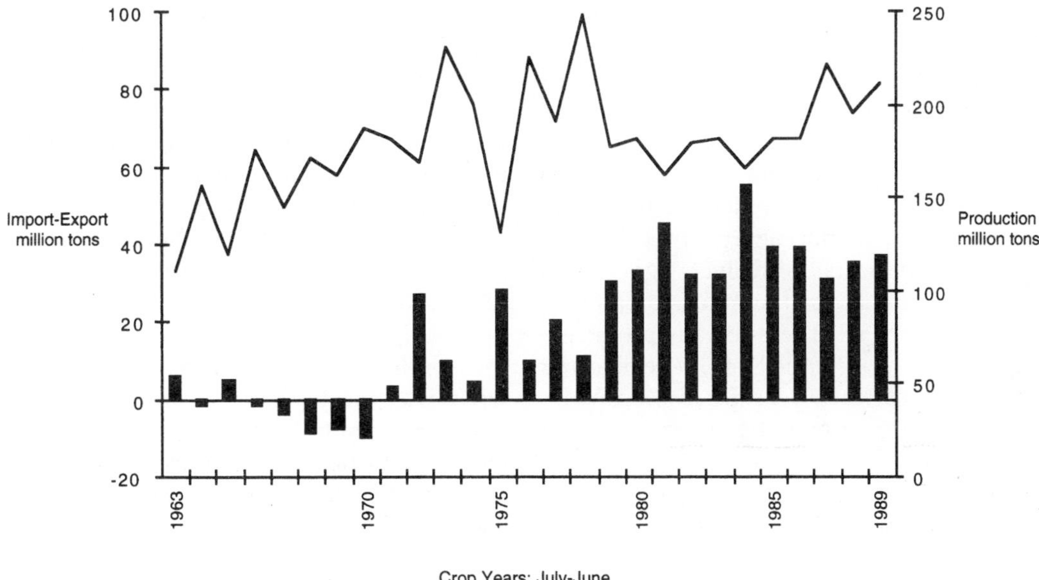

FIGURE 7.7 *Soviet-era grain production, 1963–1989*
(*Source*: Compiled from data in various issues of *Narodnoye Khozyaystvo SSSR v 1964 g ... 1989g* (Moscow: Goskomstat, 1965 ... 1990))

generally, for poor weather takes a toll of more than just the harvest of grain crops, secondly, by implication a decreasing return on the ever larger investment in agriculture registered in each five-year plan from 1965, on, and thirdly, a sizeable outlay of hard currency to pay for grain imports. It is for these reasons that grain production in particular, and agriculture in general, were regarded as the 'Achilles heel' of the Soviet economy. Lack of incentive figured prominently in the poor performance of this sector.

INVESTMENT AND INCENTIVE IN SOVIET AGRICULTURE

The first steps toward modernizing agriculture were taken by Khrushchev, but were frequently at cross-purposes. Massive campaigns to introduce new crops, or new ways to handle old tasks, were introduced only to be changed or reversed a few years later. The scientific and technical personnel in the rural sector were then inadequate in number to educate the *kolkhozniki*, who anyway often knew better what needed to be done than could be dictated by the plan. But Khrushchev's initiation of a price reform for agricultural products was a significant and beneficial step. By the early 1960s, the prices paid to collective farms had been substantially increased, the price paid for above-quota production delivered to the state was more attractive still, and of particular importance for collective farms in marginal areas, the first step was taken toward creating the regionally differentiated scale of payments. Thus, collective farms in many areas of inferior quality land were paid higher prices for the same crop produced on a collective farm in a region known for its productive soil. This was a partial recognition of the higher costs of production in the former case. Price regionalization schemes subsequently became more refined. Notwithstanding greater procurement price sensitivity to regional variations in soil, climate and agricultural potential, some collective farms remained unable to turn a profit. In many instances, relative location of the collective farm accounted for some of the red ink. Since prices paid for crops excluded the cost of transportation to the nearest railroad terminal, this had to be borne by the collective, or state, farm. The abysmal state of most rural roads simply made matters worse. The massive requisitioning of vehicles and some labour, from non-agricultural rural, as well as urban, enterprises which occurred every autumn in an attempt to move commodities from the field to the depot was necessary, but did not solve the problem. The inability of producers to deliver agricultural commodities to rail or road terminals prejudiced the full recovery of what was actually produced. Indeed, a fifth or more of the grain, vegetable and fruit crop was lost each year because of inadequate facilities for transportation, storage and distribution. The legacy of decades of little if any investment in basic infrastructure in the countryside hampers agriculture to the present day. In the early 1960s, a little more than one-fifth of total investment capital was directed towards agriculture. By the end of the Soviet era, the share was closer to one-third. Little return came from it. The absolute number of roubles spent on this sector clearly increased very considerably. Some of them went into the pockets of collective and state farm workers, and were perhaps among the most cost effective.

Higher prices paid for agricultural products brought a commensurate increase in return for the labour expended. On the *kolkhoz* the relationship was quite clear. The more profitable the operation, the larger the payment for the labour invested by the *kolkhozniki*. But until the economic reform initiated by Brezhnev and Kosygin in 1965, there was still a major disincentive for the peasant willing to work on the collective farm instead of on the private plot – annual income was not predictable. Thus, in the late 1960s a programme was introduced in which the *kolkhozniki's* annual income was averaged over a period of several years and a basic monthly wage established. The use of the labour-day concept was dropped. Annual income of collective farmers drew closer to that of state farm workers, and base levels of both increased substantially from the late 1960s. From the point of view of instilling incentive to work harder for the *kolkhoz* or the *sovkhoz*, the changes in wages and salaries initiated in the 1960s had an impact. There was a notable improvement in yields, attributable in part it seems to the attitude of farm workers. Indeed, in some regions the personal plot and livestock holdings were let slip. After all, if an adequate income could be had from the *kolkhoz* why put in the additional hours in physically demanding labour on the plot and tending livestock? Putting state and collective farms on a profit or loss basis as part of the 1965 economic reform held out the prospect of stimulating higher levels of output. However, it also raised the spectre of farm managers producing only what was most profitable and dropping that which was not. Local discretion in decision-making was held at bay during the 1970s as a result of the continuing intervention in collective and state farm management by central authorities. With the arrival of Gorbachev, however, ideas regarding deconcentration of decision-making responsibility found more favourable reception in government circles, although not much actually changed in practice. And overall, the role of personal farming was still an essential component in the agricultural scene.

Much of the increased investment in agriculture after 1965 went into large-scale amelioration projects, some of which were described in the preceding chapter. In some observers' opinion the proportion of investment directed to rural construction projects of this type was too high. At the time, it customarily ran around two-fifths of the total in the Soviet Union, compared to about one-quarter in the United States. In the latter, the balance, that is, about three-quarters of the total investment in the agricultural sector, went into increasing the stock of machinery, a substantially larger share than in the Soviet Union. Thus, despite a cultivated area at least 25 per cent larger than that of the United States there was far less equipment. In 1966, for example, the United States' stock of tractors and combine-harvesters on farms numbered 4.8 and 0.88 million respectively. In 1989, the Soviet tally was only 2.7 million tractors and 0.689 million combine-harvesters, fewer in fact than in 1986 when both had reached a peak. Trucks, as noted already, were in chronic short supply and were requisitioned annually from other sectors. Under-capitalization took many more forms than just these examples. Given that only 3 per cent of the American labour force was occupied in agriculture in the late 1980s, whereas the share was more than 20 per cent in the Soviet Union, there were clearly huge differences in the ratio of machine to farm workers. Part of the problem was that improving infrastructure and stocks of equipment necessitates even larger investment. Return on this investment, however, was unacceptably low. What was not low was the level of subsidy to agriculture.

In virtually all modern industrial states, and a good many developing ones as well, there exist extensive programmes of agricultural subsidy. But amongst industrialized states, few provided as much support for agriculture as did the Soviet Union. This was not always the case, of course. During the Stalin years, prices paid by the state for agricultural commodities were exceedingly low. For many collective farms, income received did not even cover the cost of seed, let alone provide a source of income for the *kolkhozniki*. Thus, the preoccupation of peasants with the personal sector in agriculture and the drift from *kolkhoz* to *sovkhoz* status were rational responses to an intolerable situation. From agriculture, wealth was extracted with short-term gain and little sense of long-term cost. To cheaply purchased agricultural commodities, the state applied a turnover tax. Put simply, as the principal intermediary in the process of exchange, the state charged substantially more than it paid for agricultural products. As all prices outside the free collective farmers market were determined by the state, supply and demand factors were irrelevant. Income from the turnover tax on agricultural and other commodities provided the investment capital to finance the industrialization drive. Eventually, as we have seen, adjustments had to be made in order to induce greater output from this long-exploited sector of the economy. But this came to pose a real dilemma. Although the state paid little for such products as wheat, or rye, the final cost to the consumer of bread was low even with the inclusion of turnover taxes. And it mattered little that the price of meat bore no relation to actual cost, if the volume sold by the state was small owing to short supply, if the population inured to such deficiencies was prepared to pay the going price in the collective farmers market when money was in the pocket at the same time that meat was on the counter. Cheap food was long touted as one of the benefits of socialism, just as low rent for housing and public transportation in cities were taken for granted by the populace. As the wage bill of agricultural workers rose, as the prices paid for products increased, as investment generally in agriculture jumped following the 1965 reform, the spread between cost of production and revenues received created a vast chasm into which huge amounts of money were poured. Save for the presumed benefits derived from feeding the population better and putting more spending money in the pockets of farm workers, the net result was escalating subsidy.

Between 1965 and the late-1980s, state retail price subsidies for agricultural products increased more than seventeen-fold. By the close of the Soviet era, the sum involved was close to 50 billion roubles annually. The irony was that while the demand for staples such as bread or potatoes is usually relatively inelastic, that is, only a certain amount will be consumed more or less independent of the price, these products were so cheap that peasants bought bread and potatoes to feed their livestock. On the other hand, despite the substantial increase in domestic meat production, and imports of meat and meat products, domestic demand was still not satisfied. Increased consumption of meat and dairy products in the last two decades of Soviet power was a major reason for the jump in the total retail price subsidy. In the late 1980s, it was estimated that prices charged by the state for beef, lamb and butter represented less than half the cost of production. For milk and pork the price covered perhaps as much as 60 per cent of production costs. Partly out of concern over possible negative public reaction to dramatically higher prices, the state simply absorbed the cost. Inadequate supply in the state sector enabled peasants to turn a handsome profit on privately produced meat sold at market prices in the collective farmers market. The combination of high prices for commodities sold on the collective farmers market, higher income from the *kolkhoz* owing to price and wage adjustments, and availability of bread and potatoes from the state sector at exceedingly low prices gives rise to the intolerable situation of bread and potatoes destined for humans being converted into meat. For the peasant, it was perfectly logical of course to secure feed for livestock this way if the alternatives were to invest a substantial amount of time personally supervising grazing livestock or competing with the collective farm for available fodder. Clearly, those charged with managing the domestic food supply were caught on the horns of a dilemma. While the urban household was the primary beneficiary of cheap food, we might well ask at this point, how did the dramatic increase in expenditure on the Soviet agricultural sector affect the conditions of daily life and labour of those who farmed land?

The rural household economy in the Soviet era

In the Soviet Union the word farmer was hardly ever heard. Farm workers were peasants. Strictly speaking, the term peasant was restricted to the *kolkhozniki*, or the more than 33 million members of the country's 26,000 collective farms. But to the urban sophisticate, it mattered little if the man on the street in front of him sporting knee high, dirty black boots, a tattered, padded jacket and cap and smoking a *papirosa,* was a state farm employee, and therefore a member of the country's working class, and not strictly speaking part of the peasant class. He was just another in the horde of *kolkhozniki* invading the city to buy goods probably already in short supply, or to sell something on the collective farmers market at an inflated price. And indeed he probably was right, for each day millions of peasants made the journey to cities for precisely such purposes. To be sure, they were often unsophisticated by contemporary urban standards. But by the 1980s, they were no longer likely to be literally poor country cousins. With the rapid escalation in wages paid farm workers came a huge jump in potential purchasing power. The fact that in the late Soviet period there was more money in the average rural savings account than in the urban said as much about the dearth of consumer items in rural stores as it did about the propensity to save amongst the rural population.

Rural–urban differences in household incomes were steadily eroded by the policies introduced in the 1960s. In 1970, urban incomes were on average about one-quarter again as large as those in the countryside. By the mid-1980s, if the earnings from the private plot and animal holdings of *kolkhozniki* are included, the gap was probably less than one-tenth. In peasant households close to major urban centres where agricultural commodities were easily sold, incomes frequently exceeded those of factory workers in the city. But distance from a city was not necessarily a barrier either. For many peasants, travelling by aeroplane to distant, cold climate, urban markets to sell the products of the sunny south was commonplace, and presumably profitable.

Irrespective of locational or climatic situations, much of the activity in the Soviet village was centred on personal auxiliary farming. As remarked earlier, this incorporated both the cultivation of the individual private plots, which for *kolkhozniki* usually were one-half hectare in size, and tending privately owned livestock. In regions in which irrigation was common, the area of the private plot was smaller since in theory potential yields were greater. Since the membership of the collective farm assembly was charged with monitoring the land so allocated, all had an obvious vested interest in pushing the tolerance of regulations to the limit. The typical household plot was adjacent to the home itself. Across much of European Russia and Siberia, the household agricultural enterprise comprised a detached wooden house (plastered exterior in much of Ukraine, sometimes even brick in parts of the Baltic region), with outbuildings, orchard and plot. In the Caucasus and Middle Asia, stone and adobe are the more common materials used in private rural house construction. The outbuildings provide both storage facilities and shelter for privately owned livestock. Depending on the region, animal husbandry included looking after a single cow and calf, sow and piglets and a gaggle of sheep, lambs, ducks, chickens or whatever combination the local culture dictated. Numbers owned were restricted. From time to time, the personal auxiliary farming system came under government pressure, or outright constraint. In the late 1950s, for example, Khrushchev endorsed policies which severely restricted the personal sector. Peasants could sell produce in the collective farmers markets in the cities, but only at the state-regulated price. Roadside stalls and other forms of private initiative in marketing products from the personal sector were prohibited. The result was predictable. The supply of commodities to the urban market dried up, and the resultant food shortages produced widespread complaint, and according to some reports, riots in the cities. After Khrushchev's ouster, such restrictions were lifted; indeed, in the Gorbachev period there was positive encouragement of personal auxiliary farming as part of the campaign to improve the food supply.

The millions of tiny, private plots comprised only about 3 per cent of the total cultivated area in the Soviet Union. They were not just associated with *kolkhozniki*, although they were more dependent upon them than state farm workers because for many years their annual income from the collective farm was so highly variable, or non-existent. The right to grow crops on a private plot embraced all of the rural population, including non-agricultural sector workers. In addition, urban apartment dwellers could have a summer garden, a much smaller plot than legally available to the *kolkhozniki*, and one on which no permanent habitable structure was allowed. This latter restriction was widely ignored for in the hinterland of many cities were to be found summer garden plots with what for official purposes were described as tool sheds. The so-called shed served as a poor man's *dacha*, complete with sleeping facilities and sometimes even a front porch! Produce from these plots, as well as from those tended by rural inhabitants, found its way into the collective farmers market, and hence onto the kitchen table in the typical Soviet apartment.

While making up only 3 per cent of the cultivated area, the personal sector generated more than half of the total potato production and around a third of most other vegetables. The personal sector accounted for nearly one-half of the nine million hectares sown to vegetable crops in the late 1980s, so its contribution in this regard is not surprising. Vegetables were of obviously comparatively high value. Yields from the private plots were often compared to those of the *kolkhoz* or *sovkhoz*, to the disadvantage of the latter. But it should be borne in mind that the personal sector was not entirely independent of the state sector, or the collective farm. Both the *kolkhoz* and *sovkhoz* provided machinery to facilitate ploughing and harvesting on the private plot. Indeed, under Gorbachev loaning such equipment was encouraged by the government. A little fodder was grown on private plots, but as indicated, most of this land was used for supplying the rural household with food, the surplus being marketed as opportunity permitted.

While private plots comprised 3 per cent of the total arable, or about 6.5 million hectares, there were significant regional differences. In the more bountiful agricultural regions with significant rural populations the share was quite high. In Georgia, for example, private plots accounted for nearly one-fifth of all cultivated land. In Ukraine, the figure was 7 per cent. In the Russian republic, the share was below the national average. The total area of personal plots, and its share of the total arable, declined during the last two or three decades of the Soviet period. To some degree this reflected less dependence on the household's own labour to supply food and more reliance on foodstuffs purchased from the state. In the late 1980s, the personal sector met about two-thirds of the typical rural household's food requirements. Clearly, personal auxiliary farming still accounted for a substantial share of the rural household's time budget. For women, it was a major focus of attention, but even male *kolkhozniki* probably spent a quarter of their time in the family farming enterprise.

Though the contribution of privately owned animals to the national food supply remained important, it had also steadily declined in relative terms. In the space of less than one generation, the role of private egg production dropped from more than two-thirds of the total to barely one-tenth. The development of *sovkhozy* specializing in poultry and egg production between the mid-

1950s and early 1970s accounted for the change. The state also had increased its share of other traditional markets of the private sector as well. Dairying is a case in point. In 1961 there were approximately 35 million milk cows, of which 46 per cent, or about 16 million, were privately owned. By 1990, the personal sector accounted for 13.2 million head, but the total number of cows was 41.7 million. The share of the personal sector was now about 32 per cent. Some expansion in holdings had occurred on collective farms, but the most significant growth had taken place on the *sovkhozy*. As was noted earlier, government policy in the late 1980s encouraged production in the personal sector. While the limits on the number of cows, pigs, and so forth were not increased to any real degree, other ways of tapping the skills of the *kolkhozy* households were pursued. One successful programme was the collective farm household taking piglets from the collective farm and fattening them to an agreed-upon weight. Fodder was supplied. The inducement centred on the household retaining one of the fattened pigs as payment for service rendered, or sharing in the profit realized from their sale. By this and similar schemes, including providing peasants with loans to purchase livestock, the state sought to augment meat production.

Much was often made of the fact that personal auxiliary farming still accounted for something in the order of 25 per cent of the value of total agricultural production at the time of the collapse of the Soviet Union. But recognition of the state's contribution needs to be taken into account. Privately owned livestock was often supported by fodder from the state or collective farm. Peasants often took part of the payment for their labour in kind, not cash. Privately owned livestock also grazed on state land, whether vested in the collective farm or not. And mention has already been made of the use of equipment to assist with the cultivation of the private plot. In irrigated areas, water was drawn for use on the personal plot free of charge. Notwithstanding this symbiotic nature of the relationship between the state and personal sector, the yield from the arable and livestock of the latter was extremely high. It was the comparatively higher yields which supported the argument, increasingly vocal in the Gorbachev era of *glasnost'*, that more family farming needed to be encouraged.

The huge scale of the agricultural enterprise – *kolkhoz* as well as *sovkhoz* – militated against a close link between cultivator and the land. Over the years there were experiments in which a small area of the *kolkhoz* or *sovkhoz* was turned over to a few men, or contract groups, to cultivate according to their estimation of which crops would do best. Gorbachev sanctioned more of this type of activity in 1983, before taking over as Party leader. Where such decentralized control was linked to a share of the harvest, the results were invariably the same – substantially higher yields than from similar quality land farmed by salaried *sovkhoz* employees, or regular collective farm labour brigades. In 1987 legislation was introduced which extended the concept of agricultural contracts embedded in the 1983 decree to families. This legislation permitted families to lease land for periods up to 15 years. The incentive of higher personal income, combined with autonomy in decision-making, was a potentially powerful force for change in Soviet agriculture. But these initiatives were clearly too little, too late.

As the state continued to tinker and experiment with such innovations, daily life for the majority of rural inhabitants continued in much the same manner as always. The private plot and livestock remained a major focus of activity, even though the income from auxiliary farming declined as a proportion of total peasant household income. Peasants were now not just paid more, but members of the *kolkhoz* were eligible for benefits such as state pensions. Up until the mid-1960s they had been excluded, the principal reason being that as they were 'owners' of a share of the collective farm's assets, the *kolkhoz* rather than the state should be responsible for their welfare. Thus, the private plot and livestock not only sustained families during the working life of household members, it was assumed by the state that they could support old-age pensioners as well. In the culture of villages everywhere, the aged and infirm were the responsibility of family and neighbours. The fast rising rural household incomes were spent in part on better equipping the family with consumer durables such as refrigerators, washing machine, television sets, and so on. Indeed, through the 1970s and 1980s, differences between the urban and rural sectors in the per capita ratio of these consumer durables declined dramatically. The motorcycle, with or without side-car, was commonplace in all rural areas. By 1991 there were more privately owned automobiles, the ultimate consumer durable, in the countryside than in the city.

Yet for all the signs of marked material progress, it would be a mistake to paint too rosy a picture of rural life. While the Soviet agricultural labour force comprised a fifth of the total, its composition and geography presented numerous problems. As the young and ambitious males left, especially, but by no means exclusively in the Slavic and Baltic regions, the proportion of women in the rural labour force increased. By the end of the Soviet era, about two-thirds of the agricultural population was female. Characteristically the female rural worker was less skilled than the male, and hence less productive. Given the proportion of the young and the old of the total rural population, those comprising the potential workforce were a smaller proportion than in the urban sector. But even so, it was not everywhere the case that there were too few hands for the work available. In the southern regions underemployment was rife. Indeed, the euphemism for the unemployed, the 'unengaged work-capable' population, was estimated to be in the order of one-sixth to one-fifth of the potential labour force in the Turkmenistan and Kyrgyzstan republics at the end of the Soviet period. As these were averages for the total republic populations, in the countryside where the problem was really situated it was worse still. In Uzbekistan, the number of unemployed was conservatively pegged at one million out of a total population in 1990 of just over 20 million. This huge surplus labour pool was part of the reason for the continuing intensive use of labour in the fields. For instance, about half of the cotton harvested was still picked by hand in the late 1980s. It was also the reason for the disproportionate importance of personal auxiliary farming in rural Middle Asia. The diversion of north-flowing Siberian rivers, the SIBARAL project, was very much regarded as a solution to the underemployment, indeed, unemployment, problem of rural Middle Asia.

In the Slavic realm, that is, Russia, Ukraine and Belarus', as well as in the Baltic region, a vast number of people were consigned to live in what were officially classified as 'futureless villages'. The General Scheme for the System of Settlement, described in an earlier chapter, affirmed a long-standing policy objective of consolidating the rural population. Some 350,000 of the country's 400,000 villages had too few people to qualify as a district centre, which brought with it entitlement to state investment in services ranging from day-care facilities to cultural centres. For the 15 million inhabitants of the country's 'futureless villages', isolation, if not deprivation, was portended. Out of sheer frustration with existing conditions in the 'futureless villages', some people moved of their own volition. Those left behind were further deprived as a result, since they were typically the aged, infirm or the less industrious. The human resource potential in many regions diminished accordingly.

The legacy of the Soviet agricultural system clearly is inauspicious. Each of the new post-Soviet states faces enormous challenges in adapting its collectivized agricultural sector to the new economic, social and political realities. The privatization of agriculture is proceeding in most states, even if the thorny issue of land ownership has not been resolved. The marketplace is steadily replacing state planners in determining crop mix and price. But by no means has government involvement in the sector disappeared. Nor is it likely to any time soon. Agricultural production and distribution remain major concerns for all post-Soviet states. And for good reason. Despite what a shift to a market economy may proffer by way of a more efficient agricultural system, the collapse of the Soviet Union has had negative consequences for agriculture everywhere. The most telling measure of the state of agriculture at present is how much food is being consumed on a per capita basis. Before describing the general trends in agricultural output, however, a brief overview of progress to date in reforming the institutional arrangements for this sector will be presented. The extent of privatization of agriculture in the western and southern tier post-Soviet countries will be described first. A more detailed examination of the process of agricultural reform in Russia will follow.

Agricultural reform across the post-Soviet scene

Amongst the western tier post-Soviet states, Lithuania and Latvia have made most progress in de-collectivizing their agricultural sectors. Within two years of becoming independent, both states had transferred more than half of their agricultural lands into private ownership. The formation of individual peasant farms started in the final years of Soviet power, but the process greatly accelerated after independence in 1991. In Lithuania, for example, the crop area of peasant, or private, farms increased 20-fold between 1990 and 1992 alone. In 1992, there were more than 100,000 individual farms. These farms represented 17 per cent of the total area under cultivation. By 1994, the 111,400 peasant farms accounted for 987,000 hectares, or roughly 27 per cent of the arable. Each was less than nine hectares. State farms accounted for about two-fifths of the arable in 1994. Already in 1992 in the much smaller Latvian agricultural sector, peasant farms numbered more than 52,000, and accounted for more than one-fifth of the total area under cultivation. Privatization since then has proceeded apace. Privatization of agriculture in Estonia, on the other hand, has been much slower owing to legal complications associated with the restitution of property to those people whose lands were nationalized when Estonia was forcibly brought into the Soviet Union. While restitution of former estates to heirs in Latvia has been faster, it has had the effect of distorting the land market.

Notwithstanding legislation regarding the privatization of land and property introduced shortly after declarations of independence in Ukraine, Belarus' and Moldova, the actual pace of agrarian reform has been slow. In Ukraine, for example, legal title was soon granted to the Soviet-era personal plots to peasants, but the formation of peasant farms faces many obstacles. By mid-1993, Ukraine had less than half the peasant farms of Latvia, a country with scarcely one-twentieth its population. By mid-1995, little more than one-sixth of Ukraine's agricultural land was held by the state's 33,000 peasant farms, the average size of which was 22 hectares. In Belarus', reform of the agricultural sector has been even slower than in Ukraine. A mere 3000 or so peasant farms were operating in 1995, the countryside remaining firmly under the control of the collective and state farm managers from the Soviet era. The average size of the peasant farm in Belarus' at this date was 21 hectares. Moldova was wracked with internal disorder and near civil war following the early post-independence efforts of the Moldovan majority to break all contacts with Russia and Ukraine, notwithstanding substantial localized minorities of both peoples. Thus, it had not made much progress in reforming agriculture by the mid-1990s. In 1995, there were approximately 13,000 personal farms which had been formally registered, and another 8000 in the process. But as their average size in mid-1995 was only two hectares, they had little visible impact on the rural landscape. However, Moldova received the support of international agencies for its programme to quickly privatize half of its state enterprises. Since the majority of Moldova's population is rural, conservative, and not yet out from under the thumb of Soviet-era institutions and associated managerial class, privatization of agriculture is proceeding slowly. In Ukraine and Belarus', to say nothing of Russia, there is also very strong resistance to change in the countryside. In some southern tier parts of the former Soviet Union, land reform has figured more prominently than in the Slavic core region.

In Georgia, a land reform programme that was very far reaching was introduced in early 1992 by the government. About three-quarters of agricultural lands were to be allocated to farmers and citizens without charge, the rest remaining in the state's hands. Within a matter of months, roughly half of the land to be privatized had been turned over to individuals and families. The tempo of land conversion slackened thereafter. As close to one half of Georgia's 5.4 million citizens is classified as rural, the land privatization initiative involved a quite substantial number of people. Not surprisingly, most privatization transactions involve small parcels of land. The vast majority are less than three hectares. Since the original decision to permit buying and selling of privatized agricultural land by 1994 was rescinded, there is little scope for consolidation into larger, more efficient private farming units until such time as this can occur. Meanwhile, the basis for an independent farming class has been put in place. In Armenia, which is more urbanized that Georgia, the de-collectivization process was completed more fully, and more quickly. Less than one-third of Armenian's 3.7 million citizens live in the countryside, as opposed to close to one-half in Georgia. Like Georgia, its mountainous terrain affords limited opportunity for scale economies in terms of the size of agricultural production units. In 1995, 313,000 peasant farms had replaced the few thousand Soviet-era collective and state farms. But their average size at this date was only one hectare. Unlike Georgia, there is provision in the enabling legislation which permits farmers to consolidate small plots into more viable larger scale operations. To date, however, the consolidation process has been slow owing to the devastation visited upon the domestic economy by the on-going conflict with Azerbaijan. With close to half of its 7.5 million population living in the countryside, Azerbaijan is the most rural of the three Caucasian states. Nonetheless, it has made the least effort, and in consequence the least progress, in reforming its agricultural sector. While across much of the former Soviet Union agricultural prices had been substantially freed up by the mid-1990s, in Azerbaijan the state still controlled prices of commodities whose production was deemed to be of strategic importance. This restriction applied to most of the commodities which can be exported, or which supplied state-run industries. The plan to reform agriculture was still in preparation in the mid-1990s, little having been achieved save for the formation of a mere 1300 peasant farms by mid-1995. The average size, 22 hectares, is larger than elsewhere in the Caucasus, but as there are so few of them they have had little impact. To date, the sector operates neither according to the precepts of state planning, nor those of the marketplace. In the climate of internal political instability and external military dispute with Armenia, that agricultural output has dropped comes as no surprise.

Kazakhstan has made substantial progress in reforming its agricultural sector which employed about a fifth of the country's labour force. About 44 per cent of its nearly 16.7 million population are rural. The 8000 or so state and collective farms, each with on average of 35,000 hectares of land, were the mainstay of the sector on the eve of the collapse of the Soviet Union. Several hundred of them adopted some form of private ownership within the first year of independence. The government meanwhile passed legislation privatizing the personal plots of the Soviet era, new owners possessing the right of inheritance although not sale. A number of collective farms were broken up into small scale, peasant farms. By mid-1995, there were 24,000, each with 348 hectares of land on average. All told, the private sector accounted for more than 40 per cent of the value of agricultural production. Although the average size of the private farm was larger in Kazakhstan than anywhere else in the post-Soviet scene, because of Kazakhstan's size they only accounted for a tiny share of cultivated land. While the government of Kazakhstan is widely recognized for its positive stance regarding market reform, the settlement history of the country has necessitated a rather cautious approach to creating a market for land. More than three-fifths of the cultivated area was brought into production as part of the Khrushchev-era Virgin Lands Scheme. As noted earlier, not only was there a vast requisitioning of agricultural equipment for this project, it also entailed colonization of the territory by non-Kazakhs. The colonists were principally Russians and Ukrainians. Elsewhere in Kazakhstan, there are other non-Kazakh ethnic enclaves, principally, but not exclusively, in irrigated areas. Thus, in order not to precipitate any internal political and social disorder which might arise from allowing land ownership to further ensconce minority groups within particular territories, the issue of land ownership in effect has been shelved for the near-term future. This has been accomplished by only permitting land to be held in long-term leasehold. The government, however, does allow such leaseholds to be inherited, and bought and sold. It is anticipated that this arrangement will permit consolidation of land to occur, thereby enhancing efficiency in agricultural land use, and at the same time mitigating possible controversy over land ownership by Kazakhstan's substantial regional ethnic minorities.

With more than three-fifths of Kyrgyzstan's 4.5 million total population classified as rural, any agrarian reform programme impacts the bulk of the citizenry. By the mid-1990s, close to half of the country's state and collective farms had adopted a form of ownership more compatible with a market economy, but the sector as a whole was still heavily influenced by government. For example, government agencies still controlled prices for some agricultural commodities and the state sector was completely dominant in terms of the share of cultivated area. And collective farms were accorded priority in the allocation of scarce agricultural inputs. Still, for all the lingering government involvement in agriculture, a substantial number of collective and state farms have been transformed into peasant farms, a process which began in 1991. Within two years there were already more peasant farms than in Kazakhstan. While that situation had changed by mid-1995, the 20,300 peasant farms in Kyrgyzstan at that date were of a comparatively substantial size. On average each comprised 53 hectares of land. Title to land in Kyrgyzstan is primarily vested in leasehold, not outright ownership, and for the same reason as in Kazakhstan. Colonization of rural areas characterized both the Russian Empire and Soviet eras, and sensitivity to Kyrgyz sentiments regarding land tenure is essential for domestic political stability. Indeed, after the initial flurry of privatization in which a number of collective farms were broken up into peasant farms, the issue of who controls the land became increasingly controversial. In consequence, the privatization process was temporarily suspended by the government. It was resumed later in 1993. The government has not abandoned the privatization process, however. In fact, pilot programmes of full privatization of land, including the creation of a land market, have been undertaken in two provinces.

In the other Middle Asian countries, Uzbekistan, Turkmenistan and Tajikistan, there is much less commitment to economic reform in general than is common in the other southern tier countries. Agricultural reform therefore has not been a high national priority. It has progressed furthest in Uzbekistan. In 1995, there were nearly 16,000 peasant farms, each with 14 hectares on average. Soviet-era personal plots also have been permitted to expand in area, and in fact now represent close to 15 per cent of the cultivated area. Given that 61 per cent of Uzbekistan's 22.6 million inhabitants live in the countryside, the benefit of this adjustment has been far reaching. It was also a necessary adjustment. All of Middle Asia in the late Soviet period was a food deficit region, notwithstanding the agricultural potential of the vast irrigated areas. These areas, however, were largely committed to industrial crops, most notably cotton. And cotton, of course, was not even much processed in the region, feeding instead the traditional textile production centres in Russia and Eastern Europe. Permitting larger personal plots was a logical adjustment to the post-Soviet reality of disrupted trade relations, including the inter-regional shipments of foodstuffs. However, raising the limits on the size of personal plots has not altered the relationship of individuals or households to the land. The right to a personal plot is inheritable, but not transferable, and therefore there is no basis for the creation of a market for land. The peasant farms are similarly restricted. Indeed, output from them is still subject to compulsory delivery to the state. Most of Uzbekistan's agricultural land and production, the latter still dominated by the cotton industry, is bound up with the roughly 2000 state and collective farms, which in number and function differ little from the late Soviet era. The rapid increase in the role of the personal sector in producing foodstuffs (now over half of its meat, milk and vegetables and fruits) is symbiotically related to the collective and state farms from which necessary inputs for the personal plots of their workers are derived. Even the retail prices of basic foodstuffs remain under government control. There is a measure of privatization in name, but very little of it is much related to the marketplace. In Turkmenistan the agricultural scene is also little changed. About 55 per cent of the 4.5 million total population is rural. For them life continues little changed by independence, except that the former dependence on imported foodstuffs has forced a reallocation of effort. Previously, Turkmenistan imported half or more of the cereal and dairy products, and a third of the meat requirements of the domestic market. Now there is necessarily more reliance on Turkmenistan's own 'personal' agriculture. Personal plots are now a little larger on average, but as in Uzbekistan they cannot be bought and sold, only inherited. On some collective and state farms, small areas are now leased to individuals, but output is mostly 'sold' to the collective or state farm concerned. Once government-imposed quotas have been met, above quota state and collective farm production can be sold at market prices. In 1995, there were only 300 peasant farms, each averaging only eight hectares in size. Thus, in Turkmenistan there is really little functional difference from the Soviet agricultural system described earlier. In Tajikistan, on-going civil war stands in the way of economic reform, though there is little to suggest that government agricultural policy would be much different from Uzbekistan or Turkmenistan. There were only 200 peasant farms in 1995. While small in number, they differed from their counterparts in Turkmenistan and Uzbekistan since each averaged 131 hectares. Land leasing, not ownership, is possible according to legislation adopted by the mid-1990s. As yet there is no indication of much support for the creation of market for land. Thus, nearly three-quarters of Tajikistan's 5.8 million citizens will continue to live and work in largely Sovietized countryside for the forseeable future.

The fact that cotton production remains so important to the economic well-being of much of Middle Asia is at least partially related to the reluctance of some governments to reform agriculture. High quality raw cotton remains a vitally important export crop. It is therefore a source of much needed foreign exchange. But cotton growing is dependent upon irrigation, and irrigation in Middle Asia is a vast enterprise, one complicated by serious water deficits, pollution, secondary salinization, and a seeming incapacity to reform irrigation practices in the interest of improving agriculture. The mis-use of water resources in Middle Asia is widely acknowledged. The SIBARAL project, discussed in the preceding chapter, was to have been the solution to all Middle Asia's water resource management problems. Immediately following its cancellation in the mid-1980s, charges for irrigation water were introduced on an experimental basis in the Tajikistan and Kyrgyzstan republics. Consumption was halved without adversely affecting cotton production. Indeed, it is entirely possible that greater economy of water use in irrigation projects would not only enable the current acute water deficit to be mitigated, but the annual loss of existing irrigated land to salinization could be reduced as well. Conventional wisdom has it that for every hectare added to the area irrigated one is lost, principally to secondary salinization brought about by overconsumption of water. If this is anything like representative of the actual situation, the cost of achieving the vast expansion of irrigated land during the last 25 years of Soviet power must have been colossal. As the data in Table 7.2 reveal, the Middle Asian republics accounted for the largest proportion of the irrigated land. Management of irrigation systems on the scale of those in Middle Asia requires some overriding authority. As a result of the collapse of the Soviet Union, this is now even more complicated than before. Post-Soviet political jurisdictions are not coincident with either the natural drainage systems or the irrigation networks they supply. While the question of charging for water is being addressed, by the mid-1990s there was still no rational system in any Middle Asian, post-Soviet state for pricing water consumed in irrigation systems. Water all too often is still allocated essentially free of charge to state and collective farms to manage as best they can. Until such time as a universal, and realistic, charge for water is introduced, the development of private farming, which invariably will be in large measure

Table 7.2 *Soviet-era irrigated land, 1965–1989*

Region	1965		1975		1985		1989	
	Thousand hectares	Per cent	Thousand hectares	Per cent	Thousand hectares	Per cent	Thousand hectares	Per cent
Russia	1501	15.4	3684	25.4	5805	29.1	6101	29.0
WESTERN TIER STATES								
Ukraine	503	5.1	1483	10.2	2456	12.3	2561	12.1
SOUTHERN TIER STATE								
Azerbaijan	1278	13.1	1141	7.9	1318	6.6	1390	6.6
Uzbekistan	2752	28.0	3006	20.8	3930	19.7	4164	19.8
Kazakhstan	1255	12.8	1648	11.4	2172	10.9	2294	10.9
Kyrgyzstan	861	8.8	910	6.3	1009	5.1	1034	4.9
Tajikistan	468	4.8	567	3.9	653	3.3	691	3.3
Turkmenistan	514	5.2	819	5.6	1107	5.5	1259	5.9
OTHERS	671	6.8	1228	8.5	1501	7.5	1570	7.5
TOTAL	9812	100.0	14,486	100.0	19,951	100.0	21,064	100.0

Sources: Narodnoye Khozyaystvo SSSR v 1965g (Moscow: Finansy i Statiska, 1966), 363; *Narodnoye Khozyaystvo SSSR v 1985g* (Moscow: Finansy i Statistika, 1986), 228; *Narodnoye Khozyaystvo SSSR v 1989g* (Moscow: Finansy i Statistika, 1990), 956

irrigation-water dependent and thus in competition with established rural interest groups for a scarce resource, is not likely to gain much political or public support in the countryside.

Agricultural reform in Russia

Russia has the smallest share of its total population classified as rural of all post-Soviet states. It was 27 per cent in 1995. Still, this comparatively small percentage represented almost 40 million people. However, not all of these rural folk are in fact farmers, as noted earlier. Some are urban workers unable to acquire a residence permit, and therefore obliged to live in rural settlements and commute daily. Some are workers in non-farm enterprises located in the countryside, and still others are part of the agriculture service sector. There were nearly 30,000 industrial, construction, trade and supply enterprises with a substantial labour force all geared to servicing the agricultural sector. Of course, the majority of the rural population is dependent upon the land for a living. In 1995, the on-farm labour force comprised about eight million people. They were employed in close to 27,000 farming units of one kind or another. At least one-third of these was still dependent upon state subsidy. Some estimates suggest that at least one-half of all farming units, irrespective of whether or not they are privatized, were running at a loss in 1995. The share of regional state employment in the agricultural sector in 1991, the largest part of which is employment on state farms, is portrayed in Fig. 7.8. Those regions in which agriculture is an especially important employer stand out clearly. For the most part, these are regions in which industrialization or resource extraction have played a minor part in economic development over the years.

For those people employed on collective and state farms especially, agricultural reform figured very prominently in their daily lives. The reform of Russian agricultural institutions has a long history, as earlier discussion has already made plain. The challenge facing the leadership in the late Soviet era was clearly multi-faceted. There were environmental limits to the area that could be cultivated. Massive programmes to transform the environment in an attempt to remedy this were eventually recognized as ill considered. Return on rising investment was abysmal. Aside from calling for more efficient use of existing land,

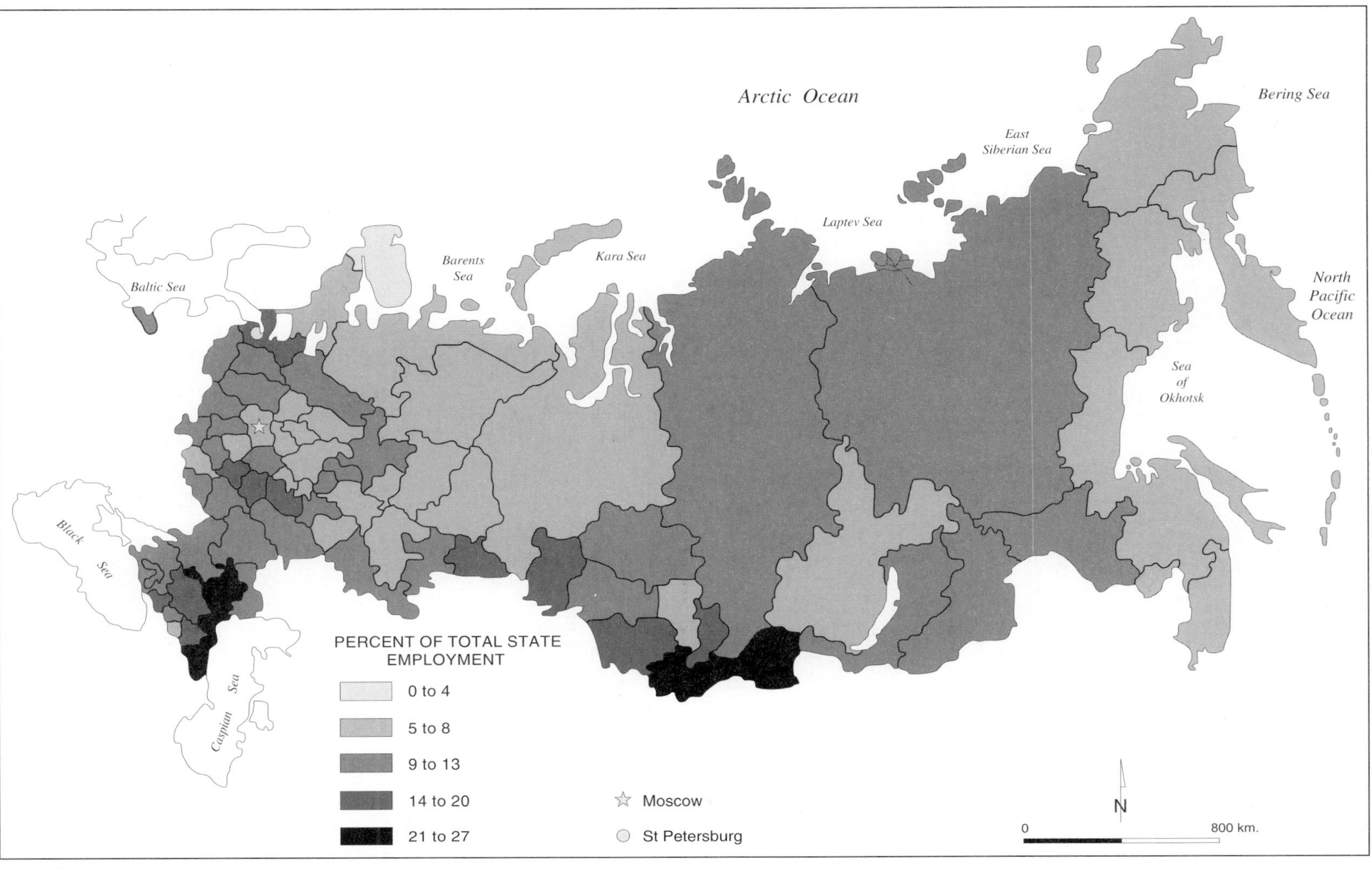

FIGURE 7.8 *State employment in agriculture: Russia, 1991*
(*Source*: T. Heleniak, *Economic Geography of the Russian Federation* (Washington: The World Bank, Country Department III, Europe and Central Asia Region, 1994), Vol. 1, No. 11)

labour and capital, there was little chance of enhancing output beyond what next year's better weather might bring. The new post-Soviet political economy offers scope for improving the management of agriculture, but it too must deal with fundamentals of soil and climate which are still far from ideal for growing crops.

While privatization in the Russian countryside to date is far less than 'market reformers' desire, it is far more than the dominant rural political elite want. Those promoting a market-oriented reform of the rural sector envision the process leading to the creation of a large, property owning, independent class of farmers, a class whose vested interests would support the further democratization of the countryside, and whose capitalist farming units would improve agriculture resource management efficiency. These objectives echo the intended outcomes of the 1906 Stolypin Agrarian Reform, described in Chapter 2. At that time, the reform movement faced stiff opposition. In the 1990s, agrarian reform faces similar resistance. Few in the farming sector are satisfied with the present state of affairs, no matter whether reformist or conservative. The dilemma is that the old system of central state planning has collapsed, agrarian reform is frequently more apparent than real, there is precious little evidence of improvement in management of the agricultural sector, and none in terms of output. The historic pattern of decentralization, then recentralization of institutional arrangements, continues, but with steadily diminishing effect. Post-independence reform of Russia's agricultural sector involves government institutions, farming organizations and land tenure. Some of the key government institutional changes will be described first.

Beginning in 1991, many thousands of government agencies dealing with agricultural supplies and output were privatized. By 1995 the process was largely complete. At first glance, this development implies enhanced competition, reduction of bloated bureaucracies, cost cutting and, in general, a more efficient use of resources. In reality, little had changed for the better. Officials and managers everywhere quickly seized the opportunity to privatize enterprises and agencies, but more often for personal profit and enrichment than for altruistic reasons such as promoting greater efficiency. And making a profit does not mean the decisions taken are subject to the rules of the marketplace. As in most parts of the Russian economy, decision-making in newly privatized enterprises and organizations occurs in a rather murky institutional and legal environment, with limited accountability to shareholders or to government, and with enormous scope for corruption. Because the federal government has not yet entirely removed itself from the business of dictating agricultural outputs and prices paid for them, coordination of the activities of these newly privatized enterprises is still necessary. But the necessary central authority required for such coordination scarcely exists at all. This situation is further complicated by the fact that since 1991 regional authorities across Russia have been busy assuming ever greater responsibility for managing agriculture within their jurisdictions. For example, already in 1993 about half of all agricultural input and output was under the jurisdiction of Russia's 89 regions. The central administrative structures are therefore increasingly irrelevant to management of many agricultural activities. Were this an intended outcome, carefully planned and executed, this could well have been a positive development. Instead, it is simply a by-product of the struggle for power between diametrically opposed political interest groups.

In 1991, a new Ministry of Agriculture for Russia replaced the Soviet-era republic level ministry. Personnel changes were made from the level of Minister on down. The new Ministry was headed by a reform-minded leader conversant with agriculture. But before long, Moscow politics gave rise to a new organization, the Federal Centre for Land and Agrarian Reform. This organization quickly became a focal point for interest groups opposed to the agrarian reforms being promoted by the Ministry of Agriculture in particular, and President Yeltsin's administration in general. There soon ensued a struggle between the Federal Centre and the Ministry of Agriculture for control of the various agricultural administrative functions, because control over them provided opportunities to influence policy implementation. President Yeltsin closed the Federal Centre in mid-1993. But by this date, agricultural reform and Moscow's control over agricultural developments across Russia already had been severely compromised. A year later, the federal government formed a new Ministry of Agriculture and Food Supplies to coordinate better the administration of the sector from Moscow's perspective. But the new Ministry does not have very much control over the day-to-day activities of the myriad of privatized former state agriculture agencies, and cannot impose its will on the regions.

Funds for agriculture are allocated from central government according to plans for output generated by all agricultural production units across Russia. Once the process of gathering and collating output data from all of the state and collective farms is completed, the Ministry of Agriculture submits the projected agricultural commodity production figures to the Ministry of Finance. Between it and the Ministry of Economics, funding requirements for the producing units are calculated, a package of financial credits is eventually approved by Parliament, and authorization for the Central Bank to release the approved funds sent back down through the system. In this respect, there is seemingly little difference from what went on during the Soviet period, save for the participation of a number of new banks, both central and regional, which handle the financial transactions. Most of these banks are still state owned. It soon became evident that the deregulation of prices for agriculture inputs, but the continuing fixed prices for agricultural commodities the state procured at 'planned' levels, made many potentially profitable farms insolvent. The cost plus basis of the calculations of Ministry officials dealing with credits for agricultural production units exacerbated inflationary pressures in the domestic economy. Put simply, with a growing share of agricultural production units going further and further into debt, the agricultural system was not sustainable. This immediate problem of what to do with thousands of insolvent agricultural enterprises was resolved, at least for the time being, when the State Duma passed a law in 1995 writing off the accumulated debt and interest owing, of all agro-industrial organizations and cooperatives. How private sector banks that were owed money were to be treated was left in abeyance. Eventually, the Russian Finance Ministry was instructed by the State Duma to issue interest-bearing notes, with maturities ranging up to ten years, to the private banks as compensation. Monies owed the government were simply forgiven, much as similar debts were handled during the Soviet era. Clearly, old habits are difficult to give up. Indeed, direct government intervention in the agricultural sector is far from waning.

Russian government subsidies to agriculture continue to be huge. Appropriations of funds for special agricultural 'needs' are multiplying, and run the gamut from support for private farmers, to price support for particular commodities, to funds for infrastructure and social services. Clearly, the financial burden of this sector on central government has not been eased as a result of the agricultural reform to date. Indeed, in some respects the situation is even less satisfactory than during the Soviet era, difficult as that may be to conceive. Central government still attempts to plan, but in an environment over which it exercises even less control than before. It still provides huge amounts of financial support in a system in which some prices are set by the market and others are not. In an era of growing regional authority, which is coming at the expense of the centre, Moscow has had to accede to demands from the regions for a larger share of taxation income. Fiscal responsibilities increase, but revenues do not. By the mid-1990s, the institutional arrangements put in place by Moscow for managing the agricultural sector were still far short of what was needed in a market economy. Reform in the other parts of the agricultural system was similarly inadequate. The privatization of existing production units, and the creation of new ones, is a case in point.

The Law on the Peasant Farm, one of the most significant pieces of legislation dealing with privatization in the agricultural sector, was adopted just before the demise of Soviet power in December 1990. This law enabled a collective or state farm worker, or indeed anyone who could demonstrate expertise in agriculture, to obtain land to farm. Some 4000 so-called peasant farms were created in the final weeks of 1990. By the end of 1991, a total of 49,000 peasant farms had been formed as a result of the legislation. Subsequently, the tempo picked up, and by late 1994, there were close to 286,000 peasant farms across Russia. The average size of a peasant farm is currently 41 hectares, a figure scarcely changed since 1990. The cultivated area of all peasant farms comprises about five of the 12 million hectares in this sector. This represents roughly 4.5 per cent of the 112 million hectares under cultivation in Russia. Obviously, the peasant farm is still a minor component of the Russian agricultural scene. But there are some interesting regional variations in the importance of peasant farms.

As Fig. 7.9 indicates, the greatest concentration of peasant farms at the beginning of 1994 was along the southern flank of the country. Six political administrative units in the North Caucasus and Volga river region accounted for about one-third of all peasant farms. This is not entirely surprising given that much of Russia's better agricultural land is located in these regions. It is also worth noting that most of these regions have very substantial non-Russian populations. In general, this pattern of peasant farm formation reflects the regional variation in the role of the personal sector in agriculture during the Soviet era. It will be recalled that during this era the personal sector was especially important in the Caucasus. Put simply, where the natural environment affords some reasonable prospects for agriculture, there is naturally greater propensity for individual initiative. As it turns out, in the southern regions of Russia the average size of a peasant farm is roughly 20 hectares more than in northern regions. There is also an east–west difference in peasant farm size. In the east, peasant farms are larger than in the west, most likely because there the rural population to land ratio is smaller, and because the agricultural potential is lower owing to the harsher natural environment. But as Fig. 7.9 clearly indicates, there are only two political administrative regions east of the Ural mountains in which the number of peasant farms is in the top category. The pattern of peasant farm formation is not simply a function of propitious conditions in which to grow crops and a pool of local entrepreneurial talent. Land availability is a key factor, and this is influenced in no small way by prevailing values and attitudes towards individual peasant farms. Land for peasant farms was obtained from two sources – collective and state farms and the so-called Special Land Funds which had to be established in each local administrative district. In the early months of 1991, the vast majority of peasant farms were established on lands from the Funds by urbanites seeking new opportunities, not by entrepreneurial farm workers who had taken their share of the land out of the collective or state farm to work independently. Urbanites continued to be a significant proportion of the proprietors of peasant farms. Indeed, in many regions of Russia there was outright resistance to the formation of peasant farms from the collective. When the Law on the Peasant Farm was successfully invoked, the land 'freed up' for the peasant farm was frequently the worst. This was a common practice in allocating land for the Special Land Funds in many jurisdictions as well. In consequence, in many areas peasant farming is handicapped by poor land, and the attitudes which gave rise to this situation often hamstring the efforts of the individual peasant farmer to purchase inputs, secure credit, or market whatever commodities produced. Such barriers have had predictable results. The tempo of peasant farm formation has dropped. By the beginning of 1995, the total number was nearly 7000 fewer than six months earlier, that is, about 279,000. Some 20,000 peasant farms have gone bankrupt, the land most often reverting back to collective or state farm organizations. Other aspiring farm entrepreneurs have quit before acquiring land. While the number of peasant farms has since increased once more, reaching 283,000 by spring sowing time in 1995, the fact remains that collectivism, not individualism, is still the prevailing ethos in the Russian countryside. The share of peasant farm arable of the total area under cultivation in 1994 is depicted in Fig. 7.10. There is a general spatial correlation of course with the number of peasant farms (Fig. 7.9). Whether peasant farms averaging only 42 hectares in size can ever be efficiently managed remains a central question. Even assuming

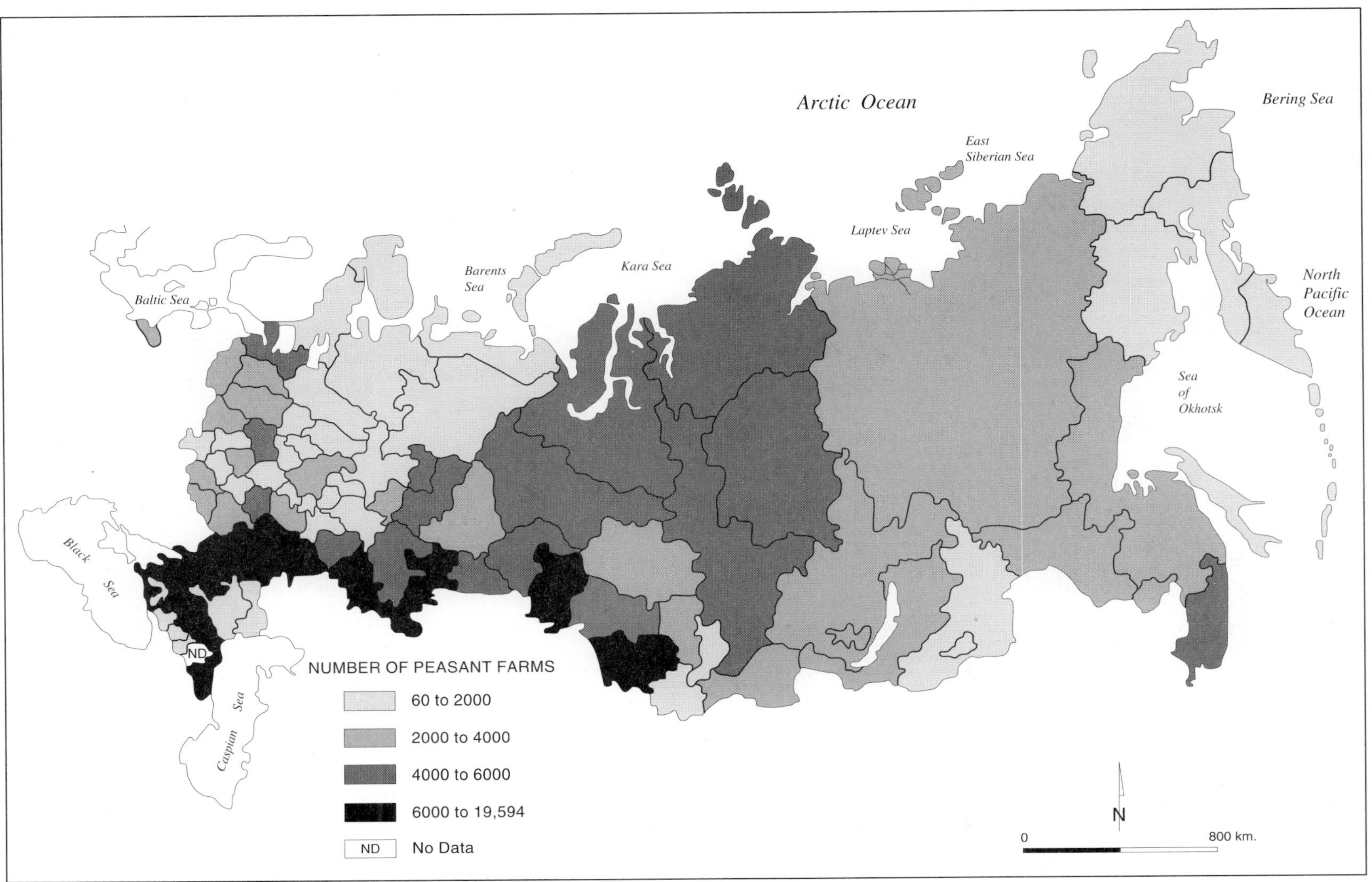

FIGURE 7.9 *Number of peasant farms: Russia, January 1, 1994*
(*Source*: P.R. Craumer, 'Regional Patterns of Agricultural Reform in Russia', *Post-Soviet Geography*, Vol. 35, No. 6, 1994, 336)

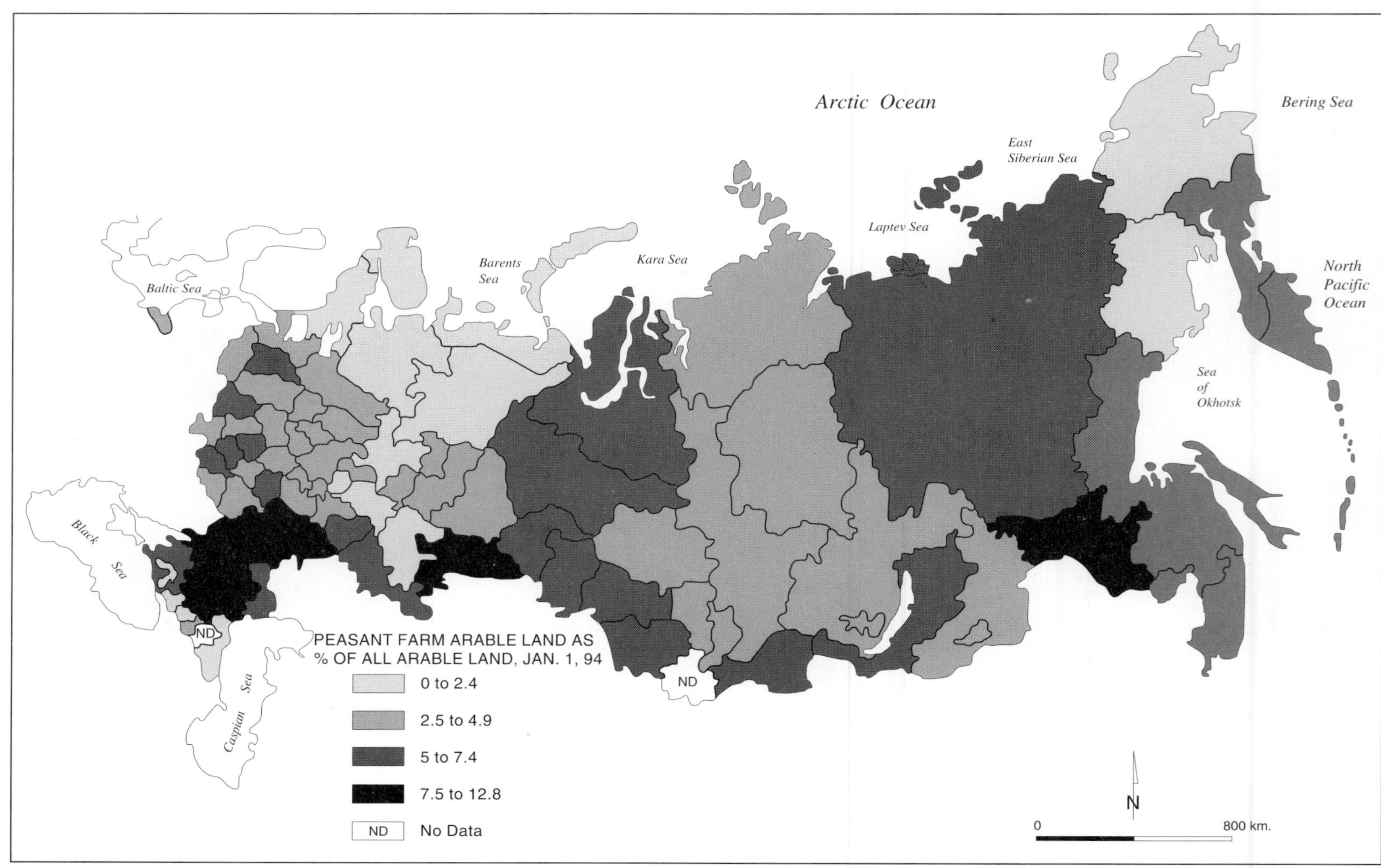

FIGURE 7.10 *Peasant farms as a per cent of arable: Russia, January 1, 1994*
(*Source*: P.R. Craumer, 'Regional Patterns of Agricultural Reform in Russia', Post-Soviet Geography, Vol. 35, No. 6, 1994, 336)

that financial and other resources were readily available, there is the not inconsiderable constraint that agricultural equipment which is domestically produced has been designed to meet the needs of very much larger agricultural operations. To reiterate, the average areas cultivated on each collective and state farm were 3600 and 5000 hectares respectively. Machinery designed for that size of operation is scarcely suited to peasant farms of only 42 hectares.

In December 1991 a Presidential decree was issued which stipulated that by the end of 1992 farm land was to be de-nationalized, that is, converted to some form of private ownership. Owing to both some confusion about the process and a lot of resistance in some regions, the deadline was later extended to March 1993. By the beginning of 1993, the privatization process had made substantial progress in many Russian regions, as Fig. 7.11 indicates. Those regions in which privatization had made least inroads by the beginning of 1993 were the non-Russian Republics or other minority group political administrative areas (see Fig. 9.12, p. 280). By 1995, almost all collective and state farms had been reorganized. In practice privatization meant that collective and state farms could opt to become a joint-stock company, a cooperative, a partnership, an association of peasant farms, or be broken up into individual peasant farms. As the foregoing discussion implies, the latter option was not very common. Indeed, amongst those collective and state farms which opted for association of peasant farms status, a great many were associations in name only as individual farms were rarely designated, or titles registered to specific parcels of land. About a third of all collective and state farms opted to retain their existing organizational form, though the land no longer belonged to the state. This response to the Yeltsin decree was especially evident in the non-Russian regions. The rest did change their status. Of this group, about three-quarters became joint stock operations, about 12 per cent opted to become cooperatives. Few elected any of the other options, which included complete liquidation of all assets.

The peasant farm phenomenon has captured a great deal of attention since the enabling legislation was first introduced in 1990. Notwithstanding rhetoric both supporting and decrying this innovation, the fact of the matter is that in Russia the peasant farm has not had a significant impact on the landscape, or on production. Because it has been seen as the market reform 'wedge' being driven into a traditionally collectivized countryside, publicity accorded the peasant farm has overshadowed the rather significant increase in the importance of the traditional personal sector, that is, the millions of personal plots and stocks of privately owned farm animals. The personal plots accounted for about 2 per cent of Russia's total cultivated land in 1990, but had increased to 3.5 per cent by 1993. Together with the peasant farm, about 8 per cent of Russia's cultivated hectarage in 1993 was represented by these two categories (Fig. 7.12). As emphasized earlier in this chapter, there has always been a symbiotic relationship between the household personal plot and the larger farming organization in which the household is located. That the personal sector derives a wide range of inputs for both the sown crops and livestock holdings from the larger farming unit, whatever its legal form, is certainly still the case. But the relative importance of the larger farms, still collective in function if not in law, has been in steady decline since independence. As Fig. 7.12 clearly indicates, their share of three key agricultural commodities, grain, potatoes and vegetables, has been eroded quite substantially since 1991, and quite dramatically if the last half of the 1980s is used as the reference point. Clearly, the peasant farm and personal sector have little importance in the production of a wide range of other crops, ranging from cotton, to soy bean, to flax. But those commodities depicted in Fig. 7.12 do figure prominently in the diet of Russia's citizenry. Including the personal household plots, collective fruit and vegetable gardens, and so forth, there were over 38 million small scale agricultural production units in rural Russia in 1993. Output from them comprised an important share of urban food supply. In addition, it is reckoned that about half the people living in cities have garden plots on which they grow vegetables and fruit to supplement their purchased foodstuffs.

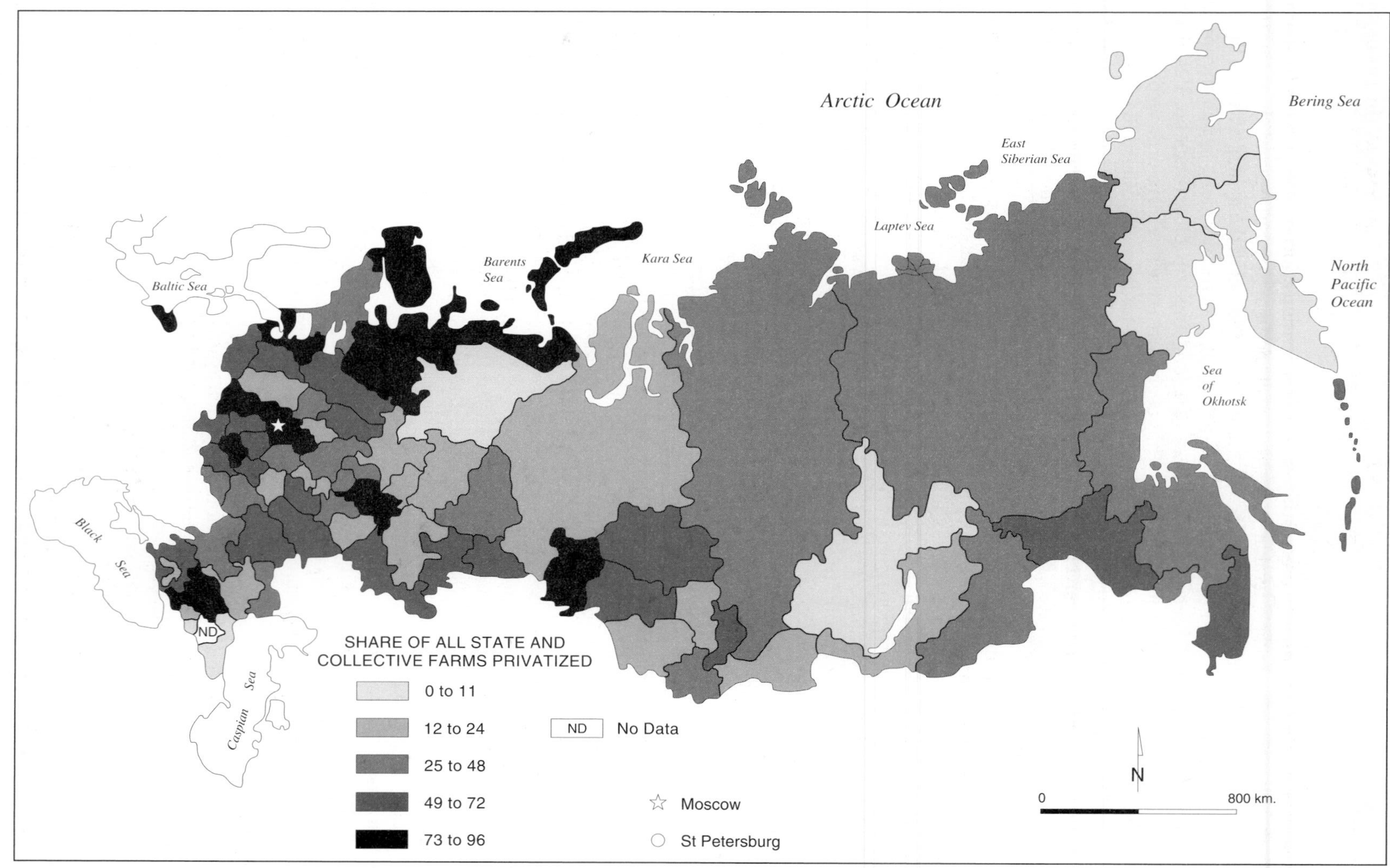

FIGURE 7.11 *Privatization of agriculture: Russia, 1993*
(*Source*: T. Heleniak, *Economic Geography of the Russian Federation* (Washington: The World Bank, Country Department III, Europe and Central Asia Region, 1994), Vol. 1, No. 21)

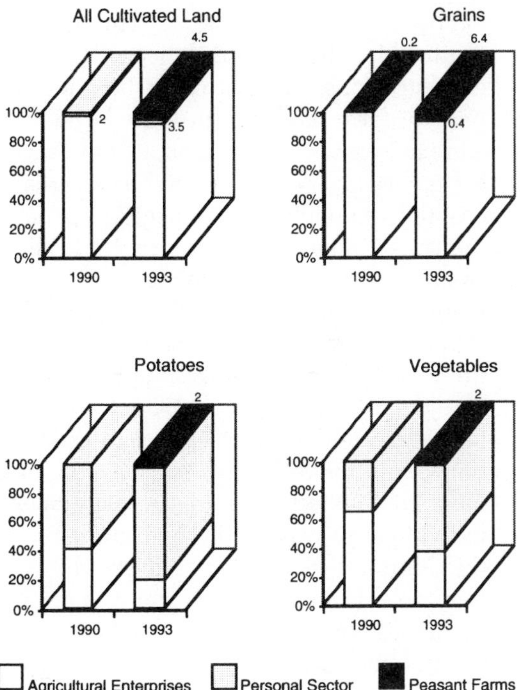

FIGURE 7.12 *State, personal and peasant farm sector agricultural commodity production: Russia, 1994*
(*Source*: Based on *Rossiyskiy Statisticheskiy Yezhegodnik* (Moscow: Goskomstat Rossii, 1994), 362)

In the countryside, the relative importance of the personal sector and the peasant farm in the overall agricultural scene needs to be placed in context. Since the late Soviet period, there has been a drop in virtually all components of the agricultural system, from area cultivated, to livestock holdings, to output. Grain production in 1994, for example, was 81.3 million tons. The target for 1995 was 82 million tons, but given the shortfall in transferring funds to the agricultural sector for the spring sowing of crops, achieving this level of production was problematic. Estimates of the 1995 harvest were around 70 million tons. As grain reserves had already been largely depleted, sizeable imports are needed to make up the difference. The grain harvest has fallen steadily since 1990 when it was 116.7 million tons. Against this rather dismal background, it is notable that area cultivated and some categories of livestock holdings in the personal sector increased. And there is now the peasant farm component which is making some progress as well. For example, not only has the total area under cultivation in personal plots increased since 1991 as already mentioned, but so have the personal sector holdings of cattle, including milk cows, and pigs. Although peasant farms are still very minor in terms of numbers of livestock, there has been some growth there as well. In contrast, in the now 'privatized' former collective and state farms, there has been a sharp decrease in livestock numbers and in output since 1991. These important trends are depicted in Figs 7.13 and 7.14. The changing relative importance of the personal sector is especially evident. The peasant farm is obviously still a minor source of agricultural output. The regional variations in these trends are predictable. For example, where conditions for animal husbandry are harsh, the impact on livestock numbers has been greatest. In the late Soviet era, livestock holdings could be maintained at the then prevailing all-time high

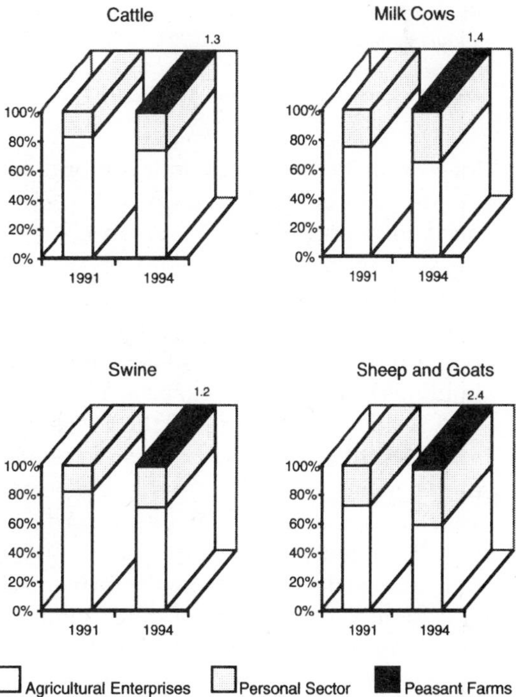

FIGURE 7.13 *State, personal and peasant farm sector livestock holdings: Russia, 1994*
(*Source*: Based on *Rossiyskiy Statisticheskiy Yezhegodnik* (Moscow: Goskomstat Rossii, 1994), 362)

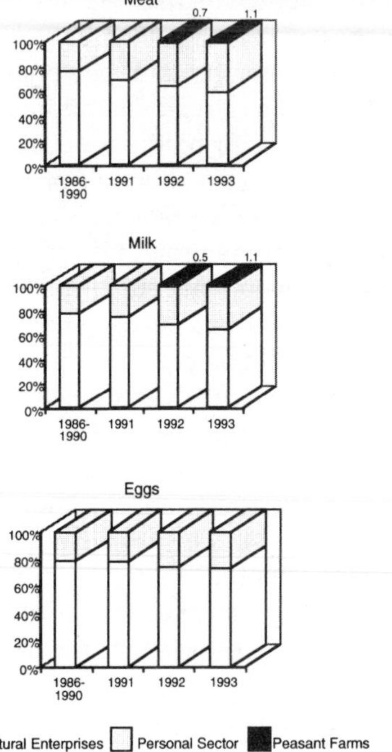

Figure 7.14 *State, personal and peasant farm sector meat and dairy production: Russia, 1994*
(*Source*: Based on *Rossiyskiy Statisticheskiy Yezhegodnik* (Moscow: Goskomstat Rossii, 1994), 365)

levels only because substantial amounts of feed grains were imported each year. Given the need to import grain for human consumption, Russia can no longer afford to import feed grain at the scale prevailing in the late Soviet era. Since domestic fodder production has dropped from already inadequate levels, livestock holdings have had to be cut back in consequence. The relative importance of the personal sector and peasant farm in the production of meat and dairy items, poultry and eggs, potatoes, vegetables and fruits is now greater to be sure (Fig. 7.14). But this is related both to an absolute reduction in the output from the former collective and state farm operations and some modest expansion in the personal sector and a small output from the peasant farm. In a functional sense, there is still as much continuity as change in the Russian agricultural scene. A case in point concerns the question of creating a market for land.

The 1990 Law on the Peasant Farm, which transcended the collapse of the Soviet Union, placed restrictions on what could be done with the peasant farm land. For instance, while peasants were to be given title to land, it could not be sold for ten years. Within the Russian Parliament, a coalition of political parties began obstructing any reformist initiative which might facilitate the creation of a rural land market. Often the arguments were couched in economic terms. The coalition opposed the break up of the existing farming units, arguing that the resultant peasant farms would be too small for efficient and profitable farming. Sometimes an appeal to Russian traditional cultural values was made – the old argument that collectivism was morally better than individualism was used. The Agrarian Party and the Communists were key players in this opposition force. But agrarian reform was not the only point of contention between reformist and opposition forces in government. Indeed, there was a growing rift between the executive, headed by Yeltsin, which was pushing for market reform throughout the economy, and the increasingly contrary-minded government. As described in Chapter 2, this rift culminated in President Yeltsin abolishing Parliament in September 1993, an act which precipitated a near civil war on the streets of Moscow in the first days of October. Elections to a new State Duma were called for December that year, with local and regional elections following in early 1994. Yeltsin took advantage of the hiatus between the dissolution of Parliament and the elections to the new State Duma to push further the agrarian reform agenda. A Presidential decree aimed at facilitating the creation of a market for land was passed in October 1993. The intent of the decree was to permit peasant shares in land to be traded, just like the privatization vouchers issued in 1992. The decree further stipulated that by January 1994 compulsory deliveries of agricultural commodities to the state at fixed prices were to end. Thereafter, the state would pay market prices for what it needed. However, since this decree would have to be endorsed by the new State Duma, and clearly challenged the well entrenched vested interests in the countryside, not surprisingly little was actually accomplished.

After the elections to the State Duma, instead of adopting the Presidential Decree a committee was set to work writing a draft Land Code which would lay the legal framework for transactions involving land, whether agricultural or not. The Duma Committee on Agrarian Questions was assigned the task of writing the legislation, notwithstanding the fact that 90 per cent of the land in Russia is not agricultural. A new Land Code was finally adopted by the State Duma in the summer of 1995, and supersedes all previous laws and Presidential decrees pertaining to land ownership in general, and the ownership of land by peasants in particular. This is only a first step as the Code has to be translated into a host of new laws. But Land Code is scarcely what the reformists hoped to see. The Code permits private ownership of land for legal entities with few if any restrictions. In essence, this protects the status quo, that is, the reorganized state and collective farm members. Individual peasant rights, on the other hand, are severely restricted. Whereas previously peasants had the right to withdraw their shares of land, now this is possible only if the farm charter permits a land allotment to be withdrawn from the farm, and only with the concurrence of the farm membership. Land can be leased, but only under very stringent conditions, conditions which can go so far as to determine which crop can be grown. Land can be inherited, however. Recourse to the courts is proposed in the event of disputes, but even this route is laden with many barriers. In effect withdrawal of land from the farm is nearly impossible, and the right to obtain compensation for a land share 'given up' would be delayed for a decade. In short, the peasant member of a now reorganized, 'privatized' state or collective farm remains firmly tied to the land in the traditional manner. Clearly, the new Land Code presents some new, and major obstacles to the formation of a peasant farm. The decline in the rate of peasant farm formation, already evident in late 1994, is likely to be accentuated as a result of the new Code. If a peasant departs the farm out of frustration, there is no compensation for property left behind. The Land Code goes further. It prohibits the sale of land in the large urban centres; there only leasehold is possible. In smaller cities, land can be bought and sold for housing, garages and so forth. Russian citizens may also buy and sell plots of land for gardens and *dachi*. If the 1995 Land Code is to be the cornerstone of agrarian reform, obviously not much of substance will be built upon it. While the apparatus of government was changed as a result of Yeltsin's quashing the Parliament and calling for elections to a new State Duma, the old coalitions reappeared, ever ready to control the political agenda in Moscow.

There remain many difficulties in Russian agriculture, and very few bright spots. One of the latter is the absolute increase in the personal sector output, something both important and necessary for ordinary Russians. For the farm workers involved, it is a quite rational response to risk. Inputs from the larger organization on which they work have always been essential to this component of the agricultural scene. From fodder, to fertilizers, to equipment, there is a dependent relationship between the personal sector and the 'host' farming organization, be it collective or state farm in the Soviet era, or the joint-stock, cooperative, or partnership farming organization of the post-Soviet period. Expansion within this framework clearly poses less risk than attempting to establish a separate peasant farm, with all the attendant problems in doing so, to say nothing of the possible subsequent retribution in terms of being denied access to necessary inputs. With the new Land Code restrictions, there is even more reason to follow this route. Without the increase in personal sector production, more food would have had to be imported. Indeed, this is already a problem in that those urbanites with money seemingly prefer imported foodstuffs, irrespective of price. For some people fearful of eating food possibly grown in the vast areas contaminated by the Chernobyl' nuclear power station disaster, health concerns are a major motivation in their decisions to buy imported foodstuffs. For others, it is simply a matter of exercising a new found freedom to choose, and there is now much more to choose from for those who have money. But for Russian agriculture, beset with problems of poor quality and high costs, imported foodstuffs are a real threat. Already there has been a Presidential decree giving Moscow Oblast' agricultural organizations preferential market rights in Moscow over imported foodstuffs. But such efforts are rarely effectual, and in any event run counter to market reform.

Even if there were support in government for far reaching agrarian reform, the demographics of the countryside are not auspicious. In general, the age–sex profile shows an imbalance between males and females. But in the predominantly ethnic Russian agricultural regions, the age–sex profile of the population is even more heavily skewed toward older women. The young, educated and ambitious males have always found reason to leave. Now there is even more reason to do so. In areas dominated by non-Russian nationalities, there has been a decidedly mixed response to agrarian reform. As noted earlier, in many of these regions privatization of agricultural organizations was resisted. Paradoxically, the formation of peasant farms in the predominantly non-Russian regions along the southern flank of the country has been the greatest in absolute terms. Across most of Russia, the natural conditions for agriculture are marginal. The simple fact is that for many people, and perhaps especially the older generations, there is greater personal security being part of a larger organization than being 'independent'. Moreover, a good many newly privatized former state and collective farms are capable of operating at a profit. Indeed, some now do, notwithstanding the scissors effect of being obliged to deliver to the state agricultural commodities at fixed prices, while having to pay market prices for inputs. Agriculture, consuming a huge share of the state budget and an excessively large share of the labour force, was long regarded as the 'Achilles heel' of the Soviet system. In post-Soviet Russia, agriculture still consumes far too much of the state's limited financial resources in the form of subsidy of one kind or another, producing in return far too little in quantity, and far too little of quality. And while some people in Russia may well be eating better now than before, they are in a distinct minority. Most probably, a large and growing proportion of what they eat comes from farms outside Russia, indeed from beyond the borders of the former Soviet Union.

Post-Soviet foodstuff consumption trends

One of the simplest measures of the overall state of agriculture in the years since independence is the per capita consumption of basic foodstuffs. The data in Table 7.3 describe the trends in per capita consumption of a selection of dietary staples in the Commonwealth of Independent States (CIS) to 1994. While the Baltic states were not members at that date, there is no compelling reason to assume that the general trends in these countries would be much different than in the CIS.

Even a cursory look at the data presented in Table 7.3 indicates that in most of the CIS the average diet was heavier in starch and lower in protein in the post-independence years than in 1990. This was especially evident in Russia and the western tier countries of Ukraine, Belarus' and Moldova. In these countries there had generally been a fairly sizeable decrease in the per capita consumption of vegetables, meat and meat products, dairy products and eggs since independence. Save for consumption of bread products in Belarus' which decreased slightly, people were eating more bread and potatoes than in 1990. Of course, the trends depicted in Table 7.3 are based on official statistics pertaining to food production and the net foodstuffs export/import balance. Thus, they do not capture the full spectrum of food consumption. As noted above, throughout the Soviet era a substantial share of urban residents had access to a garden plot and supplied part of their own vegetables and fruits. Since the post-independence economic collapse, there has been a very significant increase in urban household dependence on such ancillary 'enterprise'. In Ukraine, for example, it is estimated that most urban households were engaged in cultivating vegetable gardens. But even allowing for a much greater reliance on sources of foodstuffs which are not captured by the official statistics, the data in Table 7.3 still indicate that diet of urbanites has deteriorated, and markedly. For rural folk there has always been a much greater reliance on household-generated food supplies, including meat, dairy and poultry products. As emphasized earlier in this chapter, during the late Soviet era many rural families stopped baking bread for their own consumption since they now had the money to buy it. Indeed, store-bought bread was so heavily subsidized it was used as fodder by peasants for their livestock. Since independence, food from the personal sector probably figures more prominently on the kitchen table in the rural household if for no other reason than there has been a general reduction in supplies from the former state food production sector. Whether rural families are eating as well as before is a moot point. The situation in the cities is less ambiguous. Since urban households are less likely to be in the business of raising chickens or looking after a cow, there can be no other conclusion than that their diet has worsened, and quite decidedly in terms of protein intake. And except for Moldova, urbanites comprise the majority of citizens in the western tier countries. The health consequences of what can only be construed as a general deterioration in the diet will be explored more fully in Chapter 10, but obviously judging from the general trends evident from the data presented in Table 7.3 the consequences are not likely to be very positive.

Across the southern tier states, the trends in per capita food consumption depicted in Table 7.3 if taken at face value are even less auspicious than in Russia and the western tier states. With few exceptions, per capita consumption of all basic foodstuffs was less after independence than before it. In the few instances where per capita consumption of some foods was higher, such as in Kazakhstan and Armenia, it was carbohydrates, that is, in terms of bread products or potatoes. The decline, in some instances quite dramatic, in per capita consumption of vegetables, meat, dairy and eggs suggests certainly a less healthy, if not a quite inadequate food consumption pattern. The difficulty in drawing hard and fast conclusions from the data in Table 7.3 is related to the role of ancillary household agricultural activities, the output from which does not enter into the government's official statistics. In four of the southern tier countries the urban population comprises less than one-half the total, a bare majority in three others, and only in Armenia is it a significant majority. The bulk of the southern tier population being rural, there is no doubt a heavy reliance on food products from the personal plot, privately owned farm animals, and orchard. Perhaps this sector's output has compensated for the shortfall from the former collective and state farm operations. But the precipitous drops registered in Table 7.3 cannot be without some untoward consequences. Certainly, many urbanites can only be worse off now than before. And it is entirely possible that the output derived from ancillary household agriculture in the countryside which escapes the statisticians' purview does not fully compensate for the officially recorded drop in per capita consumption of basic foodstuffs in general, and of vegetables and protein foodstuffs in particular in the southern tier states.

Obviously, none of the post-Soviet states has much to gain from under-reporting the actual per capita food consumption situation. In the not so distant Soviet past, food shortages were enough to inflame public opinion, and in many instances resulted in riots. The present state of domestic agricultural production is clearly an issue of importance in all post-Soviet states. Where domestic output is simply inadequate to meet demand, food is sometimes imported. But not all states can pursue this option with impunity. Domestic agriculture is in need of modernization, of reform, and to simply import food is not likely to assist in the transition from a command to market economy. In most places, the prices charged for imported food simply puts it out of reach of ordinary citizens. Russia imports significant amounts of food, both basic and exotic, from outside the former Soviet realm. For example, fresh frozen meat imports in 1993 alone totalled $ 116 million. Retail prices are such as to ensure sizeable profit, and there are many consumers who can well afford to pay whatever the price demanded. But for millions of others, and not just

Table 7.3 *Post-Soviet states consumption of basic foodstuffs per person (in kilograms), 1990–1994*

	Bread products	Potatoes	Vegetables	Meat and meat products	Milk and milk products	Eggs
Russia						
1990	119	106	89	75	386	297
1991	120	112	86	69	347	288
1992	125	118	77	60	281	263
1993	124	127	71	59	294	250
1994	124	125	66	57	278	245
WESTERN TIER						
Ukraine						
1990	141	131	102	68	373	272
1991	143	116	102.5	65.5	345.5	256
1992	142.5	133	89	53.4	284.5	227
1993	146	138	94	46	275	193
1994	140	130	82	43	284	181
Belarus'						
1990	126	170	78	75	425	323
1991	126	165	78	73	415	320
1992	119	169	76	72	396	305
1993	178	178	78	70	384	306
1994	120	180	76	64	378	303
Moldova						
1990	171	69	112	58	303	203
1991	175	69	113	56	195	195
1992	170	66.5	95	46	166	166
1993	173	80	91	34	110	110
1994	136	80	80	31	161	100
SOUTHERN TIER						
Georgia						
1990	183	41	82	42	289	140
1991	169	55	76	31	135	139
1992	147	50	54	21	87	55
1993	114	48	77	19	80	46
1994	100	46	77	18	78	46
Armenia						
1990	151	58	132	44	446	163
1991	134	77	145	31	392	143
1992	150	64	132	20	122	65
1993	110	73	99	20	99	45
1994	174	80	138	22	150	57
Azerbaijan						
1990	129	27	67	32	292	143
1991	130	22	68	26	250	125
1992	114	26	53	19	204	103
1993	153	25	58	17	170	70
1994	140	23	63	14	143	78
Kazakhstan						
1990	146	85	75	71	307	222
1991	147	75	62	70.5	303	206
1992	153	86	62	61	269	175
1993	180	80	50	59	260	170
1994	217	65	42	47	253	164

Table 7.3 *Continued*

	Bread products	Potatoes	Vegetables	Meat and meat products	Milk and milk products	Eggs
Kyrgyzstan						
1990	139	69	78	54	266	154
1991	134	62	73	48	249	144
1992	135	68	75	46	206	128
1993	135	59	50	44	193	81
1994	134	56	50	40	177	42
Tajikistan						
1990	167	35	95	26	161	111
1991	131	31	83	21	124	82
1992	158	32	101	18	127	34
1993	155	31	98	14	136	23
1994	155	30	97	15	134	12
Turkmenistan						
1990	165	21	121	43	207	98
1991	167	19	123	38	176	82
1992	170	23	92	38	185	75
1993	142	33	95	33	196	75
1994	147	30	95	30	183	64
Uzbekistan						
1990	170	29	107	32	210	120
1991	167	25	114	30	196	107
1992	164	27	124	27	175	80
1993	162	25	122	27	177	74
1994	161	22	118	25	168	68

Sources: Ekonomika Sodruzhestva Nezavisimykh Gosudarstv v 1993g (Moscow: Statkomitet SNG, 1994), 39–40; *Ekonomika Sodruzhestva Nezavisimykh Gosudarstv v 1994g* (Moscow: Statkomitet SNG, 1995), 85

the elderly pensioners, the total cost of even a largely starch-based diet of home-grown products represents a very significant share of monthly income. A diet of bread, pasta products, tea, and the occasional egg or piece of sausage, is now the daily reality for literally millions of poor people, not all of them in the cities. Fresh meat of any quality is simply no longer affordable. The transition from inefficient peasant workers on a vast collective farm to entrepreneurial independent farmers portends more efficient management of the land, increased output and higher quality foodstuffs. But there were few places in any post-Soviet state in the mid-1990s where this was the case. Indeed, in no state had agricultural output increased since independence. In the agrarian reform scenario of greater efficiency, greater output and higher quality foodstuffs that ordinary people can afford to buy, the operative word remains 'portends'.

FURTHER READING

Atta, D. van (ed) 1993 *The 'Farmer Threat'. The Political Economy of Agrarian Reform in Post-Soviet Russia.* Boulder, CO, Westview Press.

Belov, F. 1955 *The History of a Soviet Collective Farm.* New York, Praeger.

Borders, K. 1976 *Village Life Under the Soviets.* Plainview, NY, Books for Libraries.

Brada, J., Wadekin, K.E. (eds) 1987 *Socialist Agriculture in Transition: Organizational Response to Failing Performance.* Boulder, CO, Westview Press.

Bridget, S. 1987 *Women in the Soviet Countryside.* Cambridge, Cambridge University Press.

Conquest, R. 1986 *The Harvest of Sorrow: Soviet Collectivization and the Terror-Famine.* New York, Oxford University Press.

Danilov, V. 1988 *Rural Russia Under the New Regime.* London, Hutchinson.

Denisova, L.N. 1995 *Rural Russia: Economic, Social and Moral Crisis.* Commack, NY, Nova Science Publishers.

Hedlund, S. 1984 *The Crisis in Soviet Agriculture.* London, Croom Helm.

Humphrey, C. 1983 *The Karl Marx Collective: Economy, Society and Religion in a Siberian Collective Farm.* Cambridge, Cambridge University Press.

Johnson, G., McConnell, K. 1983 *Prospects for Soviet Agriculture in the 1980s.* Bloomington, Indiana University Press.

Joravsky, D. 1970 *The Lysenko Affair.* Cambridge, MA, Harvard University Press.

Lewin, M. 1963 *Russian Peasants and Soviet Power.* London, George Allen and Unwin.

McAuley, M. 1976 *Khrushchev and the Development of Soviet Agriculture. The Virgin Land Programme 1953–1964.* London, Macmillan.

Millar, J. (ed) 1971 *The Soviet Rural Community.* Champaign, University of Illinois Press.

Pryor, F.L. 1992 *The Red and the Green: The Rise and Fall of Collectivized Agriculture in Marxist Regimes.* Princeton, NJ, Princeton University Press.

Reston, R. 1975 *Aftermath to Revolution: The Soviet Collective Farm.* London, Collier Macmillan.

Shaffer, H.G. (ed) 1971 *Soviet Agriculture: An Assessment of its Contributions to Economic Development.* New York, Praeger.

Stuart, R.C. (ed) 1984 *The Soviet Rural Economy.* Totowa, NJ, Rowman and Allanheld.

Symons, L. 1972 *Russian Agriculture: A Geographic Survey.* London, Bell and Sons.

8

ENERGY IN THE SOVIET AND POST-SOVIET ERAS

'sales of "black gold" are possible even as production declines'
(*Izvestiya*, August 18, 1994, 2)c

One of the characteristic features of modernizing states around the world is an annual increase in the consumption of energy. Indeed, per capita energy consumption often is used as a basic indicator of the relative level of national economic development. The importance of energy resources in the modern industrial economy was apparent to the Soviet leadership from the very beginning. Fortunately, within the Soviet Union's expansive borders were large reserves of all fossil fuels, as well as enormous hydroelectric power potential. To support the industrialization drive initiated by Stalin, high priority was given to the task of increasing output of all forms of energy, renewable and non-renewable. The energy sector therefore commanded a significant share of the total capital investment in each plan period. The energy resource scene proved to be more bountiful than even the most optimistic of early Soviet expectations. By the mid-1980s, Soviet production of coal, oil, natural gas and hydroelectricity ranked first in the world. By this date an integrated distribution system locked energy producing and consuming regions of the country together in a complex, inter-dependent fashion. Put simply, energy production and distribution were planned and managed within the context of a single space economy.

Patterns of energy production, distribution and consumption in the post-Soviet scene are still in a state of flux, but there is no denying the fundamental importance of energy supply security to the future economic viability of each post-Soviet state. Energy resources are not distributed evenly. Indeed, most new states are energy deficient, and must contend with the possibility of disruption of supplies if they cannot finance imports. Oil and natural gas are especially critical in this regard. Few states produce more energy from all sources than they consume. For the few states that do have a marketable energy surplus, a key problem is access to hard currency markets which lie beyond the borders of the former Soviet Union, markets obviously more attractive in financial terms than near bankrupt post-Soviet states. The financial stakes are high. As energy prices slowly rise from the often below cost of production level characteristic of the Soviet era, revenue potential increases accordingly. Each of the new states must also begin to change Soviet-era energy consumption practices conditioned by artificially low prices. Consumption practices typically are profligate by western standards. Long overdue demand management will take considerable time, if not money, before savings can be realized. Having control over energy supply, or energy movement, obviously brings with it some opportunities to exact political concessions from dependent states. Thus, the politics of energy resource management figure prominently in the post-Soviet scene. In the geopolitical context, Russia plays a dominant role.

Russia remains the principal source of energy supply, just as it was in the Soviet era. Russia faces three energy resource management challenges: producing enough to meet its own energy needs; producing enough to meet ongoing commitments to supply energy-deficient post-Soviet states; producing enough to maintain, or increase, exports to western hard currency markets. Energy exports, especially, but not exclusively oil and natural gas, remain the leading source of foreign exchange earnings, just as they were for the last two decades of the Soviet era. Russia, like all post-Soviet states, needs hard currency to pay for imports, to say nothing of paying off its foreign debt. Although Russia is an energy resource rich country with land and sea links to western markets, it is not spared its own unique problems in energy resource development. Constitutionally, Russia is a federation. As it happens, many important energy resources are located in non-Russian Republics. A number of these Republics have attempted to assume a proprietary position with respect to ownership and development of natural resources in general and energy resources in particular, an issue discussed in Chapter 6. Even some political administrative regions not based on ethnic minorities have tried to put regional energy resource development interests over national ones in the current climate of weakening control of the centre over the periphery.

In this chapter, the management of energy resources, both renewable and non-renewable, during the Soviet era will be reviewed first. This discussion will provide the necessary framework for an examination of the geography of energy resource development across the post-Soviet scene at the close of the twentieth century.

SOVIET ERA ENERGY RESOURCE DEVELOPMENT STRATEGIES

We will begin our examination of Soviet-era energy resource development and management by briefly describing general trends in both the renewable and non-renewable sectors. The first topic to be considered is electric power production; the second, the contribution of specific fossil fuels to the Soviet fuel balance.

While electrification is of obvious importance in all industrializing countries, the heavy emphasis on it in the Soviet era was the result of Lenin's integration of the electrification concept with his scientific theory for the formation of a socialist state. The first major regional planning initiative, the State Commission for the Electrification of Russia (GOELRO), was introduced in 1920 in order to facilitate the restoration and reconstruction of the country's industrial infrastructure following the years of world war and then civil war. Electric power grids served as the planning regions. Within and between each of them increased generation of electric power and the interconnection of the installed generating capacity were seen as essential steps in the industrialization process. Hydroelectric power development was accorded considerable publicity since it was often held up as an example of socialist man's ability to transform or conquer nature. Despite the pre-eminent position of the hydraulic engineering project in the public mind during the GOELRO, and later the Five Year Plan, era, electricity generated by thermal plants predominated from the outset (Table 8.1). In fact, the GOELRO scheme fostered the use of any and all local fossil fuel resources within each planning region. Thus, coal, shale, peat and even wood on occasion were consumed in the rapidly expanding network of small thermal power stations. Larger scale heat and power thermal stations were developed as part of the overall electrification programme after 1931. In these plants the heat produced in the process of generating electricity was converted into steam and fed into extensive urban heating networks. In this manner an otherwise wasted by-product helped to heat

Table 8.1 *Soviet-era installed generating capacity and electricity production: 1913–1989*

	Installed capacity (thousand MW)*					Electricity generated (billion kWh)**				
	Total	Hydro	%	Nuclear	%	Total	Hydro	%	Nuclear	%
1913	1.1	0.02	<1	–	–	2.0	0.04	<1	–	–
1928	1.9	0.12	6.3	–	–	5.0	0.4	8.0	–	–
1940	11.2	1.6	14.3	–	–	48.6	5.2	10.7	–	–
1950	19.6	3.2	16.3	–	–	91.2	12.7	13.9	–	–
1960	66.7	14.8	22.2	N/A	–	292.0	50.9	17.4	N/A	–
1970	166.0	31.4	18.9	0.9	0.5	741.0	124.0	16.7	3.5	0.5
1975	217.0	40.5	18.7	4.7	2.2	1039.0	126.0	12.1	20.2	1.9
1980	267.0	52.3	19.6	12.5	4.7	1294.0	184.0	14.2	73.0	5.6
1985	315.0	61.7	19.6	28.1	8.9	1544.0	215.0	13.9	167.0	10.8
1986	322.0	62.1	19.3	30.1	9.4	1599.0	216.0	13.5	161.0	10.1
1989	341.0	64.4	18.9	37.4	11.0	1722.0	223.4	13.0	212.6	12.4

*MW = megawatts
**kWh = kilowatt-hours
Sources: Narodnoye Khozyaystvo SSSR v 1958g (Moscow: Finansy i Statistika, 1959), 215; *Narodnoye Khozyaystvo SSSR v 1985g* (Moscow: Finansy i Statistika, 1986), 155; *Narodnoye Khozyaystvo SSSR za 70 Let* (Moscow: Finansy i Statistika, 1987), 161; *Narodnoye Khozyaystvo SSSR v 1989g* (Moscow: Finansy i Statistika, 1990), 375

factory and new apartment house alike. Heat and power thermal stations continue to play an important role, accounting for about half of the total urban heat supply across the post-Soviet scene, as well as generating a substantial share of total electricity.

As Table 8.1 indicates, at the end of the Soviet era hydroelectric power stations accounted for about a fifth of total installed generating capacity, and around 13 per cent of the electricity generated from all sources. Up until the mid-1950s development had been concentrated in European Russia and the Caucasus, but it is apparent from Fig. 8.1 that there was a steady shift toward the eastern regions in the development of hydroelectric energy over time. Wherever dams were built, improved navigation, flood control, expanded irrigation and fishing were presumed benefits in addition to the electricity generated. In practice, however, the management of water levels by hydrostation engineers typically was geared to the maximization of electricity produced, to the considerable cost of other users. The trend toward investment in ever larger scale hydropower projects, which is indicated by Fig. 8.1, was given a fillip with the decision in the 1950s to tap the huge water power potential of the Angara and Yenisey rivers in central Siberia. The lure was the prospect of getting large blocks of hydroelectricity at very low cost per kilowatt-hour. Favourable river regimes, ideal site characteristics for dam construction, and technical innovations in plant design and in the transmission of electric power over long distances all helped to bring about the changes in distribution depicted in Fig. 8.1.

The prime function of the large scale Siberian plants was to provide power to meet the base rather than peak load demand of the regional electric grid system. This was possible because of the unusually regular flow of the Angara and Yenisey rivers. The Angara is especially notable in this regard for there is hardly any seasonal variation in the regime owing to the fact that it is the only outlet for Lake Baykal. While small in surface area, Lake Baykal is very deep: indeed, it contains the largest volume of water of any lake in the world. The combination of a large, regular flow, steep gradient and ideal geological conditions for dam construction permitted the design of hydropower stations which could generate sufficient electricity on a regular basis to satisfy the requirements of industry. On most rivers the pronounced seasonal variations in flow make it problematic for hydropower stations to serve the industrial market, that is, to meet regular, daily base-load demand for electricity. Instead, hydropower stations are usually designed to meet part

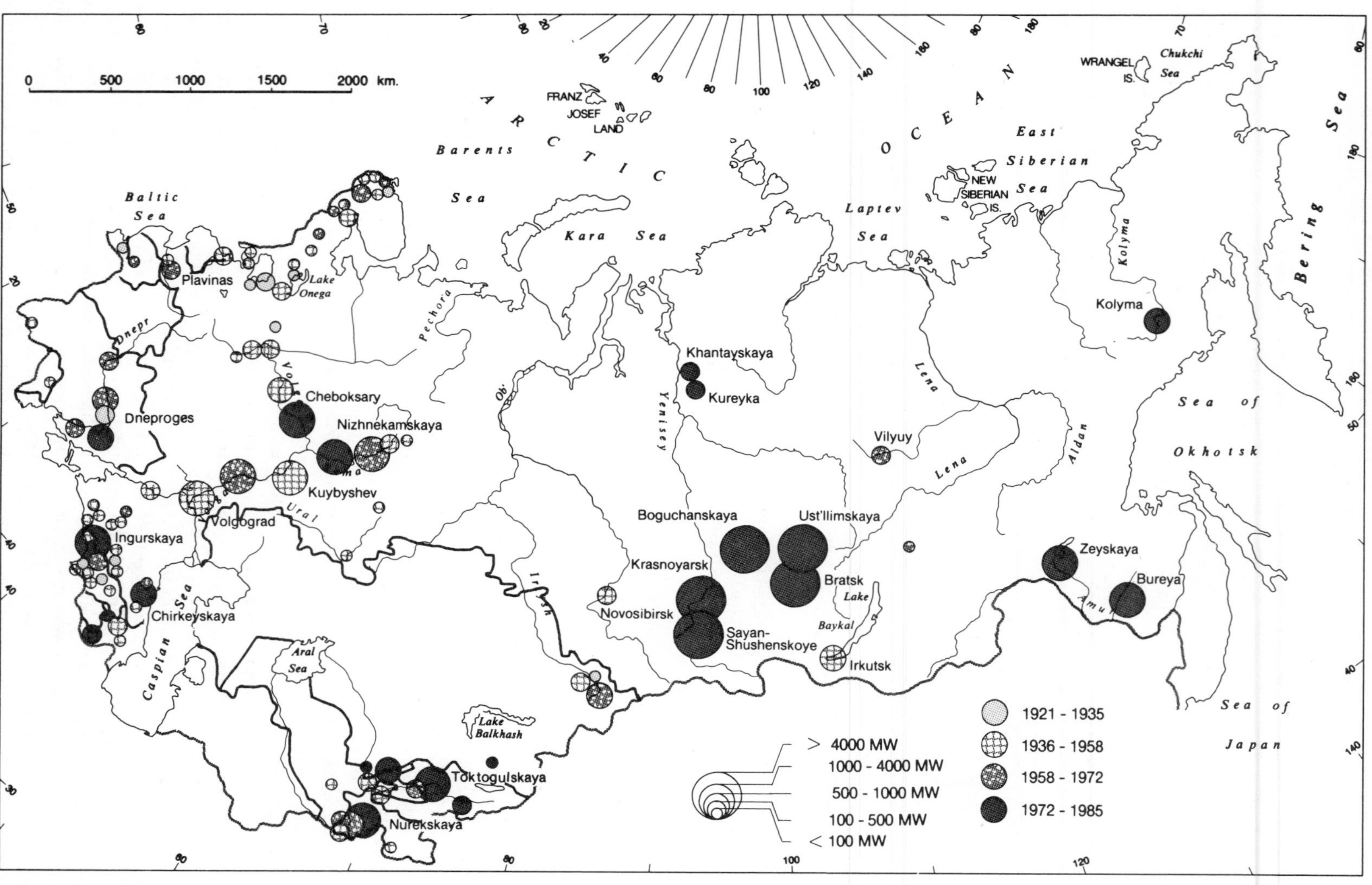

FIGURE 8.1 *Soviet-era hydroelectric power developments, 1921–1989*
(*Sources*: J.H. Bater, 'The Development of Hydropower in the Soviet Union', *Water Power*, June 1974, 216; L. Dienes and T. Shabad, *The Soviet Energy System: Resources Use and Policies* (New York: John Wiley, 1979), 135)

of the base load when ample water is available and to cover the peak load at other times. Because of the storage capacity of reservoirs and near instantaneous production of hydroelectricity, hydropower stations can more easily meet peak daily demand for electricity than thermal power stations which frequently take a few hours to 'come on stream.' Thus, around dinner time when urban-industrial electricity requirements commonly 'peak', hydroelectricity is generated to meet the demand. The fact that hydropower stations stand idle a good part of the time is the reason why they accounted for about one-fifth of the total installed generating capacity, but only 13 per cent of total electricity generated (Table 8.1). This is a relationship common the world over.

Because the large scale Siberian hydropower stations portrayed in Fig. 8.1 were built to meet base, rather than peak, load demand in the regional power system, and therefore operated on a regular basis throughout the year, substantial economies were achieved. Proponents of these projects contended that the prime cost of electricity from them was less than one-tenth the average cost of thermally generated electricity. The peculiarities of Soviet prices, the absence of a proper accounting of costs for reservoir preparation, infrastructure and so on, ensured that the projects appeared economically attractive. But the huge blocks of cheap electricity generated required markets. The water power potential of Siberia would have remained untapped had there not been a willingness on the part of Soviet authorities to invest heavily in associated energy-intensive industrial complexes, new towns and transportation systems. Hydropower developments in both Siberia, and later in Middle Asia, therefore became the catalysts in major regional economic development schemes. Were it not for the huge hydroelectric power stations in central Siberia, which together accounted for about one-third of all hydro generating capacity, and which because of their base-load function in the regional power system represented an even larger share of the total hydroelectricity generated, the role of water power in the Soviet-era energy scene would have been decidedly minor.

The notion of generating electricity from a renewable source of energy as opposed to those of finite quantity such as fossil fuels is obviously appealing. Over the years the Soviet authorities experimented with solar, tidal, geothermal and wind electricity generation. They were of local significance only. However, given the very high level of integration of regional power systems, even these minor sources of electricity contributed to overall demand. The Soviet Union spanned eleven time zones, as Russia still does at present. It is therefore technically possible to shunt 'surplus' electricity from east to west as the peak demand changes geographically. Although eminently flexible, energy in the form of electricity cannot be 'stockpiled' like coal or oil to be used later. The integration of the European Russian electric power grid with that of Eastern Europe and Finland provides even more scope for making most efficient use of electric power, and it is still 'exported' in substantial amount. While conventional fossil fuels like coal, oil and natural gas predominate in the generation of electricity in thermal power stations, nuclear power stations have played a role, albeit a tragic one, in electricity production.

In 1970 the role of nuclear electricity was quite insignificant. It accounted for a mere 0.5 per cent of all electricity generated in the Soviet Union. Thereafter, the rate of expansion of nuclear power generating facilities outstripped that of conventional fossil fuel-based thermal stations and hydroelectric plants. By 1986, 10.1 per cent of all electricity generated came from nuclear stations. As Table 8.1 indicates, total electricity production nearly doubled between 1970 and 1986. Against this background the gain registered by nuclear power is even more impressive. The attractions of the nuclear electricity to Soviet authorities were plentiful. Uranium, the basic nuclear fuel, was available in sufficient quantity to sustain a rapid escalation in nuclear power. Prior to the collapse of the Soviet Union, mining of uranium was concentrated in Middle Asia. The technology available included breeder, fusion, and controlled thermonuclear fusion reactors. The breeder reactors offered the possibility of producing more fissionable materials than they consumed, thus in terms of conservation of non-renewable energy resources they were viewed as a positive technological innovation.

Amongst the benefits of nuclear power, the portability of the basic raw materials means that plants could be located near the market, thereby reducing the loss of electricity inevitably associated with long-distance, high-voltage transmission. Save for the need for an adequate water supply for cooling purposes, there are few locational constraints on where a nuclear power station can be constructed. For these reasons as well as the fact that in the pre-Chernobyl' period nuclear power stations were officially proclaimed to be a safe technology, they were most often sited close to the major urban-industrial centres. The geography of nuclear power stations presented by Fig. 8.2 obviously stands in stark contrast to that of hydroelectric power development depicted in Fig. 8.1. Soviet-era regulations dictated specific site requirements such as soil and geological conditions, prevailing winds, and proximity of suitable land for disposal of radioactive wastes. While most plants were at least 40 kilometres from a major city, housing facilities for power station personnel were customarily within easy reach. On balance, the guidelines controlling location relative to major urban centres were less than precise. Indeed, some plants such as the one in Armenia were even constructed in earthquake prone regions. The environmental disaster produced by the meltdown at the huge nuclear installation at Chernobyl' north of Kiev in Ukraine in the spring of 1986 (Fig. 8.2) resulted in the mothballing of several installations, including the one in Armenia. More significant politically, the grass-roots anti-nuclear movement which quickly gathered momentum in the post-Chernobyl' period coalesced with nationalist, independence movements in several Soviet Republics and was a potent element in challenging central Soviet authorities. The latter had been characteristically secretive in dealing with the Chernobyl' disaster, and clearly duplicit in its dealings with the directly affected populations, notwithstanding the Gorbachev-era emphasis on *glasnost'*, or openness. Initially, the disaster was attributed to errors in management of the facility and not the technology itself. In a subsequent chapter the environmental and public health consequences of the 1986 Chernobyl' disaster will be examined in some detail. Electricity production in the remaining reactors at Chernobyl' soon resumed, a situation unchanged nearly a decade later. While in the aftermath of Chernobyl' there was a very strong anti-nuclear energy grass-roots movement, the need for electricity did not diminish. Indeed, right to the end Soviet authorities were still planning on increasing by five to six times nuclear electricity production.

As the data in Table 8.1 indicate, thermal power stations were always the mainstay of Soviet electricity production. Indeed, thermal electric power stations were a major market for Soviet coal, oil and natural gas production. Unlike the United States, however, where coal has remained the dominant fuel for the generation of electricity, the role of coal in the USSR declined sharply after 1960. Coal comprised about 70 per cent of the fuel supply for power plants in 1960, using standard calorific equivalents as the basis for measurement. By the mid-1980s coal's share was less than one-half, despite a major programme initiated in the 1970s to replace oil with coal in electricity generation, about which more will be said shortly. Across the country there were significant regional variations in the fuel mix of thermal power stations. In European Russia, local coal production was insufficient to meet the demand for electricity, hence oil and natural gas dominated. Increasingly both oil and natural gas had to be 'imported' from Siberia because of declining reserves in European Russia. In Siberia, coal was the principal fuel for thermal electric power stations. While the abundant supplies of natural gas were used for electricity generation in Middle Asia, there too coal was still the pre-eminent fuel for thermal electric power stations. Natural gas was used to generate electricity in all regions not so much because of a concern for cleaner air, which is one of the side-benefits of its use compared to coal or oil, but rather because of its comparatively abundant reserves relative to petroleum. At the end of the Soviet period, natural gas and oil each accounted for a shade more than a quarter of the fuel consumed by all

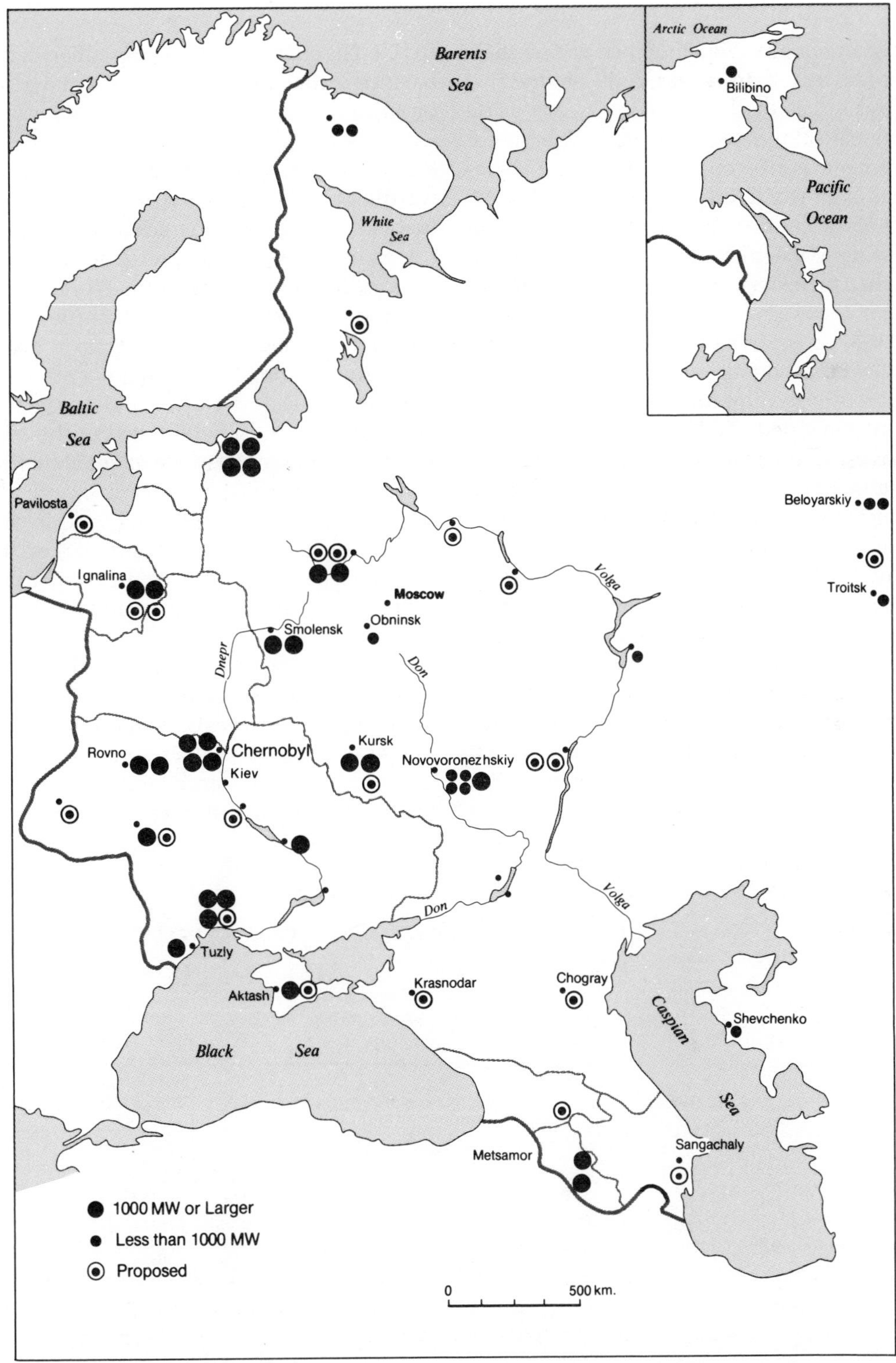

FIGURE 8.2 *Soviet-era nuclear power development*
(*Source*: L. Symons *et al., The Soviet Union. A Systematic Geography* (Totowa, NJ: Barnes and Noble, 1983), 149)

thermal power plants. Together peat and oil shale represented less than 4 per cent of the fuel consumed by all thermal stations, again using conventional calorific equivalents as a basis for measurement. However, as peat and oil shales were found mostly in the energy-deficient western part of the country, their local importance in contributing to electricity production was much greater than their share of the total fuel mix might suggest.

Electricity generation plays a vital role in modern industrial economies to be sure. And fossil fuels obviously account for the lion's share of all electricity generated (Table 8.1). However, fossil fuels have a multitude of uses in addition to generating electric power. From fuel for engines to chemical by-products such as synthetic fibre, fossil fuels are essential to the modern industrial economy. The basic indicator of the role of fossil fuels in an economy is the fuel balance (Table 8.2). The importance of individual fossil fuels in the fuel balance is determined by converting each type of fossil fuel into tons of standard fuel equivalent according to the actual energy content or calorific value of particular fuels produced. A ton of standard fuel is equal to seven million kilocalories or 27.8 million British Thermal Units (BTU). The changes in the composition of the fuel balance over time are worth noting, if only briefly.

On the eve of the revolution the fuel balance of the Russian Empire belied its industrial backwardness. Although wood was still an important fuel, it is the role of oil which stands out in Table 8.2. As was noted in Chapter 2, at the turn of the century Russia was one of the world's principal oil producing and exporting countries. Exploitation of the oil deposits around Baku in Azerbaijan had brought profits to foreign entrepreneurs, taxes to the state treasury and harsh working conditions for those who toiled in the oil fields. The importance of oil in the national fuel balance hinted at modernity, but this was more likely to be found in the industrializing states of Europe, which were the principal export markets, than in Russia itself. Within the Empire, coal and wood held sway as fuels, as they would for decades to come.

During the early Soviet period the industrialization drive was based on the proven, extensive, and accessible reserves of solid fuels, such as coal, peat and wood. With the onset of the Stalin-era Five Year Plans and forced industrial growth, coal production in particular was singled out as the

Table 8.2 *Soviet-era fuel balance: million tons of fuel equivalent (M.T.)*, 1913–1989*

	Total	Coal		Oil		Natural gas		Peat		Oil Shale		Wood	
	M.T*.	M.T.	%	M.T.	%	M.T.	%	M.T.	%	M.T.	%	M.T.	%
1913	45.9	23.1	50.3	13.2	28.8	–	–	0.7	1.5	–	–	8.9	19.4
1940	237.7	140.5	59.1	44.5	18.7	4.4	1.9	13.6	5.7	0.6	0.3	34.1	14.3
1950	311.2	205.7	66.1	54.2	17.4	7.3	2.3	14.8	4.8	1.3	0.4	27.9	9.0
1953	384.2	252.3	65.7	75.5	19.6	8.7	2.3	15.8	4.1	2.1	0.5	29.8	7.8
1965	969.0	451.9	42.9	347.3	35.9	151.3	15.6	17.0	1.7	7.5	0.8	30.0	3.1
1970	1221.8	432.7	35.4	502.5	41.1	233.5	19.1	17.7	1.5	8.8	0.7	26.6	2.2
1975	1571.3	471.8	30.0	701.9	44.7	342.9	21.8	18.5	1.2	10.8	0.7	25.4	1.6
1980	1905.7	484.4	25.4	862.6	45.3	515.7	27.1	7.3	0.4	11.9	0.6	23.8	1.2
1985	2137.3	486.9	22.8	851.3	39.8	759.9	35.5	5.5	0.3	10.2	0.5	23.5	1.1
1986	2165.7	454.8	21.0	879.1	40.6	792.7	36.6	6.6	0.3	9.6	0.4	22.9	1.1
1987	2194.0	450.3	20.5	882.3	40.2	882.7	37.5	8.1	0.4	9.5	0.4	21.1	1.0
1989	2,271.2	447.9	19.7	868.4	38.2	919.5	40.5	5.7	0.3	8.9	0.4	20.8	0.9

*M.T. = million tons of fuel equivalent. A ton of standard fuel is equal to seven million kilocalories or 27.8 million BTU.
Sources: Narodnoye Khozyaystvo SSSR v 1958g (Moscow: Finansy i Statistika, 1959), 200; *Narodnoye Khozyaystvo SSSR v 1965g* (Moscow: Finansy i Statistika, 1966), 174; *Narodnoye Khozyaystvo SSSR v 1980g* (Moscow: Finansy i Statistika, 1981), 156; *Narodnoye Khozyaystvo SSSR v 1985g* (Moscow: Finansy i Statistika, 1986), 157; *Narodnoye Khozyaystvo SSSR za 70 Let* (Moscow: Finansy i Statistika, 1987), 163. *Narodnoye Khozyaystvo SSSR v 1989g* (Moscow: Finansy i Statistika, 1990), 377

critical component in the development process. Each year invariably brought higher output targets, and higher costs of production in the country's predominantly underground, or pit, mining operations. The GOELRO plan of the 1920s had fostered the use of all local energy resources, but these were often of low calorific value, difficult to handle and bulky to transport. As the production of lower grades of coal increased in importance, as the extraction of peat expanded, as the use of wood as a fuel continued to be quite important, the burden on the transportation system increased. From the late 1920s to the mid-1950s the fuel balance became even more dependent upon these fuels, as well as oil shale (Table 8.2). The result of the vast expansion in production of low calorific value coals, peat and shale was scarcely any improvement in the average heat value per ton of fuel produced from 1928 to 1955. In other industrialized states, oil and natural gas were of growing importance. With a higher calorific value than those fuels dominating Soviet production, they were providing many economies in extraction and transportation, to say nothing of their diversified range of chemical by-products. Whether because of rigidities in the decision-making environment, a lack of investment capital, or simple inertia, the Stalin-era fixation with solid fuel production was costly. In 1953, the year Stalin died, the fuel balance was dominated by coal. In terms of tons of fuel equivalent, it accounted for 65.7 per cent of the total. Compared to 1913 oil had declined in relative importance from 28.8 per cent to 19.6, a slight improvement over 1950 when it was only 17.4 per cent. Peat, on the other hand, had tripled its share of the fuel balance by 1950, and still represented more than 4 per cent in 1953. Unlike virtually every other industrialized country, wood still figured prominently in the fuel balance at the end of the Stalin era. To be sure, the Soviet Union was industrializing, indeed modernizing, quickly. During the 1950s the economy was expanding at a singularly rapid rate. But it was growth generated by particular sectors, and often at the cost of other sectors of the economy such as consumer goods production or agriculture. It was also a tempo of growth dependent upon a narrowly defined range of inputs. The fuel balance in 1953 is a case in point.

Stalin's death brought a reassessment of many engrained features of the Soviet system. The heavy reliance on solid fuels was one of them. The decision was taken to diversify the fuel balance, to enhance the role of oil and especially natural gas. The outcome of this reassessment is readily apparent from the data presented in Table 8.2. Between 1953 and 1965 the combined share of oil and natural gas eclipsed coal, although production of the latter had by no means waned. Coal output in actual tons, as distinct from tons of fuel equivalent used in Table 8.2, rose from 320 million in 1953 to 578 in 1965. But without the contribution of oil and natural gas the sizeable gain registered in the total amount of fuel from all sources, as measured in tons of fuel equivalent (Table 8.2), would not have been possible. The basis for this expansion was laid in large part during World War II when the German invasion threatened the vital and vulnerable oil fields in the Baku region of Azerbaijan. A crash programme of geological exploration in more militarily secure regions of the country uncovered new reserves of hydrocarbons, and oil and gas in particular. The Volga–Ural oil fields, first discovered in the late 1920s, were to prove of critical importance both during the war and after when production in the Baku fields began to decline. They also comprised a major part of the resource-base inventory for the decision taken after Stalin's death to diversify the fuel balance. However, production from this region was soon to be eclipsed by that of West Siberia. Natural gas, which previously was often flared off and thus wasted when found in association with oil deposits, became a significant component of the strategy to modernize the fuel balance. The change in its share between 1953 and 1965 was perhaps the most dramatic consequence of the new Soviet energy policy. As Table 8.2 indicates, natural gas grew steadily in importance. Between 1965 and 1989 the full repercussions of the decision to change the structure of the fuel balance become apparent. The long-standing domination of the fuel balance by coal ended in 1962, and by 1965 it accounted for scarcely two-fifths of total production. Twenty-five years later its share had been reduced to barely one-fifth, despite an overall increase in production from 578 to 740 million 'actual', not fuel equivalent, tons (Table 8.2). After steadily garnering a larger share of the fuel balance until 1978, when it peaked at 45.7 per cent, the relative importance of oil then waned and represented less than two-fifths of the fuel balance at the close of the Soviet era (Table 8.2). Natural gas continued to register substantial relative increases in importance one year to the next. The remaining solid fuels – peat, shale and wood – were relatively unimportant in 1989.

The renewed interest in coal in the latter years of the Soviet period was the result of several factors. The opportunity costs associated with domestic consumption of oil, compared to selling it on the world market for hard currency, figured prominently among them. But to sell more oil on the world market for hard currency required finding a substitute for it in some domestic uses. This resulted in a government policy to replace oil with coal wherever possible in the generation of electricity. Thus, toward the end of the Soviet era the prevailing strategy for the generation of electricity entailed a significant expansion of large scale, coal-fired thermal generating stations at resource locations, principally in Siberia, the Far East and Kazakhstan. Power from stations fuelled by low-cost, strip-mined coal was to be fed into the national electric grid system to supply distant markets. Although oil and natural gas were the preferred fuels from the standpoint of ease of transport, calorific value, and comparative cleanliness when burned, the growing importance of strip-mined coal and the modernization of some pit mines combined to make coal at least appear to be a potentially viable alternative to oil in the generation of electricity. Comparative cost assessments, however, were problematic given Soviet pricing and subsidy practices. Rarely were natural resource development projects assessed in realistic benefit–cost terms. Moreover, from the late 1970s subsidies were increased substantially so as to keep the delivered price of coal competitive with other alternative, higher calorific value, and therefore more economic, fuels. Indeed, subsidies of one kind or another ensured that the price charged for coal was well below even Soviet calculations of the costs of production. The cost of transporting coal on the railroad was maintained at an artificially low level as well. This policy of production and transport subsidy was intended to guarantee coal markets throughout the country. By the end of the Soviet period, the average length that a ton of coal was being transported exceeded 800 kilometres, a staggering distance given its comparatively low average calorific value per ton.

An additional reason for trying to replace oil with coal had to do with security of supply. As will be described at a later point, oil reserves and production were in decline. The supply of coal, however, was not problematic; indeed, from the resource supply side the attraction of coal was compelling. Nearly 7000 billion tons had been estimated to lie beneath the vast expanse of the former Soviet Union, a rather significant share of the world's potential reserves. Two enormous basins in Siberia – the Tunguska with about 2000 billion tons of hard-coal potential and the Lena with about 1500 billion comprise about half of this huge potential reserve. Geologically proven reserves of hard coal total nearly 150 billion tons, almost half of which is coking coal quality and therefore of particular significance to the iron and steel industry. Just over 100 billion tons of the proven reserve are lignite, a softer, more friable and lower calorific-value coal than anthracite or bituminous, which constitute the hard-coal supply. The difference between hard and soft coal is significant beyond the simple designation these adjectives imply. For instance, while the Soviet Union was the world's largest coal producer in actual tonnage at the end of the 1980s, a smaller output from mines in the United States actually had a higher total calorific value because most of it was hard coal. At the close of the Soviet era, more than one-fifth of Soviet coal production was lignite, whereas in the United States it comprised less than one-twentieth of total output. Thus, despite the steady contraction in the role of coal in the fuel balance during the final decades of the Soviet era, it was anticipated that it would begin to replace oil in some uses, and that overall coal production in absolute terms would be maintained or even increased.

ENERGY IN THE POST-SOVIET SCENE

In the 1990s, there is little that resonates with the energy scene of the previous decade, save perhaps for the basic distribution of energy resources. And even in this context, there is some change. For example, recent exploration of what were previously listed as 'potential' areas of oil or natural gas has resulted in additions to the 'proven' column of the energy inventory ledger. This development has typically followed from foreign investment and joint ventures with multi-national oil companies, and while obviously welcome, it simply has served to widen the gap between the energy 'have' and the energy 'have not' post-Soviet states. Of the 15 post-Soviet states, fewer than one-third have an energy supply in excess of domestic demand. Energy 'have not' states remain dependent upon imports from energy surplus post-Soviet states, but the terms of trade are fast changing. The exceedingly low prices and comparatively easy terms of credit commonly available during the first year or so of independence are gone. In the case of oil, for instance, the price charged is moving quickly to the world level, and payment in hard currency or equivalent in manufactured goods is expected. Under the old Soviet system, any inter-republic trade 'debt' resulting from transactions was simply covered by transfers from central government.

In this section, the development and management of the key energy resources – coal, oil and natural gas – will be examined in the context of the new political economy of the 1990s. A brief consideration of electricity generation and the role of nuclear energy in the post-Chernobyl' period will conclude the section.

COAL

Bankruptcy is part of the vocabulary in the new economy of the post-Soviet scene, and in the energy sector it is a word often heard in the business of producing and transporting coal. Past practices involved enormous state subsidies, as noted earlier, and these have now come home to roost. However, in most coal-producing states, miners are a major political force and abolishing unwarranted subsidies has proved easier said than done. Even with continuing subsidy, the coal industry is in serious difficulty. Between the peak year of inter-regional coal shipments in 1988, and the first full year of independence, 1992, the volume shipped (excluding exports outside the former Soviet Union) dropped by nearly 50 per cent. In 1988 seven republics shipped coal to markets outside their borders. Only four post-Soviet states did so in 1992, and Ukraine and Uzbekistan played a very minor role compared to Russia and Kazakhstan (Table 8.3). Clearly, the economic disarray precipitated by the collapse of the Soviet Union figures prominently in the explanation for both the contraction in total coal production, and in inter-republic shipments of coal. In the new economy of the 1990s a number of states have placed restrictions on the amount of coal which may be exported, an initiative intended to ensure domestic needs are satisfied first.

From 772 million tons in 1988, coal production fell to 606, to 539 and to 474 million tons in 1992, 1993, and 1994 respectively (Table 8.3). Of the seven coal-producing states, only Uzbekistan maintained the 1993 level of output in 1994, but it accounted for a mere 3.8 million tons. Russia was still the largest producer, and the major importer. Ukraine which had ranked a distant second in both categories, was only the third largest producer in 1994. As Table 8.3 indicates, production there has continued to slide precipitously. Kazakhstan's production has also dropped but by nowhere the same amount, and consequently it was the second major producer in 1994. Amongst the three major producers, it was the only net exporter. With the exception of Azerbaijan, all post-Soviet states were coal consumers, although many were decidedly minor. There is no indication that the trend of declining production will be reversed any time soon given the economics of coal mining. Russia, for example, has been forced to cut back subsidies to the coal industry since it simply could not afford to continue to allocate 6 per cent of its total budget

Table 8.3 *Coal production, 1992–1994, and coal shipments between post-Soviet states 1992 (million tons)*

	Production (1992)	Production (1993)	Consumption (1993)	Production (1994)	Shipments from				Total by State
					Russia	Ukraine	Uzbekistan	Kazakhstan	
Russia	337.3	305.0	344.1	271.0		1.8	–	38.9	40.7
WESTERN TIER									
Ukraine	134.0	116.0	136.8	94.4	8.2		–	2.2	10.4
Belarus'	–	–	1.6	–	0.7	0.8	–	0.1	1.6
Moldova	–	–	4.0	–	0.9	3.1	–	–	4.0
Estonia	–	–	0.2	–	0.1	–	–	0.1	0.2
Latvia	–	–	0.4	–	0.1	0.1	–	0.2	0.4
Lithuania	–	–	0.9	–	0.6	0.1	–	0.2	0.9
SOUTHERN TIER									
Armenia	–	–	0.1	–	0.1	–	–	–	0.1
Azerbaijan	–	–	–	–	–	–	–	–	–
Georgia	0.5	0.5	0.6	0.5	0.1	–	–	–	0.1
Kazakhstan	127.0	112.0	87.1	104.0	3.7	–	–	–	3.7
Kyrgyzstan	2.2	1.7	3.6	0.8	0.3	–	–	1.1	1.4
Tajikistan	0.2	0.2	0.3	0.1	–	–	0.1	–	0.1
Turkmenistan	–	–	0.2	–	–	–	0.1	0.1	0.2
Uzbekistan	4.7	3.8	5.6	3.8	1.0	–		0.1	1.2
Total	605.9	539.2	585.5	474.1	33.9	7.6	0.2	43.6	85.3
Total Exports					18.1	1.7	–	0.6	20.4

Sources: M. J. Sagers, 'Inter-republican Coal Trade in Sharp Decline', *Post-Soviet Geography*, Vol. 34, No. 10, 1993, 662; *Ekonomika Sodruzhestva Nezavisimykh Gosudarstv* v 1993 g (Moscow: Statkomitet SNG, 1994), 43; *Ekonomika Sodruzhestva Nezavisimykh Gosudarstv* v 1994 g (Moscow: Statkomitet SNG, 1995), 88

to this sector, as it did in 1992. A Presidential decree was passed in 1993, the intent of which was to free coal prices, enabling them to move up from a bureaucratically determined level to one determined by the market place. To assist coal producers and consumers in the transition from heavy state subsidies to market conditions, a support fund was created through a special federal tax on all goods and services. Mines which are unable to adapt to the new economy will close. Some in fact have done so already. However, as noted above, miners are well organized and have in the past been able to exact concessions from government, thus the full impact of this 1993 decree is not yet evident. Ukraine and Kazakhstan are not spared the problem of dealing with the high coal mining subsidy legacy of the Soviet era, but by 1995 had not yet taken action comparable to Russia's. Clearly, as coal prices are forced up, as subsidies are removed, coal inevitably will be less competitive, if for no other reason than many customers will be unable to pay the higher prices even if they were prepared to do so. Indeed, there are already serious problems of non-payment for coal deliveries to industrial customers, a situation which higher prices will just exacerbate. Rising costs to ship coal by rail are also inevitable as transport subsidies are removed, and commodity movements are rationalized amongst and within post-Soviet states. Unlike oil and natural gas, the prospects for selling coal on the international market are not auspicious. Global recession and industrial restructuring during the early 1990s have reduced demand for coal on the world market. Despite all the problems associated with the coal industry in the mid-1990s, deposits in Russia and Kazakhstan especially are significant by any standard of comparison, and will afford some domestic energy resource management and strategic planning opportunities in the future, if not at present because of the problems of economic transition.

The most valuable deposits are the hard-coal basins outlined in Fig. 8.3. The largest anywhere in the post-Soviet scene is the Donets basin in southeast Ukraine and the adjacent part of the Russian republic. This region was the mainstay of the industrialization drive of the late imperial era. By the late 1880s, coking coal from the Donets basin supported most of the iron and steel production in the Russian Empire. A century later it still dominates hard-coal production, and supports a massive array of urban and regional heavy industrial complexes. Ukraine accounts for a quarter of the total hard-coal output, which was more than 400 million tons from all coal-producing post-Soviet states in the early 1990s. But there are problems associated with the basin. Many of the accessible coal seams have been exhausted. Ever deeper pit mines to reach the remaining thin, discontinuous coal seams have pushed the already heavily subsidized costs of production even higher.

The second most important source of hard coal is in Russia. It is the Kuznetsk basin in Siberia (Fig. 8.3). First developed in the late nineteenth century to supply the coal requirements of the newly developed Trans-Siberian railroad, it was a major component of Stalin's industrialization drive during the First Five Year Plan. Unlike the Donets basin, coal seams here are continuous and thick, with a high proportion of the coal which must be pit-mined lying close to the surface. Nearly two-thirds of proven reserves are within 400 metres of the surface. In the Donets basin in contrast, about half of the remaining proven reserves are more than 600 metres below ground. Moreover, unlike the Donets Basin, a significant proportion of the Kuznetsk basin reserves can be strip-mined. Coking and bituminous steam coal from the Kuznetsk basin not only sustain the region's own huge industrial demand, it supplies as well the bulk of the demand of the Urals industrial complex and a portion of the requirements of electric power generating stations even further west in Russia. The Kuznetsk Basin alone accounts for more than a third of total Russian coal output. The remote location and harsh environment pose problems for further development of Russia's other major hard-coal deposit near Vorkuta in the northern Ural mountains region. Indeed, its very future is seriously compromised by the fiscal requirements of Russia's new economy. Elsewhere are a number of small scale coal mining operations, the combined output from which accounts for only about

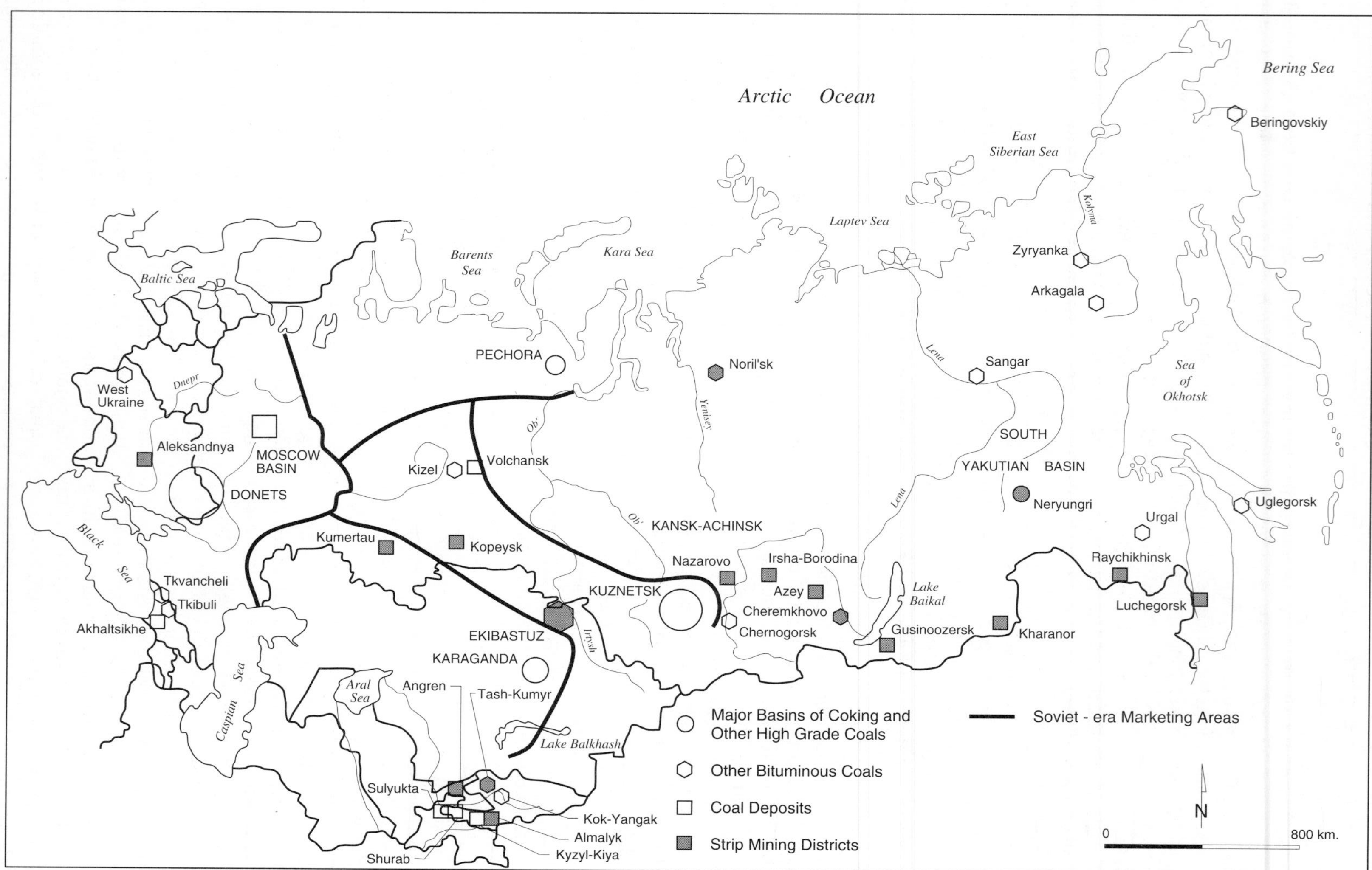

FIGURE 8.3 *Coal resources*
(*Source*: L. Dienes and T. Shabad, *The Soviet Energy System: Resource Use and Policies (New York: John Wiley, 1979), 109)*

one-sixth of total production (Fig. 8.3). Some of these mines yield coal of coking quality, but most output ends up fuelling thermal power stations. The South Yakutian basin, for example, possesses fairly extensive reserves of coking, and reasonable quality steam, coal, both of which can be strip-mined. Owing to the existence of the Baykal-Amur railroad, the centrepiece of a late Soviet-era, large scale regional development project, coal from this basin could supply the Russian Far East market, as well as export markets such as Japan and Korea, should domestic and international demand warrant it. For this to occur, however, money will have to be spent on upgrading the railroad itself as it was poorly constructed in many sections.

The Karaganda region in Kazakhstan is an important source of hard coal also. As Fig. 8.3 indicates, it is located roughly half way between the Ural mountains and the Kuznetsk Basin. The creation of the Ural–Kuznetsk Combine, an early Soviet-era regional development project described in Chapter 3 which involved an exchange of iron ore and coking coal, had ended up putting an unsustainable burden on the existing rail line across Siberia. Development of the Karaganda fields first began in the 1930s in order to supply the Urals steel industry, thereby reducing the huge westward flow of coking coal from the Kuznetsk basin to the steel mills of the Urals. Inasmuch as Karaganda coal is pit mined, it is more expensive than coal produced in the Kuznetsk basin in Russia, where geological conditions favour strip mining. While proven reserves are substantial, they are incapable of sustaining significant annual increases in output. Thus, the Karaganda region has not been able to augment production at the high rates which characterized its early years of operation. Kazakhstan, however, still exports a substantial amount of coking coal to the steel plants in the Urals and elsewhere in central Russia, and to Ukraine (see Fig. 8.3).

Soft, lower calorific value, coal, principally though not exclusively lignite, is found in many post-Soviet states. The Ekibastuz deposit in Kazakhstan is the largest in terms of production, and has sufficient proven reserves to sustain a very substantial increase. Its low quality, high ash content sub-bituminous coal supplies local thermal stations and meets a portion of the demand of power plants in other post-Soviet states as well (Fig. 8.3). The Kansk-Achinsk lignite reserves in Russian central Siberia are extensive, but only produce about half the annual output of the Ekibastuz mines. Coal from the Kansk-Achinsk region is of such low calorific value it is not economic to transport it any significant distance. But because Kansk-Achinsk comprises the lion's share of Russia's lignite reserves, and because the thick seams are readily strip-mined, it has already become the centre of a complex of very large-scale thermal power stations which supply electricity to other parts of the country through the interconnected grid system. Russia's other major soft-coal-producing area is the Moscow basin. Production here has been declining in recent years. Of very low calorific value, the Moscow basin lignite is further disadvantaged since most of it comes from expensive pit-mining operations. But it does sit amidst one of Russia's principal industrial regions, and thus coal from the Moscow basin mines supplies some local thermal power stations. Increasingly, however, these plants must rely on steam coal from more distant sources, notably those in Siberia.

Oil and natural gas import dependencies

Oil and natural gas figure prominently in the energy needs of all modern industrial states. To the extent that a state can itself meet the demand for these essential sources of energy, obviously the better off it will be. The dependence of former Soviet Republics on 'imports' to satisfy their requirements for oil and natural gas is illustrated by Fig. 8.4. These data provide a rough benchmark for assessing only one facet of the energy scene, of course, but it is a very important one. The basic configuration of oil and natural gas import dependencies in 1989 affords some insight into the nature of the challenge confronting Russia and the other newly independent states in the 1990s.

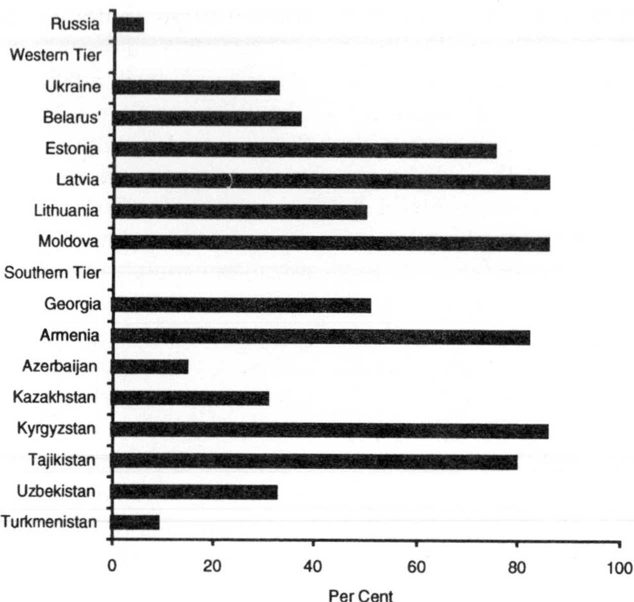

FIGURE 8.4 *Oil and natural gas import dependence: Soviet republics, 1989*
(*Source*: Based on R.A. Watson, 'Interrepublic Trade in the Former Soviet Union: Structure and Implications', *Post-Soviet Geography*, Vol. 35, No. 7, 1994, 378–79)

According to the information presented in Fig. 8.4, in 1989 Russia was in the most enviable position of all republics in terms of oil and natural gas supplies. In fact, Russia accounted for well over four-fifths of all oil and natural gas produced at that date, a position of absolute dominance unchanged to the present. All of what are now the western tier states were heavily dependent upon oil and gas imports to sustain their economies in 1989. In a strategic sense, most can be described as potentially vulnerable to external manipulation of basic energy supplies. Ukraine and Belarus', even though meeting more than half of their 1989 combined oil and natural gas requirements themselves, are no exceptions in this regard. For example, given Ukraine's very large complement of heavy industries, dependence on substantial volumes of imported oil, petroleum products, and natural gas is a significant liability in both the financial and geopolitical sense. The energy import dependency in the Baltic region and Moldova is also a very serious issue. Political relations between these countries and the principal supplier, Russia, are not always harmonious. Amongst the southern tier states only Turkmenistan and Azerbaijan come close to Russia in terms of self-sufficiency. In the Caucasus region, Armenia's energy import dependency has been used to advantage by Azerbaijan in its dispute with Armenia over the Nagorno-Kharabakh enclave. Armenia has been cut off from outside supplies of oil delivered by railroad across Azerbaijan territory. The natural gas pipeline from Russia crosses Georgia. However, Georgia has been unstable politically since independence. Beset with periodic outbreaks of civil war in at least two of its non-Georgian regions, South Ossetia and Abkhazia, the smooth operation of anything on its territory has been exceedingly problematic. Indeed, the natural gas pipeline has been the object of terrorist attack. Additionally, the gas pipeline itself has been plagued with technical problems and could not be an assured source of energy supply irrespective of Georgia's internal difficulties. The Armenian population has had to endure unheated buildings during the winter months, and limited electricity supply year round since the dispute with Azerbaijan began. Industrial production has been severely impacted as well. While this is only one illustration of the geo-political dimension of energy import dependence, it should be noted that such deprivation has not brought about Armenian compliance with Azeri demands for the return of Nagorno-Kharabakh. In Middle Asia, Turkmenistan, Kazakhstan and Uzbekistan are, in that order, the most favourably positioned in terms of meeting domestic oil and natural gas requirements from domestic supplies. Obviously, the situation in Kyrgyzstan and Tajikistan is one of extreme dependence on imports. Most energy-deficient post-Soviet states have little real opportunity to switch to energy suppliers outside the former Soviet Union because of their relative locations. In the 1990s, geo-politics figure prominently in the oil and natural gas industries. We will consider first some aspects of the political geography of oil production.

OIL

Oil production in the former Soviet Union peaked in 1988, the same year that coal output did. At that date, 624 million tons, including gas condensates, were produced. In Russia, which alone accounted for well over four-fifths of the total, production had in fact peaked a year earlier, its output then being almost 570 million tons. By 1991, Russian oil production had slipped to 462 million tons (Table 8.4), of which domestic demand took up 270 million. Net exports to other post-Soviet states accounted for another 80 million tons. Exports elsewhere totalled nearly 100 million tons, the balance of output that year going into reserve stock. Clearly, oil exports were a very significant source of hard currency income. In 1994, Russia's output was 316 million tons, barely two-thirds that of 1991. Since the peak year, 1987, Russian oil production has fallen by almost one-half. The precipitous drop in output is obviously linked to the collapse of the Soviet system. This has not just resulted from a drop in demand from industrial consumers, but to departure of key technical staff, and fast deteriorating infrastructure, from wellhead, to pipeline, to refinery. In some instances, enterprises have reduced output in anticipation of higher prices. The decline from peak output in the late 1980s, however, is also related to a more fundamental issue – depletion of major oil fields. Put simply, by the late 1980s the prospects for the oil industry were problematic just in terms of maintaining production, let alone increasing it.

Given the importance of the oil industry, plans were soon adopted by the Russian government to ameliorate the situation. Investment in the industry was to be substantially increased, in part by opening the door wider to foreign oil company participation in the industry, something already started in the final year or so of the Soviet period. Existing oil well facilities were to be upgraded to improve oil recovery rates. Oil exploration activity was augmented, again through a combination of domestic and foreign investment. The plan was to stabilize oil output by 1995, and then return annual production to approximately 350 million tons by the end of the century. An earlier production target of about 400 million tons for the year 2000 was soon jettisoned because it was unrealizeable, even if highly desirable. Indeed, this level of production is not likely to be regained until well into the next century, if at all. The economic crisis which has enveloped most Russian industry in one sense has a positive side for the government because a higher volume of exports for hard currency can be managed than would otherwise be the case. Once industrial demand picks up, however, unless oil supply increases accordingly there simply will be less available for sale abroad. And although energy wastage, including over-consumption of oil, is everywhere a serious issue, conservation and demand management measures cannot be introduced quickly enough to have a fully offsetting impact either. With the notable exception of Uzbekistan, in the other states producing in excess of a million tons of oil annually declines in output also characterize the 1991 to 1994 period (Table 8.4). But none experienced a relative decline comparable to Russia's. Uzbekistan's output while not large, almost doubled over this period.

The importance of Russia's oil exports, and the reliance on them on the part of some western countries, make developments there of more than

Table 8.4 *Post-Soviet oil production, 1991–1994*

	Million tons				Kilograms per person			
	1991	1992	1993	1994	1991	1992	1993	1994
Russia	462.0	399.0	352.0	316.0	3109.0	2686.0	2370.0	2129.0
WESTERN TIER								
Ukraine	4.9	4.5	4.2	4.2	94.7	85.7	81.4	80.9
Belarus'	2.1	2.0	2.0	2.0	200.0	194.0	194.0	193.0
SOUTHERN TIER								
Georgia	0.2	0.1	0.1	–	32.9	23.8	16.5	–
Azerbaijan	11.7	11.1	–	–	1620.0	1512.0	–	–
Kazakhstan	26.6	25.8	23.0	20.3	1576.0	1523.0	1354.0	1207.0
Kyrgyzstan	0.1	0.1	0.1	0.1	31.9	25.1	19.5	19.7
Tajikistan	0.1	0.06	0.04	0.03	19.6	11.0	7.5	5.7
Turkmenistan	5.4	5.2	4.4	4.1	1431.0	1290.0	1021.0	922.0
Uzbekistan	2.8	3.3	4.0	5.5	133.0	153.0	182.0	246.0

Sources: Ekonomika Sodruzhestva Nezavisimykh Gosudarstv v 1993g (Moscow: Statkomitet SNG, 1994), 42; *Ekonomika Sodruzhestva Nezavisimykh Gosudarstv v 1994g* (Moscow: Statkomitet SNG, 1995), 87

passing interest. Indeed, from the mid-1970s on, oil production trends were the object of intense scrutiny by intelligence agencies in several western countries. There was good reason for this. Oil well productivity declined each year from the late 1970s, such exploratory drilling as occurred more often than not came up dry, and the total proven reserves declined in consequence. Despite a flurry of decrees issued by the Soviet government exhorting greater effort to produce more, the level of production attained in the late 1980s simply could not be sustained. As it was, oil recovery practices were invariably geared to meeting annual targets rather than maximizing output over the longer term, and in consequence much oil remains in the ground, unrecoverable at present prices and technology. Given the state of the Russian economy since independence, and the competing, indeed, conflicting private sector and state interests in the oil industry, it is no easy task to accurately assess what oil production in the near and mid-term future might be. It would be appropriate at this juncture to describe in broad brush fashion the changing geography of oil reserves and extraction, and the part Russia plays in these developments.

It has been noted already that the importance of oil in the fuel balance at the turn of the last century was related to the output from the Caucasus, primarily but not exclusively from the fields around Baku. Until the German invasion in 1941 forced rapid exploration and production from other known reserves, the Caucasus accounted for at least four-fifths of the total production. While the Caucasian oil fields were certainly vulnerable to the German advance, in fact they remained in Soviet hands throughout the war. In the post-war period, the proven reserves of the Volga–Ural fields, roughly the region stretching from Kuybyshev to Perm (Fig. 8.5), seemed to increase substantially with each exploratory well sunk. By the early 1950s, the output from the Volga–Ural fields was already more than three times that of the Caucasus, despite a vigorous programme of off-shore exploration and production around Baku after the war. With the government's decision to promote oil in the fuel balance, production from the Volga–Ural wells soared. By the end of the Khrushchev period in 1964, nearly three-quarters of a vastly expanded oil output came from the Volga–Ural region alone. Although the absolute volume of production from the Caucasus increased from 1940 to 1965, owing to the vigorous off-shore programme and further development of the fields in Chechnya near the city of Grozny to the north of Baku, its relative importance had obviously changed dramatically. In the mid-1960s Caucasian oil

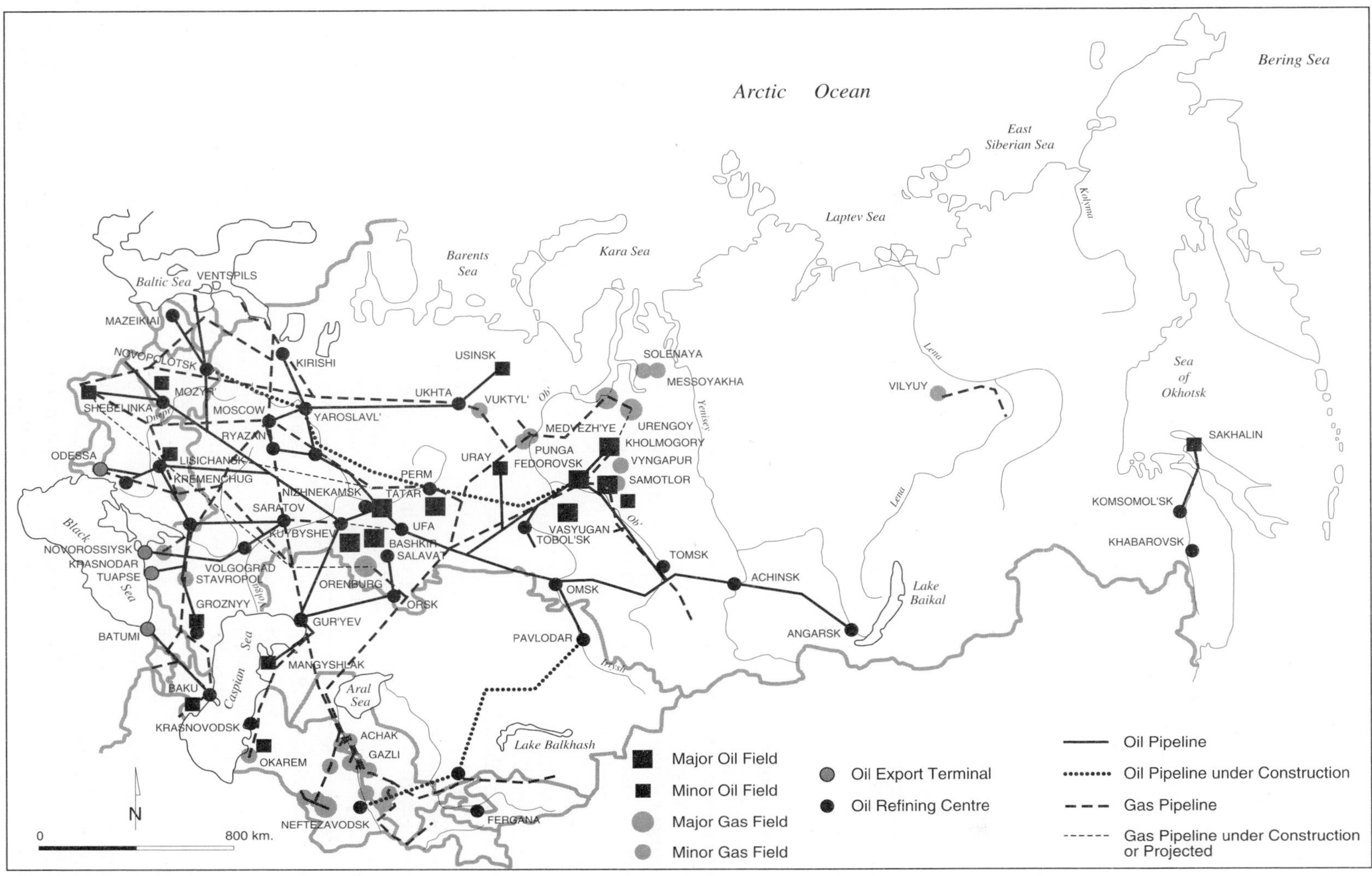

FIGURE 8.5 *Oil and natural gas resources*
(*Source*: L. Dienes and T. Shabad, *The Soviet Energy System: Resource Use and Policies* (New York: John Wiley, 1979), 49, 73)

represented only about one-fifth of the total. The balance of production at this time was came mostly from a few small scale operations in Ukraine, Belarus' and in Middle Asia.

The locational advantage of the Volga–Ural oil fields in relation to the major market areas is apparent from Fig. 8.5. Roughly equidistant from the Moscow region, the industrial regions in Ukraine and in the Urals, for many years it satisfied the bulk of domestic oil needs with comparative economy in terms of transport cost. The network of pipelines extended from the Volga–Ural region in all directions, fed as well of course by the output of the Caucasus and the other smaller oil fields wherever this was feasible. By the mid-1950s, the pipeline system had penetrated Siberia and oil from the Volga–Ural region moved east to the refinery established at Omsk. The small scale production from the Middle Asian fields on the eastern flank of the Caspian Sea for the most part was transported to consumers in European Russia and the Urals. In the mid-1960s, oil production from the vast region to the east of the Urals, including Middle Asia, accounted for little more than one-twentieth of total output. However, the distribution of proven reserves, if not actual production, was fast shifting to the east.

The West Siberian lowland (see Fig. 1.3, p. 13) was long presumed to be a storehouse of oil and natural gas. But geological exploration of this vast tract of forest, swamp and tundra had been slow because it was so difficult to penetrate. It was also some 2000 or more kilometres further east than the Volga–Ural oil fields. The first productive wells were developed in the Ob' river valley a few hundred kilometres to the east of the confluence of the Ob' and Irtysh rivers. A complex of oil wells quickly developed, the oil first being transported by river tanker to Omsk. A major refinery complex had already been built here to process Volga–Ural oil brought east by pipeline. River transport of West Siberian oil to Omsk first occurred in 1964, but was soon replaced by pipeline. This was also the year in which the pipeline carrying Volga–Ural oil finally reached a new refinery in central Siberia at Angarsk, not far from Irkutsk (see Fig. 8.5). The West Siberian oil and natural gas fields soon transformed the energy scene. At the close of the Soviet era, oil production from the West Siberian fields accounted for more than three-fifths of total production. About one-third came from the Volga–Ural wells, and the balance primarily from the Caucasus. Oil from West Siberia quickly replaced Volga–Ural oil in the pipeline to Angarsk. Indeed, by the early 1970s the flow from the Volga–Ural fields to Omsk had stopped. The pumping stations on the pipeline were put to work moving oil in the opposite direction. As Fig. 8.5 indicates, a network of large-diameter oil pipeline penetrates much of West Siberia. Oil is transported to the markets of European Russia, to western tier post-Soviet states and to Europe. The pipeline also feeds oil to the markets of Siberia, particularly the Kuznetsk basin, and moves oil to Kazakhstan and elsewhere in Middle Asia.

For a long period more than half of all of the oil pumped through the pipeline network came from one giant deposit – the Samotlar field in the Ob' river valley. By the end of the 1970s, however, production from the Samotlar wells began to taper off, prompting exploration in the more remote parts of West Siberia and elsewhere. In most regions the strata in which geological conditions were propitious for oil discovery lay well below the average depth to which wells had been sunk to date. Existing technology for oil drilling was generally outmoded, and certainly not suitable for drilling to the depths now required. At the close of the Soviet era about three-fifths of all drilling rigs were still equipped with turbo-drills, the balance comprising some form of rotary drill. Turbo-drill technology evolved very largely in the pre-World War II era, and has been more or less the same since the early 1950s. Basically the technology involves injecting water and mud under high pressure to weaken and soften the material through which the drill bit then penetrates. As the drill pipe itself does not rotate, it is under less pressure and therefore can be manufactured from lower quality steel. The adoption of the turbo-drill technology, according to some observers, was related to the state's inability to produce enough high quality steel to meet the anticipated demand for drilling equipment when the decision to rapidly expand the role of oil in the fuel balance was originally taken in the late 1950s. As a technology, it is quite satisfactory to depths of about 3000 metres. Below that the frequency of breakdown jumps substantially. Rotary drill rigs, in contrast, must be manufactured from much higher quality steel since revolving drill pipe under pressure, and the bit, do the digging. With this technology wells are more easily sunk to much greater depths than by turbo-drilling, and equipment failures are less frequent. The basic problem confronting the Soviet, and now Russian, oil industry is that most remaining potential oil reserves are found at depths in excess of 3000 metres. At these depths the rate of breakdown of turbo-drill equipment is extremely high. The deficiencies of turbo-drill technology are well recognized, and since 1991 Russia has relied heavily on foreign technology and foreign drilling crews in efforts to revamp the supply side of the oil industry.

The Russian oil industry is now obliged to look to more remote locations for oil, including potential off-shore fields. Invariably, these prospects for oil production are investment intensive because of remote location, difficult geological conditions associated with the fields, or because the resource potential lies off-shore. Estimates of the amount of recoverable oil remaining in the fields in European Russia, that is principally those in the Volga–Ural and northern Caucasus regions, range from 10 to 30 per cent of the potential maximum volume available. Thus, there is small chance that European Russia will figure very prominently in oil production in the future. While there is always an unknown element in prospecting, few observers expect that there will ever be another discovery equivalent to the giant Samotlor field in West Siberia, which for so long sustained total production at a very high level. Production from West Siberia is therefore judged to be in permanent decline. Hectic exploration activity in the northern reaches of West Siberia occasionally results in oil finds, but most are in geologically problematic fields, as well as being remote. Most often the exploratory oil drilling is adding instead to the proven natural gas reserves, for which the basic geology in the northern part of West Siberia has long been proven a rich source. According to recent surveys, about two-thirds of the potential oil in all of West Siberia have now been discovered, and close to half of it has already been pumped. What remains is more remote, and because it is found in many smaller fields, invariably will be higher cost than oil pumped to date. Exploration in other regions of eastern and northern Siberia holds out some promise. The same is true of off-shore regions in the Arctic and Pacific coast regions. The small production on Sakhalin Island led to preliminary exploration off-shore, which reputedly has considerable promise. But as in the Arctic, the environment is harsh in the extreme. In the Arctic, ice is an obvious environmental hazard. Potential oil fields off Sakhalin Island are not spared such problems, as there is a serious risk of oil rigs being damaged by ice-bergs. And on Sakhalin Island itself, seismic conditions are a hazard. Indeed, in May 1995 there was a serious earthquake which not only destroyed part of the oil production and pipeline infrastructure, but resulted in almost two thousand deaths and enormous damage to the oil field settlement of Neftegorsk. The harsh environment of the Samotlor field in western Siberia is positively benign compared to some of the hazards associated with some of the other potential oil areas. In 1995, across Russia fewer than one-third of producing oil wells were capable of increased output. To reach the projected target of 350 million tons by the year 2000, it is estimated that more than 500 new oil fields will have to be brought on stream. How, indeed whether, this can be done, remains to be seen. A basic set of prerequisites for this to happen would include a stable political situation, a functioning economy, a facilitative legal and institutional environment for both domestic and foreign oil industries, and some progress in domestic oil demand management. As yet none of these prerequisites is firmly in place.

Management of the oil industry in the Soviet era was relatively straightforward, even if the results fell short of the prescribed mark. Annual and five-year plan targets were set by central planning agencies in response to policies set by the Communist Party hierarchy. These were endorsed by the government with little, if any, discussion. The All-Union Ministry responsible for the oil industry instructed its enterprises accordingly. There is little about the management of the energy industries in general, and the oil industry in particular, in Russia since independence that is straightforward. By the mid-1990s, there was in fact little evidence of coherent management of the industry, let alone of central government control. It is yet another example of the market reform process spawning a private sector industry in which there are few laws and less enforcement. Into this murky environment has stepped a wide variety of players, not all of whom have national, regional or even corporate interests as a priority. Put simply, corruption and personal enrichment have figured prominently in the Russian oil industry in the immediate post-Soviet period. A conglomerate of associations, trusts, and joint-stock companies has emerged as a result of the privatization process described in Chapter 3. Most of them are controlled by former bureaucrats whose management decisions often are not subject to outside scrutiny or question. Thus, a mix of private sector and state agencies are now involved in production, distribution and marketing of oil. The emergence in Russia of Republics and regions as owners/managers of oil resources and industrial infrastructure has only served to complicate further institutional arrangements. The Ministry of Fuel and Energy has little if any control over the more than 30 state-owned regional production associations presently operating in Russia. Until 1993 most of these associations belonged to the Russian State Oil and Gas Corporation (Rosneftgaz), which ostensibly provided some overall policy and management direction for oil extraction. It was not effective, partly because it had no ability to enforce policy decisions regarding, for example, the rate of oil extraction, where and how much oil would be refined, and so on. Indeed, dissatisfaction with Rosneftgaz was a major reason for oil-producing firms to band together to create such organizations as LUKoil in Siberia and YuKOS in European Russia, to work out their own contracts with refineries. Rosneftgaz was abolished by the Russian government in 1993 in order that some greater degree of direct control from Moscow might be imposed on the industry. Clearly, the potential profit from the oil industry is attracting interest from both the private and public sectors. Most government agencies are in dire financial straits owing to cutbacks and non-payment of salaries. Some agencies with connections to the oil industry have already found ways to be less dependent on government finances by charging for services rendered and/or securing a place in the oil extraction, transport or refining process from which income might be derived. The Russian geological survey agency, for example, is seeking to participate in the production of oil. Small wonder all levels of government are looking for additional ways to tax the oil industry. By the mid-1990s, the oil industry was subject to more than a score of special taxes.

The domestic price of oil in Russia in 1995 was still barely one-third the world price, the direct consequence of government policy. While perhaps understandable in terms of protecting industries struggling to adapt to the new economy from even more financial difficulties, price controls have had numerous negative consequences. An important one for Russian firms is that there is little incentive to expand oil production if the rapidly escalating operating costs cannot be covered by revenues. Perhaps even more critical for the future of the oil industry in Russia, interest in oil exploration activities was dampened by price controls. Since domestic industrial production has declined by 40 per cent since 1991, there is strong incentive to export. But export quotas were used by the federal government in the early years of independence to ensure that before a permit to export was issued a certain proportion of oil extracted went to the underutilized refineries in Russia. Export quotas were also used to control the total volume exported, and to try and balance the interests of Russian oil export firms and the many foreign companies which had invested in Russia for the express purpose of producing oil for the more lucrative export market. In 1995, all export quotas were dropped, largely in response to international aid agency insistence that the Russian government stop intervening in the oil industry marketplace. Removal of export quotas was anticipated to further increase oil prices. To contain oil price rises, and to protect the comparatively inefficient Russian refining industry, the government instead introduced legislation requiring all oil production enterprises to refine domestically roughly two-thirds of petroleum extracted. Once this requirement was met, any producer would be free to export. The probable outcome of such an initiative is that exports will decline since incentive to produce for the external hard currency market would be dampened. It follows that for Russia hard currency income generated by taxing oil sales abroad would decline. Thus, as one form of government control over the oil industry is dropped, another one is imposed. The reason is simple – in Russia there remains widespread fear that the economy cannot sustain a rapid shift to world level oil prices any time soon, notwithstanding advice to the contrary from international agencies such as the International Monetary Fund (IMF). Complete freedom to export, it is contended by many Russian authorities, would soon push domestic oil prices to the world level because domestic consumers would be forced to compete with foreign markets simply to get supplies.

Across Russia a host of foreign independent and multi-national oil firms is involved in exploration, retrofitting existing wells, and enhancing pipeline and refinery technology. Since many non-Russian Republics possess producing wells and potential oil fields, contract negotiations are extremely complex. Foreign firms in some cases are having to negotiate with regional authorities, as well as those in Moscow. In the political environment of post-Soviet Russia, it is often the case that there are disputes between the two levels of government regarding export possibilities, royalty payments, and taxation. Domestic and foreign firms alike must deal with reality, but clearly the domestic firms at least have the advantage of appreciating better the political and institutional environment. There has been some progress in clarifying respective responsibility and authority for natural resource development between the federal government and some of the Republics. After several years of negotiation, Tatarstan was given oil resource rights, ownership of oil industry infrastructure, and discretion to set export quotas independent of Moscow. Similar 'deals' have been worked out with other regions. For foreign oil industry firms in particular, however, the institutional and government relations remain far from clear cut, or stable. And for such companies, Russia is not the only country in the post-Soviet scene which offers opportunities.

As the second most important oil producer, Kazakhstan moved quickly to create an economic and political environment conducive to investment by foreign oil companies. And it has the carrot to attract their interest – extensive areas in which geological conditions for petroleum and natural gas discoveries are extremely promising. Since 1980, when output from its wells was nearly 19 million tons, the production trend has been one of slow but steady annual increase. As Table 8.4 indicates, in 1991 output from Kazakhstan's wells was barely one-twentieth that of Russia. Production has dropped since independence, but the decline has been moderate in relative terms. About two-thirds of output to date has come from the fields around Mangghystau (Mangyshlak) on the east coast of the Caspian Sea (Fig. 8.6). These wells first came on stream in the early 1960s. But it is not the Mangghystau fields which are the object of so much interest. It is those first discovered in 1979 further to the north around Tengiz which are exciting foreign multi-national oil companies. Tengiz is at present one of the largest known potential reserves of oil and natural gas in the world, and accounts for the bulk of Kazakhstan's proven reserves. Production began on a small scale in 1991. Output from this field is anticipated to enable Kazakhstan's production to exceed 40 million tons by the year 2000. Leading the exploration and production in the Tengiz region is a joint venture between the multi-national corporation, Chevron, and the national oil company of Kazakhstan. Chevron has agreed to a multi-billion dollar investment commitment in exchange for half the oil output. Because the proven reserves lie at depths of 4500 to 5500 metres in geologically complex strata, and are heavy in sulphur content, they posed serious problems for the technologically handicapped Soviet oil industry and thus were largely ignored. It is believed that even more oil will be

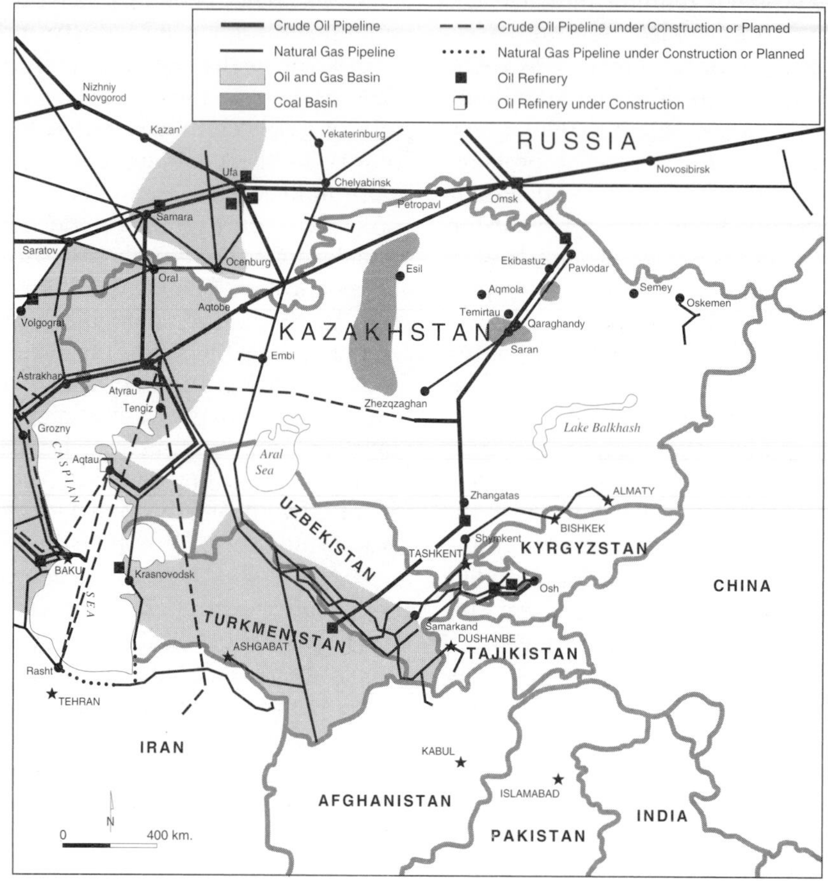

Figure 8.6 *Southern tier post-Soviet states oil and natural gas resources*
(*Sources*: Based on J.P. Dorian, S.F. Zhanseitov and S.H. Indriyanto, 'The Kazakh Oil Industry. A Potential Critical Role in Central Asia', *Energy Policy*, Vol. 22, No. 8, 1994, 686; J.P. Dorian, I. Sheffield and S.T. Indriyanto, 'Central Asia's Oil and Gas Pipeline Network: Current and Future Flows', *Post-Soviet Geography*, Vol. 35, No. 7, 1994, 416–17)

discovered once exploratory drilling penetrates to greater depths.

Several problems confront the nascent Tengiz-based oil industry. The first is that the capacity of existing de-sulphurization plants in Kazakhstan is limited. As the oil in untreated state is highly corrosive, it cannot be moved through pipelines for any prolonged period without causing some damage. Russia has also placed restrictions on the volume of oil from this field which can be moved through its pipeline network, arguing that such restrictions are necessary in order to avoid contaminating oil already in the system. As new de-sulphurization plants are put into operation, more oil will become available for export. Other problems are technical and geo-political. At present the throughput capacity of the pipeline network limits the quantity of oil which can be moved. Kazakhstan itself offers only a limited market, and one that is principally located in the eastern part of the country. There is no efficient way of moving oil to this market at the present time. And even if there was, it could not possibly absorb the huge potential output from the Tengiz area. Thus, external markets must be found, and from the standpoint of the joint venture with Chevron, hard currency markets are obviously essential. From Fig. 8.6, however, it will be clear that this will not be easy since Kazakhstan does not have direct access to the hard currency markets of the west. Like many post-Soviet states, it is essentially landlocked in this respect.

The Soviet oil pipeline system was designed to take advantage of complementary locations of oil reserves and markets in different regions. Oil from West Siberia was directed south into Kazakhstan to be refined at Pavlodar, for example. The same pipeline extends further south to Shymkent (Chimkent) and terminates at Seili in Turkmenistan (see Fig. 8.6). Some Siberian oil supplements the small scale domestic production from Kyrgyzstan and Tajikistan which supplies the refineries in the Fergana region. To offset this supply of oil to the Middle Asian refineries, oil from the fields in western Kazakhstan near the Caspian Sea was moved north into Russia. Since no pipeline links producing and consuming areas in Kazakhstan, this long-standing swap arrangement first devised for domestic Soviet purposes has been maintained. This permits some oil produced by the joint venture between Chevron and the national oil company of Kazakhstan from the Tengiz fields to reach the markets in Europe. Other alternatives for shipping Tengiz oil to markets have been under active study. Kazakhstan is part of the Caspian Pipeline Consortium which is seeking to build a pipeline for exporting oil to the west. The original signatories included Russia, Azerbaijan, the Gulf state Oman, as well as Kazakhstan. Chevron and a number of other international oil companies with business interests in Kazakhstan and Azerbaijan have an investment interest in the Consortium. Amongst the many possible pipeline routes to western markets considered by the Consortium, one involving a pipeline extending under the Caspian Sea, and then overland to Grozny, the capital of Chechnya and an oil centre itself, and from there to the Black Sea port of Novorossiysk, which offers access by oil tanker to the west, seemed a strong contender in the early days. But civil war and Russian military intervention in Chechnya in 1995 underscored the importance of securing a route through territory with some measure of political, if not economic, stability. Russia consistently pushed for an overland, all-Russian route to Novorossiysk which would pass north of Chechnya. This was not Kazakhstan's first choice, since dependence on a pipeline through Russia to move its oil to western markets continues its vulnerability to political pressure from Moscow. As well, the Turkish government is opposed to such a development because, they claim, the increased oil tanker movement through the narrow Bosporus and Dardanelles straits to the Mediterranean Sea would increase the risk of accidents and serious environmental damage. But Turkey has its own agenda. It has actively promoted a pipeline route from Azerbaijan across its territory to the Mediterranean port of Ceyhan. Such a development would further establish Turkey as benefactor to the newly independent, market-oriented, Muslim Middle Asian states. But there is no part of the Caucasus, through which such a pipeline must pass, which is immune to internal political instability. Still, oil from Kazakhstan could be shipped by tanker to Baku, even if a pipeline under the Caspian Sea was not built. Some oil does move over the Caspian Sea by tanker from Kazakhstan already. Azerbaijan and Turkmenistan also ship oil by tanker to Iran, but the volume is small and clearly pales in comparison with what could be moved by pipeline. A pipeline through Iran might also provide a possible outlet for Kazakhstan oil, but the US government, to which Chevron is to some degree accountable, is not supportive since it would give even greater oil supply leverage to a country the US does not regard as an ally. Thus, even if Chevron were willing to assist in underwriting the huge capital cost of building the pipeline, it would face political pressure from the American government not to do so. Exports to China remain a possibility of course, but once again there is as yet no trans-Kazakhstan pipeline to facilitate tapping a very large and logical market (Fig. 8.6). The necessary oil field infrastructure to increase production in Kazakhstan is steadily being put in place. Refinery capacity will be expanded from the three Soviet-era facilities to four, with the construction of a large scale plant at Aqtau on the Caspian Sea coast near Mangghystau (Mangyshlak) (Fig. 8.5). This latter project is being undertaken by a Japanese–Kazakhstan joint venture. While Kazakhstan is the only post-Soviet state which does not refine all of its oil output, and presently delivers more crude oil to Russia than it receives owing to the relative difference in domestic demand requirements between the two states, there are limits as to how much oil can be marketed in, or moved through, Russia. Put simply, Kazakhstan's oil industry needs as much access as possible to external, non-Russian markets in order to ensure profit to the private sector and revenue for state coffers. In late 1995, it was finally decided to build a pipeline to the western markets through Russia. After considerable political pressure and lobbying by Russia, Azerbaijan also agreed to send some of its surplus oil to the west through Russia as well. The rest will move across Georgia to the Black Sea port of Batumi, a proposition supported by US oil interests wary of too much dependence on Russia as the intermediary. Turkey also continued to push for some Caspian Sea oil to be shipped via its territory to the Mediterranean. But while a route through the Caucasus has some appeal, there are other problems in the region.

Azerbaijan is the third largest oil-producing country, one which faces most of the same problems confronting Kazakhstan's oil interests. In 1980 its output totalled nearly 15 million tons. Production in 1992, the last date for which reliable data are available, was 11.1 million tons, less than 3 per cent of Russia's output (Table 8.4). But its oil is high quality, and there is some opinion that output from existing wells is down at least in part owing to equipment failure in particular, and technological deficiencies in general. But of greater interest to foreign firms than re-equipping existing wells is that the adjoining Caspian Sea shelf, over which Azerbaijan lays exclusive claim, is reputed to contain extensive, untapped reserves. Off-shore production accounts for about four-fifths of the total at present. Following the collapse of the Soviet Union, major international oil firms were soon signing contracts with the new government of Azerbaijan to conduct feasibility studies and to do off-shore geological exploration. Since the change in government in 1993, political relations with Russia have improved, and those with Turkey have cooled. But for western firms it is still difficult to do business in Azerbaijan and by 1995 no significant oil field development had taken place. Put simply, geo-politics have confounded western oil business opportunities.

The first geo-political issue has to do with the Azerbaijan government itself. In short, it is still far from stable. The current government under President Heidar Aliyev replaced a strongly anti-Communist and pro-Turkish government. While relations with Russia are better, this means that Russia now has more influence on government policies. Democratic processes in Azerbaijan are far from inviolable, and government of any political persuasion, whether elected or imposed, is under constant threat of armed resistance. Thus, the risk of civil war remains very high. International oil companies find little comfort in such a turbulent political environment. But even if the domestic political situation were stable, the region of which Azerbaijan is part obviously is not.

The second geo-political issue is that the claim of Azerbaijan, and a number of other Caspian Sea states, to exclusive rights over resources found within self-proclaimed off-shore domains is not endorsed by all states abutting the Caspian. Russia especially has argued in a number of international organizations that the resource wealth which lies below the Caspian Sea should be developed by all states abutting it in such a manner that trans-national environmental risks are minimized. All states would therefore have to agree before development of off-shore oil and gas fields, for example in Azerbaijan territory, could proceed. Some Caspian Sea states (which include Russia, Kazakhstan, Uzbekistan, Turkmenistan, Azerbaijan, and Iran) view this as simply a ploy on Russia's part to retain control over the whole area. Interestingly, Russian diplomatic resistance to such off-shore territorial claims did not stop the major Russian oil company, LUKoil, from joining an international oil company consortium formed to develop Azerbaijan's off-shore Apsheron oil fields!

Regional geo-politics, the third critical issue, also play a major part in the debate over how best to move oil out of Azerbaijan to western markets. The Caucasus region has witnessed civil wars and military interventions on a regular basis since independence. From Nagorno-Kharabakh, to Chechnya, to Abkhaziya, there are few areas which have been spared the consequences of military action. Thus, a pipeline through any part of the Caucasus is more than a little problematic since no government can guarantee unimpeded flow of oil through it despite the obvious attraction to near bankrupt states of the revenues which would be part of such a venture. Azerbaijan was an original signatory to the Caspian Pipeline Consortium, but later dropped out as a result of an agreement reached between Turkey and the international and state oil consortium to build a 1600 kilometres pipeline south and west to the Mediterranean port of Ceyhan. While longer than some of the other pipeline options under consideration, it was argued it would be cheaper because part of the pipeline through Turkey from Iraq, mothballed since the United Nation's oil export embargo imposed on Iraq following its defeat after invading Kuwait, could then be put to use again. As noted earlier, it was proposed that this pipeline option would allow Kazakhstan access to western markets, should it wish to use it. But following the change in government in Azerbaijan, political pressure from Moscow to have Azerbaijan oil move over Russian territory by pipeline, and then to western markets by oil tanker from the Black Sea port of Novorossiysk, escalated and eventually brought results as noted above. But up until the end of 1995 there had been much discussion, but little actual development of Azerbaijan's oil reserves.

Elsewhere in the former Soviet Union oil output was limited. Turkmenistan, Ukraine and Uzbekistan each produce about four to five million tons annually (Table 8.4). In Turkmenistan, there are plans to increase output based on some newly discovered deposits on land and off-shore in the Caspian Sea. A number of international oil companies are already doing business there. A new oil refinery is to be supplied by augmented domestic output and oil imported from neighbouring Iran. In Ukraine, about 90 per cent of the oil refined, and consumed, within the country is imported, mostly from Russia. Payment problems and politics have combined to reduce the volume sent to Ukraine from Russia, thus Ukraine's refineries have been underutilized over the past few years. Ukraine pumps a little oil itself, but most of its output is natural gas condensate. It has also concluded contracts with a number of foreign firms to explore potential oil-bearing territories within Ukraine and in off-shore Black Sea areas. Every effort is being made to reduce the oil import dependence on Russia. But Russia is likely to remain the principal source of oil imports for some years to come. Uzbekistan was alone in increasing oil production in the 1991–1994 period, though as Table 8.4 indicates its total output is not very large. The quest for greater output has brought some new oil wells into production, principally in the Fergana Valley region, but the bulk of the increased production is gas condensate based. In keeping with the government's approach to the new economic era, it has placed all oil and gas exploration, production, distribution and refining enterprises in the country into one state corporation. Foreign investment is being sought, but the problem of finding a secure route for export remains a problem, just as it does for all of the southern tier post-Soviet states.

NATURAL GAS

In the final years of the Soviet era, the relative importance of natural gas in the fuel balance overtook oil (Table 8.2). In 1989, close to 800 billion cubic metres were produced, nearly twice the output in 1980, and not far off twenty times more than in 1960. And in terms of tons of fuel equivalent, natural gas represented about 40 per cent of the fuel balance, up from 27 per cent in 1980. In 1989, oil accounted for 38 per cent of the fuel balance, down from nearly 46 per cent in 1980. As Table 8.5 indicates, natural gas production in the 1991–1994 period has declined somewhat in most states, but the downturn has been far less severe than in the case of oil. In Russia, which accounts for 86 per cent of the total natural gas output from all post-Soviet states, 1994 output was off less than 5 per cent from 1991, a striking

Table 8.5 *Post-Soviet natural gas production, 1991–1994*

	Billion cubic metres				Cubic metres per person			
	1991	**1992**	**1993**	**1994**	**1991**	**1992**	**1993**	**1994**
Russia	643.0	641.0	618.0	607.0	4327.0	4311.0	4158.0	4092.0
WESTERN TIER								
Ukraine	24.4	20.9	19.2	18.3	468.0	400.0	368.0	353.0
Belarus'	0.3	0.3	0.3	0.3	28.6	28.3	28.1	28.4
SOUTHERN TIER								
Georgia	0.04	0.01	0.01	–	7.3	1.8	1.8	–
Azerbaijan	8.6	7.9	–	–	1190.0	1074.0	–	–
Kazakhstan	7.9	8.1	6.7	4.5	465.0	478.0	393.0	267.0
Kyrgyzstan	0.1	0.1	0.04	0.04	18.5	16.1	9.3	8.7
Tajikistan	0.1	0.1	0.05	0.03	16.9	13.0	8.7	5.8
Turkmenistan	84.3	60.1	65.2	35.6	22144.0	14908.0	15144.0	8084.0
Uzbekistan	41.9	42.8	45.0	47.2	1975.0	1995.0	2052.0	2105.0

Sources: Ekonomika Sodruzhestva Nezavisimykh Gosudarstv v 1993g (Moscow: Statkomitet SNG, 1994), 43; *Ekonomika Sodruzhestva Nezavisimykh Gosudarstv v 1994g* (Moscow: Statkomitet SNG, 1995), 88

contrast with its oil production. Ukraine and especially Turkmenistan have experienced substantial reductions in gas output over this period. Uzbekistan managed to increase natural gas output, just as it did with oil.

Natural gas production, unlike oil, has been sustained during the uncertainties and disruptions of the market reform era for a number of reasons. Firstly, a substantial share of its domestic market comprises electric power generating stations which have had a more consistent demand schedule than have oil-consuming industries. Secondly, natural gas production is also more sustainable than oil because over the past few decades proven reserves have more than kept pace with the rather remarkable increase in output. Moreover, the natural gas industry infrastructure, from well-head to consumer, is in much better condition than is that for oil. What the natural gas industry does have in common with the oil industry, however, is the steady shift from the western to eastern regions in the locus of proven reserves and output. In this shift to the east of proven reserves of natural gas, Russia's West Siberian lowland energy storehouse has figured prominently, and in contrast to oil reserves, it will continue to do so for years to come. For all former Soviet territories taken together, Russia's proven natural gas reserves account for almost 90 per cent of the total. Turkmenistan ranks second in terms of proven reserves, but these amount to only 5 per cent of the total. Kazakhstan, Uzbekistan, Ukraine and Azerbaijan have, in that order, the next largest reserves of natural gas, obviously all minuscule in comparison with Russia's.

Russia's natural gas industry is completely dominated by the Gazprom concern. This is a vast conglomerate of enterprises involved in production, processing and distribution of natural gas. It was formerly headed by career bureaucrat and Communist V. Chernomyrdin, who was appointed Prime Minister of Russia in 1993. Privatized early in the post-Soviet period, many former Soviet-era key personnel have assumed positions of authority in the new structures. The state has kept a substantial minority role as a stockholder in Gazprom, though how much influence it exerts through its 40 per cent share holding position is not entirely clear. At any rate, while the links with government are close, this does not mean that the industry has been given free rein to accumulate large profits for its share holders, government included. In fact, there is growing concern over the seeming imbalance between the industry's contribution to Russia's total gross domestic product, which along with the petroleum industry is about 12 per cent, and the taxation and royalty revenue, which is just under 9 per cent. Gazprom's operating expenses are, in the opinion of some observers, excessive. Employees of Gazprom are the highest paid, on average, of all workers. They also receive in addition to the highest average pay, a singularly rich benefits package, much in the style of the former Soviet era. While gas prices were increased early in 1992, and have been adjusted periodically since, gas is still substantially underpriced. In Russia in the mid-1990s, natural gas was less than one-half the world market price. This has posed serious problems for the industry since operating costs in terms of wages and materials, for instance, have escalated dramatically. The growing unpaid bill of industrial consumers such as electric power plants (which account for over one-half of all gas consumed), iron and steel factories and so forth, is placing further pressure on the gas producers. Many operations in Russia are therefore faced with a real cost-price squeeze. While Gazprom has a monopoly position and close ties to decision-makers in government, thus far its appeals for a substantial price hike have been denied. Gazprom has been more successful in protecting its monopoly position in Russia's potentially lucrative natural gas industry by keeping foreign companies at bay. To date, very few have set up joint venture operations in Russia, in contrast to the situation in the Russian oil industry. The huge scale of Gazprom's reserves and their relatively straightforward recovery mean that foreign technology and skills are not necessary. The natural gas industry has always been bountiful in that reserves were large and associated with geological formations which did not pose huge technological challenges in terms of recovery. Indeed, for many years during the Soviet era it was not even regarded as a resource to be husbanded and developed. As the following brief overview indicates, the evolution of the industry reflects largely happenstance developments.

Until the decision was taken during the late 1950s to start developing the proven reserves of natural gas, output was frequently just ancillary to oil production. All too typically, natural gas was simply flared off at the oil well since its recovery was not part of the plan. Most natural gas production occurred in the Baku region in Azerbaijan. A small amount came from the Dashava natural gas fields in western Ukraine. Once natural gas production became an integral part of Soviet energy policy, exploration for natural gas proceeded apace and output increased substantially (Table 8.2). From the fields near Stavropol' in the north Caucasus, natural gas flowed in ever larger volume from the late-1950s on. By the mid-1960s, it accounted for just a shade less than 30 per cent of total production. However, reserves were insufficient to sustain this volume of output for long. The Stavropol' gas fields soon began to decline in both relative and absolute importance. The giant Shebelinka reserves in eastern Ukraine took up the slack. They were better located in terms of accessibility to the major urban-industrial markets of European Russia. Combined with the output from Dashava further west, Ukraine was the Soviet Union's main source of natural gas for nearly two decades. By the mid-1960s, a pipeline system integrated these two production regions with several other fields brought on stream in the 1960s (see Fig. 8.5). It was used mostly to fuel thermal electric power stations. Beginning in the late 1970s Ukraine's production waned in relative and absolute terms. Gas produced in Ukraine satisfied a substantial share of domestic demand, as well as the export market in both eastern and western Europe. With the pipeline system in place in European Russia, natural gas from deposits deeper in the Soviet interior could be drawn on as the reserves of the north Caucasus, and then Ukraine, were depleted.

Natural gas was a common occurrence at the well-head in the oil fields of the Volga–Ural region, but as elsewhere across the former Soviet Union, it too was frequently flared off. By the mid-1960s gas production was still quite insignificant since few reserves separate from oil deposits had been discovered. This changed with the appearance of data from exploratory drilling near Orenburg. While the gas itself was impure, and thus had to be processed before it could be used, the location of this reserve was strategically significant since it was close to major industrial markets. Natural gas deposits were being discovered and tapped in other locations at this time as well, but invariably the reserves were distant from markets, in a harsh physical environment, or both. Production from the Orenburg gas reserves soon played a major role in supplying thermal power stations throughout European Russia, and was of sufficient scale that expansion of exports to both eastern and western Europe could be entertained. In the former case, exports were required as part of the bilateral arrangements which led to the development of the Orenburg gas fields in the first instance. Finances, technical personnel and labour from eastern Europe were employed in developing the Orenburg deposit, and building additional pipeline facilities. At about the same time Soviet authorities were busy linking by pipeline the considerable natural gas reserves of Middle Asia with European Russia (Fig. 8.6). By the mid-1970s, production in Middle Asia accounted for about one-third of the Soviet total. Approximately one-half of production came from the various fields in European Russia and the north Caucasus. The remainder came from the as yet not fully explored West Siberian gas fields. These were known to be substantial already in the late 1960s, but eventually the major gas fields in the West Siberian lowland depicted in Fig. 8.5 accounted for more than half of the proven reserves. Most of these fields are in an exceedingly harsh environment, and are even more remote than Russia's oil fields. Thus, access to the reserves and transport of natural gas to the major domestic and foreign markets pose major logistical problems. But as Fig. 8.5 indicates, the natural gas pipeline system penetrates this region, and in so doing makes Russia's future energy situation more secure. Gas production from West Siberia is dominated by the huge Urengoy deposit. In the early 1990s it alone accounted for close to one-half of Russia's substantial natural gas production. While output from this field peaked in the late 1980s at 300 billion cubic metres, it has been around 280 billion cubic metres since, and therefore has not experienced a significant decline. Output from the Yamburg deposit, like Urengoy also located in Tyumen' Oblast' in the northern reaches of West Siberia, has grown steadily, notwithstanding the turmoil of the transition to a market economy. In 1992, output from this field was nearly 180 billion cubic metres. Other fields in Siberia contribute to natural gas production, but in minor fashion. The Urengoy and Yamburg fields together make up nearly three-quarters of Russia's total natural gas production. West Siberia in total accounts for more than 90 per cent of Russia's gas output. And to reiterate, amongst all post-Soviet states Russia accounts for more than four-fifths of the natural gas production.

While West Siberia's reserves of natural gas can sustain a steady and significant increase in annual production for the foreseeable future, the industry is far from immune to the problems confronting the petroleum sector. There is some evidence that in the most productive gas fields of the West Siberian lowland output from some wells exceeds the operating norm which will maximize the long-term yield, all in the quest to increase output in the short-term. Clearly this does not augur well for maximizing recovery, and thus making most effective use of a non-renewable resource. The really gigantic scale of proven reserves, and the propitious geo-physical conditions for expanding the reserve potential in many other regions in Russia, tend not to encourage conservative, and hence conservation-minded, management.

To create the existing network of pipeline put huge pressure on the steel industry, amongst others. Producing high quality steel pipe and efficient pumping station technology has been a major problem in the past, and in the current economic situation in Russia remains a critical issue. Indeed, leakages from poor quality, and inadequately maintained, pipelines are commonplace. In the case of oil pipelines the resultant environmental degradation is extremely serious, as discussion in a later chapter will demonstrate. Oil is moved over long distances in pipelines whose average diameter is less than 800 mm. There are some 1200 mm oil pipeline trunk systems, but they do not comprise a significant part of the total system. More than half of the natural gas pipeline system, however, is in excess of 800 mm, and a sizeable share of the total pipeline length comprises 1200 mm and 1400 mm diameter pipe. As reserves are discovered in more remote regions (and as Fig. 8.5 indicates, there are production facilities in more easterly locations than the West Siberian lowland), distribution costs will inevitably increase. That said, there is already a substantial gas pipeline infrastructure in place, one which has for more than two decades enabled gas to be exported to eastern and western Europe in substantial volume. Current plans call for much augmented exports to the hard currency markets of western Europe. And to ensure greater control over gas being exported, Russia wants to build a new pipeline 'around' Ukraine. Ukraine has in the past simply diverted Russian gas from pipelines across its territory which connect up with the network in eastern Europe. Russian natural gas exports are vulnerable to this tactic as nearly 90 per cent move through pipelines which traverse Ukraine.

Turkmenistan was the second largest natural gas producer. It is now third after Uzbekistan, but ahead still of Ukraine. As the data in Table 8.5 indicate, output dropped quite substantially since independence, and especially in 1994. This is largely due to price resistance on the part of external markets. Turkmenistan moved quickly to adjust gas prices, pushing them substantially above the price of Russian gas, and prejudiced exports to cash-starved post-Soviet states in consequence. Ukraine was the major export market and was determined not to pay the price demanded. Turkmenistan has been obliged to compromise, at least for the time being. To access western hard currency markets for its gas, Turkmenistan is planning a joint venture pipeline construction project through Iran and Turkey (Fig. 8.6). While Iran and Turkey have contracted to take some gas, so too has Germany which thus far still relies very heavily on imports from Russia. A much larger joint venture project involves the construction of an 8000 kilometre long natural gas pipeline to Japan. The pipeline would follow the route of the ancient Silk Road to China, and then underwater from China to Japan. Japanese interest in Turkmenistan's very substantial reserves is high, and several companies have signed joint venture contracts to develop them. The natural gas industry in Turkmenistan is still very much an arm of the state, thus domestic private sector interests are limited. Notwithstanding the drop in natural gas output since 1991 in Kazakhstan, it is endeavouring to increase domestic production so as to eliminate the now very costly gas imports from Turkmenistan. Joint ventures with western companies are playing a key role in this energy development strategy. Uzbekistan now ranks second in natural gas production, and consumes most of its steadily growing output itself. Ukraine's gas production, which is fourth largest amongst post-Soviet states (Table 8.5), has dropped since 1991 despite the urgent need to reduce its costly energy import dependence. The decline is related to depleting reserves and less to economic disruption, although that is not without its consequences to be sure. There is some expectation that recently initiated exploration both in western Ukraine and in off-shore Black Sea areas will increase proven gas reserves, and permit production to grow. Output in the other states is not large, and in the main is used to meet domestic needs.

ELECTRICITY

Per capita consumption of electricity is often used as a measure of economic development, as was noted at the outset of this chapter. On this basis it is clear from the data presented in Table 8.6 that there are no positive signs anywhere in the post-Soviet scene. While in some countries, notably Russia, there were small additions to the installed generating capacity, in most cases the previous trend of rapidly expanding power generating facilities ground to a halt. There has been some privatization of electricity generating and distribution facilities, again notably in Russia where a company controlling about half of the electricity generated and all of the facilities for transmission of electricity was created in 1992. The government holds a 49 per cent share in the company, and presumably is influential in determining policy initiatives. Elsewhere in the former Soviet Union the trend typically has been one of continuing state ownership of the electric power generation and distribution systems.

Table 8.6 *Electricity production, post-Soviet states, 1989–1994 (in billion kilowatt-hours)*

	1989	1990	1991	1992	1993	1994
FORMER USSR (TOTAL)	1721.7	1726.0	1683.1	1574.1	–	–
Thermal	1285.7	1281.0	1236.2	1126.2	–	–
Hydro	223.4	233.0	234.8	240.0	–	–
Nuclear	212.6	212.0	212.1	207.9	–	–
Russia	1076.6	1082.2	1068.0	1014.6	957.0	876.0
WESTERN TIER						
Ukraine	295.3	298.5	278.7	252.6	230.0	201.0
Belarus'	38.5	39.5	38.7	37.6	33.4	31.4
Moldova	17.0	15.7	13.2	11.1	10.4	8.2
Estonia	17.6	17.2	16.2	15.9	–	–
Latvia	5.8	6.6	5.6	3.9	–	–
Lithuania	29.2	28.4	29.4	28.2	–	–
SOUTHERN TIER						
Armenia	12.1	10.4	9.6	6.9	6.3	5.6
Azerbaijan	23.3	23.2	23.5	20.0	19.0	17.5
Georgia	15.8	14.2	13.4	11.5	9.7	–
Kazakhstan	89.7	87.4	86.0	81.3	77.4	65.1
Kyrgyzstan	15.1	13.4	14.2	11.8	11.0	12.7
Tajikistan	15.3	18.1	17.5	16.8	17.7	17.0
Turkmenistan	14.5	14.6	15.0	13.1	12.6	10.5
Uzbekistan	55.9	56.3	54.2	50.9	49.1	47.7

Sources: Matthew J. Sagers, 'The Energy Industries of the Former USSR: A Mid-Year Survey', *Post-Soviet Geography*, Vol. 34, No. 6, 1993, 404; *Ekonomika Sodruzhestva Nezavisimykh Gosudarstv v 1993 g* (Moscow: Statkomitet SNG, 1994), 42; *Ekonomika Sodruzhestva Nezavisimykh Gosudarstv v 1994 g* (Moscow: Statkomitet SNG, 1995), 87

As the bulk of electricity generated comes from thermal power stations, the foregoing discussion regarding fossil fuel production trends is obviously relevant. Simply put, declining fossil fuel production has disrupted supplies to some thermal power stations and hence reduced electricity output, rather dramatically in some post-Soviet states such as Latvia, Moldova and the three Caucasian states, Georgia, Armenia and Azerbaijan (Table 8.6). Clearly, the dismal state of the industrial sector of the economy across the post-Soviet scene also had an impact on demand. Since electricity exports are relatively small scale, output by country tends to be highly correlated with consumption. In all cases total output and per capita output were lower in 1994 than in 1991, in some instances substantially so. Hydroelectricity stations and nuclear power installations, while together counting for a relatively small share of electricity produced, nonetheless have proven to be more dependable sources of electricity in these difficult times. Output from hydroelectric power stations across the former Soviet Union is now about 250 billion kilowatt hours (kWh), which is greater than at any time in the past. Russia accounts for 70 per cent of this total. An annual increase in hydroelectricity generation has been maintained in Russia despite the economic disruption of the past few years. Nuclear power station generation of electricity was only marginally less in the mid-1990s than it was in 1988 when it peaked at 216 billion kWh. Indeed, the nuclear power industry appears to have gained some momentum in the new economy of the post-Soviet era, notwithstanding the legacy of Chernobyl'.

The widespread grass-roots resistance to the nuclear power industry which the Chernobyl' disaster of 1986 spawned forced the closure of a plant in Armenia and brought to a halt installation of about 60 nuclear reactors at sites across the country. The international condemnation of certain aspects of Soviet nuclear technology in particular, and the management of the industry in general, simply underscored public unease over the industry. For a while it appeared that the nuclear industry's long-term programme of substantially increasing the number and scale of nuclear power stations would be shelved permanently. The collapse of the Soviet Union and the ensuing energy shortages have caused some states, as they grapple with how best to deal with this situation, to revisit the basic nuclear industry development strategy put forward in the early 1980s. The plan, if realized, would have seen nuclear electricity account for 30 per cent of total generation by the year 2000. In 1992, it accounted for only 13 per cent of the total from all former Soviet territories (Table 8.6). In Ukraine and Lithuania electricity from nuclear power stations would have represented more than one-half of the total generated. In the post-Soviet era of rapidly escalating prices for oil and natural gas, disruption of electricity supply from thermal power stations, and limited domestic energy resources, nuclear energy has a certain appeal, notwithstanding the continuing deep suspicion of it on the part of the general public. In Russia, Ukraine, Armenia, and even in Belarus' which does not have nuclear installations, the nuclear 'lobby' has the ear of government. In Lithuania where there was a huge public outcry in the wake of Chernobyl' over its own nuclear station, there now is grudging recognition that the domestic economy simply cannot afford to do without the electricity produced by the Ignalina nuclear installation (see Fig. 8.2). This one plant accounts for about one-third of all electricity generated. Of course, the nuclear industry has an alternative technology to the discredited system which failed at Chernobyl'. But the electricity supply situation is so serious in many places that nuclear installations which use the risky graphite-moderated reactor technology employed at the Chernobyl' facility continue to operate. Lithuania is a case a point.

Russia retains the 'distinction' of having the largest number of nuclear power stations and generating capacity (see Fig. 8.2). Total capacity in all of its stations in the mid-1990s was about 20,000 megawatts. This was a shade more than 50 per cent of all nuclear power installed generating capacity across all the former Soviet territory. One version of Russia's nuclear power industry's future envisions doubling electricity generating capacity by the year 2010. Given Russia's current economic situation, however, it is unlikely it has the financial resources to carry out such a large expansion programme. However, there are a number of installations which were mothballed in the midst of construction in the wake of Chernobyl', and these could certainly be brought on stream. Indeed, one of these reactors was commissioned in 1993, the first since the collapse of the Soviet Union. Several other reactors are currently being completed despite local opposition of the general public. While these new reactors are based on a technology deemed to be safe, the fact is that a number of nuclear installations in Russia with the Chernobyl' technology continue in operation. In the opinion of western nuclear energy experts, all such facilities should be closed. But Russian authorities argue that aside from their claim that these facilities are in fact managed well and pose no risk, the closure of such nuclear installations would not only diminish much needed electricity production, it would represent a huge financial loss to the state. According to government estimates, closure of these stations would entail a loss of some US$ 60 billion! While international opinion strongly endorses closure, to date there has been little indication that Russia would receive any significant financial compensation for doing so.

The safety of nuclear power stations is an ongoing concern in most regions although there are some exceptions. In the Kola peninsula and Karelia, for example, modernization of existing plants and their continuing operation is apparently something the general public endorses. But elsewhere the prevailing view is one of considerable scepticism regarding public pronouncements from the nuclear industry. Minatom, the Russian nuclear power ministry, has responsibility for both the civilian and military installations. But given the seeming lack of government control over, indeed, in some cases knowledge of, its activities, there is little reason to be sanguine about its official statements regarding safety in particular, and management objectives in general. Minatom which entered into an agreement to sell nuclear technology to Iran did so without government sanction. The reaction of western states to this initiative was negative, since if this sale of nuclear technology went ahead it portended the possibility of yet another state developing nuclear arms. While Russia defended its right to sell technology to Iran, it would seem this was more in defence of national self-interest than it was in support of a policy adopted by government. This particular issue served to underscore the widespread view that in Russia the nuclear industry essentially operates outside the aegis of government, and therefore without proper public accountability for decisions taken.

In Ukraine, the issue of public accountability in decision-making concerning the nuclear power industry is a high priority because of what happened at Chernobyl'. Thus, the Ukrainian government put in place a State Committee with authority over nuclear energy use. As noted earlier, Ukraine is to some degree obliged to examine carefully nuclear energy options owing to its continuing energy import dependence. While the government has plans to close the remaining operating reactors at Chernobyl', it has not done so as yet. Moreover, it is now expected that completion of several nuclear projects mothballed in the late 1980s will proceed. Grass-roots opposition to bringing more nuclear reactors on stream remains strong irrespective of the particular technology to be employed. Expansion of the nuclear generating capacity can be achieved at comparatively low cost since construction of the reactors to be brought on stream was well advanced at the time they were mothballed. The problem of what to do with the nuclear waste from the stations operating now is not resolved. Additional nuclear facilities will clearly exacerbate the waste disposal problem. Given the widespread public concern over, if not outright opposition to, nuclear energy in Ukraine the fact that expansion is planned and approved by government simply underscores how the serious is the problem of supplying the country with electricity in the most cost effective manner. Environmental and public health concerns have not abated, their resolution seemingly has just been postponed.

The trade-off between environmental well-being and economic well-being in relation to the nuclear power industry is perhaps nowhere so palpable as in Armenia. A beleaguered Armenia government, confronted with an economy reduced to a standstill, has determined that the only way the country can survive is to re-open the Metsamor station, located about 30 kilometres from the capital Yerevan, to supply electricity for both domestic and industrial use. Contrary to logic, the Metsamor nuclear power station was built in a high risk seismic zone during the Soviet period (see Fig. 8.2). It was shut down in 1989 because of safety concerns over the reactor technology, and one year after a serious earthquake occurred close by. While the technology used is not the same as at Chernobyl', there is a problem with containment of radioactive substances should there be an accident. International agencies remain convinced the Metsamor facility is faulty in design and obviously located in the wrong place. As no financial or technical assistance could be obtained from international agencies, Armenia enlisted the help of Russia instead. This clearly is in the self-interest of Russia politically, and serves to bolster the position of the nuclear power industry within Russia as well. The Metsamor station resumed operation in 1995. But in view of the ongoing disputes between Armenia and Azerbaijan, and the terrorist attacks on other electricity generating facilities and pipelines which have already occurred in the region, there are more than technical and natural hazards associated with putting such a facility into operation once again.

The problems confronting energy sector planners across the post-Soviet scene are complex to be sure, but in Russia especially they are made even more so by the geography of energy. As the more accessible reserves of non-renewable fossil fuels and renewable sources of energy are consumed or developed, the fundamental imbalance between resource location and the market is exaggerated. In other states, the sheer distance between the location of energy resources and the market is not such a critical problem. Indeed, some post-Soviet states would welcome such a problem for it is far less daunting than trying to cope with essentially no domestic energy resource supplies. For such energy import-dependent states, there is pressing need to establish a comparative advantage in some other economic sector. As the discussion in the following chapter will make plain, all post-Soviet states are actively engaged in the same search process.

FURTHER READING

Campbell, R.W. 1980 *Soviet Energy Technologies: Planning Policies and Research*. Bloomington, Indiana University Press.

Dienes, L., Shabad, T. 1979 *The Soviet Energy System: Resource Use and Policies*. New York, John Wiley.

Dienes, L., Dobosi, I., Radetzki, M. 1994 *Energy and Economic Reform in the Former Soviet Union: Implications for Production, Consumption and Exports, and for the International Energy Markets*. New York, St. Martin's Press.

Ebel, R.E. 1994 *Energy Choices in Russia*. Washington, DC, Centre for Strategic and International Studies.

Goldman, M.I. 1980 *The Enigma of Soviet Petroleum: Half Full or Half-Empty*. London, George Allen and Unwin.

Hewett, E.A. 1984 *Energy, Economics and Foreign Policy in the Soviet Union*. Washington, Brookings Institution.

Marples, D.R. 1987 *Chernobyl and Nuclear Power in the USSR*. New York, St. Martin's Press.

Raykher, E. 1984 *The Economics of the Soviet Gas Industry*. Leesburg, Delphic Associates.

Sagers, M.J., Green, M.B. 1983 *The Transportation of Soviet Energy Resources*. Totowa, NJ, Rowman and Littlefield.

Young, K.E. (ed) 1986 *Decision-Making in the Soviet Energy Industry: Selected Papers with Analysis*. Falls Church, Delphic Associates.

9

INDUSTRIAL RESTRUCTURING AND MARKET REFORM

'No country conducting privatization has ever spent that kind of money maintaining privatization agencies'
(*Izvestiya*, September 2, 1995, 2)

Since the collapse of the Soviet Union in 1991, a general crisis has enveloped most categories of industrial production. Output has declined dramatically. Millions of industrial workers are on reduced workweeks, hundreds of thousands are irregularly paid, many are simply unemployed. The dilemma facing post-Soviet states is the same everywhere – consumer goods-oriented industries are frequently technologically outdated, workers' productivity is typically quite low, the quality of product is generally poor, and in the former defence industry, demand has shrunk. Industrial enterprises under the Soviet autarchic development model were traditionally guaranteed a market for their products or services, irrespective of quality. In short, the country was to all intents and purposes closed to foreign competition. These same enterprises now must overcome the disruption in supplies of raw materials, semi-manufactured inputs, and services of one kind or another as a result of the collapse of the Soviet centrally planned system. At the same time, they must compete with foreign companies for a share of their own domestic markets. That this is proving difficult should come as no surprise. Industrial restructuring is challenging at the best of times. Having to restructure amidst the economic chaos of the immediate post-Soviet period has been enormously complex. Changing the ownership of former state sector enterprises, that is privatizing them, is one thing. Making them internationally competitive is quite another. Not all of the industrial organizations, plant and equipment inherited from the Soviet system will be restructured or will continue to operate in the new economy. Individual enterprises, indeed, in some cases entire single industry towns or industrial regions are at risk. And the longer the restructuring process takes in each post-Soviet state, the more difficult it will be to recapture the growing share of domestic markets being won by foreign companies from outside the former Soviet Union. Many of these foreign firms have already restructured so as to survive the economic recessions of the early 1980s and 1990s. Put simply, the competition for markets in the newly emergent post-Soviet states is frequently very intense.

This chapter will first provide an overview of the distribution of the principal mineral and forest resources for industry. The level of output of a number of important raw materials for industry at the close of the Soviet era will be briefly described. Production trends since independence are also discussed. This overview sets the stage for a description of the distribution of the major industries, and of the transportation system which services them. The balance of the chapter will be devoted to an examination of post-Soviet industrial production, privatization and restructuring. Developments in Russia will figure prominently in this discussion.

Natural resources for industry

Virtually all of the energy, mineral and biotic raw materials required by a modern industrial state were found within the vast expanse of the former Soviet Union. The only cloud in this picture of comparative abundance was that created by the geography of natural resources. In many instances, the locations of resources of sufficient scale to sustain long-term development were rather far removed from where they were most needed. Thus, in addition to the infrastructure and other costs associated with resource development in increasingly remote frontier regions were ever higher costs of transportation. The history of the development of iron ore reserves illustrates well the basic imbalance between the geography of supply and of demand.

During the late eighteenth century an iron industry emerged in the southern Ural mountain region far from the manufacturing centres of the Russian Empire (Fig. 9.1). The development there was based on very large, and very rich, deposits of iron ore. The technology which enabled Russia to become the world's largest producer of iron at the time was based on smelting this high grade iron ore with charcoal, not coking coal. While the Urals provided adequate forest resources from which charcoal could be obtained, there was no locally available supply of coking coal. Russia was soon eclipsed by production in a number of West European states which was based on the more advanced technology of smelting iron using coking coal as fuel. Iron production in the Urals, already handicapped by virtue of being very far from the major market in central European Russia, could not compete with higher quality and cheaper iron imported by sea from Europe, and thus steadily declined until the late nineteenth century. At this time rich iron ore deposits were discovered near Krivoy Rog (Kryvyy Rih) in south Ukraine. They were soon linked by railroad to the then enormous coking coal reserves of the Donets basin, some 300 kilometres to the east. This integration of iron ore and coking coal deposits by railroad gave rise to the Russian Empire's first modern iron and steel industry. Supported by the introduction of a wide range of protective tariffs and guaranteed state orders for steel, a very substantial industry soon emerged in southeastern Ukraine. The iron and steel industry which developed there spawned a host of related metalworking industries, all geared to the principal markets in European Russia. The demise of the Russian Empire in 1917 and the emergence of its Soviet successor bent on rapid industrialization within a socialist, planned economy brought new initiatives and goals to the fore. The vast iron ore reserves of the southern Urals soon attracted attention notwithstanding the fact that they were located more than 1500 kilometres east of the existing iron and steel industry of southeast Ukraine. It was decided to integrate the high-quality coking coal of the Kuznetsk basin in central Siberia, 2000 kilometres east of the Urals, with the Ural iron ore reserves. This decision led to the creation of the Urals–Kuznetsk Combine. The genesis of this vast regional development complex was discussed in Chapter 3. Over the years, a vast array of iron- and steel-using industries were developed by the state in the southern Urals and in the Kuznetsk basin in central Siberia. However, the average distance over which the basic raw materials for the iron and steel industry, and the production from it, had to be transported still increased over time. The dilemma was real enough. Where the market was, many of the basic resources required by the iron and steel industry were not.

There are at present more than 100 billion tons of recoverable reserves of iron ore in all post-Soviet states taken together. By far the largest part of these are in Russia. During the Soviet era, geological exploration resulted in a number of new iron ore reserves being added to the inventory. Siberia figured prominently in these discoveries, but by no means was it the only region to

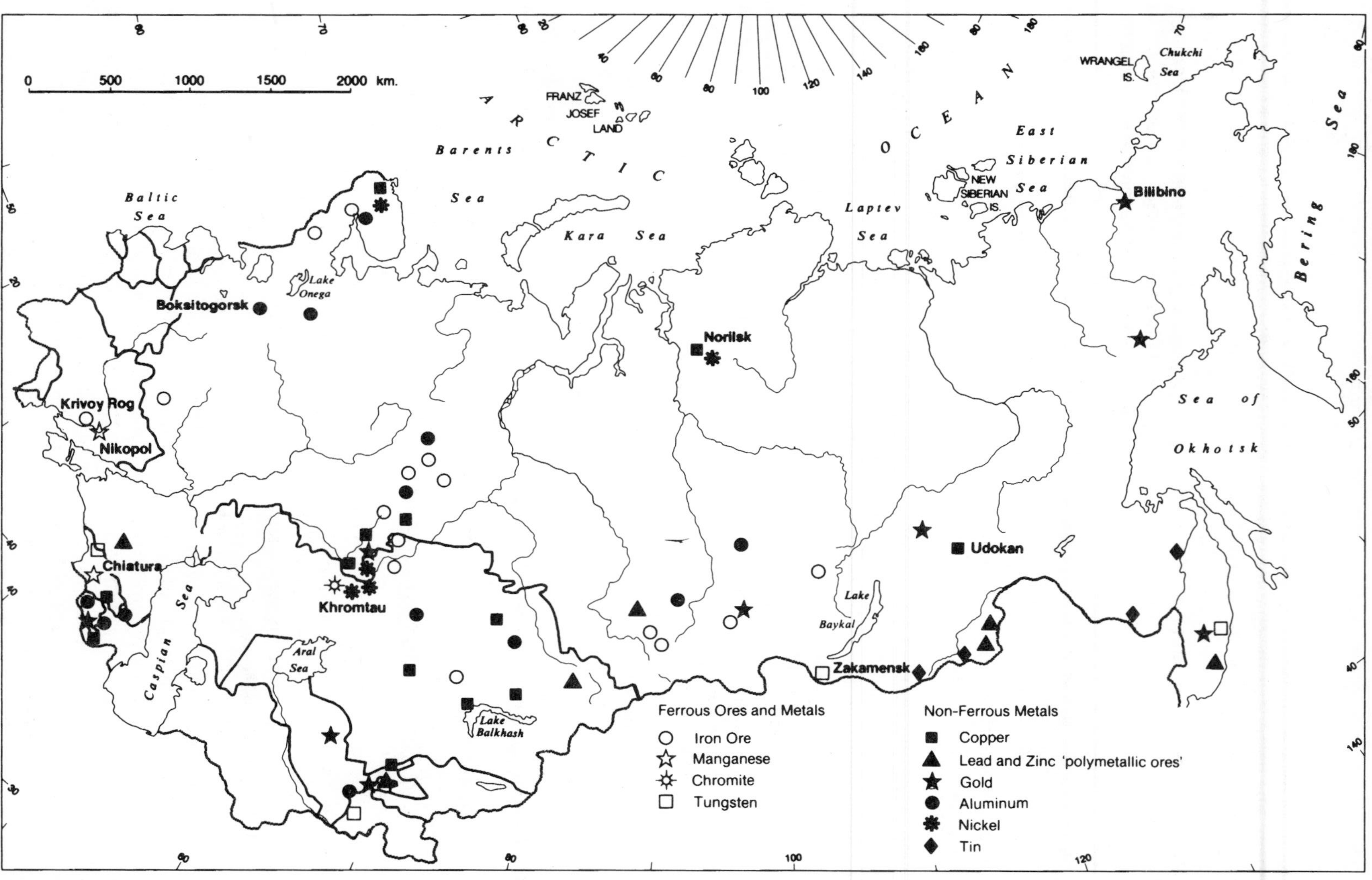

FIGURE 9.1 *Ferrous and non-ferrous metals resources*
(*Source*: Leslie Symons *et al.*, *The Soviet Union. A Systematic Geography* (Totowa, NJ: Barnes and Noble, 1983), 155).

play a part in the changing geography of resource supply (Fig. 9.1). Indeed, from the point of view of strategic location and scale of reserves, the huge magnetite and iron-bearing quartzite deposits associated with the so-called Kursk Magnetic Anomaly (KMA) south of Moscow in central European Russia were of obvious importance. It has about 50 billion tons of reserves. Given that geological exploration of potential iron ore reserves is far from complete, it is entirely possible that the geography of reserves could change again. The most likely consequence of further exploration is that the share of reserves in the eastern regions of Russia beyond the Ural mountains would increase. Russian iron ore reserves are something of a spatial anomaly amongst industrial raw materials, inasmuch as more than three-quarters are found in European Russia and the Urals. For most industrial raw materials, the pattern is reversed. Krivoy Rog in Ukraine is reputed to have about 25 billion tons of reserves. Kazakhstan also has significant iron ore reserves at Sokolvo-Sarbay in the north west near Kustanay. Reserves here are estimated to be about 25 billion tons.

While Fig. 9.1 shows the location of the major iron ore deposits, it does not indicate their relative importance in terms of production. On the eve of the collapse of the Soviet Union, Ukraine was the largest producer of iron ore, roughly 130 million tons annually. The bulk of this came from the Krivoy Rog deposit. Russia ranked second in iron ore output. The mines of the southern Urals region produced about 25 million tons. KMA production in the late 1980s was about 45 million tons. Ore from the KMA has long been transported to the major iron and steel mill at Lipetsk a few hundred kilometres to the east, and even to the steel mills in the Urals more than 1500 kilometres away. As is the case in the other major iron mining operations, output from the KMA is of relatively low iron ore content, and thus must be concentrated before shipping. The Kuznetsk Basin (KUZBAS) region, and Zheleznogorsk further east near Lake Baykal, are the major Siberian iron mining regions. Together their output was around 18 million tons. In Kazakhstan, which came third in production, the mines in the northwest region produced about 27 million tons. The general pattern of iron ore flow from mine to mill on the eve of the collapse of the Soviet Union is portrayed in Fig. 9.2. The average iron ore content of remaining reserves across the post-Soviet scene is a shade more than 30 per cent iron per ton of ore. Owing to various concentration technologies widely in use, ore shipped any distance is customarily about 60 per cent iron content.

While many post-Soviet states have substantial coal deposits, not all of them possess coal of coking quality (see Fig. 8.3, p. 230). Coking coal reserves are abundant in Russia and Kazakhstan, but problematic in Ukraine because of depletion of the richest, most accessible seams in the Donets Basin. In Ukraine, there are some options, however. With the assistance of foreign technology introduced through joint ventures with western firms, millions of tons of coking coal are now being reclaimed annually from previously abandoned mines, slag heaps and mine tailings. This coal is for the most part destined for West European export markets owing both to drastically reduced domestic demand in Ukraine itself, and to Ukraine's urgent need for hard currency revenues. While substantial amounts of high grade coal still remain to be recovered, this is a short-term strategy to compensate for depleted, easily accessible coal seams in the Donets Basin underground mines. For the long term, domestic supplies of coking coal are not auspicious. Compounding the resource supply problem for Ukraine's steel industry is the fact that the most accessible reserves in the long exploited Krivoy Rog iron ore mines in south Ukraine are being depleted, even though substantial reserves remain. Thus, ferrous metallurgy in the post-Soviet scene is secure with respect to adequate domestic supplies of essential iron ore and coking coal resources only in Russia and Kazakhstan. However, in view of the world-wide glut of modern steel making capacity in the western economies, whether there is a future commensurate with the past for the iron and steel industry in any post-Soviet state is very much a moot point.

As Fig. 9.1 indicates, deposits of the other important ferrous metals, manganese, chromite

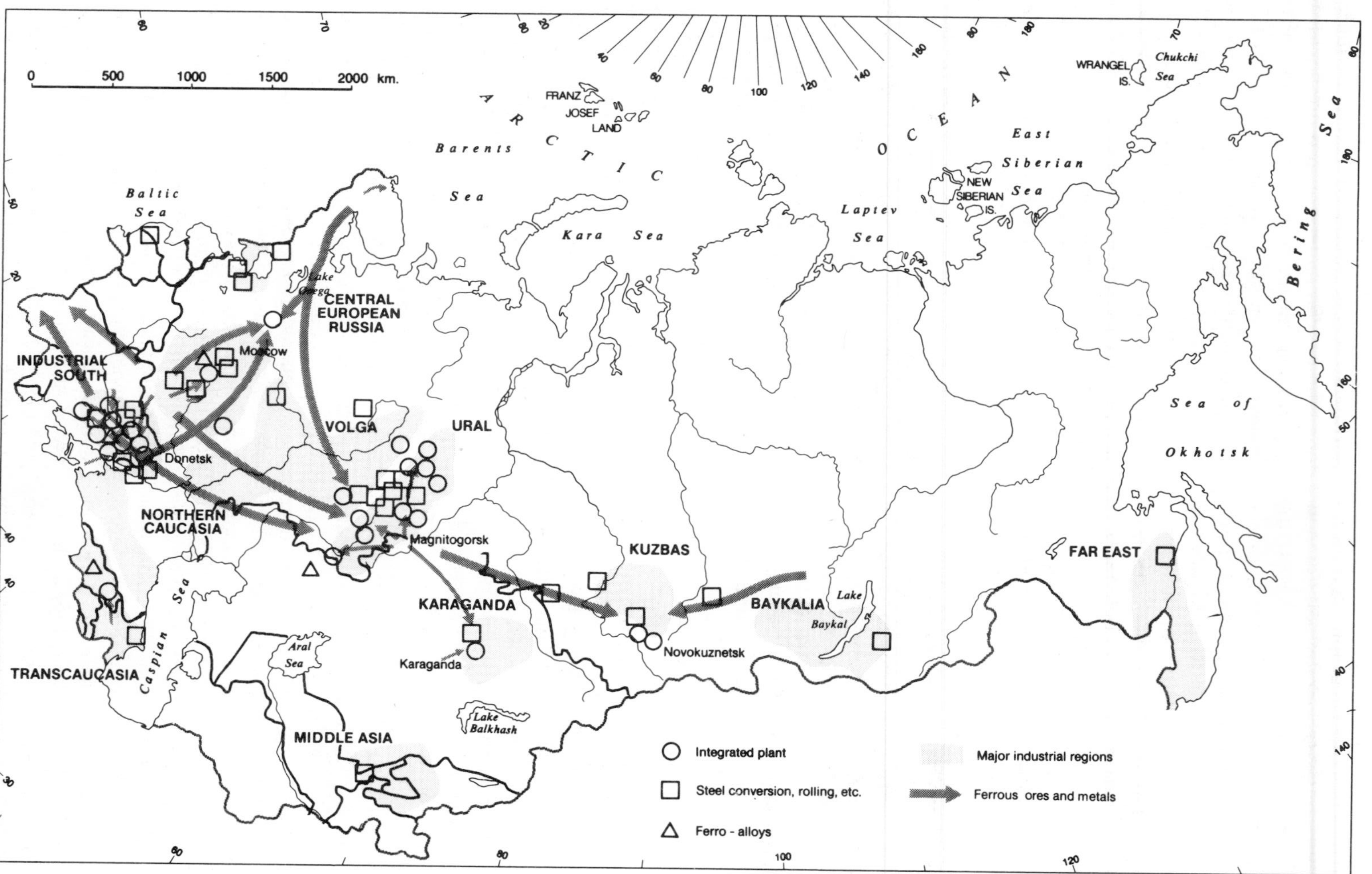

FIGURE 9.2 *Ferrous metallurgical complexes*
(*Source*: Leslie Symons *et al.*, *The Soviet Union. A Systematic Geography* (Totowa, NJ: Barnes and Noble, 1983), 155)

and tungsten, are far flung in distribution. These ferroalloys are essential for the production of steel with specific properties such as strength and conductivity. The major reserve of manganese is located in Ukraine at Nikopol'. Mines here accounted for more than three-quarters of the total annual output of about eight million tons at the end of the Soviet period, down from the peak output of nearly ten million tons in the mid-1980s. The other main deposit is the Chiatura basin in Georgia. First mined in the late nineteenth century, its highest quality ores have long been depleted. It accounted for about one-fifth of output on the eve of the collapse of the Soviet Union. Russia is dependent upon sources in the 'near abroad' for supplies of this vital raw material for the steel industry. However, further exploration of previously marginal reserves is now underway, and much emphasis is being accorded to reducing consumption of manganese during the steel making process. The huge chromite deposit at Kromtau in northwestern Kazakhstan accounted for virtually all of the late Soviet-era production of this important metal, which in the late 1980s ranked first in the world. Total chromite output then was approximately four million tons annually. In the late 1980s, tungsten production ranked second in the world after China. Tungsten is found in several regions. In Russia, which is the largest producer amongst post-Soviet states, it is mined at Zakamensk near Lake Baykal and in the Far East at Vostok-2. Less important deposits are in Georgia, Kazakhstan and Uzbekistan. Since independence, output of these important metals has fallen, but domestic demand has fallen even more. The irony is that exports are now larger than in the past.

The bulk of Soviet-era platinum and nickel output came from the massive non-ferrous mineral reserves near Noril'sk in Russia. With a peak population of more than 180,000 inhabitants, the very size of this mining town located north of the Arctic circle was testimony to its importance in non-ferrous mineral production. Noril'sk has fallen upon hard times since independence, mineral production and population having both declined. In addition to platinum and nickel, mines here also yield substantial amounts of copper, cobalt, gold and silver. Total annual copper output at the end of the Soviet era was approximately one million tons, and production was dominated by output from the Noril'sk mines and those in the Kola peninsula. Some production comes from the smaller mines in central Siberia near Salairskoye, and the older, but still important copper mines in the Urals. Amongst all post-Soviet states, the largest copper deposits are in central Kazakhstan. Substantial output comes from the huge mines near Zhezqazghan (Dzhezkazgan). There are some smaller copper mines in Georgia, Azerbaijan and Uzbekistan (Fig. 9.1). Massive reserves of lead and zinc are located in Kazakhstan near Oskemen (Ust-Kamenogorsk). At the close of the Soviet era, lead output was close to 600,000 tons annually, and ranked first in the world. Over half of the output came from Kazakhstan alone. Production also occurs in Uzbekistan at Almalyk. In Russia, production occurs at Vladikavkaz (Ordzhonikidze) in the northern Caucasus. Potential lead resources to the east and north of Lake Baykal in Russia are apparently very large. Total zinc production in the late 1980s was around 850,000 tons. In 1993, Russia produced about 40,000 tons of primary lead and 160,000 tons of primary zinc, both up slightly from the previous year but still well below peak output in the Soviet era. Tin reserves across the former Soviet Union were uncharacteristically meagre. Tin production ranked second in the world after Malaysia before independence. However, to meet domestic needs, tin had to be imported. Tin oxide is mined in Eastern Siberia and the Far East economic regions of Russia. Since independence, production of tin has declined, but again not as much as domestic demand. Thus, Russia now exports some tin. More significant gaps between production and consumption of metals such as copper and nickel have appeared, thus affording additional export opportunities.

Raw material for aluminium production is found in a number of post-Soviet states. However, substantial volumes of bauxite ore and alumina (derived from bauxite and from which aluminium is smelted) were imported for much of the Soviet period. Close to two-fifths of total aluminium production in the late Soviet era, which was second in the world after the United States, was based on imported raw materials. Bauxite is mined in a number of places, but most notably in northwestern Russia at Boksitogorsk (Fig. 9.1). Import dependence was in some respects a matter of choice rather than necessity, for alumina may be derived from materials other than bauxite. While nepheline, a by-product of apatite production, and kaolin, for example, can be substituted for bauxite, the technology involved is more complex. Production of aluminium is further complicated since the reduction of alumina requires very substantial amounts of electricity. A number of major hydroelectric power stations were developed on the premise that alumina reduction plants would comprise a major part of the industrial demand for their output. This is especially so for the Siberian hydroelectric power stations discussed in the preceding chapter. At its Soviet-era peak, annual production was close to three million metric tons. Amongst non-ferrous metals, aluminium ranked first in terms of tonnage, and accounted for about a third of exports by value. Since independence, costs of electricity have been increased substantially, as have costs of transportation. For the huge aluminium plants in central Siberia, these alone are very serious problems, but they are compounded by the fact that much of the domestic demand in the Soviet era was defence industry related and this has dropped precipitously in the post-Soviet period. Producers have turned to the export market. Output of aluminium in Russia was 2.9 million tons in 1993, up from 2.7 million in 1992. This is much in excess of domestic consumption, thus exports now exceed one million tons annually. The appearance on the world market of such large amounts of aluminium in recent years has led to charges that Russia is dumping, that is, selling commodities on the world market at prices which do not reflect real costs of production. Of course, Russia is a major exporter of aluminium in order to earn hard currency. But it is now having to export more to earn the same income. Between the late 1980s and 1993, the world price per ton of aluminium has dropped by more than half, in large measure because of ascending Russian exports.

Gold was long a major source of hard-currency earnings, and annual output was reckoned to be in the order of 325 metric tons at the end of the Soviet era. As Fig. 9.1 indicates, it is found in a number of locations. The mines in northeast Siberia in Russia, of which Bilibino is the major one, accounted for as much as a quarter of total production. Historically, mining gold underground and extracting it from alluvial gravels was linked to the use of forced labour, especially in the more remote regions of northeast Siberia. Placer operations on many of the rivers of Eastern Siberia continue to yield substantial amounts of gold. In 1993, Russian gold production was 130 tons, down a shade more than 10 per cent compared to 1992. Amongst the other post-Soviet states, Kazakhstan ranks second in gold production. In Kazakhstan, gold production has been almost entirely related to lead, zinc and copper output. It was produced as one of several by-products. In recent years, exploration for gold has resulted in some potentially significant finds. For instance, in 1995 a new gold mining operation was set up in remote eastern Kyrgyzstan as a result of an international joint venture. This is only one example of the international interest in the mineral resource potential of post-Soviet states.

Of all post-Soviet states, Russia's reserves of non-metallic materials are by far the greatest. For example, very large deposits of apatite are found in Russia's Kola peninsula. Apatite is an important input for the phosphate fertilizer industry. Substantial quantities are exported as well. The former Soviet Union was a major diamond producer, second only to Zaire. Diamond mines in Russia, for the most part located in remote and difficult to access areas in Sakha Republic in Eastern Siberia, accounted for most of the production. Prior to the discovery of these deposits in the late 1940s, diamonds came mostly from placer operations in the western Urals. Russia now is part of the international South African DeBeers cartel which plays a major role in controlling the supply of diamonds on the world market. In 1993, Russia produced about eight million carats of gem quality diamonds. Asbestos is the other major non-metallic material produced in significant quantity in Russia. Mines in the central Urals dominate output. Indeed, total production at the end of the Soviet period was the largest in the world, and roughly a quarter of it was exported. Substantial deposits of asbestos have been found in Eastern Siberia.

Aside from chemical raw materials derived as by-products, for example from the petro-chemical industry, there are some important elemental chemicals which are produced in significant amounts. In this category of industrial raw materials, Russia, uncharacteristically, does not figure prominently. For example, phosphate is principally produced in Kazakhstan. Plants here accounted for more than four-fifths of the estimated total annual output of around 600,000 metric tons at the close of the Soviet period. The principal market for Kazakhstan's present-day production is the phosphate fertilizer industries. As fertilizer output across the post-Soviet scene has plummeted, so therefore has production of phosphate. However, given the need to enhance yields in agriculture in all these countries, demand will no doubt increase in the years ahead. Potash also figures prominently in the fertilizer industry. Demand, and hence production, have similarly dropped since independence. Potash is mined primarily in the Urals region of Russia, and in Belarus'. Along with ammonia and urea, potash ranked among the top three chemical-based exports in terms of value in the late Soviet period.

Russia's forests represent nearly one-quarter of the world's total stock, and about half of the world's coniferous forests. As Fig. 9.3 indicates, with the exception of the eastern flank of Belarus', and most of northern and eastern Russia, all post-Soviet states are classified as regions with limited commercial forest, and inadequate local timber supply. Of course, not all regions possess the requisite natural conditions to support forest growth. But in many places where conditions are appropriate for growth, the forest cover has been much reduced, or removed entirely. This is the result of decades, if not centuries, of overcutting and inadequate reforestation. Even in Russia, the more accessible forest reserves can no longer sustain past timber harvesting practices. Over the years, logging operations have been pushed further north and east into zones of more marginal stocks and harsher climate. In the absence of adequate reforestation programmes during most of the Soviet era, environmental degradation in the form of erosion of soil and depletion of natural habitat for fauna is widespread. Toward the end of the Soviet period, there was concerted effort to reduce waste in logging operations, and to pursue reforestation programmes more seriously.

Forest products have long played a role in generating hard currency income. Exports traditionally were predominantly unprocessed roundwood and sawnwood, as opposed to higher value products such as plywood, particle board or paper. There was an attempt to shift the emphasis from low value to higher value forest products for export. Indeed, as several huge mills in Siberia, notably at Bratsk and Ust-Ilimsk, came on stream, exports of sulphate pulp surged. This development was a significant step towards a higher value product for export. Nonetheless, as late as the mid-1980s at least a quarter of all exports by value was still in the form of roundwood, sawnwood, pit props and pulpwood. Much of this raw material was destined for hard-currency markets, especially in Western Europe and Japan.

The collapse of the Soviet system has reduced the domestic demand for wood products, and exports have dropped off as well. The total volume of wood harvested in Russia in 1989 was about 290 million cubic metres. In 1993, it was 165 million cubic metres. Wood exports to the new states of the former Soviet Union in 1993 were barely one-tenth of the 1989 level. Exports abroad have similarly shrunk from 15 per cent of the larger harvest in 1989, to 10 per cent of a smaller one in 1993. The economics of wood harvesting are much dependent upon the cost of transportation. Since 1991, rail haulage rates have increased significantly. The higher cost of transportation has made some logging operations in European Russia unprofitable in terms of exporting output even to nearby Finland. More distant forestry operations are even more high cost, and thus competitively disadvantaged. Perhaps the one positive side to this new economic reality is that a reduced harvest may have some beneficial consequences in terms of forest resource conservation. However, some observers fear that in the current period of economic hardship the temptation will be to strip forest resources for short-term gain, rather than adopt more conservation-oriented management practices. Ensuring a sustainable yield from Russian forests over the coming years remains a basic problem. The market reform has already brought more realistic prices for forest resources, and should help to reduce waste in harvesting which has been deplorably high. More scientific management of forests, and continuing improvements in the technology of the forest products industry itself, can bring other benefits. These are important initiatives because most industrial raw materials are non-renewable resources, whereas forests properly managed are sustainable. As Fig. 9.3 indicates, the eastern regions of Russia will figure prominently as a source of raw material for the forest products industry for years to come.

Patterns of industry

As Fig. 9.2 indicates, ferrous metallurgy is an integral part of the post-Soviet industrial scene. But nowhere are these plants producing anywhere near the output of the earlier era. For example, from 1991 to 1994, output from the plants in Russia and Ukraine has dropped from 55 to 36, and from 33 to 17 million tons of rolled steel respectively. As yet there is little indication that there will be a quick turn around. The concentration of plants around Moscow in the Central European Russia region reflects market orientation, since the bulk of the raw materials are imported from other regions. The main centres of iron and steel production are Tula and Lipetsk, while fabricating plants are found in most major centres in the region. St Petersburg boasts a fair complement of steel fabrication plants. Most of the steel comes from the Cherepovets mills which were built in the mid-1950s to supply the St Petersburg and Baltic region. Located adjacent to the Rybinsk reservoir, Cherepovets is one of the country's largest plants. It relies on iron ore from deposits in Karelia and the Kola peninsula, and coking coal from the Pechora region in the European north. The iron and steel plants in the Ural region, which up until recently produced about two-fifths of total steel production, are now dependent upon both imported coking coal and iron ore. As was noted earlier, the deficiency in coking coal was of historical significance. The need to rely on non-local reserves of iron ore is a more recent development. The principal movements of ore are depicted in Fig. 9.2. Most of the iron and steel mills are located along the highly mineralized eastern flank of the Ural mountains. It is only in the steppe environment of the southern Ural region that adequate fresh water supply has posed a problem. Hydraulic engineers produced a technological solution to the problem of inadequate or unpredictable water supply. The solution was to dam the Ural river thereby creating a reservoir of sufficient size to meet the requirements for steel production. Further east, the iron and steel complex in the KUZBAS was a key element in the Urals Kuznetsk Combine, as was noted in an earlier chapter. Notwithstanding the dependence on imported ore (Fig. 9.2), the region possesses some of Russia's largest mills. Both the integrated iron and steel plant and the sheet mill in Novokuznetsk rank amongst the world's largest. Smaller iron and steel mills have been in operation for many years at Petrovsk-Zabaykalskiy near Lake Baykal and at Komsomol'sk-na-Amure in the Far East. They convert local scrap and imported pig iron into steel.

Ukraine accounted for the second largest share of Soviet-era steel production. In the mid-1980s, its steel mills based on Donets basin (Donbas) coking coal and Krivoy Rog iron ore turned out close to two-fifths of total output. Although Fig. 9.2 suggests a spatially continuous complex of iron and steel plants in the 'industrial south', that is, Ukraine, there are complexes both in the east in the Donbas coking coal region, and in the west centred on the iron ore deposits of Krivoy Rog. There are some intra-regional specializations in steel fabrication, but perhaps the most significant factor distinguishing the western and eastern flanks of Ukraine's iron and

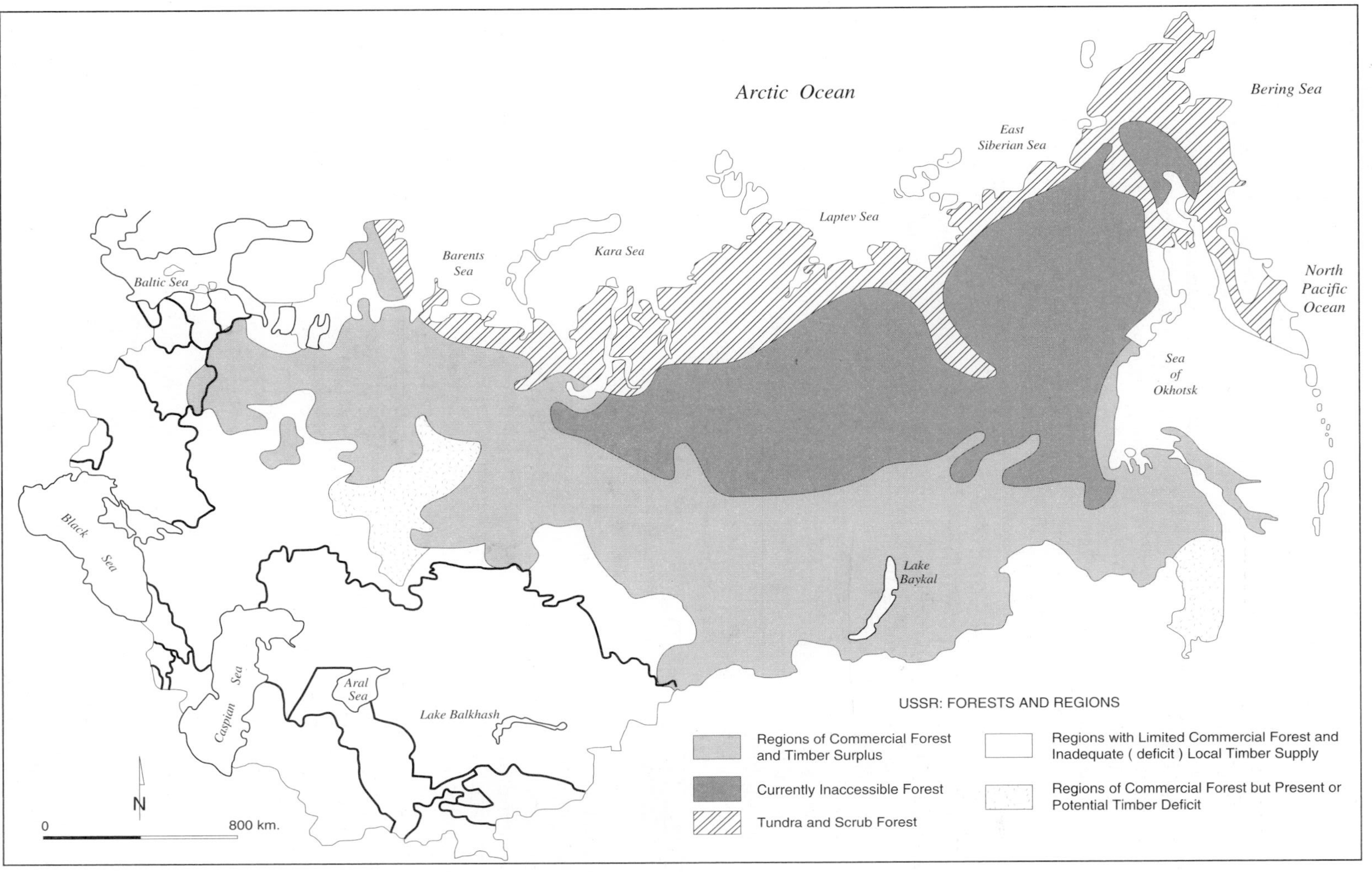

FIGURE 9.3 *Forest resources*
(*Source*: B.M. Barr, 'Regional Dilemmas and International Prospects in the Soviet Timber Industry', in Jensen, R.G., Shabad, T. and Wright, A.W. (eds), *Soviet Natural Resources in the World Economy* (Chicago: University of Chicago Press, 1983), 428)

steel complexes is water supply. Fresh water is a vital input in the iron and steel industry, and supplies are increasingly problematic in the steppe region of the eastern part of Ukraine, whereas in the west the Dnepr' river provides an assured source. This was rather simpler than the solution adopted in Kazakhstan where a several hundred kilometres long canal was created to bring Irtysh river water to the mills located in the arid steppe around Karaganda.

The existence of small steel mills in other parts of the post-Soviet scene is the legacy of the self-sufficiency principle of Soviet era regional development planning. A small plant at Liyepaya on the Baltic Sea coast in Latvia converts scrap and imported pig iron into steel. In Georgia, the mill at Rustavi near the capital city of Tbilisi began operation in the mid-1950s. Aside from supplying local needs, steel from Rustavi was shipped to the rolling mills at Baku in Azerbaijan in order to complement the output from the small steel plant nearby at Sumgait. In 1994, production of rolled steel in Georgia was below 100,000 tons, an eighth of what it was at independence. Azerbaijan's output was just 30,000 tons, a mere one-fifteenth of the 1991 figure. In Uzbekistan, steel is produced at Bekabad near Tashkent. The Bekabad plant was designed to convert local scrap and imported pig iron into steel. Production in 1994 was about one-third the level of output in 1991. At the huge mills at Karaganda and Temirtau in Kazakhstan rolled steel production is less than a half of 1991. As the world's largest steel producer, the Soviet Union's output of rolled steel was about 130 million tons in the mid-1980s. In the mid-1990s, total production in all post-Soviet states was less than 60 million tons.

The production of non-ferrous metals differs from ferrous metallurgy in at least two important respects. Firstly, total annual output of all such metals is but a small fraction of the millions of tons of steel produced, even at its present-day much reduced level. Secondly, the mining of non-ferrous metals (Fig. 9.1) is seldom coincident with the production of metals, the latter process being market-oriented or, in the case of aluminium, locationally tied to major sources of electricity (Fig. 9.4). There is a closer spatial link between the location of basic raw materials and the production of steel (see Fig. 8.3, p. 230, and Fig. 9.1). While there was some attention given to non-ferrous metallurgy in the industrialization drive of the late nineteenth century, for the most part domestic needs were met by imports from abroad. The Stalin-era industrialization programme attached high priority to the development of a non-ferrous metallurgical industry. Copper refining in the late nineteenth century was localized primarily in the Urals and in the Caucasus. Production in the Urals was given a fillip in the early Five Year Plans, but after World War II the locus of production shifted to Kazakhstan, especially to Zhezqazghan (Dzhezkazgan) where a major smelter and refining complex came on stream in the early 1970s (Fig. 9.4). Copper is also produced in Uzbekistan and Armenia, where there are both mines and smelters. But the largest quantities of copper are obtained from non-ferrous metallurgical complexes in Russia at Noril'sk and in the Kola peninsula, and in Kazakhstan at Oskemen (Ust-Kamenogorsk).

Nickel smelters are located in a number of places in Russia, but the largest production comes from plants at Noril'sk and in the Kola peninsula. Lead and zinc ores are widespread, but smelting has been traditionally associated with the metallurgical complexes in Russia and Ukraine. Smelting technology has seen coal and coke replaced by natural gas and electricity, and thus the production of lead and zinc metals has shifted to different locations. Among the major refineries outside Russia are those in Kazakhstan at Oskemen (Ust-Kamenogorsk) and Shymkent (Chimkent), and at Almalyk in Uzbekistan. Tin smelting occurs near Moscow and at Novosibirsk in Russia. The very promising reserves in the Far East have resulted in the further development of tin mining, concentration of ore and construction of a new smelting complex at Solnechniy near Komsomol'sk-na-Amure. Aluminium refining, as noted, has been closely associated with major electric power installations, especially hydro, since the reduction of alumina is so heavily electricity intensive. Production of aluminium occurs in a number of locations, as Fig. 9.4 indicates. But the bulk of output comes from the Siberian refineries, notably Shelekhov, Bratsk, Krasnoyarsk

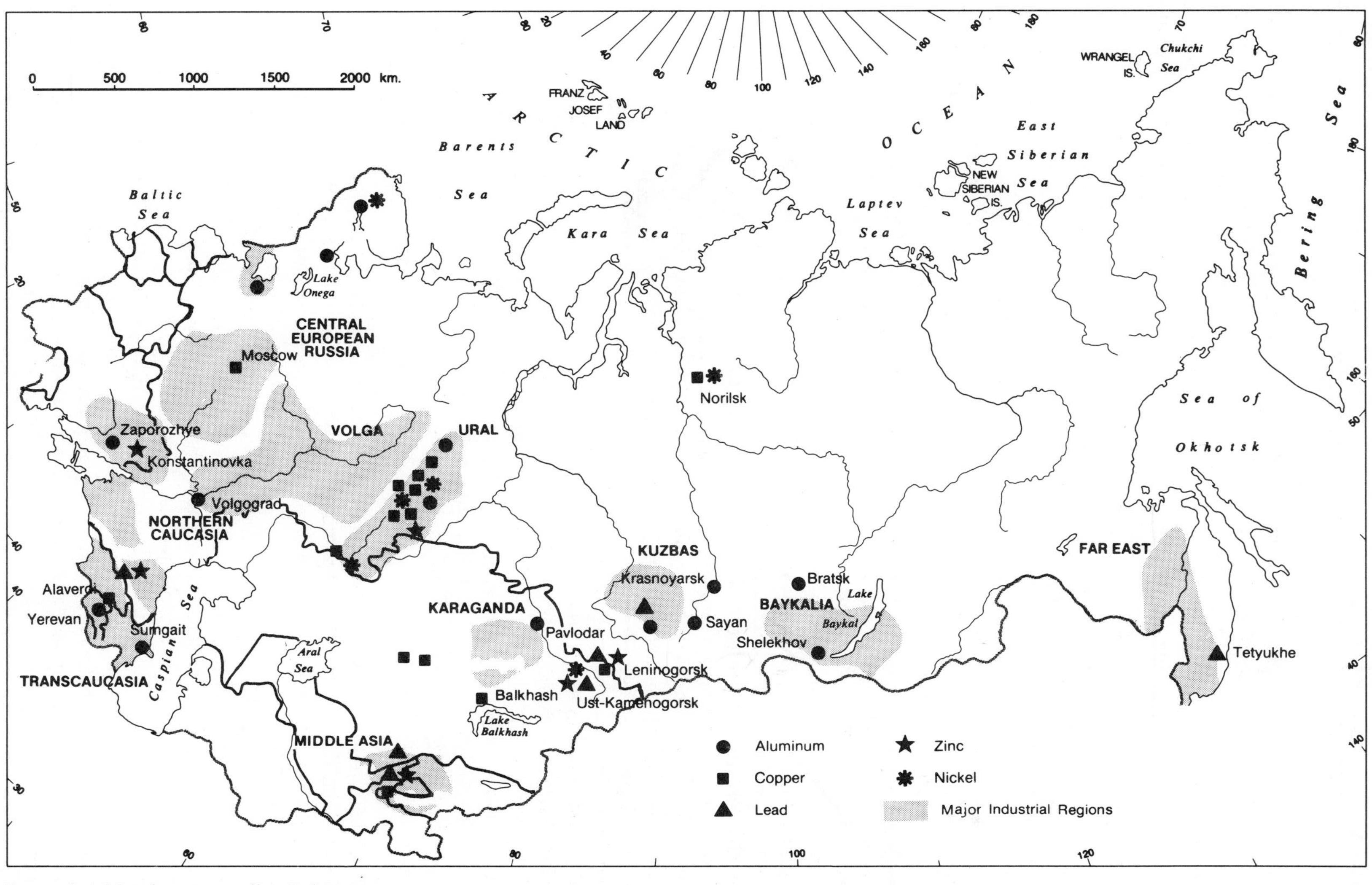

FIGURE 9.4 *Non-ferrous metallurgical centres*
(*Source*: Based on J.C. Dewdney, *The USSR* (London: Hutchinson, 1978), 93)

and more recently the Sayan plant on the upper Yenisey river. Each is associated with a major hydroelectric power station. To date, Siberian production of alumina is inadequate to meet the aggregate demand of these refineries and therefore long rail haulage from sources in European Russia has been required.

Ferrous and non-ferrous metals are themselves inputs for a vast array of production processes, not least of which are the engineering industries. These have been developed in many regions of the country, as the data on machine tools, transport and agriculture equipment portrayed in Fig. 9.5 clearly indicate. Some of the key chemical and related synthetic products industries are also shown on this map. Taken together Fig. 9.2 showing ferrous metallurgy, Fig. 9.4 depicting non-ferrous metallurgy, and Fig. 9.5 indicate the locations of the major concentrations of heavy industry. What is missing are the many industries which comprise the consumer goods sector, a sector which has consistently received only a small share of the total investment capital allocated to industry. From food products, to textiles, to toys, literally tens of thousands of goods are produced each year for the consumer. In terms of location, these industries tend to be market-oriented, that is, close to the major centres of population concentration.

The most important region in terms of the total value of all industrial production is that roughly delimited by Central European Russia and the St Petersburg regions (see Fig. 9.5). Much of the industry in these regions is market-oriented and relatively labour-intensive. In the late 1980s, they accounted for almost one-quarter of the total value of Soviet industrial production. Although production has been negatively impacted by the prevailing economic crisis in the years since independence, the Central European Russia and St Petersburg regions remain very important. As well as having a reasonable complement of those industries turning out capital goods, industrial raw materials and energy, these two regions have a disproportionate share of Russia's consumer goods industries. Industrial development in the Urals was given a fillip with the emergence of the Urals–Kuznetsk Combine in the 1930s. Industry in the Volga and Baykalia regions is largely the result of post-World War II development programmes and is heavily skewed toward their energy bases, respectively oil and gas and electricity. Consumer-oriented industries are also commonplace in the Baltic zone (Fig. 9.5). Ukraine is principally a region of heavy industry, and in the Soviet period accounted for roughly one-fifth of the total industrial output by value. The remaining industrial regions in post-Soviet states shown in Fig. 9.5 typically have a reasonable mixture of consumer goods and heavy industries, though it should be noted that the southern tier countries tend to have a higher share of consumer goods manufacturing than they do heavy industry.

The functional integration of different regions of the former Soviet Union in 1989 is highlighted by the degree of import dependence for a selection of key industrial sectors depicted by Figs 9.6–9.8. For light industry, not surprisingly, import dependence was not especially large. However, from machinery, to ferrous metals, to non-ferrous metals, to chemicals, the dependency steadily increases. Clearly what these diagrams imply is that there was a very substantial movement of materials within the former Soviet Union. In large measure this reflects the imbalance between the location of basic industrial resources and the location of centres of demand. But it also is a reflection of the basic economic development strategy employed by the state which entailed concentrating some types of production in particular locations in order to achieve economies of scale. The consequences for the transport system are obvious enough. Huge volumes of materials moved between regions. Figure 9.9 depicts the importance of trade between the Soviet republics as a per cent of the gross domestic product in each republic in 1989. The Soviet Union in its final years was a highly integrated economic system.

The composition of industry in all regions represents some degree of regional self-sufficiency and specialization. But since none were, or are now, self-sufficient, complementarity of industrial structures requires transportation systems to facilitate inter-regional commodity flows. The

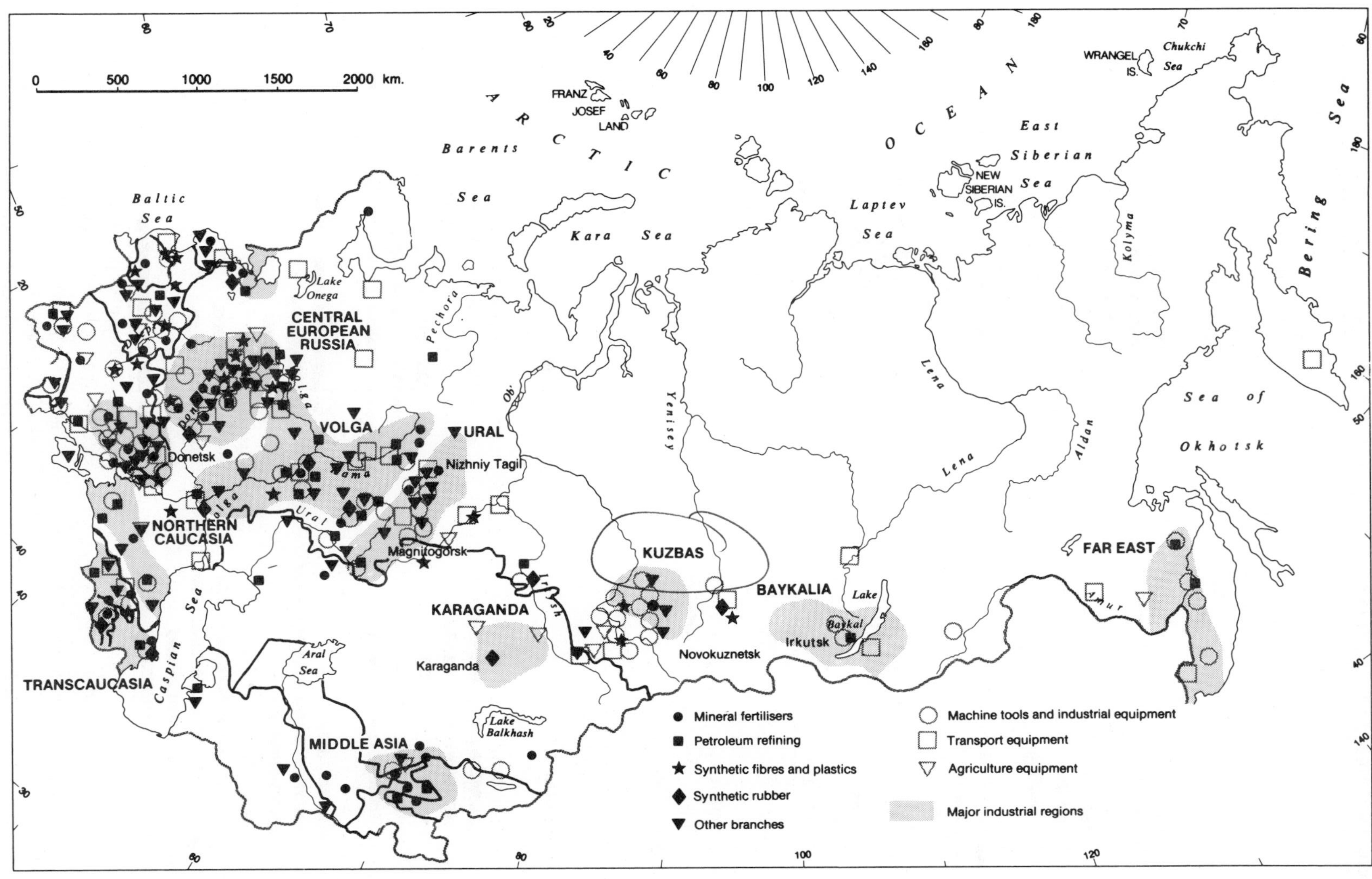

FIGURE 9.5 *Engineering and chemical industries*
(*Source*: L. Symons et al., The Soviet Union. A Systematic Geography (Totowa, NJ: Barnes and Noble, 1983), 170, 173)

Light Industry Import Dependence, 1989

Machinery Import Dependence, 1989

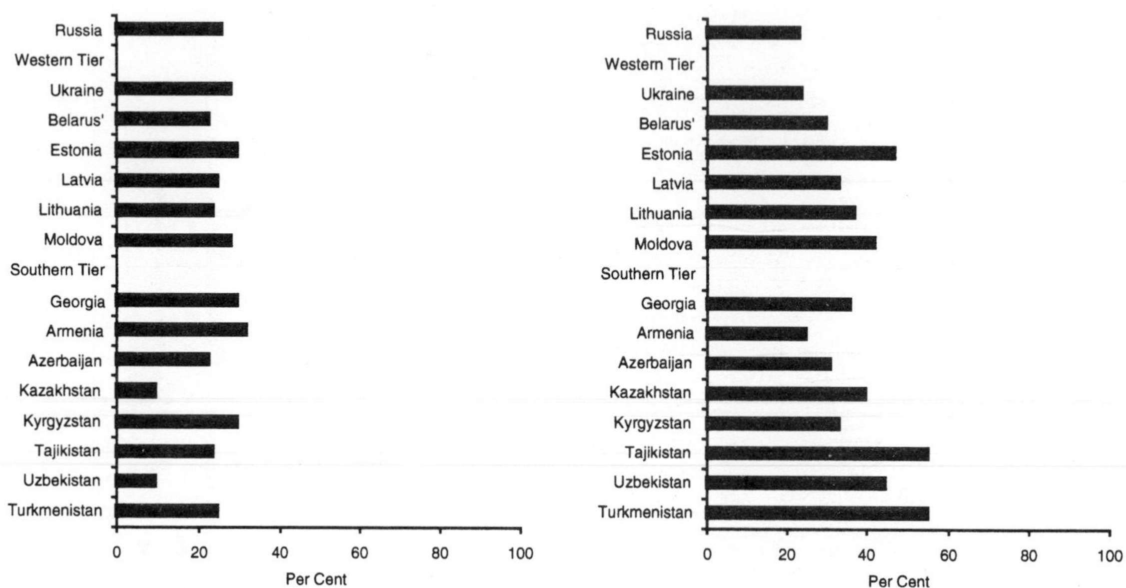

FIGURE 9.6 *Import dependence by Soviet republic, 1989: light industry and machinery*
(*Source*: Based on R.A. Watson, 'Interrepublic Trade in the Former Soviet Union: Structure and Implications', *Post-Soviet Geography*, Vol. 35, No. 7, 1994, 378–79)

Nonferrous Metals Import Dependence, 1989

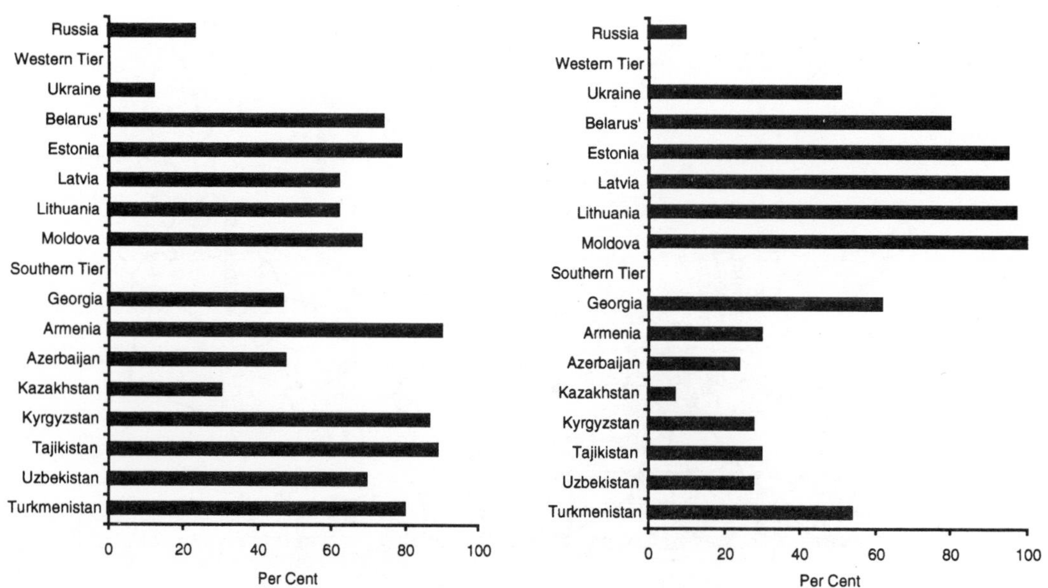

FIGURE 9.7 *Import dependence by Soviet republic, 1989: ferrous and non-ferrous metals*
(*Source*: Based on R.A. Watson, 'Interrepublic Trade in the Former Soviet Union: Structure and Implications', *Post-Soviet Geography*, Vol. 35, No. 7, 1994, 378–79)

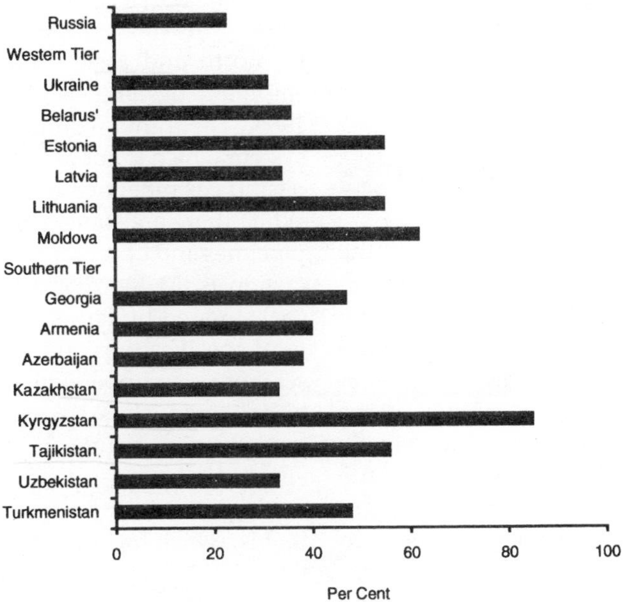

FIGURE 9.8 *Import dependence by Soviet republic, 1989: chemicals*
(*Source*: Based on R.A. Watson, 'Interrepublic Trade in the Former Soviet Union: Structure and Implications', *Post-Soviet Geography*, Vol. 35, No. 7, 1994, 378–79)

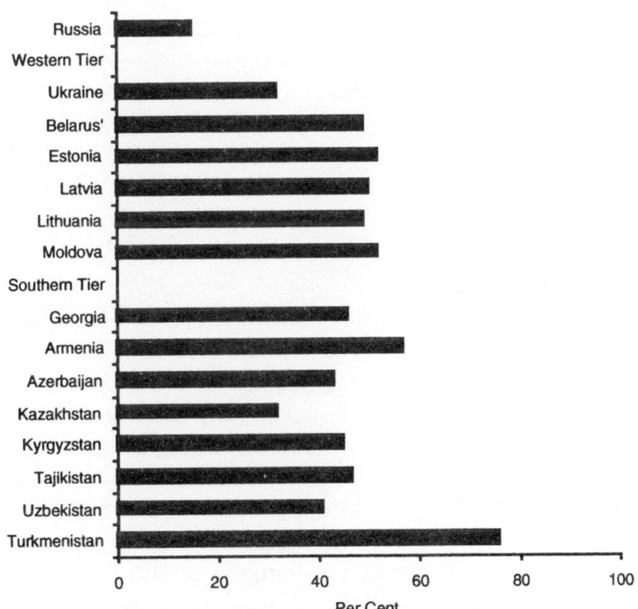

FIGURE 9.9 *Trade between Soviet republics as a percentage of Gross Domestic Product (GDP), 1989*
(*Source*: Compiled from data in R. Sakwa, *Russian Politics and Society* (London: Routledge, 1993), 208

transport system had to be designed to accommodate these substantial internal flows of raw materials and manufactured goods. The geography of transportation merits separate, if brief, consideration at this point.

The geography of transport

To integrate the many different regions of the Russian Empire and then the Soviet Union, required more than just political power. An efficient transportation system was an essential prerequisite for both political, and economic, integration. In the nineteenth century, the coming of the railroad permitted for the first time comparatively efficient movement of goods and people. But the railroad was not at first welcome. It was comparatively slow to develop in Russia, as noted in Chapter 2, because of concerns of the government and elites in society over the potentially disruptive consequences of facilitating the movement of people across the country in general, and from the countryside to the town in particular. Nonetheless, in regions where the railroad did develop, traffic on the pre-existing system of post roads, or *trakty*, the rivers, and the canals of the Russian Empire era was soon eclipsed in relative terms. On the eve of the Russian revolution, the country had a fairly extensive rail system, but movement of goods and people in many regions remained completely dependent on outdated modes of transport. After the revolution, transport system development was a more important national priority than it had been in the preceding era. The transport system which emerged was part of a national economic development strategy.

Something of the spatially complementary nature of the road, rail and water transport systems developed during the Soviet era is conveyed by Fig. 9.10. Clearly, much of the country did not have ready access to transport by land or by water. What is missing from this picture is, firstly, the extensive network of air routes. These connected some 3600 settlements of all sizes in all regions. In the remote far north and east air freight still plays an important part in supplying settlements, though to be sure the absolute volume pales in comparison to that moved by land. In the late Soviet era, about three-fifths of all air freight was still represented by shipments in the far north and east. Average distance of goods shipment by air was about 2000 kilometres in 1989. The other component of the transport system which is not represented in Fig. 9.10 is the oil and gas pipeline network (see Fig. 8.5, p. 235). While there are some possible tradeoffs between pipelines and other forms of goods transport, discretion is limited. For example, oil may replace coal shipped by rail or water, but once the technical specifications and route of the pipeline are set there is an upper limit on the quantity that can be delivered. Thus, in creating a unified transport system very careful consideration had to be given to the best use for each mode, and to the determination of the most effective modal splits, that is, which form of transport to use when and where.

Unlike Western Europe or North America where duplication of modes, and competition between them, provides a type of safety net in ensuring that products and people do get to their destinations more or less on time, during the Soviet era there was much less choice as to how shipments might occur. The whole transport system was for example 'inward looking', in that it was developed with a view to meeting domestic needs. Little attention, or certainly less attention, was accorded linkages with other countries. This feature of the Soviet transportation system further complicates the efforts of post-Soviet states to realign their domestic economies, to reform their economies. For example, there has long been discussion of developing a land bridge across Russia. This would facilitate the rapid transport of containerized goods between the Atlantic and Pacific realms. However, realization of the land bridge concept has been frustrated because there are inadequate facilities to handle containers at the major ports, and more critically perhaps, because there is inadequate infrastructure, including rail equipment and trucks, to handle container shipments across the country. The trans-Siberian railroad has long been one of the most heavily used railroads in the world. Adding an unpredictable

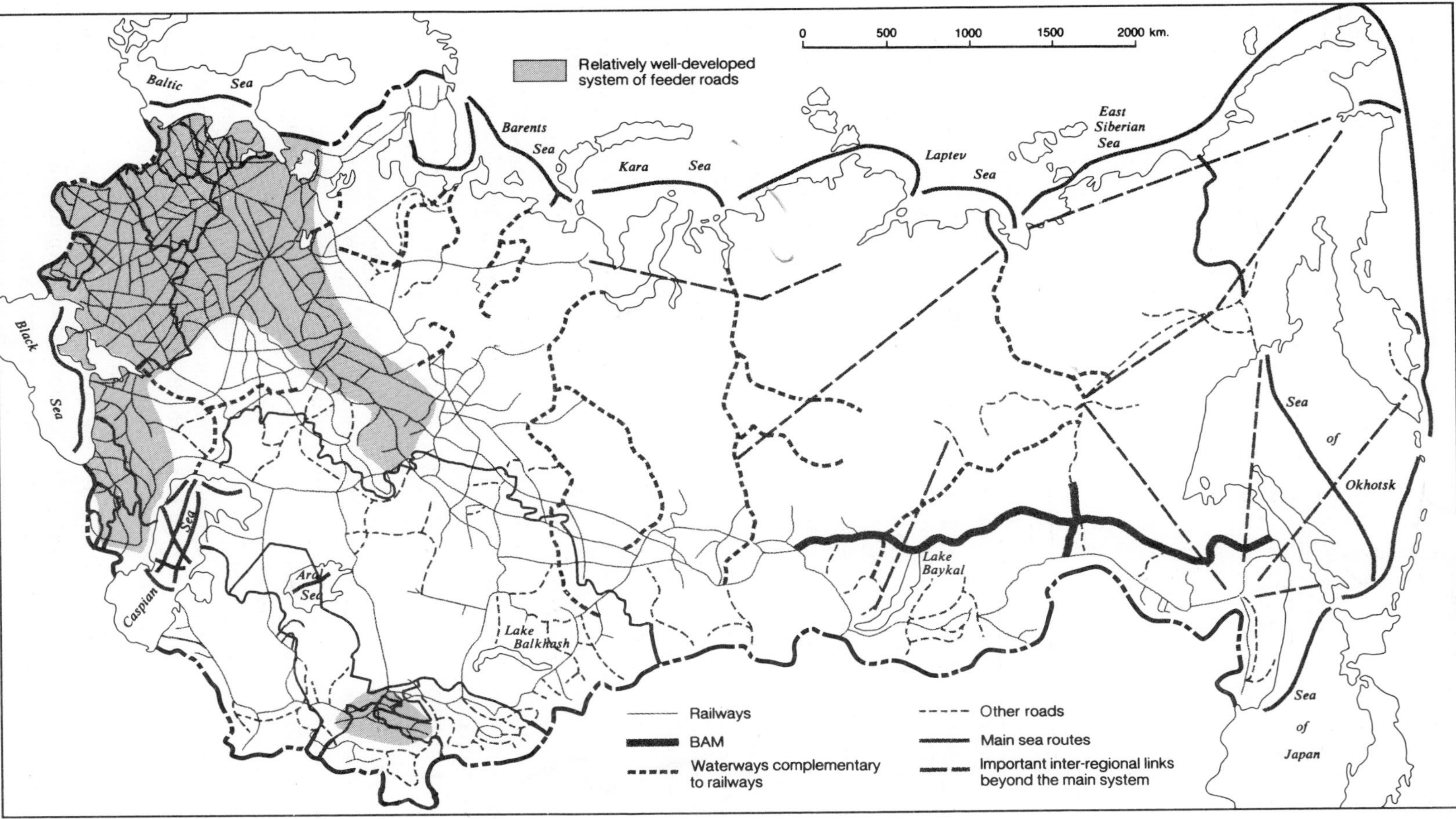

FIGURE 9.10 *Transportation system*
(*Source*: A. Brown *et al.* (eds), *The Cambridge Encyclopedia of Russia and the Soviet Union* (Cambridge: CUP, 1982), 351)

trans-national volume of traffic to a system which in the past has been barely able to cope with domestic traffic volumes, obviously poses problems. The completion of the Baykal-Amur railroad late in the Soviet era was supposed to relieve the pressure on the rail system in the Russian Far East, but it is plagued by technical problems related to both the harsh environment and shoddy construction.

As Fig. 9.10 indicates, the European Russian core is especially favoured in terms of the density of transportation systems compared to the situation in most other post-Soviet states. Given the distribution of population and economic activities, the greater development of the transportation system in European Russia is scarcely surprising. But the transportation system in the eastern regions of Russia should not be underestimated in terms of its impact on the country as a whole simply because it is so sparse in comparison with European Russia. As has been emphasized, it is from these regions that so many of the essential raw materials for industry must now be obtained. While in Russia it is obvious that there is a major cost associated with overcoming vast distances, in all post-Soviet states there is the compelling need to properly assess the real costs of goods transport in the new economy. Irrespective of the territory of each state, this adjustment from the heavily subsidized transportation rates of the Soviet era to the higher rates necessary in market-based economies must still be absorbed. This remains a major challenge at the present time.

Within the Soviet scheme for transport of goods, the road system was viewed as providing a feeder service to rail and water routes. The average distance of goods haulage by road on a per ton basis was only about 21 kilometres in 1989, less than one-twentieth the figure for the United States. Rail and waterborne shipments were 962 and 346 kilometres respectively at this date. While trucks were supposed to dominate in short haul goods shipment, in fact probably close to one-sixth of all rail shipments were only 100 kilometres, or less. This is partially explained by the general shortage of goods-hauling vehicles, especially multi-axle, long distance transport trucks, a shortage which persists to the present day. Thus, what was supposed to be a rational modal split between rail and road short-haul goods haulage was confounded by deficiencies in the supply of the necessary equipment. The general dearth of trucks of any description is typically exacerbated at harvest time when there is a widespread requisitioning of all suitable vehicles from rural and urban enterprises to assist in moving crops from farms to rail terminals. A second reason for the feeder, as opposed to long distance, transport role of trucks is related to the comparatively underdeveloped state of the system of roads. For instance, with more than two and one-half times the territory of the United States, the Soviet Union in 1991 still had less than one-sixth of its hard-surface road network. Moreover, there was no comparable system of high quality, multi-lane, interstate highways which so greatly facilitates rapid shipment of goods within and between regions in the United States, and with Canada. While over the years Soviet road and air transport became more important in moving goods and people, they were viewed as complementary rather than competitive modes. The principal exception to the planned short-haul focus of goods movement by road was in the far north and eastern regions of Russia. For example, the Aldan Highway links the trans-Siberian railroad at Never to Yakutsk, a distance of roughly 1000 kilometres. In terrain such as the Siberian upland, the cost of railroad construction is clearly prohibitive. While the comparatively limited system of all-weather roads is complemented by river steamers and barges in the summer, the harsh climate severely limits the shipping season. And the harsh climate renders most dirt roads, which comprise a very sizeable share of the total system, impassable for certain times of the year as well. In the northern regions a costly but necessary adjunct to the transport system is the extensive network of winter roads which ensures a lifeline to the numerous remote northern settlements. With the collapse of the Soviet system, continuing and undiminished government support for maintenance of this system is no longer guaranteed. Satellite beamed television now informs people even in the most remote corners of the post-Soviet scene, but the abysmal condition of the road network which each state has inherited is such that the centuries-old phenomenon of *bezdorozh'ye*, or roadlessness, can still bring the movement of goods and people to a standstill during the spring and autumn rains.

That the average shipment of a ton of freight by rail should be so long, 962 kilometres in 1989, is partially explained by the fact that coal makes up about one-fifth of the total tonnage. As noted earlier, the integration of iron ore and coking coal deposits separated by long distances adds substantially to the traffic on the railroad. Construction materials also comprised a large share of rail cargo. Although the share of goods movement by water has dropped in relative terms, it has continued to increase in absolute volume. In Russia, the geographic orientation of the major inland water routes also complements the rail system since the former tend to be north–south and the latter east–west. Historically this was an important factor in keeping a sizeable volume of goods shipment on the Volga river system. Indeed, the Volga and related canal/river systems are still the most important routes for water-borne goods movement, accounting for the lion's share of the total inland haulage. While coastal goods movement is important, it is the Northern Sea Route which has traditionally garnered most publicity. Despite the short ice-free season, there is a sizeable volume of goods moved each season from the west to the Yenisey river. At Igarka transshipment of goods from the ocean going, ice-breaker freighter fleet occurs. From here goods can move by river steamer to Dudinka where a short rail line extends to Noril'sk, or further upstream. But the main movement of goods, principally forest products and ore from Noril'sk, is outbound. In the former Soviet Union, goods haulage by sea route averaged just over 4000 kilometres per ton in 1989.

There are several ways of assessing the comparative importance of different modes of transport, and these have been used in the foregoing discussion. The absolute tonnage shipped, for example by road, rail and pipeline, is the simplest one. Near the end of the Soviet era, motor transport was most important in terms of absolute tonnage carried. In 1989, it accounted for about 6.8 billion, or 52 per cent, of the nearly 13 billion tons total commodity movement on all transportation systems. Rail, pipeline, water and air goods transport were 31, 9, 7 and a fraction of a per cent respectively. But how far each ton is hauled by each mode throws a different light on the transport system usage. The data in Table 9.1 illustrate comparative freight movement in terms of billions of ton-kilometres. At the close of the Soviet era, the railroad, not motor transport, accounted for almost one-half of the total. Indeed, between 1917 and 1960 its relative importance actually increased slightly, largely at the expense of water-borne freight movement. In 1989, the latter accounted for less than 3 per cent of all ton-kilometres. In view of the limited shipping season on inland waterways and along most of the coastal zone, this change perhaps is not too surprising. As Table 9.1 reveals, between 1960 and 1989 there were several significant changes in the pattern of freight movement. The dramatic increase in the importance of pipeline movement of oil and natural gas is readily apparent. Between 1960 and 1989, shipments in terms of ton-kilometres increased fifty-fold. At the latter date, more than one-third of all freight shipment as measured in ton-kilometres was accounted for by pipelines.

In terms of passenger movement measured in billion passenger-kilometres, railroads traditionally dominated. The pattern began to change in the 1960s when Soviet planners turned their attention to the potential role of cars and especially buses, for short-haul journeys. In 1960, the automobile/bus share of passenger movement was less than 25 per cent; by 1991, it was close to one-half. In the same period the share of rail passenger movement fell from 69 per cent to about one-third. In absolute terms, traffic increased more than four times in this period, and was well in excess of one billion passenger-kilometres in 1991. There was only a marginal change in water-borne passenger traffic. In 1960, some 12.1 billion passenger-kilometres were chalked up by air travellers; by 1991, more than 200 billion were recorded. This rate of change exceeded that of any other mode. Given the time required to travel across the Soviet

Table 9.1 *Soviet freight movements (billion ton-kilometres)*

Mode	1917	1928	1940	1960	1980	1986	1989
Rail	63.60	93.4	420.7	1504.3	3439.9	3834.5	3851.7
Sea	7.7	9.3	24.9	131.5	848.2	970.0	991.2
Inland waterway	15.0	15.9	36.1	99.6	244.9	255.6	239.6
Pipeline – oil	0.01	0.7	3.8	51.2	1812.9	2641.3	2944.4
Pipeline – gas	–	–	–	12.6			
Road	0.1	0.2	8.9	98.5	131.5	141.3	143.2
Air	–	–	0.02	0.6	3.1	3.4	3.3
Total	85.8	119.5	494.4	1898.3	6480.5	7846.1	8173.4

Sources: *Narodnoye Khozyzystvo SSSR za 70 Let* (Moscow: Finansy i Statistika, 1987), 341; *Narodnoye Khozyaystvo SSSR v 1965g* (Moscow: Finansy i Statistika, 1966), 457; *Narodnoye Khozyaystvo SSR v 1989g* (Moscow: Finansy i Statistika, 1990), 555

Union by rail, it is scarcely surprising that so many chose to fly. Since independence, however, the steady deterioration in the maintenance of aircraft has resulted in escalating numbers of casualties related to air travel. On short distance journeys, prudent travellers now go by train.

The economic disintegration which the demise of the Soviet Union occasioned not surprisingly has plunged all forms of commodity movement into disarray. Nowhere in the post-Soviet scene is the volume of goods movement in 1994 greater than in 1991. Excluding pipeline shipments, only Turkmenistan and Uzbekistan at 70 and 72 per cent respectively had a reasonable share of the 1991 goods shipment volume in 1994. In both countries there is continued emphasis on maintaining the area under cotton cultivation, and on maintaining past levels of cotton output and export, and this government policy figures prominently in explaining the continuing relatively large commodity flows. In the rest of the post-Soviet states, goods shipments in 1994 were less than 40 per cent of 1991 levels. Indeed, in Moldova, Azerbaijan, Armenia, Georgia, Kyrgyzstan and Tajikistan commodity flows were less than a quarter of those in 1991. The impact of the changes since independence in terms of total volume of commodity flow by mode of transport on the Russian system is reflected in Table 9.2. Of course, commodity flows are simply one expression of the level of industrial activity. It is patently obvious that there has been a drastic change. It should be borne in mind that what has happened in Russia, is in fact in relative terms less dire than is the situation in many other post-Soviet states.

Table 9.2 *Freight shipments: Russia, 1990–1993 (million tons)*

Mode	1990	1991	1992	1993
Rail	2140	1957	1640	1348
Road	2941	2731	1862	1110
Inland waterway	562	514	308	215
Sea	112	104	91	83
Air	2.5	2.2	1.4	0.9
Oil and gas pipeline	1101	1042	947	873
Total	6858.5	6350.2	4849.4	3629.9

Source: *Rossiyskiy Statisticheskiy Yezhegodnik*, 1994 (Moscow: Goskomstat, 1995), 391

INDUSTRIAL OUTPUT IN THE NEW ECONOMY

The overall drop in industrial raw materials output obviously means fewer final goods are being produced. Charting actual production levels of basic industrial inputs and consumer goods is the most obvious way of acquiring some sense of how post-Soviet economies are faring in the transition to a market economy. From earlier discussion, there is little to suggest that these economies are very robust. Tracking the output of cement and the manufacture of metal cutting machine tools gives a reasonable picture of what is happening in the industrial economy in general. Cement production, of course, is closely linked to demand generated by the construction industry. All states produce cement, and given the nature of the product importing it from outside the former Soviet Union is unlikely. The metal cutting machine tool, a producer not a consumer good, is perhaps best thought of as an essential tool for producing things which are then used in the manufacture of other goods. In contrast to cement, importing foreign-made machines is possible, but from available data appears not yet to be very widespread.

From the data presented in Table 9.3, it is patently obvious that demand for cement from 1990 to 1994 dropped rather dramatically right across the post-Soviet scene. Output peaked in most states in 1989; in some the decline started a year earlier. In 1994, production of cement was about 45 per cent of output in 1990. In the Baltic states, recovery in demand is likely to begin first. In Russia, there were preliminary signs in 1995 that suggest the collapse of industrial production had bottomed out. As the restructuring of the economy proceeds, new construction will pick up and the demand for cement will rise. But it will be several years before production even comes close to the 1990 output level. In the southern tier states of Uzbekistan and Turkmenistan, the continuing dominant role in the economy played by the state sector has helped to prevent a collapse in the demand for basic construction materials, such as cement, comparable to that which has occurred in the other states. Elsewhere in the post-Soviet scene, it is not at all clear when economic recovery will start. A total of about 148,000 metal cutting machine tools were produced in the Soviet Union in 1990, down from 216,000 a decade earlier. In the new economy of automated production techniques, computer driven numeric

Table 9.3 *Post-Soviet states manufactured industrial inputs: selected indicators, 1990–1994*

State	Cement (million tons)		Metal cutting machine tools (thousands)	
	1990	1994	1990	1994
Russia	83.0	37.2	74.2	18.2
WESTERN TIER				
Belarus'	2.3	1.5	15.5	6.2
Ukraine	22.7	11.4	37.0	8.5
Moldova	2.3	0.04	–	–
SOUTHERN TIER				
Georgia	1.3	0.3*	1.6	–
Armenia	1.5	0.1	8.6	0.8
Azerbaijan	1.0	0.5	0.6	0.2
Kazakhstan	8.3	2.0	2.6	0.4
Kyrgyzstan	1.4	0.4	1.3	0.1
Tajikistan	1.1	0.2	6.7	1.1*
Uzbekistan	6.4	4.8	0.02	0.01
Turkmenistan	1.1	0.7	–	–

*Data for 1993

Source: Ekonomika Sodruzhestva Nezavisimykh Gosudarstv v 1994g (Moscow: Statkomitet, 1995), 89, 92

control devices and robots are the equivalent of metal cutting machine tools of the old economy. Certainly some enterprises in the industrial economy are reasonably well equipped in terms of this new economy-era technology, but in the main these were part of the former Soviet military industrial complex. Such equipment is not yet very common in consumer goods factories, save for those which have been converted from military to civilian production. Although the annual output of metal cutting machine tools had been steadily declining in the latter decade or so of the Soviet era, since independence production has plummeted. In 1994, a total of 35,500 such machines were made in all post-Soviet states, less than one-quarter the number in 1990. The data presented in Table 9.3 imply that production of other goods, both producers and consumers, has declined since independence as well. The production trends for the manufactured goods listed in Tables 9.4 and 9.5 do not contradict such a conclusion.

Materials used to build automobiles, trucks and tractors in 1994 were but small fractions of the amounts consumed in 1990 (Table 9.4). Most post-Soviet states had inherited some machine building capacity, as is reflected in the fairly wide distribution of plants which made metal cutting machine tools. But fewer states produced trucks and tractors, and only Russia and Ukraine had an automotive industry. Despite different industrial structures, all post-Soviet states had one thing in common – output in 1994 was very much smaller than in 1990. Notwithstanding the numerous problems in agriculture which are in part at least a function of an inadequate supply of equipment, trucks and tractors being but two examples, the output of both in 1994 will do little to ameliorate the deficiencies noted in Chapter 7. During the 1980s, production of tractors averaged more than 550,000 units annually. In 1990, 495,00 were produced. In 1994, a mere 92,500 tractors left the factory production lines in all post-Soviet states. The near-bankrupt position of former collective and state farms everywhere has impacted demand for much needed equipment. On farms hard pressed to find money to buy diesel fuel to run existing machinery, buying a new tractor is indeed unlikely. An additional problem is that the tractors manufactured are not designed for the peasant farms which have appeared in the past few years. Even if a proprietor had the money to buy one, there are none suited to the needs of the peasant farming operation which presently aver-

Table 9.4 *Post-Soviet states industrial production: selected indicators, 1990–1994*

State	Cars (thousands)		Trucks (thousands)		Tractors (thousands)	
	1990	1994	1990	1994	1990	1994
Russia	1103.0	798.0	665.0	146.0	214.0	28.7
WESTERN TIER						
Belarus'	–	–	42.0	21.3	101.0	42.9
Ukraine	156.0	93.6	27.7	23.1	106.0	16.0
Moldova	–	–	–	–	9.8	1.2
SOUTHERN TIER						
Georgia	–	–	5.7	0.1	–	–
Armenia	–	–	6.1	0.2	–	–
Azerbaijan	–	–	3.1	0.01	–	–
Kazakhstan	–	–	–	–	41.1	2.0
Kyrgyzstan	–	–	24.3	0.2	–	–
Tajikistan	–	–	–	–	–	–
Uzbekistan	–	–	–	–	23.3	1.7
Turkmenistan	–	–	–	–	–	–

Source: *Ekonomika Sodruzhestva Nezavisimykh Gosudarstv v 1994g* (Moscow: Statkomitet, 1995), 90

Table 9.5 *Post-Soviet states consumers goods production: selected indicators, 1990–1994*

State	Television sets (thousands)		Refrigerators/freezers (thousands)		Washing machines (thousands)		Footwear* (millions pairs)	
	1990	1994	1990	1994	1990	1994	1990	1994
Russia	4717.0	3987.0	3774.0	2631.0	5419.0	2107.0	385.0	66.7
WESTERN TIER								
Belarus'	1302.0	465.0	728.0	742.0	33.1	76.9	46.8	24.8
Ukraine	3774.0	1874.0	903.0	650.0	788.0	422.0	196.0	37.6
Moldova	138.0	167.0	133.0	53.2	298.0	80.8	23.2	2.2
SOUTHERN TIER								
Georgia	51.3	0.9**	–	–	–	–	17.0	1.4**
Armenia	–	–	–	–	110.0	0.3	18.7	1.5
Azerbaijan	5.9*	2.3	330.0	96.9	–	–	15.3	3.0
Kazakhstan	–	44.4	–	33.1	367.0	87.7	36.5	7.7
Kyrgyzstan	–	42.6	–	–	234.0	17.1	11.6	1.6
Tajikistan	–	–	167.0	3.2	–	–	10.9	0.9
Uzbekistan	–	48.3	201.0	19.8	–	7.8	46.7	27.9
Turkmenistan	–	–	–	–	–	–	5.1	1.8

*1992
**1993
Source: *Ekonomika Sodruzhestva Nezavisimykh Gosudarstv v 1994g* (Moscow: Statkomitet, 1995), 93–95

ages only 42 hectares in size. The cultivation of thousands of hectares on the average collective and state farms resulted in standardization of product lines during the Soviet era. Tractors and other farming equipment were designed and built to operate on large scale, not small, farms. In the automotive industry, standardization was also commonplace, as were huge car factories in only a few locations. Since independence, the economic chaos has greatly reduced the institutional demand for domestically produced automobiles. Put simply, fewer enterprises and departments have the financial resources to add cars to their fleets. For the newly rich, the ability to exercise some choice in what they buy has also impacted the automobile industry in Ukraine and Russia. Since independence there has been an influx of new and used foreign-made automobiles, customarily of the up-market variety, into all corners of the former Soviet Union. Cars made in Russia and Ukraine cannot compete, except perhaps in price. When a Mercedes or BMW is on offer, discerning buyers with money seem little interested in purchasing a Volga or Zhiguli (Lada) given their deserved reputations for poor design and worse construction. About all that is guaranteed when purchasing such a car is the need for regular repair. During the Soviet era, car parts such as brakes, drive shafts and engine blocks were manufactured as far afield from the automobile manufacturing centres in European Russia and Ukraine as Georgia, Kazakhstan and Uzbekistan. Import substitution has not been a major obstacle since independence, however. As Table 9.3 indicates, car production has dropped by more than a quarter in Russia, and by more than a third in Ukraine since 1990. The production of trucks has fared even worse than automobiles. While there are some industrial enterprises which have adapted well to the market economy, are profitable, and thus have the requisite financial resources to purchase new trucks, they quite clearly are far fewer than the number of citizens in the post-Soviet states who can afford to purchase a car. Indeed, in all countries with an automotive industry far fewer trucks than cars are produced in any event. In short, demand for new trucks has been hard hit by the events of the past few years.

A selection of some basic consumer goods is included in Table 9.5. During the 1980s, total production of each increased annually. Clearly, the same cannot be said of the post-independence period to date. In a few states, there are instances of some consumer goods being produced in greater quantity in 1994 compared to 1990. For example, in Belarus', output of refrigerators and washing machines was higher, and Moldova produced more television sets. In Kazakhstan, Kyrgyzstan and Uzbekistan, the manufacture of television sets first began following independence. The total number of sets produced in these three states reached 135,000 in 1994. But overall, the 1994 output of television sets was fully one-third fewer than the nearly 10 million units produced in 1990. Kazakhstan also began producing refrigerator/freezers as well. While these examples of import substitution are important in their own right, they do not offset the prevailing trends. In the case of each consumer good listed in Table 9.5, overall output in 1994 was less than in 1990. But demand for these and other consumer goods has not shrunk in parallel fashion. Across the former Soviet Union, there has been a sizeable influx of foreign-made consumer goods. Electronic wares from around the world are prominently displayed in some stores in all of the major cities, and in a good number of provincial centres as well. Major appliances with brand names familiar to Europeans, and in some cases to North Americans, may be purchased with comparative ease if price is not the most important issue in the decision. To be sure, domestic production continues to satisfy the bulk of demand for such wares. But as the post-Soviet consumer becomes more affluent, and more concerned with quality as opposed to just price, the consumer goods industries inherited from the Soviet era inevitably will be forced to adapt, and improve, their products. The quite marked decline in shoe production is perhaps an indication of what is in store for many post-Soviet manufacturers of clothing, personal care items and so on. Shoes from developing countries are already competing with domestic output on price, if not quality. Amongst higher order personal items ranging from clothing to jewelry, European and American styles, design, quality, and marketing strategies will put enormous pressure on domestic consumer goods enterprises grown accustomed over the years to supplying not selling. Already there is enormous interest in foreign consumer goods of all descriptions amongst the general population. This is evidenced in part by the volume of such products now being sold, and in part by the appearance of a host of magazines catering to the tastes and pocket books of the growing middle class as well as the rich. Doubtless, western consumer goods will continue to make substantial inroads into the markets of domestic enterprises which over the years turned out items of poor design and worse quality.

Privatization and Restructuring of Industry

The presumed panacea for both poor quality producer and consumer goods was market reform, since this would place all enterprises on the same footing. To stay in business, to expand production, would depend not on the plan but on how the market responded to what was being offered for sale. In Chapter 3, the path to market reform and the degree to which enterprises across the post-Soviet scene have been privatized, was described in general terms (see Fig. 3.5, p. 78). The retail trade sector was the first to undergo extensive privatization, in large measure because the typical retail establishment was small scale, possessed limited infrastructure and equipment, and therefore was of comparatively limited value. Privatization of retail trade was endorsed by all levels of government and a large segment of the general public as well (see Fig. 3.8, p. 84). Industrial enterprises posed rather different challenges. Capital intensive, large scale, and consequently of substantial worth, how to share out their asset value equitably was a major problem. Moreover, a substantial share of the former Soviet Union's most important industrial infrastructure was associated with the military industrial complex (MIC). Privatization of parts of the MIC is now underway.

The MIC is no longer as high a national priority in any post-Soviet state as it was during most of the Soviet period. The raw materials, semi-manufactured goods, and technical-service linkages between the huge and often very specialized military-industrial production components located in different parts of the former Soviet Union have been fundamentally fractured. Even if one post-Soviet state wished to maintain production from the MIC remaining on its territory, it would be hard pressed to do so given the high degree of functional integration between the production complexes which had evolved during the Soviet era, and which are now often located in different countries. For example, important components of the former Soviet Union's space industry were located in Kazakhstan. This part of the MIC remains important in present-day Russia. Following independence, Russia assumed administrative and financial responsibility for some of the Kazakhstan space and defence-related installations, associated workforces and dependent population, as part of a formal agreement with the government of Kazakhstan. Clearly, this is less than an ideal situation given the theoretical possibility of Kazakhstan exerting political leverage on Russia by virtue of the fact that an important part of Russia's defence complex is located on its territory, and some distance from the Russian border at that. Political relations to date have been amicable, but this can always change. To set the privatization and restructuring of the MIC in context, it is necessary to first briefly review how industry as a whole has been impacted. This discussion will focus on the situation in Russia.

It was noted in Chapter 3 that Russia's voucher privatization process initiated by Presidential decree in mid-1992 was put on a fast track in that all state enterprises employing more than 1000 people, or having a book value of 50 million roubles or more, were obliged to submit privatization plans to the federal government by November 1992. Most enterprises opted for the privatization plan that gave employees the right to acquire 51 per cent of the stock of their enterprise at a fixed price. Even though the programme only began in November, already by the end of December 1992 about 14 per cent of all industrial workers were employed in newly privatized firms. Across Russia there was considerable regional variation in the percentage of industrial employment that was now in the privatized, as opposed to the state sector, as Fig. 9.11 indicates. In parts of the Urals economic region, and in Kemerovo Oblast' in West Siberia for example, there were many large enterprises to which the Presidential decree applied. The North and Far East economic regions also stand out in terms of the degree of privatization in part because of the dominant role of a few very large enterprises in the regional industrial economies there. (Figure 9.12 provides the names and locations of Russia's 89 regions, while Fig. 9.13 details the 12 major economic regions). In the Central Chernozem economic region, smaller scale enterprises dominated the industrial structures, and therefore they were not obliged to participate in the first wave of privatization. As Fig. 9.11 indicates the percentage of industrial employment in the private sector there was comparatively low. Another factor helps to account for the regional variations presented by Fig. 9.11. In some areas, for example in Moscow and St Petersburg and the surrounding oblasts, enterprises which were part of the MIC were important employers. In the early stages of privatization, many of these huge and specialized defence industry plants were exempted from the process owing to lobbying on the part of their managers and some members of government. The rationale for their exclusion at the time was that Russia needed to maintain at least part of its MIC in a state of readiness in the event of a threat to national security. It was contended that this was better done by having the state retain control as opposed to private sector interests. But by early 1993, a growing number of managers of MIC enterprises appreciated some of the advantages of privatization in terms of restructuring and converting part of the operation to the needs of the market. Put simply, there was the possibility not just of corporate profit, but of personal gain as well. The parts of the MIC which the government then permitted to privatize were those which had the best prospects of converting production to the market. Thus, between 18 and 30 per cent of the MIC plants in communications technology, electronics and space technology were allowed to privatize from 1993 to early 1994. By way of contrast, only

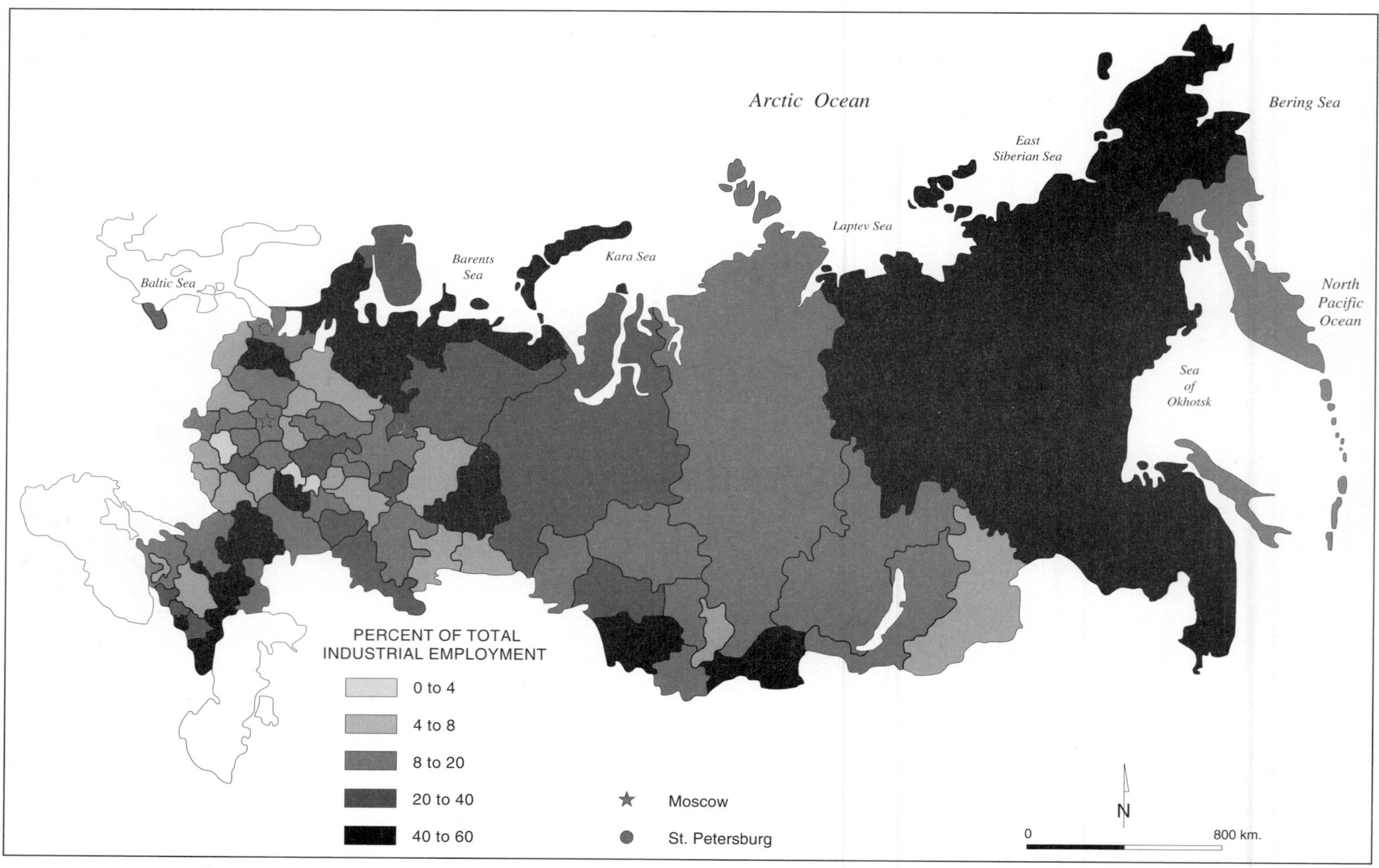

FIGURE 9.11 *Privatization of industry: Russia, 1992*
(*Source*: T. Heleniak, *Economic Geography of the Russian Federation* (Washington: The World Bank, Country Department III, Europe and Central Asia Region, 1994), Vol. 1, No. 15)

5 and 8 per cent respectively of the MIC enterprises producing armaments and warheads were given permission to privatize. By the end of the voucher privatization programme in July 1994, 13.3 million industrial workers out of a total of just over 18 million were employees of privatized enterprises, or about 73 per cent. By mid-1995, they accounted for over 80 per cent of total industrial employment. This relative increase resulted from additional conversions of state enterprises to some form of private ownership, and because total industrial employment in Russia continued its steady decline from nearly 21 million in 1990. Workers in the MIC comprised a substantial share of the privatized and remaining state sector employment categories in 1995.

Table 9.6 outlines the employment in privatized industry in June 1994, according to industrial category. Several features of the nearly 9000 enterprises listed stand out. The first is that the average employment per enterprise, 1498 workers, was extremely large. With 24,000 workers each, the 26 enterprises in the oil and gas industry obviously led the pack in average size, having twice the average number of employees of the next largest group, fuel and power industry enterprises. Average monthly wages in the gas industry were also high. They exceeded those in all other industrial sectors in 1994. Of the ten industrial groups listed in Table 9.6, only three had fewer than 1000 employees on average. Of course, when the privatization drive began in 1992, the largest enterprises were singled out for early privatization in the aforementioned Presidential decree. In the years since, the privatization drive embraced smaller industrial enterprises as well. Thus, what the data in Table 9.6 also underscore is the legacy of the Soviet-era preoccupation with scale economies in factory production.

In the majority of the enterprises listed in Table 9.6, at least one-half of the company shares were held by the employees at the end of June 1994. Thereafter, and for the first time in Russia's market reform process, it was possible to purchase shares in companies with cash. As shareholders can now at least consider exchanging their shares for cash, doubtless there will be some adjustment in the ownership patterns. In many of the larger enterprises, the Russian government itself is a major, albeit minority, shareholder. With government representation on the Boards of Directors of some of the most important private enterprises, implementation of government policies of one kind or another can be monitored more easily, and government priorities can in theory at least influence corporate production and marketing decision-making. With government bureaucrats as directors of companies working together with former members of the *nomenklatura*, who often are now corporate managers and

Table 9.6 *Privatization of Russian industries, June 1994*

Industry	Number of privatized companies	Number of employees in private enterprises (thousands)	Average employment per privatized enterprise
Oil and gas	26	616	23,692
Fabrics and clothing	731	877	1200
Chemicals and petrochemicals	1316	1509	1147
Steel production	389	1655	4255
Electronics and engineering	1767	3587	2030
Fuel-and-power complex	68	816	12000
Transport equipment	175	1515	8657
Construction	2122	1385	653
Transport	1345	510	379
Timber and pulp/paper	875	735	840
Total	8814	13,205	1498

Sources: Igor Filatochev, Roy Bradshaw, 'The Geographical Impact of the Russian Privatization Program', *Post-Soviet Geography*, Vol. 36, No. 6, 1995, 374

FIGURE 9.12 *Regions of the Russian Federation*
(*Source*: T. Heleniak, *Economic Geography of the Russian Federation* (Washington: The World Bank, Country Department III, Europe and Central Asia Region, 1994), Vol. 1, No. 1)

1 Northern economic region
- 1 Kareliya Republic
- Komi Republic
- 2 Arkhangel'sk Oblast'
- 3 Nenets AOkr
- Other Arkhangel'sk Oblast'
- 4 Vologda Oblast'
- Murmansk Oblast'

2 Northwest economic region
- 5 Leningrad
- St. Petersburg city
- Leningrad Obast'
- 6 Novgorod Oblast'
- 7 Pskov Oblast'

3 Central economic region
- 8 Bryansk Oblast'
- 9 Vladimir Oblast'
- 10 Ivanovo Oblast'
- 11 Kaluga Oblast'
- 12 Kostroma Oblast'
- 13 Moscow
- Moscow City
- Moscow Oblast'
- 14 Orel Oblast'
- 15 Ryazan' Oblast'
- 16 Smolensk Oblast'
- 17 Tver' Oblast'
- 18 Tula Oblast'
- 19 Yaroslavl' Oblast'

4 Volgo-Vyatskiy economic region
- 20 Mariy El Republic
- 21 Mordoviya Republic
- 22 Chuvashskaya Republic
- 23 Kirov Oblast'
- 24 Nizhegorod Oblast'

5 Central Chernozem economic region
- 25 Belgorod Oblast'
- 26 Voronezh Oblast'
- 27 Kursk Oblast'
- 28 Lipetsk Oblast'
- 29 Tambov Oblast'

6 Povolzhskiy economic region
- 30 Kalmykiya Republic
- 31 Tatarstan Republic
- 32 Astrakhan' Oblast'
- 33 Volgograd Oblast'
- 34 Penza Oblast'
- 35 Samara Oblast'
- 36 Saratov Oblast'
- 37 Ul'yanovsk Oblast'

7 North-Caucasus economic region
- 38 Adygeya Republic
- 39 Dagestan Republic
- 40 Kabardino-Balkarskaya Republic
- 41 Karachay-Cherkesskaya Republic
- 42 North Ossetiya Republic
- 43 Chechnya & Ingushetiya Republics
- 44 Krasnodar Kray
- 45 Stavropol' Kray
- 46 Rostov Oblast'

8 Urals economic region
- 47 Bashkortostan Republic
- 48 Urmurtskaya Republic
- 49 Kurgan Oblast'
- 50 Orenburg Oblast'
- 51 Perm' Oblast'
- 52 Komi-Permyat AOkr
- Other Perm' Oblast'
- 53 Sverdlovsk Oblast'
- 54 Chelyabinsk Oblast'

9 West Siberian economic region
- 55 Altay Republic
- Altay Kray
- 56 Kemerovo Oblast'
- 57 Novosibirsk Oblast'
- 58 Omsk Oblast'
- Tomsk Oblast'
- 59 Tyumen' Oblast'
- Khanty-Mansi AOkr
- 60 Yamalo-Nenets AOkr
- Other Tyumen' Oblast'

10 East Siberian economic region
- 61 BuryatiyaRepublic
- 62 Tuva Republic
- 63 Khakasiya Republic
- Krasnoyarsk Kray
- Taymyr AOkr
- Evenki AOkr
- Other Krasnoyarsk Kray
- Irkutsk Oblast'
- 64 Ust'-Ordo Buryat AOkr
- Other Irkutsk Oblast'
- Chita Oblast'
- 65 Aga AOkr
- Other Chita Oblast'

11 Far Eastern economic region
- Sakha Republic
- Primor Kray
- Khabarovsk Kray
- 66 Yevrey AO
- Other Khabarovsk Kray
- Amur Oblast'
- 67 Kamchatka Oblast'
- 68 Koryak AOkr
- Other Kamchatka Oblast'
- Magadan Oblast'
- Chukotka AOkr
- Other Magadan Oblast'
- Sakhalin Oblast'

12 Kaliningrad Oblast'

AO – Autonomous Oblast'
AOkr – Autonomous Okrug

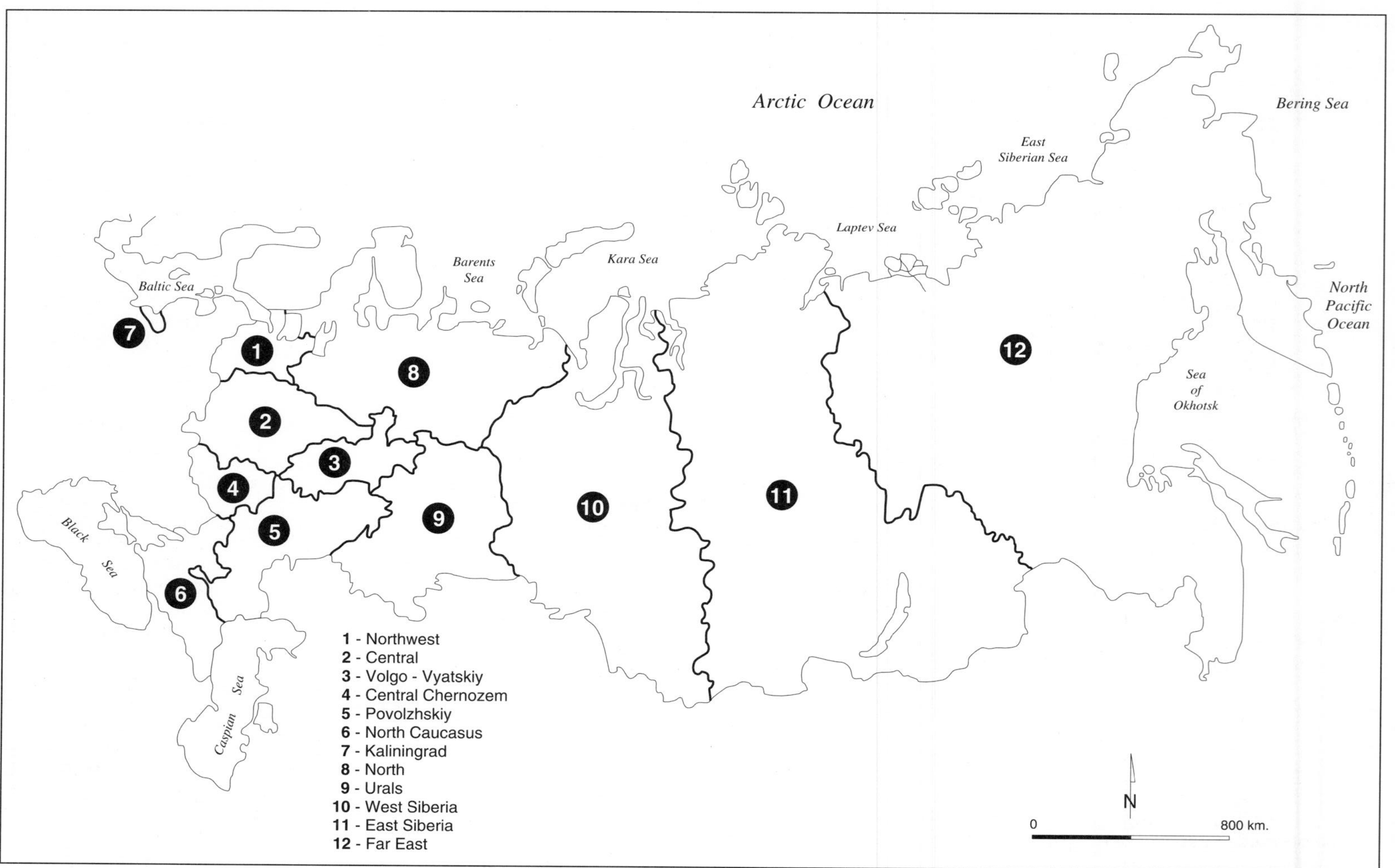

FIGURE 9.13 *Economic regions of Russia*
(*Source*: T. Heleniak, *Economic Geography of the Russian Federation* (Washington: The World Bank, Country Department III, Europe and Central Asia Region, 1994), Vol. 1, No. 3)

significant company shareholders, there is ample opportunity for these elites to share potential private profit as well. Government as shareholder is typical in the privatized, and profitable, energy sector industrial enterprises. It is also a standard arrangement in those MIC enterprises which have privatized. Strategically, a continuing presence of government in the corporate management of the military industrial complex is a sensible arrangement given the chaos which the market reform process has visited upon the Russian economy since independence.

In light of the MIC's strategic importance and the historic Soviet penchant for secrecy with respect to statistical data of often the most innocuous kind, it should come as no surprise that detailed information on defence industries remains exceedingly difficult to come by. It has been estimated that in the mid-1980s defence employment was around eight million people, that number being in addition to the armed forces themselves. In the cities of Moscow and St Petersburg, and in Sverdlovsk Oblast' in the Urals economic region, were more than 300,000 defence industry sector workers. In each of Nizhniy Novgorod, Perm', Samara and Moscow Oblasts between 200,000 to 300,000 more defence industry employees were employed. While more is now known about the MIC than during the Soviet era, much of the available information is far from precise. Indeed, this is immediately evident from a quick perusal of the data presented on the Russian and former Soviet MIC in Table 9.7. In most instances, estimated ranges rather than specific totals are given for enterprise numbers and employment. Nonetheless, what emerges is an interesting picture, albeit one painted in broad brush strokes, of a very important element in Russia's contemporary industrial economy. As Table 9.7 indicates, the largest share of the MIC of the former Soviet Union was located in Russia.

Soviet-era defence industry facilities were under nine different ministries, and military equipment components were produced in a wide range of other enterprises which lay outside the defence industry's immediate ministerial jurisdiction. The exact number of industrial enterprises under the direct administration of the Soviet MIC is not known, but was probably in the order of 3000 to 4000. The number of enterprises working on defence industry products included many outside direct jurisdiction, and this is the basis for the upper end of the ranges listed in Table 9.7. As a share of total Soviet industrial employment, the MIC figured prominently. While the MIC's precise share of Soviet gross domestic product was also never known, that it was disproportionately large in comparison with, for example, the United States, is not subject to dispute. Thus, the continuing large scale production from the defence sector industries was for the Soviet population a real opportunity cost in terms of the state's resultant inability to meet the domestic demand for consumer goods. Therefore, already in the late Soviet era converting some of the bloated MIC's plants to the production of consumer goods was part of Gorbachev's economic reform programme. Beginning in 1989, there was a planned reduction in the Soviet armed forces which reinforced the decision to convert part of the MIC from military to civilian production. There were some positive results attributable to the conversion process in that there was an increase in consumer goods output. By 1991, a large share of the country's electronic consumer goods such as television sets, VCRs, and tape recorders were produced by MIC enterprises. And so too was a substantial share of large appliances. In the spirit of the times, the former Soviet Union's largest missile plant reputedly added sausage making machines, microwave ovens and umbrellas to its product line! From the point of view of reducing government expenditures, the conversion process was less successful. Many defence enterprises built new plants in order to produce consumer goods, rather than convert existing ones. This additional plant capacity apparently consumed fully three-quarters of the funds the government had set aside to facilitate the conversion of existing MIC enterprises from defence to civilian production. From tanks to tin pots is not necessarily an easy step, nor necessarily a logical one.

As Table 9.7 shows, the number of enterprises and workers in Russia's MIC is substantial. It has been suggested that a conservative estimate of

Table 9.7 Military industrial complex: Russia and Former Soviet Union (FSU)

	Russia	FSU
Number of industrial enterprises	731 – 4000	1100 – 5000+
Employment in industrial enterprises (million)	3.2 – 8.0	5.0 – 9.0+
Number of research and development and design enterprises	714 – 1125+	920 – 1500+
Employment in research and development and design enterprises (million)	0.8 - 1.7	1.7 – 1.9

Source: James H. Noren, 'The Russian Military Industrial Sector and Conversion', *Post-Soviet Geography*, Vol. 35, No. 9, 1994, 499

the employment in the defence industry might be six or seven million, or more than a third of Russia's total industrial employment in 1994. The machine building and metal working sector is the most important, accounting for between two-thirds and three-quarters of the total MIC employment. Defence industry employment has declined in the past few years, just as employment in the non-defence industries has. In part this is a function of reduced demand, something which began with Gorbachev's efforts to diminish the role of the defence sector in the overall economy. In the final two or three years of Soviet power, procurement of tanks, aircraft and armoured personnel carriers from MIC factories in Russia dropped by 30, 50, and 80 per cent respectively. Production from the MIC in Russia has continued to decline since independence. But the relative rates of decline in MIC output in the military and civilian sectors have not always been what the government wants. In the important machine building and metal working industrial sector, for example, defence industry output between 1992 and 1994 seems to have declined less than output from MIC plants geared to the civilian market. Stockpiling is commonplace since government cutbacks in procurement have reduced demand for the products of the MIC sector. Wages in the MIC are now lower than those paid in privatized plants geared to the civilian market, whereas not many years ago MIC employees were comparatively privileged. In the machine building and metal working sector, for example, average monthly wages were about three-quarters of the average for all Russian industry in 1994. In the 1980s, workers in this sector earned more than the industry average. The Russian MIC figures very prominently in this sector and is impacted accordingly. Ironically, it is sometimes the case that the more educated, specialized and skilled the workforce in a defence sector plant is, the more difficult is the conversion of it to the civilian market. The material circumstances of daily life for the employees are often quite difficult in these situations.

Part of the difficulty in converting MIC enterprises to the civilian market is related to the nature of the production process, of course. But the very nature of the MIC gave rise to other problems, not least of which was the creation of single industry towns. During the Soviet era there were about 50 cities which were not just single industry, but completely closed to all outsiders as well. A number of these closed cities have been perpetuated in post-Soviet Russia, about ten of them related to the nuclear armaments industry. With the reduction in government funding for the MIC system, of which these cities are a part, their populations have had some special problems in adapting to the market economy. Denied ready access to the outside world, limited in opportunity to convert the basic economy from defence industry products to those wanted in the civilian market, and traditionally very well provided for in terms of social services and infrastructure, the reality of the 1990s has been painful. Some fortunate people have left for jobs elsewhere. But many have no such opportunity. They may now own their apartment, but as there is no market for it the inhabitants are tied to these cities unless wealthy enough to simply walk away, and buy or rent another apartment somewhere else. Few people are in such an enviable financial position. In a climate of frustration, growing deprivation and personal hardship, criminal activity is likely to mushroom. Indeed, already there is alarm in the international community over the clandestine movement of nuclear materials out of Russia, apparently destined for countries intent upon building a nuclear armaments arsenal. As the government funding of systems for control over these MIC closed cities shrinks, little improvement in internal security can be expected. As well, some former MIC technical specialists have departed Russia for related employment in countries currently not possessing a nuclear industry, but actively trying to develop one.

Single industry cities in the non-defence sectors of Russia's economy are more numerous, but face many of the same problems. Figure 9.14 depicts all cities of more than 100,000 population which have more than 30 per cent of the total employment concentrated in a single enterprise. The Urals economic region has a very substantial number of such cities, many of which are both dominated by a single enterprise and heavy industry at that. It is so-called heavy industries such as ferrous and non-ferrous metallurgy, chemicals and machine building which have been especially hard hit by the economic reform and reduction in, or outright loss of, guaranteed government funding. Attempting to adapt to changing conditions is much more difficult in cities with a narrow economic base. On the other hand, the broader the array of industries and occupations in a particular place, the greater is the potential for finding opportunities in the new economy, for creating new business opportunities, simply because of the greater mix of employment skills and aptitudes. Not all of the single industry cities depicted in Fig. 9.14 are facing severe economic crises, but a good many are. In many of them the bulk of the social services, housing, and maintenance of municipal infrastructure was supported by the dominant enterprise. As many of these have come upon hard times, the city's social and other services as well as wage income have been hard hit. For those still employed wages are frequently in arrears. Unemployment, though still a small share of the workforce, is rising. In such circumstances social tensions increase, and understandably so.

During the Soviet era there had been much effort to develop the peripheral regions of the country, to spread, geographically, the benefits of socialism. In this context, industrial development was inextricably linked to urban development. In theory, resource-rich regions on the periphery – principally the European Russian north, Siberia and Middle Asia – were not to be exploited for the benefit of the core region. The state ownership of natural resources and their planned development in accordance with socialist principles would, it was claimed, preclude any core–periphery economic dependent relationship. Over the years some progress in reducing regional disparity was in fact achieved. But as the Soviet economy became more technologically complex, indeed more modern, the attraction of the core region over the periphery increased. Balancing the need to achieve a measure of regional equality in economic development with the compelling logic of enhancing efficiency in all sectors of the national economy proved impossible. Thus, investment in the core region increased in relative terms in the final years of Soviet power. In the interplay of forces for national efficiency or regional equality, the former was ascendant in the final analysis. Since independence, there has been limited investment in new industrial infrastructure. Where it has occurred, it has been driven by market considerations and is mostly in the European Russian core. As prices in Russia come to reflect more the marketplace and less the values of the Soviet-era planners, doubtless more of the industrial plant in the eastern and northern regions will have greater difficulty in competing in the new economy. In Russia, the periphery will increasingly be regarded as a resource base to be exploited, not developed irrespective of the cost as so often was the case under Soviet regional industrial development planning principles. Migration now is predominantly from periphery to core. This is a harbinger of things to come. In short, regional disparity between core and periphery exists, and it is intensifying. The quality of life is impacted in consequence. This fundamentally important issue is the subject of the next chapter.

FIGURE 9.14 *Single enterprise cities: Russia, 1991*
(*Source*: T. Heleniak, *Economic Geography of the Russian Federation* (Washington: The World Bank, Country Department III, Europe and Central Asia Region, 1994), Vol. 1, No. 1)

FURTHER READING

Akopian, A. (comp.) *Industrial Potential of Russia: Analytical Study Based on Fixed Assets Statistics to 1992*. Commack, NY, Nova Science Publishers.

Ambler, J., Shaw, D.J.B., Symons, L. (eds) 1985 *Soviet and East European Transport Problems*. London, Croom Helm.

Bahry, D. 1987 *Outside Moscow. Power Politics and Budgetary Policy in the Soviet Republics*. New York, Columbia University Press.

Barr, B.M., Braden, K. 1988 *The Disappearing Russian Forest. A Dilemma in Soviet Resource Management*. Totowa, NJ, Rowman and Littlefield.

Blandon, P. 1983 *Soviet Forest Industries*. Boulder, CO, Westview Press.

Boycko, M., Shleifer, A., Vishny, R. 1995 *Privatizing Russia*. Cambridge, MA, MIT Press.

De Melo, M., Ofer, G. 1994 *Private Service Firms in a Transitional Economy: Findings of a Survey in St. Petersburg*. Washington, DC, World Bank.

Hauner, M. 1992 *What is Asia to Us? Russia's Asian Heartland Yesterday and Today*. New York, Routledge.

Hunter, H., Kaple, D.A. 1983 *The Soviet Railroad Situation*. Washington, Wharton Econometrics Inc.

Kaser, M.C. 1995 *Privatization in the CIS*. London, Russian and CIS Programme, Royal Institute of International Affairs.

Koropeckyj, I.S., Schroeder, G.E. (eds) 1981 *Economics of Soviet Regions*. New York, Praeger.

Kosals, L., Crowfoot, I.P., Sedova, V. 1994 *Why Doesn't Russian Industry Work?* London, I.B. Tauris.

Lewis, R. 1979 *Science and Industrialization in the USSR*. London, Macmillan.

McFaul, M., Perlmutter, T. 1995 *Privatization, Conversion, and Enterprise Reform in Russia*. Boulder, CO, Westview Press.

North, R. 1978 *Transport in Western Siberia: Tsarist and Soviet Development*. Vancouver, University of British Columbia.

Rees, E.A. 1994 *Stalinism and Soviet Rail Transport, 1928–41*. Basingstoke, St. Martin's Press.

Rowen, H., Wolf Jr, C., Alotnick, J. (eds) 1995 *Defense Conversion, Economic Reform and the Outlook for the Russian and Ukrainian Economies*. Basingstoke, Macmillan.

Swearingen, R. (ed) 1987 *Siberia and the Soviet Far East: Strategic Dimensions in Multinational Perspective*. Stanford, Hoover Institution.

Wood, A. (ed) 1987 *Siberia. Problems and Prospects for Regional Development*. Totowa, NJ, Rowman and Littlefield.

ZumBrunnen, C., Osleeb, J.P. 1986 *The Soviet Iron and Steel Industry*. Totowa, NJ, Rowman and Allanheld.

10

QUALITY OF LIFE IN THE ERA OF THE NEW ECONOMY

'the rich have become richer, while the poor have become even poorer'
(*Trud*, August 19, 1994, 1)

From what already has been said in the foregoing chapters there can be little doubt that for very significant numbers of people the quality of life has become materially worse since independence. Real incomes are falling, employment for many people has become a part-time affair, personal savings have been eroded or have disappeared altogether. Yet it is also fair to say that there are no post-Soviet countries in which an overwhelming majority of the population wants to return to the previous political and economic system. Nonetheless, there are some features of the former system which are particularly attractive at the present time, cradle to grave economic and social security being prominent amongst them. The economic crisis of the immediate post-independence period has made it extremely difficult for the new states to maintain the Soviet-era social safety net. All social services, but especially health care, have deteriorated. For those people who have successfully carved out a niche for themselves in the new economy, of course money can buy all the social and health care services they might need or want. But money cannot completely insulate the wealthy. In many regions of the former Soviet Union, the health and well-being of the rich and the poor alike is negatively impacted by a common legacy of the past era – an unsatisfactory, if not hazardous, ecological environment. With all these problems, it is small wonder that some people yearn for what they perceive to be better times. However, even the former Communists, who are once more acquiring the reins of political power across the post-Soviet scene, seldom call for a complete rejection of the changes which have been set in motion. Rather, slowing the transition to a market economy is a common goal, and so too is the maintenance of some parts of the former social contract such as adequate health care and a minimum standard of living. But talking about improving the quality of life is one thing, finding the resources to actually do so is quite another.

In this chapter, some basic indicators of personal income and household expenditure patterns at the end of Soviet era will first be examined. This will provide the necessary benchmark for determining where each state was at the time of independence, and for assessing what has happened since. In providing a rough sketch of quality of life changes since 1991, the focus of the discussion will be on the state of public health. In most regions there are clear indications that the health of the average citizen is deteriorating. The poor quality of the environment frequently figures prominently as a factor in public health deterioration, although the link between poor health and quality of environment is more often than not based on intuition. Thus, in the final section of this chapter an overview of environmental degradation across the post-Soviet scene will be provided.

Personal income and expenditures during the late Soviet era

The declining rate of economic growth was highlighted as one of the key problems confronting Soviet economic planners. Official statistics indicate that the growth of national income dropped from a shade less than 8 per cent per year in the latter half of the 1960s to less than 3 per cent in the late 1980s (see Fig. 3.4, p. 73). During the Gorbachev era of *glasnost'* even more pessimistic assessments were put forward by a number of leading Soviet economists. For example, A. Aganbegyan, one of Gorbachev's closest economic advisors during the late 1980s, contended that the decline was from a high of only 5 per cent per annum in the late 1960s to zero in the early 1980s. Whatever the real situation, any improvement in the standard of living of the average citizen must be linked to a net addition to the real wealth of the state, a major realignment of existing priorities in the allocation of the state's resources, or a combination of the two. For example, defence-related expenditures in the late Soviet era were reckoned to consume about 15 per cent of the gross national product, roughly twice the share in the US economy at the same time. The US economy was close to twice as productive, so actual military production was not in fact smaller. Thus, the burden of military expenditure was far greater in the former Soviet Union than in the United States.

The productivity of a workforce is directly tied to its skill level. In the late Soviet era, labour productivity was notoriously low in relation to the US, a reflection of differences in basic skills of the respective workforces and their sectoral allocation. The more skilled the labour force, the greater is the potential for successful innovation in an economy, the fewer the workers needed to produce a specific quantity of output, and the greater the opportunity to create new products and new markets. While workers in some sectors of the Soviet system, and especially those embraced by the military industrial complex, were certainly well trained and highly skilled, overall the Soviet-era labour force did not compare very favourably with other western industrialized states. In 1985, more than a third of all Soviet industrial workers were classified as unskilled. Less significantly perhaps, more than one-half of all construction workers were similarly classified. But in agriculture, which typically employs less than one-twentieth of the labour force in Western Europe and North America, more than two-thirds of the Soviet Union's vastly larger rural labour force was unskilled. Not only was the US economy more than twice as productive, it was in consequence characterized by important differences in the structure of the labour force, as the data for 1985 depicted in Fig. 10.1 indicate. The role of quaternary and quinary employment in the US, that is, the white collar, administrative and research employment of the service sector, dwarfed that of the Soviet Union. Obviously, inventions and their application, or innovation, will find much more fertile ground when there is less reliance on muscle and more on strategic planning. The latter is dependent upon widespread access to information systems. In the mid-1980s, it was difficult enough to find a city telephone book in the Soviet Union, let alone someone owning a personal computer and modem. Control over information and its manipulation or transmittal was one of the many ways of keeping potentially contentious elements in society under the thumb of the authorities. The cost in terms of civil rights was substantial. But the economic costs were huge, since innovation was stifled. Ten years later, personal computers and open information systems access are found in offices and homes alike. As well, access to fax and e-mail is increasingly taken for granted. But these communication technologies are dependent upon a telephone system which, following years if not decades of inadequate investment, is deficient in every respect. International direct dial telephone communication via satellite from

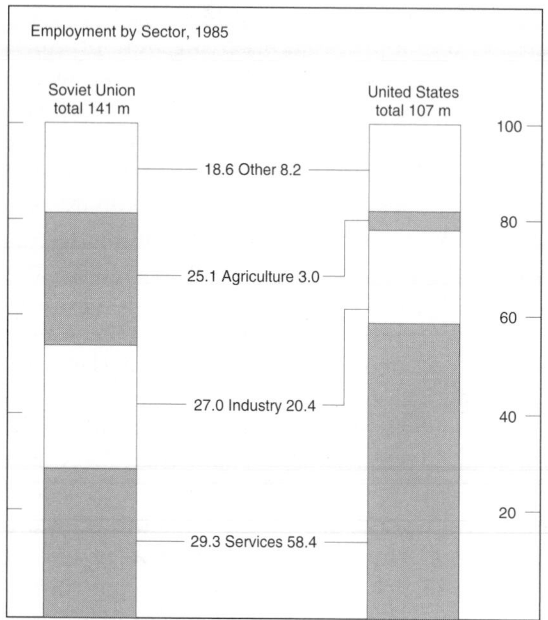

FIGURE 10.1 *Employment structure: Soviet Union and United States, 1985*
(*Source*: Based on *The Economist*, 9–15 April 1988, 11)

Moscow to the capitals of the rest of the world is frequently far more reliable and clearer than calling someone across the city via the overloaded and inadequate local telephone exchange, to say nothing of trying to place a call to someone elsewhere in the former Soviet Union. Innovation is still possible in such an environment, of course, but is handicapped on an international competitive basis. In the late Soviet era, and at the present time, competetive disadvantage related to communication and information systems has a price in terms of what society can produce by way of wealth. And upon the creation of wealth, the average wage and salary are dependent. How they are allocated amongst society's constituents is a matter of social policy. In the Soviet era, removing income differences between groups within society and between the regions across the country was a publicly stated objective.

Between 1940 and 1986, Soviet per capita national income as measured by wages and social benefits increased by 663 per cent, according to official statistics, and the differences amongst groups of workers narrowed. The tempo of change from the late 1960s on was especially notable. It was pointed out in Chapter 3 that the economic reform initiated by Brezhnev and Kosygin in 1965 was premised on enhancing the incentive to work and on improving the quality of output. Enterprises were to be judged on their profitability, which would reflect their ability to sell what they produced. Controls on prices were relaxed somewhat. Enterprises were permitted to retain a portion of profits for the benefit of workers. At the same time, basic wage rates improved, particularly in the agricultural sector. Not only was the gap between the average income of collective farm (*kolkhoz*) and state farm (*sovkhoz*) workers significantly narrowed, the gap between the farm and the urban industrial employee was also eroded and stood at less than 10 per cent in the late 1980s. While these basic relationships were changing, the average monthly wage of all non-agricultural workers climbed steadily. In 1960, it was 81 roubles per month; in 1986, 196 roubles; in 1989, 240 roubles. These figures are not adjusted for inflation, however, which in the late 1980s was of growing concern to the public at large. Paralleling the increase in wages was a steady growth in absolute expenditure on social benefits such as education, health, housing, family allowance payments, pensions, and so on. On a per capita basis, social benefits amounted to 127 roubles a year in 1961. By 1989, these social benefits were close to 600 roubles a year. At a rough reckoning, wage payments accounted for about three-quarters of what was described as the 'real income' of the typical Soviet urban family. The balance comprised the wide array of state benefits. For the two-wage earning, non-agricultural family in 1989, this worked out to just over 530 roubles income per month from wages and state benefits.

Soviet statistics indicate that per capita national income rose nearly two-fold between 1970 and 1989. Expressed in terms of average income per family member, the changes were as follows. In 1970, 18 per cent of the population belonged to families with monthly incomes in excess of 100 roubles per member. By 1989, close to two-thirds of the population were in this category. Indeed, by the latter date more than 30 per cent of the

Soviet population belonged to families whose monthly income per member exceeded 150 roubles. There was stratification in wage levels according to state priorities. Thus, skilled workers in the energy sector, metallurgy or machine construction earned as much as two and one-half times more than their counterparts in food processing or textiles. And the system of regional wage coefficients simply served to accentuate the differences in the monthly wage packet. At the other end of the skill level continuum, in the trade, catering or custodial occupations for example, wages were quite low. Compared to the highest paid workers they earned as little as one-third, or in extreme cases, one-quarter as much. But wage differentials in the mid-1980s were considerably smaller than 30 years earlier. The narrowing income differential between groups of workers, however, was thought by some people to remove incentive to work and thus thought to be partially responsible for the declining level of workers' productivity. After years of reductions in income differentials resulting from government social policies, the government took note of the argument linking incentive and income. Developments in the last few years of the Soviet power resulted in growing wage and salary disparity. Attempts during Gorbachev's regime to stimulate output and productivity by giving more discretion to enterprises in terms of how much workers were paid, and the legalization of cooperatives, soon created a class of comparatively wealthy citizens, and produced widening income differentials not just amongst employees with different skills in the same enterprise, but across regions as well.

Non-agricultural wage rates varied very markedly across the republics, as the relative data presented in Fig. 10.2 indicate. In 1989, the average monthly pay packet for a non-agricultural worker was 240 roubles, as noted above. But there were significant regional differences; indeed, in the majority of republics monthly non-agricultural wages were well below 240 roubles. In Estonia, the average was 270 roubles, nearly 90 roubles or 50 per cent more than in Azerbaijan. In general, there was a north-west, south-east gradient in average wages. To some degree this reflected differences in the structure of the regional economies. But there is more to the differences portrayed in Fig. 10.2 than simply a core–periphery

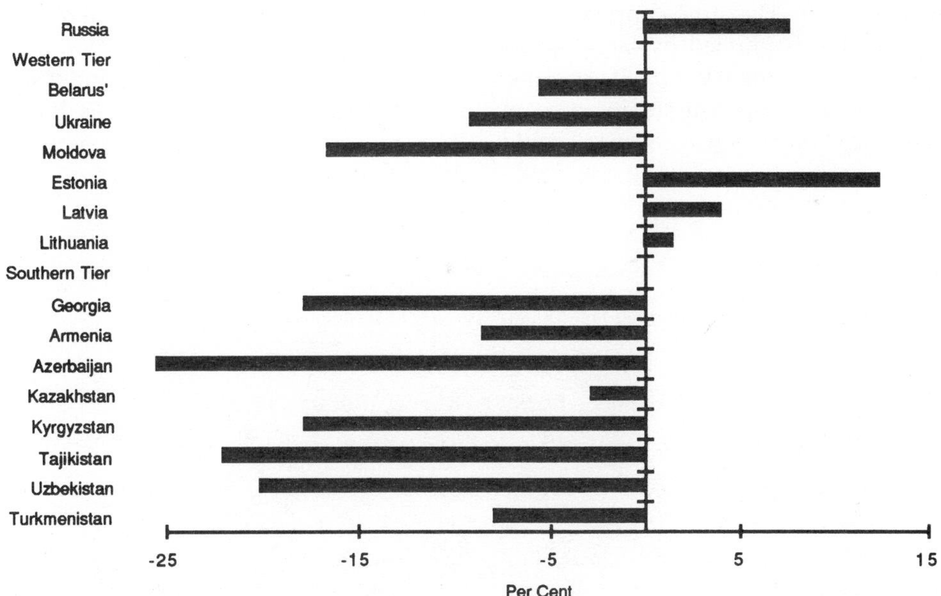

FIGURE 10.2 *Ratio of average monthly non-agricultural wages by republic to Soviet average of 240 roubles, 1989* (*Source*: Compiled from data in *Narodnoye Khozyaystvo SSSR v 1989g* (Moscow: Goskomstat, 1990), 78)

relationship. Estonia, for example, was at the forefront in the economic reform promoted by Gorbachev. It was the first republic to switch from an administrative supply allocation to something akin to a wholesale-trade system. In 1988, Estonia moved a step further along the *perestroyka* track by petitioning to have All-Union ministerial control, which emanated from Moscow, turned over to republic-level decision-makers. Cultural factors played a part in the ready adoption of innovations in economic management in Estonia. In short, in 1989 Estonian enterprises more than anywhere else in the former Soviet Union were market oriented. In other regions, cultural values may also have played important roles in the economy, but sometimes in ways that the state did not support. The populations of the southern tier republics were comparatively poor in terms of the official average monthly wage presented in Fig. 10.2. But interviews with Soviet émigrés conducted in the late 1980s suggest that substantial earnings were derived from the second economy in many southern tier republics, income which of course did not figure in the official statistics. Unreported income was not restricted to the southern tier republics to be sure, but it no doubt served to reduce somewhat the official average wage differentials highlighted in Fig. 10.2. Improvement in the quality of life is implied by larger wage packets and resultant higher per capita income. But this assumes a commensurate increase in goods and services on which consumers could spend money.

There is no question that in the late 1980s the average Soviet household was better off in a material sense that at any earlier time. The data on consumer durables presented in Table 10.1 certainly make this clear. In 1989, television sets were ubiquitous, though the demand for colour receivers remains strong despite their high cost. Major appliances were commonplace in both the urban and rural households, a rather dramatic change compared to 1970 in the latter instance. Indeed, the growing material prosperity of the rural population was perhaps best reflected by the ratio of private automobiles per 100 households in 1989. But basic statistics such as those presented in Table 10.1 convey only part of the picture. In many, if not most, rural areas repair services for the growing volume of household appliances were seriously deficient. This facet of the consumer scene was exacerbated by the customarily low quality of such durables. Urban households were similarly disadvantaged in respect of the quality of goods available. However, they had somewhat better prospects of obtaining repairs when necessary, if not from state-run agencies then from the urban *shabashniki* of the second economy, that is, the legion of individuals who provided such repair and other services on the side, often at sizeable personal profit, or from the growing number of cooperatively owned consumer service shops permitted under the Gorbachev reforms. In the final analysis, however, the gains registered in per capita income, in ownership of consumer durables, in general material

Table 10.1 *Soviet-era consumer durables per 100 households, 1970–1989*

	1970		1986		1989	
	Urban	Rural	Urban	Rural	Urban	Rural
Refrigerators	43	13	101	78	101	81
Washing machines	64	26	78	58	78	63
Sewing machines	57	54	63	70	61	68
Vacuum cleaners	16	3	50	21	56	24
Television	61	32	102	94	107	101
Private car	–	–	15	17	18	20

Sources: *Narodnoye Khozyaystvo SSSR za 70 Let* (Moscow: Finansy i Statistika, 1987), 473; *Narodnoye Khozyaystvo SSSR v 1989g* (Moscow: Goskomstat SSSR, 1990), 122

well-being, while substantial, failed to satisfy consumer demand in the late 1980s.

While the roubles in circulation increased more than three-fold between 1970 and 1989, production of consumer goods barely doubled. As basic household durables became more commonplace, as opportunities to spend comparatively large sums of money on other consumer goods remained limited, savings, not surprisingly, accumulated. Between the late 1960s and the late 1970s, registered savings rose four-fold. During the late 1980s the trend continued. Savings grew by one-third between 1984 and 1989 alone. Too much money chasing too few goods and services clearly produces price inflation. Since the state controlled most prices, and could only increase charges for basic items slowly because of the long-standing, implicit social contract with the people, if not concern over adverse public reaction, inflation was most evident in the second economy and in the price of foodstuffs sold in the free market environment of the collective farmers market and the cooperative. Again, not surprisingly, consumer reaction to escalating prices for goods and services not readily available from the state was negative.

Household expenditure patterns provide some additional insights into the relationship between the rising per capita national income and general well-being. In 1980, about 36 per cent of the average non-agricultural household budget was spent on food and another 4 per cent on alcohol. About 30 per cent was spent on non-food items, and the balance on other expenses. By 1989, the allocations were 31, 3 and 32 per cent on food, alcohol and non-food items respectively, again the balance going toward other household expenses. Rent remained at more or less the same rate. Clearly, a larger share of a larger family income was going toward clothing, consumer durables and social–cultural activities than ever before. While expenditures on food apparently declined in relative terms, there was a steady improvement in the diet of the population. These data are, of course, national averages, and therefore in some categories of expenditure cultural values play an important role. This is certainly true of alcohol. In many southern tier republics the traditional Muslim interdiction concerning consumption of alcohol was widely observed. Thus, in the Slavic and Baltic households the level of outlay for alcohol was substantially higher than suggested by the foregoing average figures. In summary, notwithstanding some important regional variations in household expenditure patterns, a larger share of a larger average household income was devoted to non-food expenditures, a trend which certainly suggests a material improvement in the quality of life. But improvement in the standard of living was by no means universal.

The data cited above were based upon a sample survey of 62,000 household budgets in various representative regions, and included an estimate of state payments and benefits to the average family. At the time these data were first gathered and the results published there was considerable discussion as to how close the relative allocations by category squared with reality. In the late 1980s, *glasnost'* provided an opportunity for the readers of *Izvestiya*, then the main newspaper of central government, to take issue with the official view of the average family budget. The size of the 'average family' budget, estimated to be just over 530 roubles per month when state benefits and payments were included, was felt by many people to be completely unrealistic. It offended the sensibilities of lower-income families to be lumped together with those who had more. Put simply, families with average or higher than average monthly incomes comprised part of the privileged sector of Soviet society. And privilege afforded opportunities to purchase goods at state prices in special, restricted entry shops. Thus, the proportion of the family budget needed to purchase food, officially 31 per cent in 1989, was widely disputed. Deficient supplies in state shops obliged ordinary households to purchase foodstuffs from the collective farmers markets or cooperatives where prices reflected supply and demand. For many families unable to access special shops where foodstuffs were usually available, and at state prices, the additional cost was quite substantial. Official surveys corroborated this state of affairs. It was estimated that those who earned low or even average wages had to pay almost 50 per cent more for a kilogram of meat than those with high monthly incomes who typically had access to special shops. In a similar fashion, that segment of the population with above-average incomes also received more by way of state benefits simply because they had far greater opportunities to take advantage of them.

The attempt to compile a representative household expenditure budget is problematic in any society. Few people relate well to a statistical average. But there were clearly more difficulties in managing household affairs in a chronically goods and services deficient society than in a more bountiful one, irrespective of how closely one's own circumstances fit the average. Thus, as remarked in earlier chapters, privilege in the Soviet Union was more often expressed in terms of access to goods and services than in terms of money alone. What compounded the difficulties in household management for many Soviet women was the attitude of Soviet men toward domestic chores.

While in the typical Soviet family both husband and wife had jobs, there was an additional burden imposed on working women. Not only did Soviet women living in cities work nearly as many hours as men each week, their contribution to household chores was between two to three times as great. As a rule, the higher the level of education the greater was the male contribution to household duties, but the distinction is relative. The substantial improvement in the ratio of consumer durables per 100 households described in Table 10.1 certainly made housework easier, but largely for the wife and mother. Thus, compared to the 1960s when on average 30 hours a week were taken up on household duties – in addition to the 50-odd hours of work and work related activity – the Soviet housewife of the late 1980s was much better off. The work week was reduced in the late 1960s to 41 hours, the six-day work week was reduced to five, thereby providing a free weekend for most workers, and the simple chores of washing clothes, cleaning house and food preparation had been much aided by an improved supply of washing machines, vacuum cleaners and refrigerators. But attitudes obviously change less rapidly than the material circumstances of the household. Married men spent substantially more time on personal pursuits and asleep each day than women. As perhaps might be expected, the situation for the woman of the rural household was decidedly less favourable than her urban counterpart's. Thus, the Soviet Union shared with many other countries considerable inequality in gender roles in the home and workplace.

While the data on household expenditures go some way toward enhancing our appreciation as to how the average Soviet family lived, they tell only part of the story. To reiterate, despite real improvements over the decades the standard of living in the Soviet Union in 1989 was considerably lower than in the major western industrial states. Only in education were per capita outlays consistently greater in the Soviet Union than in western states. Housing and transportation expenditures on the other hand, fare especially poorly in a comparative context. In 1989, millions of Soviet citizens still awaited private accommodation while living with relatives, while renting a room or part of one, or while existing in communal apartments or dormitories. The data provided earlier on per capita allocations of living space in 1989 demonstrated that there remained significant regional disparities (see Fig. 5.3, p. 127). And within cities of course, there was substantial variation in housing space allocation between the poor and privileged (see Table 5.5, p. 148) And, of course, all of the data mentioned thus far mask important regional variations in quality of life resulting from factors such as climate, cultural values and economic opportunity. By the late 1980s, the void between Soviet rhetoric and the reality of daily life was palpable, and everywhere the cause of growing public discontent and resentment.

Standards of Living Across the Post-Soviet Scene

The demise of the Soviet system was bemoaned by some people to be sure, but across the post-Soviet scene they were doubtless a distinct minority. New countries, new economic opportunities, a new and better life were all portended by the arrival of political independence. But in the early stages of the transition to a market economy, a higher standard of living remains an elusive goal for most people. In the absence of a common currency and high and variable rates of inflation, comparisons of per capita income trends across the former Soviet Union are problematic in the extreme. Indeed, even acquiring reliable basic data for some post-Soviet states is often difficult. Data on real income and its distribution within Russian society, and data on average monthly income for the 89 regions in Russia in August 1993 are therefore of particular interest since the basic trends in Russia may be indicative of what is perhaps happening more generally in the post-Soviet scene.

Personal income in Russia has fallen and become more stratified by social class group since 1991. As the data in Fig. 10.3 indicate, real income in the early part of 1995 was only 70 per cent of what it was in 1990. While there were some signs of economic recovery in 1994, the first half of 1995 witnessed a drop, in large measure owing to the government's tight money policy aimed at keeping inflation in check. At the end of 1994, some 31 million people, or about 21 per cent of the total population were defined as poor. The threshold defining poverty at the end of 1994 was a monthly income of less than 145,000 roubles. This was an improvement over mid-1993, when roughly 31 per cent of the population fell below the official poverty line. As many people are involved in the underground economy, the actual level of poverty may not be as extreme as these percentages suggest. But millions of people are still suffering extreme economic hardship. Toward the end of 1995, indicators were suggesting that the economic situation was again improving. But as the economy is slowly restructured and resuscitated, the distribution of income between the rich and the poor is becoming more imbalanced. In Fig. 10.4 the Gini coefficient is used as the measure of income inequality. A coefficient of 1.0 would mean that all income is concentrated in the hands of just one person in society. As Fig. 10.4 indicates, the trend since independence has been one of steady concentration of a larger share of total income in the pockets of fewer people. At the end of 1994, it stood at 0.42. In the US in 1992, the comparable figure was 0.40, up from 0.35 in 1969. In Britain in 1991, the Gini coefficient was 0.34. This particular measure suggests that incomes were more equally distributed in the US and Britain than in Russia. Between 1993 and 1994, the real income of the top one-fifth of the Russian population grew at more than 30 per cent. At the other end of the income scale, the poorest one-fifth saw their incomes decrease by between 3 to 5 per cent over the same period. That this has produced different perceptions of the future according to socio-economic groups is not surprising. In a survey conducted in September 1995, it was reported that 40 per cent of entrepreneurs interviewed thought that their material circumstances would be better the next year. Of the managers, office employees, workers and pensioners who were sampled, only 15, 9, 8 and 3 per cent respectively thought their personal material circumstances would improve the following year. Similar trends in income differentiation no doubt characterize the other post-Soviet states actively pursuing market reform of their economies. Indeed, in the Baltic states, it is entirely possible that income disparity may even exceed that of Russia. Where economic restructuring has been more moderate, for example in Turkmenistan and Uzbekistan, the income disparity may be less, but there are no data available to assess the income gap by group. Available data do reveal a trend of growing income disparity by region, as well amongst social class groups.

Average monthly wages by region in Russia were surveyed in August 1993. As Fig. 10.5 indicates, there was considerable variation. The highest average wages were in the West Siberia and Far East economic regions, more specifically in the energy and resource rich parts of them (see Fig. 9.12, p. 282). Wages there ranged between 154,000 and 225,000 roubles per month. The lowest average monthly salaries were in the six republics of the north Caucasus – the Karachay-Cherkessiya, Kabardino-Balkariya, North Ossetiya,

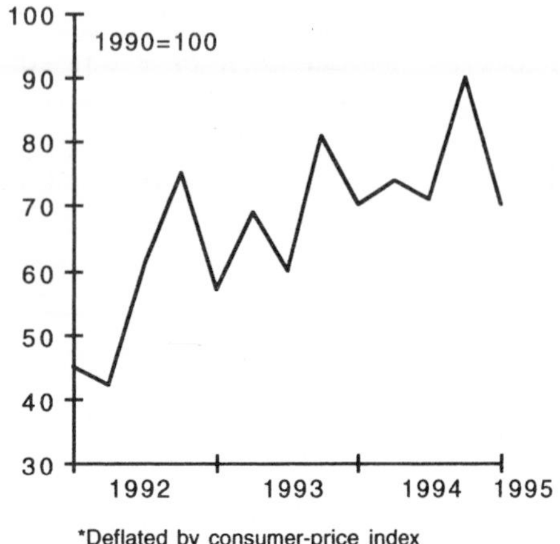

FIGURE 10.3 *Real income change: Russia, 1991–1995* (*Source*: Based on *The Economist*, 25 April–May 5 1995, 63)

FIGURE 10.4 *Income inequality: Russia, 1992–1994* (*Source*: Based on *The Economist*, 25 April–May 5 1995, 63)

Chechnya, Ingushetiya, and Dagestan Republics. Average wages in these troubled parts of the Russian Federation ranged between about one-tenth and one-fifth of those in the Siberian resource frontiers. In the late Soviet era, the maximum average regional income differential was in the order of 1:3, and this mostly reflected the basic core–periphery dichotomy in the country, that is, differences between European Russia and some of the republics in Middle Asia. In August 1993 when the wage survey was undertaken, the official threshold for poverty in Russia was about 50,000 roubles per month. While thresholds of this type must be assessed in terms of regional variations in the actual cost of living, Fig. 10.5 nonetheless indicates that in many of Russia's regions in mid-1993 there was likely considerable economic hardship. Indeed, even the comparatively high incomes in those resource rich but northerly parts of Siberia have to be seen in the context of the local cost of living. Clearly, food expenditures alone would have been very much greater in the highest average monthly wage regions in the climatically harsh Siberian realm than in the poorest regions in the more equable north Caucasus simply because of the relative differences in agricultural productivity. Moscow's workers fell into the 48,000 to 85,000 rouble average monthly wage category. Following the price liberalization programme introduced by President Yeltsin in 1992, Muscovites with average monthly incomes of less than 50,000 roubles would have had difficulty making ends meet. Most of the city's pensioners, a group comprising about a quarter of the city's total population, received pensions of less than 48,000 roubles per month, and were impacted accordingly. The average monthly wage data for Russia in August 1993 depicted in Fig. 10.5 certainly suggest growing regional disparities. In those regions dominated by heavy industry of the old economy, iron and steel, chemicals and the like, wage earners are at considerable risk since the demand for what they can produce has shrunk or evaporated entirely. Already in 1993, shortened work weeks had significantly reduced incomes. Agricultural workers on the now privatized collective and state farms are at most risk in terms of being poor or very poor. This is because a very large share of such enterprises is in fact bankrupt. Fortunately, many of these people can produce a substantial part of what they eat. In short, the regional income disparities which had narrowed during most of the Soviet era have widened since independence, and very significantly at that. Even though there are no comparable regional average salary data available for the other post-Soviet states, one measure of standard of living for which there is information, household budget

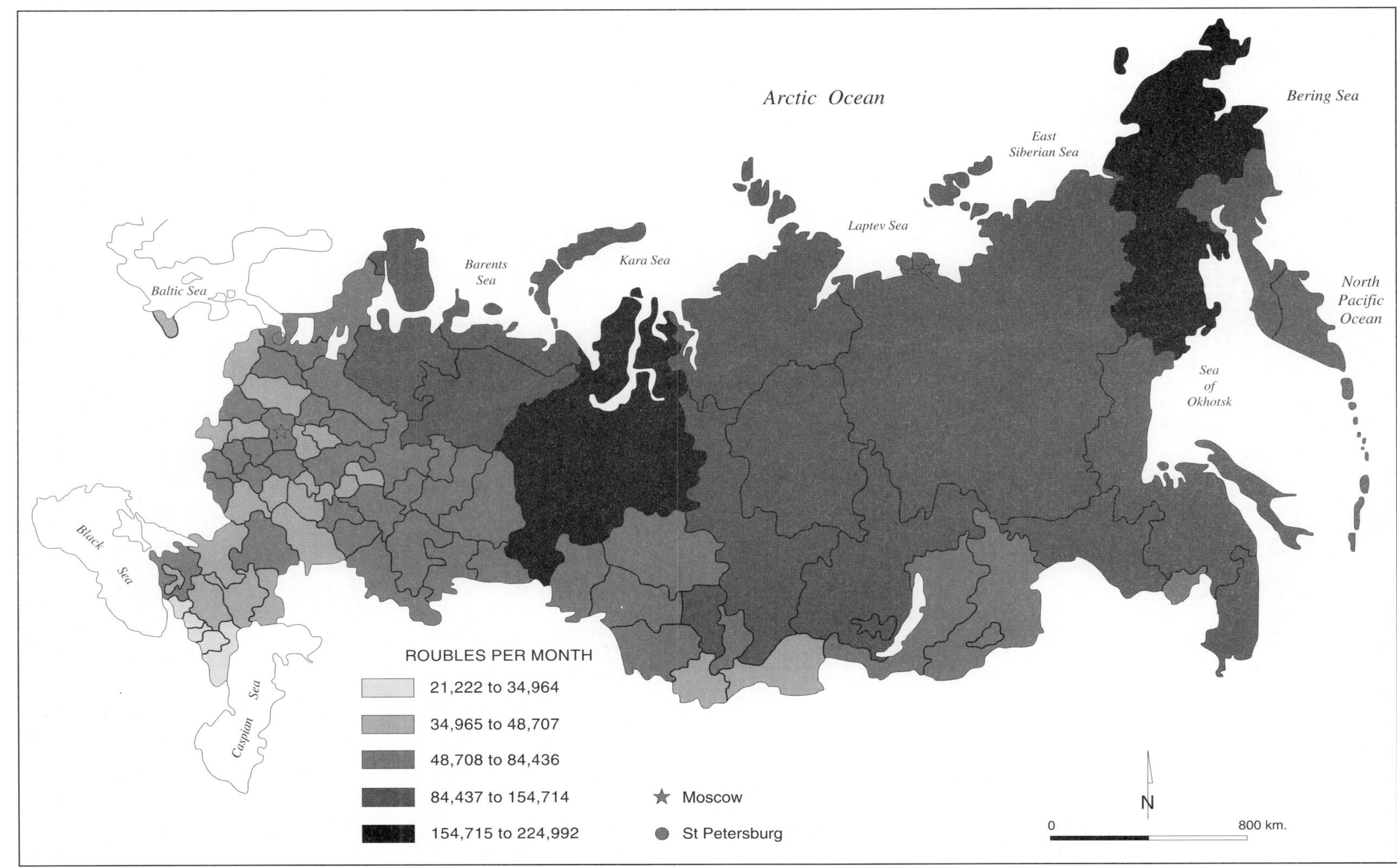

FIGURE 10.5 *Average monthly wages: Russia, August 1993*
(*Source*: T. Heleniak, *Economic Geography of the Russian Federation* (Washington: The World Bank, Country Department III, Europe and Central Asia Region, 1994), Vol. 1, No. 25)

allocations, confirms that there are more similarities than differences in standard of living trends since 1991.

Table 10.2 provides a measure of temporal change in the relative importance of the basic household budget allocations across all but one of the post-Soviet states. While data for Estonia were not available, the basic trends in Latvia and Lithuania are most likely reasonably representative of the situation there as well. The average expenditure pattern in Soviet households in 1989 provides the first point of reference, while the republic average income for non-agricultural workers offers some additional insights into what the relative differences in household allocations might mean in monetary terms (Fig. 10.2). Per capita consumption of basic foodstuffs also affords some insights into the quality of the diet which the expenditures on food give people across the post-Soviet scene (see Table 7.3, p. 214). The three Baltic states had above average monthly incomes in 1989. Thus, the proportion spent

Table 10.2 *Post-Soviet states household expenditures: 1990–1994 (per cent)*

State		Food	Non-food	Alcohol	Consumer services, etc.
Russia	1990	36.1	46.6	5.0	12.3
	1994*	46.4	40.6	2.7	10.3
WESTERN TIER					
Belarus'	1990	33.7	47.3	7.4	11.6
	1994*	57.2	28.4	4.7	9.7
Ukraine	1990	38.4	46.1	3.7	11.8
	1994	50.6	37.1	1.9	10.4
Moldova	1990	34.4	50.2	2.6	12.8
	1994	45.5	37.1	2.0	15.4
Estonia	1990	–	–	–	–
	1992	–	–	–	–
Latvia	1990	29.4	37.6	4.4	28.6
	1992*	48.1	24.1	2.9	24.9
Lithuania	1990	33.9	47.5	6.1	12.5
	1992	59.7	27.4	10.0	2.9
SOUTHERN TIER					
Georgia	1990	49.7	38.6	1.3	10.4
	1991*	55.5	35.2	1.1	8.2
Armenia	1990	52.1	35.8	4.0	8.1
	1994	68.6	23.1	1.3	7.0
Azerbaijan	1990	53.4	36.5	1.3	8.8
	1994**	72.3	21.0	1.0	5.7
Kazakhstan	1990	38.6	44.5	5.0	11.9
	1994	54.4	30.2	3.4	12.0
Kyrgyzstan	1990	37.8	47.3	3.9	11.0
	1994	55.4	31.3	2.7	10.6
Tajikistan	1990	44.4	44.1	1.8	9.7
	1993	70.3	22.5	2.2	5.0
Uzbekistan	1990	48.6	37.5	2.9	11.0
	1994*	63.9	27.0	2.6	6.5
Turkmenistan	1990	43.5	43.3	3.4	9.8
	1994	56.7	34.2	4.0	5.1

*First nine months
**October
Sources: Ekonomika Sodruzhestva Nezavisimykh Gosudarstv v 1994g (Moscow: Statkomitet, 1995), 82–83; *Latvia Today, 1992* (Riga: State Committee for Statistics, 1993), 41; *Lithuania's Statistics Yearbook* (Vilnius: Lithuanian Department of Statistics, 1993), 58

on food in 1990 in Latvia and Lithuania, perhaps not surprisingly, was certainly less than in many other states. But it did not differ much from the relative allocation in Belarus' and Moldova, both of which had below average monthly wages (Fig. 10.2). Obviously the price of food plays an important role in explaining differences in household expenditures. But it is especially notable that, with the exception of Latvia, in no state in 1990 was the share of household budget expenditures for food below the official 1989 Soviet average of 31 per cent, a figure angrily refuted by many ordinary people as noted earlier. Indeed, in six states, more than 40 per cent went on food, and in Armenia and Azerbaijan more than one-half. To be sure, varying classifications of food items, to say nothing of the accuracy of the data collected and published, may well mean that the figures in Table 10.1 are at best approximations of the real situation. But the temporal trends are consistent within individual states, and they unequivocally indicate a substantial shift in domestic spending habits which have direct impact on the standard of living.

A careful examination of Table 10.2 reveals that for the period from 1990 to 1994, expenditures on food increased dramatically. Even in the historically better off Baltic region, there was in excess of a 50 per cent increase in the share of the household budget expenditures on food in both Latvia and Lithuania, and this had already occurred by 1992 (the most recent year for which information was obtainable). It is not unreasonable to assume a similar trend prevailed in Estonia. By 1994, in at least 11 of the 15 states buying food took more than one-half of the household budget. In Armenia, Azerbaijan and Tajikistan where civil wars have devastated large parts of the domestic economy, expenditures on food represented more than two-thirds of the budget. And as the discussion of post-Soviet consumption of basic foodstuffs indicated, nowhere has there been an improvement in the diet since independence (Table 7.3). About the only positive feature of the current spending patterns is that with the exceptions of Tajikistan, Turkmenistan and Lithuania, alcohol comprises a smaller share of the budget. Why these states should be apparent exceptions is not clear. At any rate, it is evident that it was necessary to spend more on food in 1994 than in 1990 without improving the diet. Obviously, real incomes for most people have declined since independence, a conclusion fully in keeping with many of the trends in economic activity discussed in the preceding chapters. However, it is not just real income which has dropped, so too has the number of people with jobs. With economic reform has come unemployment.

Official statistics for unemployment across the post-Soviet scene are widely acknowledged to be somewhat off the mark since they customarily only reflect the number of people who have registered as unemployed, who are actively seeking a job through the state employment bureaus. Many unemployed workers simply do not bother registering. In 1994, average unemployment across the CIS was officially just 2 per cent. Armenia had the largest share of the workforce unemployed, 5.6 per cent, or nearly 92,000 people in total. In absolute numbers Russia's 1.6 million officially unemployed topped the list, accounting for more than three-quarters of the total for all CIS countries, that is, all post-Soviet states save for the Baltic states. In Latvia, nearly 50,000 workers were unemployed in April of 1993. This was 3.4 per cent of the labour force. Lithuania's unemployment was only 1.1 per cent at the end of December 1992, or about 21,000 people. In Lithuania, as in all other post-Soviet states, women predominate amongst the unemployed. In December 1992, they accounted for more than three-fifths of the total. In Russia, women comprised about 70 per cent of the registered unemployed in 1993. In Uzbekistan, the figure was 64 per cent in 1993. When redundancies occur, it is women who are usually the first to be dispatched since their jobs are seen by the typical male decision-maker, and by some women as well, as being less valued than those held by the male 'bread winner' of the family. In the Soviet era, most urban women worked at a job outside the home out of necessity. Two incomes were customarily required in order to achieve a reasonable standard of living. A common refrain in the post-Soviet era is that decisions to terminate women ahead of men simply reflects the deeply ingrained inequality in gender opportunities which characterized the Soviet era, something which was reflected in the previously mentioned mal-distribution of hours spent on household chores between men and women. It is difficult to know with any precision just how many people have lost their jobs in the new economy. The official unemployment level in Russia in mid-1995 was just under 3 per cent, or a shade more than two million workers, but this is obviously an underestimation of the real situation. One estimate was that real unemployment in Russia at this time was about 9.6 million, or 13 per cent of the country's labour force. Job losses are not everywhere of the same order. Regions dominated by heavy industry, and especially those regions and cities with a narrow economic base, have been especially hard hit. The distribution of official unemployment across Russia in late 1993 is depicted in Fig. 10.6. Areas of heavy industrial development, such as the Urals economic region and parts of Siberia, stand out as having higher than average levels of unemployed.

The old Soviet-era adage that the state just pretended to pay a decent wage being matched by workers just pretending to work has been taken to a new level in the post-Soviet era. Since many enterprises are now in arrears in salary and wage payments, many people who remain on the books as employees do not even bother going to work regularly. For those with marketable skills, especially but scarcely exclusively members of the *intelligentsia,* multiple 'full-time' jobs are not uncommon. Indeed, this is often necessary in order for members of the *intelligentsia* to sustain their former standard of living since salaries in many previously well-paid government and scientific occupations have been eclipsed by inflation. But not all who require a certain level of income are able to earn it. In restructuring the economy, jobs are being shed, permanently. In many enterprises, the official employment has not been reduced, largely through fear of creating widespread social unrest. Being employed, but not working regularly or being paid regularly, is commonplace. For example, in 1993 and 1994, only about two-fifths of the labour force in Russia was being paid fully and on schedule. In the first year or so of independence, many people were able to sustain their standard of living in the face of significant levels of inflation and worsening employment prospects by relying on savings, which as noted already had escalated during the last decade of Soviet power owing to the shortages of consumer goods on which to spend higher salaries and wages. However, this was only a short-term response to a longer term problem of economic dislocation associated with the market reform process.

Throughout the Soviet era, comparatively low wages and standard of living were at least partially compensated for by the prevalence of a social safety net. People did not live as well as in many western industrial economies, but they were at least spared the worry of what to do if a member of the family got sick. To put the changes in health care and public health after independence into context, the broad features of the Soviet system will be described first.

Soviet-era health care and quality of life issues

According to the Soviet Constitution all citizens had the right to free, qualified medical care. Thus, the state began to assume responsibility for health care from the outset. During the years of Lenin's New Economic Policy (1921–1928), a fair measure of free enterprise was tolerated, as noted in Chapter 3. While the medical profession was not prohibited from providing services for fees, doctors were encouraged to play a role in developing the nascent state socialist medical care system by becoming state employees. In 1922, there were about 5000 hospitals, 193,000 hospital beds and just over 21,000 medical doctors of all specializations for a population of 136 million. By 1926, the state itself had established polyclinics which charged moderate fees and offered services in competition with doctors in private practice. By the end of the New Economic Policy period, some progress in expanding the public health care delivery system had been made. The

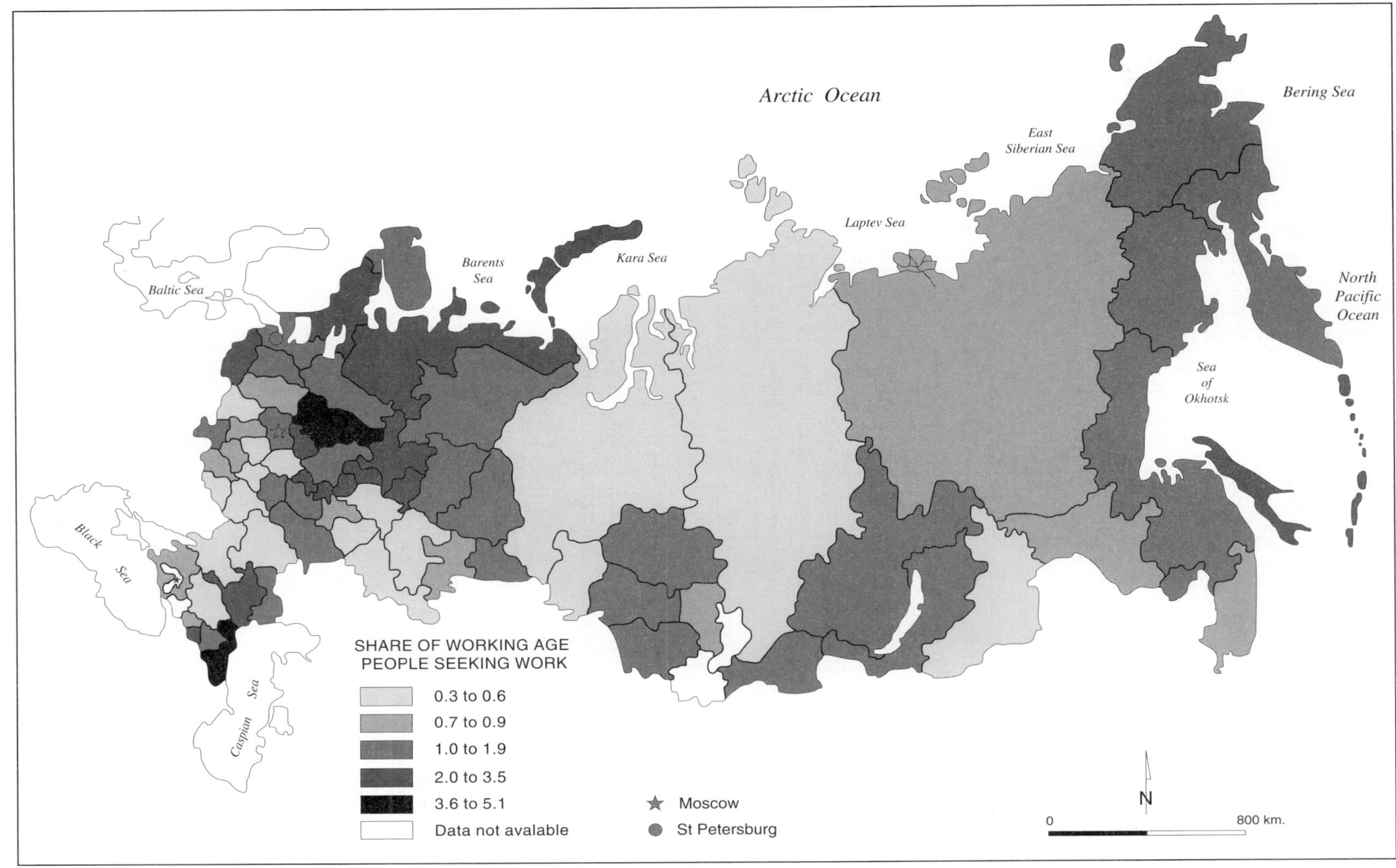

FIGURE 10.6 *People seeking employment as a share of total working age population, August 1993*
(*Source*: T. Heleniak, *Economic Geography of the Russian Federation* (Washington: The World Bank, Country Department III, Europe and Central Asia Region, 1994), Vol. 1, No. 24

number of doctors per 10,000 population stood at four, still clearly inadequate but a four-fold improvement over the situation in 1917. The number of beds per 10,000 population increased from ten to sixteen over the same period. In summary, the 1920s witnessed a continuation of the pre-revolutionary-era practice of fee for service medical care, the beginnings of a system of special facilities for the privileged, and the attempt to provide free health care for the masses.

The Stalin era produced a vast expansion of the health care system, but one which was more highly differentiated than the relatively simple system which preceded it. Under Stalin, privilege was permanently embedded. The delivery of medical care was first of all divided into what were often labelled open and closed systems. The first included all those free facilities and services which were available to the general public according to place of residence. Additionally, since medical doctors were still able to have a limited private practice, those ordinary folk with the necessary wherewithal could avail themselves of private care as they saw fit. The second, closed, system was designed to deliver free medical care to special groups. Generally speaking, the more important the group the higher the quality of medical attention and facilities provided. So much for the egalitarianism heralded by the 1936 proclamation that the Soviet Union was now officially a socialist state.

By 1940, the number of doctors per 10,000 population stood at nearly eight; the number of hospital beds had increased to 40 per 10,000. A more illuminating statistic in terms of the quality of life is how long the average person could expect to live. At the turn of the century life expectancy was around 32 years. By the late 1920s it was already at 44. During the 1930s it improved still more, but there were considerable regional variations, and the difference between men and women was becoming quite pronounced. For millions of people, the 1930s were scarcely years of a good, or even adequate, standard of living. Rationing of food was introduced in cities at the beginning of the decade. In the countryside collectivization took a huge toll in human life. The late-1930s purges put all citizens at potential risk. Nonetheless, officialdom could and did point to the growth in medical care personnel and facilities. For the privileged and the emerging *meshchanstvo*, or middle class, life and, statistically, more of it, had never been better. For the urban masses, and the impoverished peasantry on the collective farms it was a moot point whether the same could be said.

Over the next 40 years the advances registered in medical care infrastructure and personnel were quantitatively impressive. For instance, the number of medical doctors increased from roughly 155,000 in 1940 to nearly one million in 1980. By 1989, the number was nearly 1.3 million. From 1940 to 1989, the number of hospitals doubled, and the number of beds grew from about 790,000 to 3.8 million. Measured in terms of doctors and hospital beds per 10,000 population the changes were 8 to 44 and 40 to 133 respectively between 1940 and 1989. In terms of crude measure such as these, the situation in the Soviet Union compared very favourably with other industrial states. Indeed, on the basis of these two measures alone, in 1989 no other country was so well off on a per capita basis. However, the health care system in the 1980s was far from homogeneous on a qualitative basis, and far from satisfactory when measured in terms of the health care delivered to the average Soviet citizen. In terms of the distribution of health care facilities, it was also far from equitable.

As Fig. 10.7 reveals, in the period from 1940 to 1981 there was general improvement in the ratio of doctors per 10,000 population across all regions. Most authorities agree that the relative difference between the best and worst-off regions was ameliorated in the post-war era, as indeed were most other measures of the material conditions of daily labour and life. Still, in 1981 the Middle Asian realm trailed the other regions. For example, Tajikistan with 24 doctors per 10,000 people had less than one-half of the number serving the population of the Centre region. Since the Centre included the Moscow agglomeration that there should be a difference is not surprising. However, as Fig. 10.7 indicates, there were many other regions outside the Moscow area where access to medical personnel was substantially

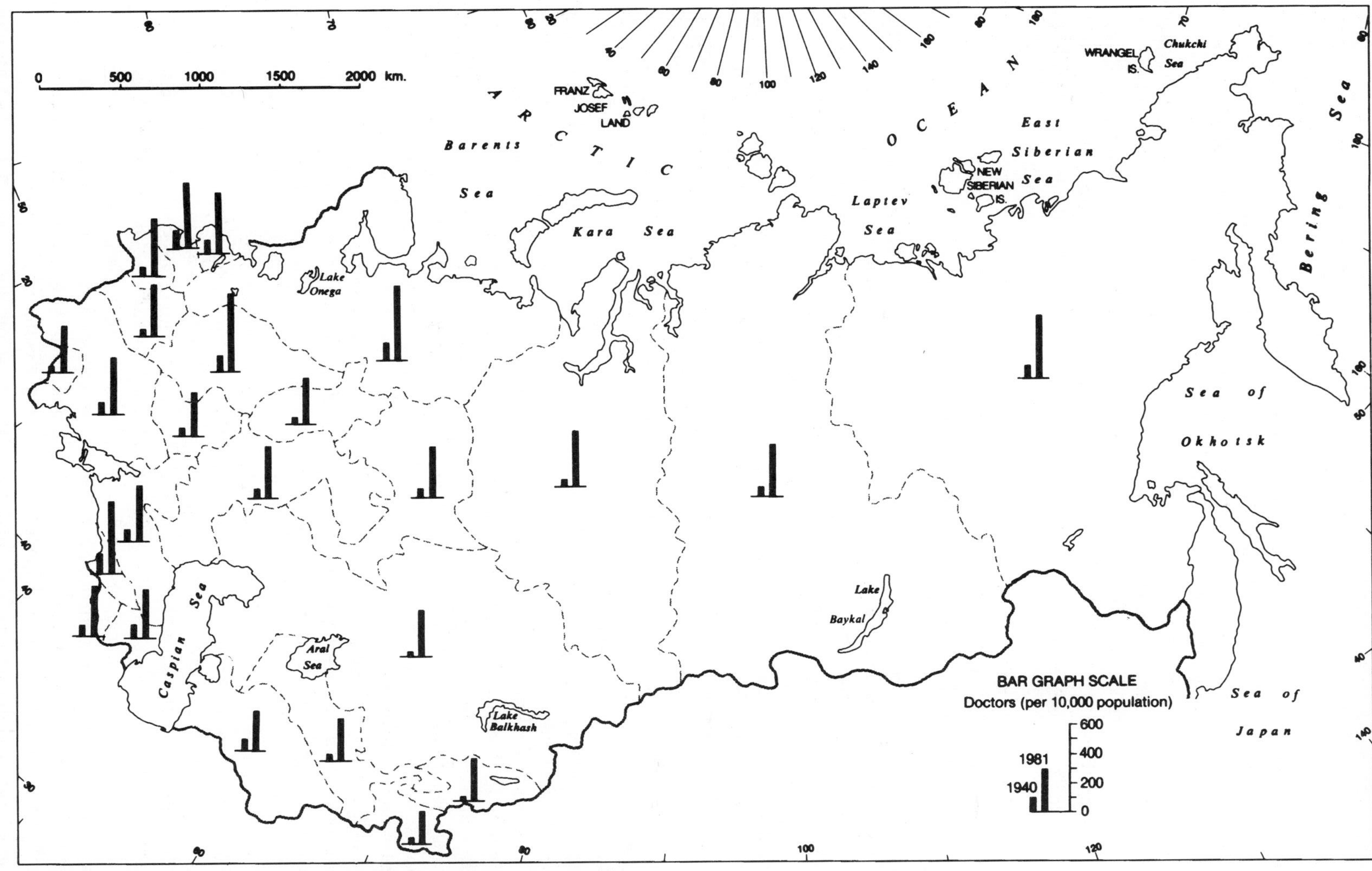

FIGURE 10.7 *Soviet era medical doctors per 10,000 population, by republic*
(*Source*: T. Buck, J. Cole, *Modern Soviet Economic Performance* (Oxford: Basil Blackwell, 1987), 165)

better than in Tajikistan. For the most part these regions were part of the Baltic and Slavic culture realms. With the notable exception of Georgia, the ratio of doctors per 10,000 people in the southern tier republics was about 20 per cent below the national average of 38.5 in 1981. Oddly enough, the number of hospital beds per 10,000 Georgians was below the national average (107 as compared to 126), as it was in Armenia and Azerbaijan as well. There were suspicions at the time that admissions to the Georgia's medical schools were less than rigorous, and that children of those in positions of authority and privilege seemingly were predominant amongst beneficiaries of admissions policies. Bribes were also presumed to have played some part in determining who, and how many, were admitted. The net result was that the number of trained medical personnel in Georgia actually exceeded the number of positions available to practice, while in other regions there were critical shortages. In any event, numbers of doctors, hospital beds and so forth neither speak to the question of accessibility of the population at large to health care, nor to its quality.

In most western industrial states, medical doctors enjoy prestige and incomes commensurate with high status amongst the professions. During the Soviet era, in contrast, most medical doctors earned less than skilled workers in priority industrial sectors. Indeed, salaries were generally below those for white-collar workers until 1987 when the discrepancy was tackled by a decree which raised the average salary of all medical personnel by more than one-third. Much was made of the fact that women played an important part in the expansion of medical staff during the Soviet period. However, those sectors of the economy in which a female labour force predominated were invariably amongst the lowest paid. The medical profession was no exception. Women comprised about 70 per cent of the medical doctor category during the 1980s, a decline from earlier years when their share had reached 76 per cent. The relative size of the wage bill for Soviet health care therefore was traditionally lower than might otherwise be expected based on western experience.

More significant than the relative size of the wage bill was the fact that the share of the state budget allocated to health care was declining for much of the post-Second World War period. About 5.2 per cent of the total budget was devoted to health care in 1950. By the early 1960s, it was nearly 7 per cent. However, by the end of the decade it had dropped to 6 per cent. And ten years later, it had fallen a further percentage point. By the close of the Soviet era it had dropped to about 4 per cent, about half what many authorities reckoned was required. In most western industrial countries, the growing proportion of the elderly has been responsible for the steady escalation in the cost of health care. The Soviet Union was no exception in having to deal with an ageing population, as was noted in Chapter 4, but the state did nothing to augment the health care system in order to deal with the growing demand of the elderly.

In the final few years of the Soviet era, the number of fee-charging health care institutions began to increase. In 1987, visits to fee-paying hospitals and polyclinics represented about 3 per cent of the 123 million visits paid to state-supported institutions. The share rose each year thereafter until the collapse of the Soviet system. Fee for service medical care simply expanded on the existing system of differentiated health care. But apparently about three-quarters of the population were already making 'payments' to medical personnel in the open system with the expectation of better treatment in return. Thus, the publicly accessible system was long 'distinguished' by requiring payment of unofficial fees in the hope of receiving better service. The so-called closed system of medical care clearly provided a more sanitary environment and, one presumes, less busy if not more proficient medical personnel. Highly differentiated itself in accordance with the importance of ministry, enterprise, or special interest group represented, it was not insignificant in terms of its contribution to health care of the Soviet population. Such closed facilities employed about 40 per cent of Moscow's physicians. But in the final years of Soviet power, money as well as the time honoured privilege was sufficient condition for access to better health care. The expansion of fee for service medical care was not introduced without controversy, however. With the Gorbachev era phenomenon of *glasnost'* came public criticism directed at all forms of privileged access to something which people firmly believed was everyone's right as part of the unwritten social contract with the state. Notwithstanding public sentiment about the inequities of the system of health care, the abysmal quality of the facilities available to the average person assured any new fee-paying enterprise a steady clientele. For example, of the nearly 23,500 hospitals in the country in 1986 about 3900 were district-level institutions. Of the latter, 65 per cent lacked a hot-water supply. Perhaps more serious, in excess of a quarter of them were dependent upon septic tank systems of one type or another. About one-sixth of them did not even have running water! The condition of the 18,000-odd rural polyclinics may be imagined. Medical facilities being the source of infectious disease is less than surprising in light of such statistics. Quality of the staff was also becoming a serious issue in the late Soviet period. Of some 350,000 doctors assessed for medical competence in the mid-1980s, nearly a tenth were found to be deficient in one way or another.

With the emphasis accorded to free medical care after the revolution, infant mortality rates dramatically improved compared to the levels obtained in late imperial Russia. In 1913, it is reckoned that 273 out of every 1000 infants did not survive the first year, but this is likely a conservative figure given contemporary methods of reporting such deaths. By 1940, the ratio stood at 184 per 1000. During the 1950s it was more than halved, and by 1960 was just 35 per 1000. By 1971 health care improvements had reduced infant mortality to 22.9 per 1000. During the 1970s the situation deteriorated, however, as was briefly mentioned in Chapter 4. In 1980 the ratio had risen to 27.3, which was cause for widespread concern. Indeed, the actual figure might have been higher than 25.4; for example, 30 per 1000 has been mentioned in some official statements. The latter figure put the Soviet Union at around fiftieth in the world in terms of combatting infant mortality at that time.

As the data in Table 10.3 indicate, the pattern of infant mortality also varied markedly from one republic to another. Moreover, the general improvement noted above was not consistent by year in all republics. For example, in Estonia and Turkmenistan infant mortality was higher in 1989 than in 1985. The Turkmenistan figure was about the same as Guatemala and Madagascar, and was higher than Mexico or the Philippines. The ratios in Middle Asia were especially poor, but there were many other regions where there was also reason for concern.

Table 10.3 *Soviet era, infant mortality by republic: number of deaths before one year of age per 1000 births*

Republic	1970	1980	1985	1989
Russia	23.0	22.1	20.7	17.8
WESTERN TIER				
Ukraine	17.2	16.6	15.7	13.0
Belarus'	18.8	16.3	14.5	11.8
Estonia	17.8	17.1	14.0	14.7
Latvia	17.9	15.4	13.0	11.1
Lithuania	19.4	14.5	14.2	10.7
Moldova	23.3	35.0	30.9	20.4
SOUTHERN TIER				
Georgia	25.3	25.4	24.0	19.6
Armenia	25.3	26.2	24.8	20.4
Azerbaijan	34.8	30.4	29.4	26.2
Kazakhstan	25.9	32.7	30.1	25.9
Kyrgyzstan	45.4	43.3	41.9	32.2
Tajikistan	45.9	58.1	46.8	43.2
Uzbekistan	31.0	47.0	45.3	37.7
Turkmenistan	46.1	53.6	52.4	54.7

Sources: *Narodnoye Khozyaystvo SSSR za 70 Let* (Moscow: Finansy i Statistika, 1987), 408; *Narodnoye Khozyaystvo SSSR v 1989g* (Moscow: Goskomstat, 1990), 41

Amongst the reasons for the abysmal statistics for Middle Asian republics, the poor, indeed, unsanitary state of local hospitals and clinics figured prominently. Overcrowding of facilities simply compounded problems raised by unhygenic conditions. In the late 1980s in Uzbekistan, for example, an additional 29,000 paediatric beds were required to fulfil existing norms. In Turkmenistan three-fifths of the maternity clinics, maternity wards and paediatric hospitals did not provide hot water. Septic tanks serviced about two-thirds of the republic's hospitals. Inadequacies in these systems for handling waste were thought to be a major reason for pollution of the water supply for a substantial number of the 127 hospitals in Turkmenistan which still had to make do without piped water at the close of the Soviet era. Intestinal infections, viral hepatitis, toxaemia and septicaemia all took an inordinate number of the newborn in Middle Asia. Deficiencies in training medical personnel were also acknowledged to figure in the exceptionally high levels of infant mortality. In short, these and other regions compared poorly with most western industrial states. The health care system legacy for the post-Soviet states was far from auspicious, and regonally variable.

Inasmuch as these data are averages across a differentiated health care system – poor facilities for the masses, better facilities for special groups – there was much scope for bitterness on the part of the disenfranchised segment of society. The situation in the capital was atypical, but by the same token Soviet society comprised a substantial number of special interest groups. Under Gorbachev's policy of *glasnost'* such facilities came under close scrutiny and public discussion. Comparisons were frequently made with publicly accessible facilities, which were not always physically separate, but which were invariably poorer.

POST-SOVIET HEALTH CARE AND QUALITY OF LIFE ISSUES

The notion of cradle to grave security was the centre piece of the social contract the Soviet government had with the people. The continuation of the health and welfare benefits and social services not surprisingly is widely endorsed by people across the former Soviet Union. While maintaining some form of social safety net at a level consistent with the needs of the population remains a common goal, in practice the present state of post-Soviet economies makes it nearly impossible to do so. From hospitals, to schools, to pensions, to cultural programmes, the monies available are insufficient to meet Soviet-era standards, let alone what might be optimally required. During the final years of Soviet power, the health care system was under growing pressure. Indeed, the central government had begun the process of both opening the door to fee for service, as well as off-loading financial responsibility for it to the republic and local level governments. On an international comparative basis, supply side indicators such as number of doctors or hospital beds put the Soviet Union in a quite favourable position. What the system delivered in terms of quality of health care was another matter, as already noted. In this section, the discussion will focus on selected facets of the health care system. For the post-Soviet scene, trends for a selection of supply side indicators will be considered. But more emphasis will be given to qualitative measures of the state of public health such as rates of infant mortality and causes of death amongst selected post-Soviet populations. Finally, the issue of life expectancy discussed in Chapter 4 will be briefly revisited. While there are many other aspects of the health care system which affect the quality of life, these are deemed to be representative indicators of what is happening in the new economies of the post-Soviet period.

Across the former Soviet Union health care has deteriorated. In most states, the share of total budget allocated to this sector has declined, and given the fast rising costs of equipment and medicines this means that what is being delivered is less than during the Soviet era. In Russia in 1993, the share of enterprise budgets dedicated to social–cultural expenditures, which included health care, varied quite markedly by region as Fig. 10.8 indicates. While the map does not indicate the absolute scale of the budgets involved, and hence the actual sums spent, it does suggest growing regional disparity in provisionment of both hard and soft services, that is, facilities and staff complements. Basic supply side indicators are slipping. A few examples of prevailing trends will suffice. In Russia, the ratio of doctors dropped from 47.3 per 10,000 population in 1989 to 45.2 in 1993; the same ratio of hospital beds from 138.7 to 129.4. In the comparatively well-off Baltic states of Lithuania and Latvia, the supply of doctors and hospital beds per 10,000 population also declined from the late Soviet era to 1992, the latest date for which data were available. In 1989, Lithuania's ratio of doctors per 10,000 ulation was 45.7; by 1992 it was 42.4. The ıber of hospital beds had dropped by nearly ·r cent over the same period. Latvia was the provisioned of all Soviet republics in 1989 in ıs of these two indicators. From 1989 to 1992, ratio of doctors in Latvia slipped from 50 to he ratio of hospital beds, from 147 to 130 per 00 population. Similar declines are common ughout the other western tier and southern states. The attrition in the number of medical ors is certainly caused in part by low month- alaries, salaries which at present are little ve the official threshold for defining poverty. ıies allocated to health care as a proportion of, nstance, gross domestic product are not only erally less now than in the Soviet era, given escalation in prices for such essentials as rmaceuticals, the impact on health care is ally much greater than implied by the ive change in share of budget allocated to this or. Throughout the former Soviet Union ss to necessary medicines is hampered both ost and supplies. Whereas previously pres- ions were essentially free of charge, fees are charged in many countries. Where prescrib- rugs can be had, the cost is frequently well nd the means of the patient. Often what is ired is simply unavailable. Ukraine used to luce about a third of all Soviet-era pharma- ical output. The disintegration of the Soviet em, however, has left many of the pharma- ical factories in Ukraine without key inputs. any such plants the equipment is outdated. ıort, production of prescription drugs has psed in consequence. Hard currency is cus- rily demanded for what is for sale. Owing to disruption in production of prescription s since independence, pharmaceutical sup- are being imported from the west in grow- quantities. International aid agencies are ged in most post-Soviet states, but the c health problems being addressed are in y places far greater than such agencies can do ı about with available monies.

e health care delivery system everywhere e post-Soviet scene is undergoing change. ı of this change is related to a reduction in ing for health care services, facilities and materials. To be sure, there was much in the former health care delivery system that was distorted and in need of rationalization. For instance, the total number of medical doctors probably exceeded what was actually required. Ukraine, for example, is now attempting to manage the supply of doctors by restricting entry to medical schools. Moreover, the distribution of medical doctors in the former Soviet Union was imbalanced. Rural regions in poor republics were the worst off, as the foregoing discussion indicates. Reducing the days spent in hospital beds, reducing the ready and free access to a full range of health care services, all of these measures and more besides could be implemented without necessarily negatively affecting the general well-being of the population. These are policy and management issues in many, if not most, developed countries as well. In many western, developed countries the health care systems which have evolved over the years have come under intense scrutiny in terms of improving performance because of rising health care costs associated with, for instance, ageing populations. But the number of doctors, hospitals, the beds available and so on are only very crude measures of what is being provided in terms of the quality of health care. It is in this context that the post-Soviet situation is distinctly different from western industrialized states. Irrespective of the current scale and cost of health care inputs, what is being delivered is seriously deficient. In all respects there are deficiencies in the state of public health, and these reflect a deteriorating quality of life in all post-Soviet states.

Infant mortality rates were presented for the final two decades of the Soviet era in Table 10.3. The data presented in Table 10.4 describe the changes during the short period of time since independence for all members of the Commonwealth of Independent States, as well as for Lithuania and Latvia which are not member states. The picture presented by these data is broadly the same in most parts of the post-Soviet scene. Save for Armenia, Kyrgyzstan and Uzbekistan, where infant mortality has continued to drop, the trend elsewhere is the opposite. The rate has increased most dramatically in Tajikistan,

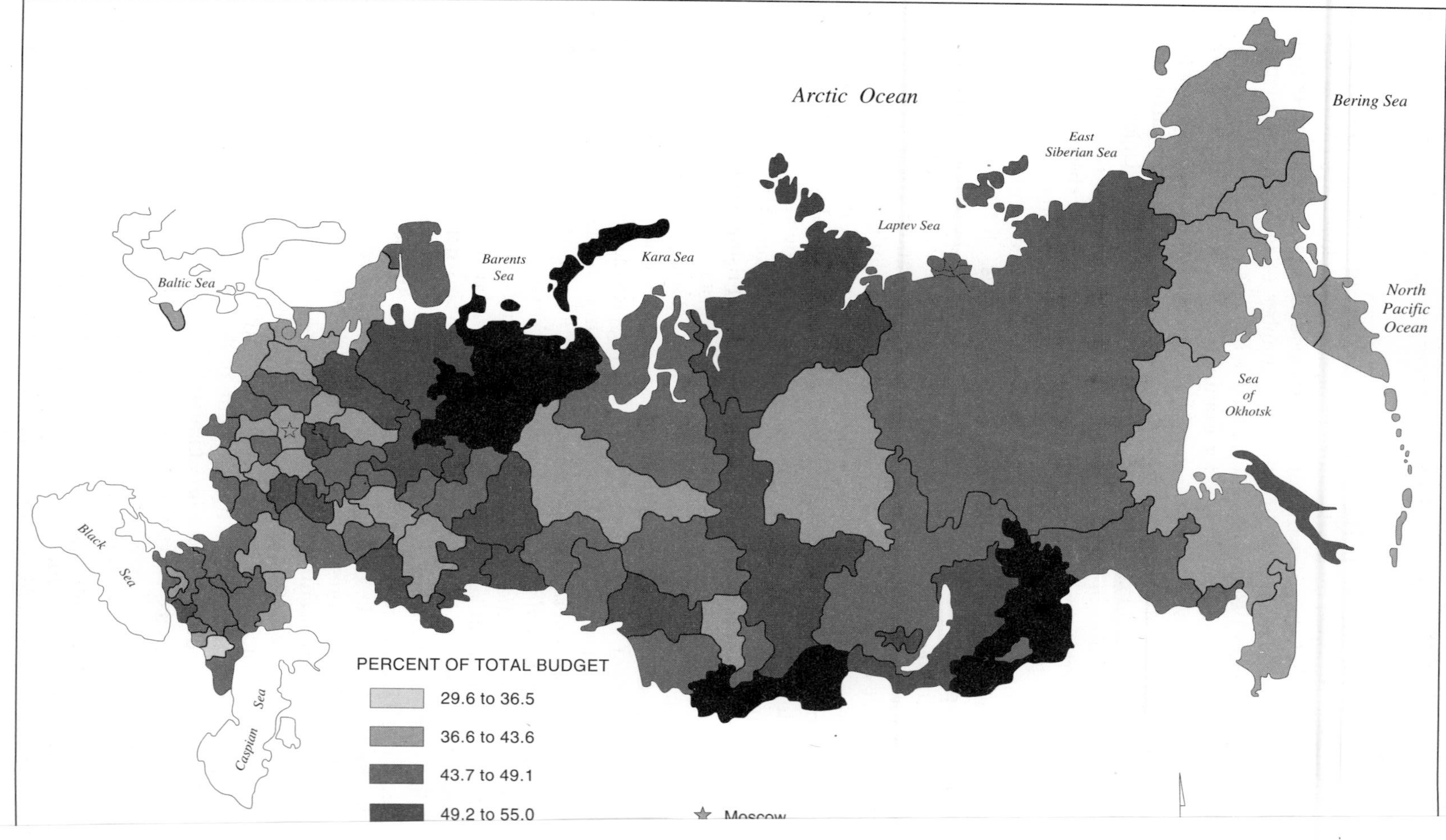
Arctic Ocean
Bering Sea
East Siberian Sea
Laptev Sea
Kara Sea
Barents Sea
Baltic Sea
North Pacific Ocean
Sea of Okhotsk
Black Sea
Caspian Sea
PERCENT OF TOTAL BUDGET
29.6 to 36.5
36.6 to 43.6
43.7 to 49.1
49.2 to 55.0
Moscow

Table 10.4 *Post-Soviet states infant mortality: number of deaths before one year of age per 1000 births*

State	1990	1992	1994
Russia	17.4	18.0	18.7
WESTERN TIER			
Ukraine	12.9	14.0	14.3
Belarus'	11.9	12.3	13.2
Estonia	–	–	–
Latvia	13.7	17.4	–
Lithuania	10.3	16.5	–
Moldova	19.0	18.4	22.3
SOUTHERN TIER			
Georgia	15.9	12.4	18.3
Armenia	18.6	18.5	17.1*
Azerbaijan	23.0	25.5	26.9
Kazakhstan	26.4	26.1	27.4
Kyrgyzstan	30.0	31.5	29.4
Tajikistan	40.7	45.9	47.0*
Uzbekistan	34.6	37.4	28.3
Turkmenistan	45.2	43.6	46.4

*1993

Sources: Ekonomika Sodruzhestva Nezavisimykh Gosudarstv v 1994 (Moscow: Statkomitet SNG, 1995), 63; *Latvian Statistics in Brief 1992* (Riga: Baltika, 1993), 6; *Lithuania's Statistics Yearbook* (Vilnius: Lithuanian Department of Statistics, 1993), 23

Latvia and Estonia. Tajikistan has been the scene of civil war which may partially account for the increase in the infant mortality rate. But Latvia and Lithuania have not had to contend with such a devastating domestic upheaval, and therefore the sizeable increase in the rates in those countries over such a short period of time must be a very disturbing trend for authorities. Overall, typical infant mortality rates in the early 1990s in the post-Soviet states were at least twice those in the industrial countries of Western Europe and North America. In post-Soviet states with infant mortality rates of 20 per 1000 births or higher, newly industrialized southeast Asian states such as Thailand and Malaysia are more appropriate reference points than are the industrialized nations of the west. But in the typical developing country of southeast Asia, the infant mortality rate trend is down, not up as is characteristic of 12 of the 15 post-Soviet states.

The death rates by cause data for Russia presented in Table 10.5 serve to underscore the dimensions of the health care problem confronting all post-Soviet states. The percentage changes in the number of deaths from the various ailments listed in Table 10.5 are indicative of deteriorating public health conditions. Infectious diseases are in the ascent in most regions, are frequently related to impure water supply, and run the gamut from dysentery to cholera. Respiratory ailments have long taken a sizeable toll of middle aged men in particular. In general, there has been a marked increase in deaths due to diseases of the heart and circulatory system. Rising levels of stress produced by chaotic economic and social conditions of the past few years no doubt figure prominently. And the deterioration of diet noted in an earlier chapter can do little to help the general state of wellness in Russia and the other post-Soviet states.

Table 10.5 *Death rates by cause: Russia, 1990–1994*

	1990	1994	Absolute change, 1990–1994	Per cent change, 1990–1994	Per cent of increase explained by cause
TOTAL DEATHS (THOUSANDS)	1656.0	2301.4	645.4	39.0	100.0
Infectious and parasitic diseases	17.9	29.5	11.6	64.8	1.8
Cancer	284.4	300.6	16.2	5.7	2.5
Heart and circulatory system	915.5	1230.4	314.9	34.4	48.8
Respiratory system	88.0	118.7	30.7	34.9	4.8
Digestive system	42.5	64.8	22.3	52.5	3.5
External causes	198.3	368.4	170.1	85.8	26.4
Other	109.4	189.0	79.6	72.8	12.3

Source: Timothy Heleniak, 'Economic Transition and Demographic Change in Russia', Unpublished paper, 1995

Overall, the lifestyle of the typical Russian, male and female, is far from ideal. Alcohol consumption is high on a per capita basis. Indeed, alcohol still figures quite prominently in the household budget expenditure data cited earlier. The average life expectancy of alcoholics is 15 to 20 years less than the norm, and while alive alcoholics invariably suffer ailments of other kinds. Alcoholics account for as much as one-sixth of the total population, and people who would qualify as heavy drinkers make up anther 10 per cent. Studies conducted in the late 1980s indicated that about two-thirds of the population did not regularly exercise, about one-quarter smoked, and among young people the share was even higher. About half the population is overweight. The deterioration in the diet in the past few years and the associated rise in consumption of starch and carbohydrates does little to help in this regard. The anti-smoking campaign started in 1988 has been completely overshadowed by the more immediate concerns of jobs and wages.

Although customs vary from one region to another, most urbanites enjoy being able to get out into the countryside. In the major cities sizeable tracts were set aside for this purpose during the Soviet era. The concept of a green belt with fingers of open space penetrating the built up area was widely employed as one means of ensuring ready access to the 'outdoors' for urban inhabitants seeking some respite from the urban condition. But now the pace of private sector urban development overrides town planning principles and environmental considerations. Around all major cities across the post-Soviet scene wealthy people are invading former green belts in the quest for a suburban lifestyle, detached house and all. The process of land conversion is having a negative impact on such zones formerly set aside for public recreation public recreation. For example, at the end of the 1980s the pace of urban development in the Moscow green belt beyond the ring road had already taken its toll of green or open space. Similarly, satellite data provide an opportunity to assess the land conversion process by the changes in general land use in the greater St Petersburg area between 1975 and 1987 (Fig. 10.9). Since independence the land use conversion process has accelerated. For many people access to a *dacha* is an important adjunct to the system of sanatoria, rest homes, camps, hotels and so forth. The *dacha*, or cottage, is scarcely a new phenomenon. In the late imperial era suburban trains provided access to thousands of *dachi* in the outskirts of St Petersburg and Moscow. Private car ownership has made weekend travel to a *dacha* easier for thousands of the better-off post-Soviet citizens. Today urbanites everywhere carry on the tradition, renting a room in a *dacha* for part of the summer when, as is most often the case, they do not own such a luxury themselves. But new economy wealth is producing large scale *dacha* development which like private home construction is leaving an indelible impress on the countryside around most cities.

There is little to indicate that anywhere in the post-Soviet scene physical fitness and recreation programmes are commanding much attention or government financial support. Indeed, many recreational and wellness facilities have been shut down for want of money to continue their operation. For ordinary people the possibilities for vacation are now much more restricted than ten years ago. Almost without exception, sanatoria and rest homes have been reduced in number and capacity since independence. In Russia, the number of beds in such facilities has dropped from a peak of 1.3 million in 1988 to 991,000 at the end of 1993. In Lithuania, where there was also an above average number of such facilities on a per capita basis, the number of beds in sanatoria and rest facilities plummeted from 71,400 in 1989, to just 44,400 at the end of 1992. In short, when opportunities to rest and relax for the general public are fewer, when the health of the population has worsened, as no doubt it has quite considerably over the past few years, disease not surprisingly takes a heavier toll of all ages, but especially of the newborn, ill and aged.

In the 'external causes' of death category listed in Table 10.5, which has seen the largest percentage increase, a number of factors are responsible. On-the-job accidents have increased dramatically since independence, in large measure because safety standards and regulations have fallen by

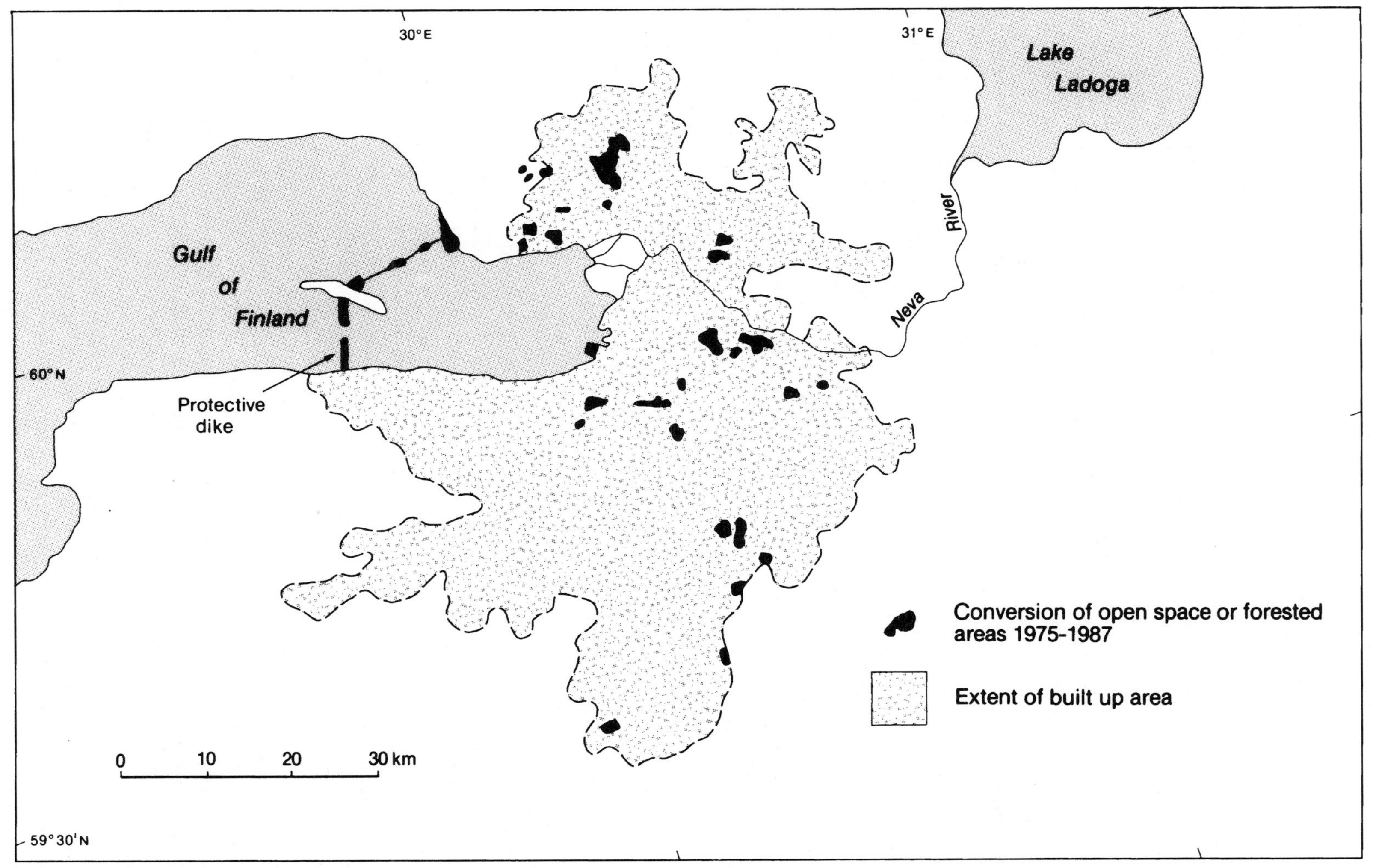

FIGURE 10.9 *Land use change in the St Petersburg urban area, 1975–1987*
(*Source*: Compiled from Landsat images, 1975, 1987)

the wayside in the quest to maintain production at lower cost. Investment in modernizing plant and equipment has all but dried up leaving workers in high risk environments. The deterioration of the health care system has been matched by the demise of long-standing social structures and networks. All of the uncertainty and turmoil has produced much stress, and a wide variety of anti-social behaviour both within families and in society at large. Suicides and murders have also risen sharply since 1991. In 1991, it was reported that 5300 women were murdered in Russia. By 1993, the number had jumped to 13,000, putting the per capita rate ahead of the US figure, and much above the rates common throughout Europe. Increased criminal causes of death are by no means peculiar to Russia. In Lithuania, the number of murders jumped from 281 in 1990 to 394 in 1992; suicides rose from 969 to 1294 over the same period of time. Murders in Latvia in 1990 totalled 165; in 1992, there were 293.

The health care data cited in this discussion pertain to just a few aspects of the system. Additionally, they mostly have been drawn from records of those states which had the most auspicious health care indicators in the late Soviet era. The health and well being of a very sizeable share of the citizens of these countries obviously has worsened since independence. The health care systems in those parts of the former Soviet Union which were disadvantaged to begin with, and which in some cases have endured much civil unrest since, can only be worse. The most telling measure of the quality of a society's well-being is life expectancy. This topic was discussed in Chapter 4, but it is worth reiterating that in Russia and all western tier states life expectancies were longer before independence than after (see Table 4.3, p. 99). Amongst the southern tier states, the most recent available data were for 1990 or 1991, and so throw little light on the mid-1990s situation. Over this short period, seven states evidenced a slight increase compared to 1989. Only in strife ridden Tajikistan did life expectancy decline. It is entirely possible that at present some of the seven states have joined the ranks of those in which life expectancies are declining. Regional variations in life expectancy are to be expected, if only between the urban and rural sectors. But in Russia, which experienced the most precipitous drop, there is no respite from the bleak picture of widespread and steep decline in life expectancy. With males newly born having a life expectancy of barely 57 years on average, it matters little where in Russia one lives because life expectancies across all regions are declining, the only variation being by how much. Close to one-half of all men aged 19 in 1994 will not reach 60 years of age at the then prevailing rates of mortality. This is symptomatic of a staggering failure of health care systems, as well as being indicative of the fundamental problems in contemporary Russian society. And if dim prospects for a reasonably long life span were not problem enough, those who do live longer will be doing so in a society not just lacking an adequate health care system, but in one increasingly impacted by criminal activity of one kind or another. Criminal activity is everywhere on the rise in the post-Soviet scene, just as death and disease are. And a growing share of deaths attributable to 'external causes' are the direct result of crime.

In Russia, the role of the so-called *mafiya* is growing. Crime and gangs are not simply a product of the new political economy, of course. They have always been a presence, and began to mushroom in number and importance in the Brezhnev era of stagnation and growing corruption. By the late 1980s, the Soviet Ministry of the Interior, which was responsible for policing the state, estimated that half the income of the typical government bureaucrat came from bribes. Crime of one kind or another has flourished in the new era of limited funds for the police, and the demise of the former agencies of government charged with superintending the population at large, and dissident and criminal groups in particular. Thus, organized crime with extensive operations in Russia, in other post-Soviet states, and abroad, now flourishes. Of course, it is difficult to be very precise when dealing with the impact of crime. However, it is reckoned that there were perhaps 4000 to 5000 gangs operating in Russia in the mid-1990s. Perhaps as many as 10 per cent operate internationally. Nearly 50 are reputed to operate in the Moscow region alone. Some 40,000 businesses and industrial enterprises are said to be controlled by organized crime. Banking has been an obvious target. In a state with few laws and less enforcement, there are many opportunities to make very significant amounts of money in business. Such an environment is fertile ground for criminal organizations. Its influence is extensive. Close to one-half of the members of the Congress of Peoples' Deputies, which was abolished by President Yeltsin in September 1993, was thought to be corrupt. Many were presumed to have links to criminal groups of one kind or another. Some members of the present State Duma are suspected of having connections to organized crime or gangs. The solution to disputes in criminal society is often violent, of course. About two score top officials of the Russian private banking industry have been murdered since 1991. By late 1995, four of the 450 members of the State Duma elected in December 1993 had been murdered, the motives being unclear but certainly suspect. But these numbers pale in comparison with the carnage on the streets. It is estimated that in 1993 alone more Russians died from criminal violence than soldiers died in Afghanistan during the nine years the Soviet Union waged war there. The very rapid rise in the number of deaths from 'external causes' is certainly partly a function of the escalation in crime. Muscovites often joke that the city is just like Chicago was in the 1930s, an impression seemingly got from movies. Most probably Chicago was nowhere near as bad. In all the other states, crime is also on the rise, as noted earlier.

The pages of statistical handbooks provide only a very clinical perspective on what is happening to public health and safety. On the streets and in the apartments, life is impacted in ways not covered by the rows and columns of numbers. Steel doors being added to individual apartment entrances, large guard dogs being added to already crowded apartments, apprehension amongst the elderly, less willingness to be on the streets after dark, all of this and more besides is very much more a part of the new reality than it was of the old order. Then corruption and crime were part of the scene of course, but criminal activity was certainly less violent. The quality of life for all members of the new post-Soviet societies is being steadily diminished in more than one way. Indeed, even where crime may not be much of a threat, simply breathing the air, drinking the water or eating fresh produce may by injurious to one's health and well-being.

Environmental degradation and the quality of life

As Fig. 10.10 makes abundantly clear, serious environmental degradation afflicts all parts of the post-Soviet scene. That public health has been negatively impacted in consequence is obvious even if the exact correlation between death or disease and the quality of the environment sometimes remains an elusive statistic. But in the worst affected regions people have well founded reason for believing that the poor quality of the environment accounts for many common diseases. While political, economic and social structures can be altered quite quickly, restoring heavily damaged ecosystems is not amenable to a 'big bang' or shock therapy approach. Much of the environmental degradation inherited from the Soviet era is probably not amenable to repair over several generations let alone years even if there were the will to do something. The range of problems is enormous, and runs the gamut from high levels of radiation to regional climate change. What might before have been addressed within the confines of a single state with a command economy, must now often be addressed by a number of independent states, each with different economic development priorities, but all having in common inadequate financial resources to do very much by way of environmental management and amelioration. The economic crisis which has enveloped all post-Soviet states might at first glance be seen in a positive light from the perspective of reduction of industrial pollution. Probably more likely is the relaxing of existing standards of pollution control in the attempt to keep industrial plants operating in the new economy. While Fig. 10.10 provides a

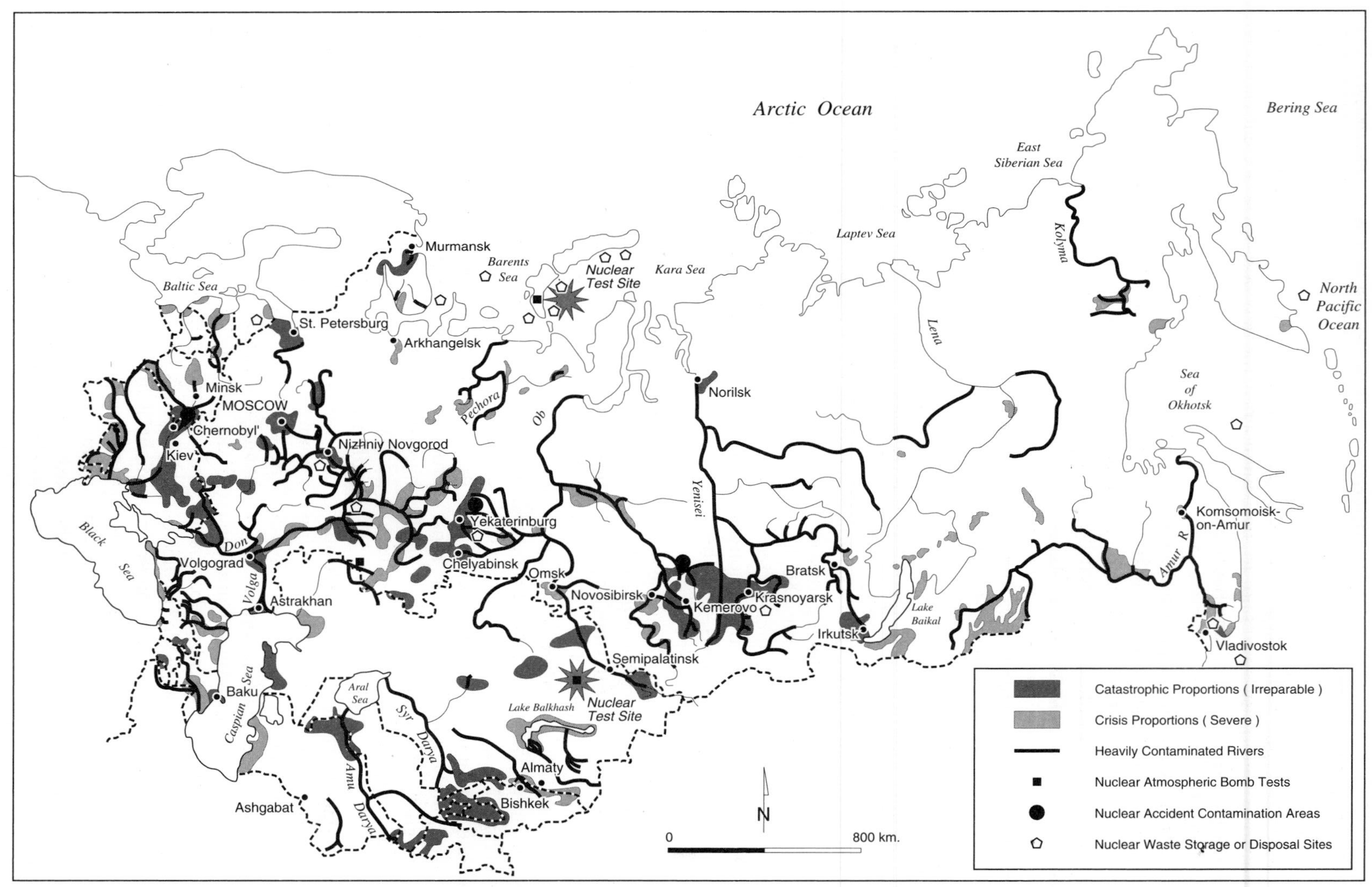

FIGURE 10.10 *Environmental degradation, post-Soviet states*
(*Sources*: Based on D.J. Peterson, *Troubled Lands. The Legacy of Environmental Destruction* (Boulder: CO, Westview Press, 1993), 8–9; P.R. Pryde and D.J. Bradley, 'The Geography of Radioactive Contamination in the former USSR', *Post-Soviet Geography*, Vol. 35, No. 10, 1994, 564)

synoptic, if depressing, view of the state of the environment bequeathed to the post-Soviet states, how it came to be so degraded requires brief discussion. Some typical post-Soviet environmental health problems and reactions will then be reviewed.

External diseconomies of industrial development are not specific to any one political–economic system. The fact that land and water were at best only nominally accounted for in the Soviet price system and accounting practices made accurate determination of the benefits and costs of a particular development more difficult to be sure. But the mere existence of a set of market-determined prices for resources is certainly no assurance of easy and accurate benefit–cost calculations, or sound environmental management, as evidence of environmental degradation in many western industrial states makes plain. Irrespective of the nature of the political economy there is still the fundamental question, from whose perspective should benefits and costs be calculated? Simply contending that the best choice is that of the common good does not resolve the problem of determining whether cleaner air is preferable to a higher standard of living. The systemic weakness of the Soviet approach to environmental management and protection was the absence of an environmental interest group within the upper echelons of the Communist Party apparatus where fundamental policy decisions were taken, and the absence of a single authority at the level of an All-Union Ministry which might have served as a countervailing influence to the economically and politically important industrial ministries.

One of the key obstacles in implementation of Soviet-era environmental protection legislation was that for a long time responsibility for environmental management was turned over to the ministries themselves. Internal, subordinate administrative units were assigned the task of implementing and monitoring environmental protection legislation. Fragmentation of authority often resulted in vital standards being ignored if they impinged on the overriding ministerial goal of meeting production quotas. Thus, antediluvian factories continued to produce and pollute, and non-conforming locations of factories in cities were overlooked when closing or relocating them could jeopardize production goals. From the standpoint of industrial managers, the choice was perfectly rational *within* the context of their priorities. Additionally, the proliferation of environmental monitoring agencies within different ministries, and on an *ad hoc* basis amongst various public interest groups, led to a fragmentation of effort at best and outright conflict of interest amongst groups with ostensibly identical objectives at worst. These systemic problems did not pass unnoticed, but it was some years after the Lake Baykal *cause célèbre* before an institutional arrangement was introduced to ameliorate the problem. Lake Baykal caught the imagination and evoked widespread public concern as an environmental problem in the 1960s.

The Lake Baykal controversy emerged in the late 1960s. For the time it was a quite singular example of public debate in an otherwise tightly controlled Soviet media, though the 'public' was largely restricted to the scientific community, ministry officials, the Writers' Union and a handful of other official environmental bodies. The latter were clearly of little potential clout, though the various official organs of the cultural establishment, such as the Writers' Union, were important. Put simply, the controversy arose over the decision to locate two huge pulp and paper mills on the southern shore of the lake. Because of its particular geological history, Lake Baykal supports a vast array of unique flora and fauna. It represents the largest single volume of fresh water in the world. But the very purity of Baykal's water, plus the local and regional forest resource wealth, were the principal locational advantages for the proposed pulp mills. The threat to the purity of the lake was recognized immediately, if for no other reason than across the lake from the proposed plants was situated a major institute of limnological research. Ministry officials at first rejected the scientific community's contention that effluent from the plants would threaten the existence of many unique species of life associated with, and dependent upon, the pristine quality of Baykal water. Eventually the debate became a national issue. Indeed, it attracted international attention. The construction of the plants went ahead, modifications were made to the effluent treatment system, the government responded by legislating a zone around the lake to be immune to the forester's axe and by banning the movement of logs on the lake itself. The latter were a particular hazard since Siberian larch, the predominant species, is heavy per unit of volume. Estimates of potential losses during rafting of logs to the mills were high. Limnologists argued that the amount of oxygen required to decompose the anticipated volume of wood and bark added to the lake through rafting would seriously alter the chemical composition of the water and thereby put at risk much of the unique vegetative and aquatic life. Legislation introduced in 1966 and in 1987 specifically to protect the Lake have probably helped to avoid the worst possible negative environmental impacts of the pulp mills in particular and other sources of polluted effluent in general, but parts of the Lake are now seriously polluted nonetheless. Equally serious as the mills as sources of pollution are the forestry, agricultural and urban industrial enterprises in the Lake's watershed. Most urban settlements treat only a small portion of their domestic sewage, and small-scale industrial plants, of which there are a great many, do even less as a rule. As Fig. 10.9 indicates, areas adjoining Lake Baykal are seriously degraded. Thus, environmental protection legislation has been of only limited assistance in the Lake Baykal region. The problem was the traditional case of production taking priority over preservation and conservation.

Soviet authorities did have some success in protecting the natural environment, however. Shortly after the revolution they took steps to secure parcels of the natural environment as protected areas. *Zapovedniki*, or nature preserves, pre-dated the revolution, but were created in many parts of the country during the early years of Soviet rule to facilitate research in the natural sciences. The Soviet government inherited a handful of *zapovedniki*, but added another 20 or so in the 1920s alone. By the 1960s the number had increased to nearly 70. While serving as scientific research areas, tourism in and around *zapovedniki* increased steadily. Notwithstanding the attraction of *zapovedniki* to tourists, these natural areas are not the same as North American or European national parks. The demands of tourism eventually resulted in a Soviet version of the national park. They were first created in the 1960s. By 1985, 14 national parks had been added to the list of *zapovedniki*, which then totalled 144 across the country. Since independence, financial resources to support such natural ecosystems have dwindled and their protection has diminished accordingly.

By the late Soviet era new institutional arrangements for managing the environment had been put in place. The creation of the USSR State Committee on Hydrometeorology and the Environment was really too little, too late. Lacking the status of an All-Union Ministry, it did not have the necessary clout to radically change practice simply by enforcing existing regulations. Its specialist staff worked with other agencies in an effort to ensure compliance with legislation. From the Ministry of Public Health, Ministry of Land Reclamation and Water Conservancy, amongst others, evidence of violation of state laws was procured. While it was not without some success as an institutional arrangement, it was simply overwhelmed by the nature and scale of the problems of environmental degradation. Fragmentation of responsibility in monitoring and regulating continued to be regarded as one of the reasons for environmental degradation. On the eve of the demise of the Soviet Union another administration arrangement was introduced to address environmental problems. The formation of a USSR State Committee for Environmental Protection, along with Union-Republic equivalents, was aimed at ensuring more rational use of natural resources and better environmental protection. Separating environment from the State Committee on Hydrometeorology was no doubt a step in the right direction, but as before it begged the question of ministerial status and authority. Mandated to improve both the monitoring and protection of the environment, it had an impossibly large task. Environmental concerns were to be fully integrated into future economic and social development plans for the country as a whole. From gathering data, setting new standards, revising existing standards, goal setting, establishing education programmes, supervising existing nature preserves – the list of responsibilities rolled on without end. Ostensibly its decisions were to be binding, irrespective of ministry, department, enterprise or geographic location. It was to have the right to levy charges in order to build up the financial resources to expand its operation as need arose. Research departments were to be transferred to it from other agencies, ministries and enterprises in order to facilitate its scientific work. In conjunction with the USSR Ministry of State Law, the new state committee was soon involved in preparing a new law on environmental protection. In 1991, the new State Committee for Environmental Protection was elevated to a Ministry. Notwithstanding the potential importance of this development, as the Soviet system disintegrated during 1991 so too did the prospects for better environmental management. In the post-Soviet era the situation has worsened, and so too have the human consequences.

Institutional arrangements to deal with transnational issues of environmental degradation are an obvious first priority if some of the worst aspects of the environmental degradation legacy of the Soviet era are to be addressed. By the mid-1990s, however, there was precious little that had been accomplished. The Caspian Sea is heavily polluted, but the five adjoining states cannot agree on off-shore territorial limits for the purpose of resource development. Russia has taken the high moral road arguing that before any off-shore oil resources are developed, the four post-Soviet states and Iran need to establish binding environmental protection measures. This is regarded as a strategy for Russia to control resource developments in the Caspian Sea, thus pollution of the Caspian Sea continues unabated. The catastrophic consequences of the falling level of the Aral Sea are evident now, yet the Middle Asian states continue to petition Russia to resurrect the scheme to divert the north flowing rivers to the south. In the autumn of 1995, a formal request was made by these states yet again. Cotton mono-culture continues as a national priority in Uzbekistan and Turkmenistan, irrigation waters are still allocated largely free of charge since the still state-controlled agricultural enterprises cannot afford to pay for irrigation waters, thus over-consumption of water is not being effectively challenged. The estimated cost of such a project is currently estimated to be in excess of $115 billion, and therefore is prohibitive. Agreement to petition for external aid is easier than addressing entrenched interest groups within these states. Indeed, there is considerable potential for conflict over access to water in the region. Indeed, there have already been outbreaks of violence in some of the most densely settled areas, such as the Fergana Basin, which is shared by 14 million people from three states, Uzbekistan, Kyrgyzstan and Tajikistan. Comparatively fast growing populations and inter-ethnic tensions simply add other layers of tension to the Syr Darya river water resource access issue. Trans-boundary air and water pollution is especially serious among the western tier states simply because those states are more heavily industrialized than the southern tier states, but by no means are the latter spared such problems as Fig. 10.10 implies. Put simply, the prospects for effective international cooperation in addressing environmental problems is far from auspicious, and thus mitigation of related human health problems in the near-term future is bleak.

There are major deficiencies in the water and sewage treatment facilities upon which many polyclinics and hospitals in southern tier Middle Asian states are still dependent. However, these alone do not account for the disproportionate number of infants who do not survive the first year, or for the nature of the health problems which afflict the population at large. Environmental hazards have been contributing factors. Widespread use of toxic chemical pesticides in labour-intensive agricultural systems continues to take a toll since the intensive cultivation of cotton as a cash crop has been perpetuated in the post-Soviet era. Its export abroad remains a vitally important source of hard currency. While much of the cotton crop in Middle Asia being harvested by hand is perhaps not an entirely unreasonable use of labour in a region distinguished by a surplus of it in the countryside, such a practice does have negative health consequences. The combination of widespread employment of women as cotton pickers and regular application of exceedingly toxic defoliants is positively correlated with the high level of infant mortality throughout the region.

On a more general level, questions arise as to both the appropriateness of standards to be followed in the application of chemical substances in agricultural production and to the efficacy of their enforcement. Given the penchant for overconsumption of water in the irrigated areas of Middle Asia, excessive application of chemical substances in the quest for higher levels of output is not surprising. Serious pollution of drinking water has occurred in consequence, and is linked to a wide variety of health disorders. The general desiccation of the Aral Sea region has given rise to dust storms which move toxic materials deposited in soil over considerable distances, and are reckoned to be the cause of unusually high levels of respiratory ailments amongst all age groups. In some of the worst affected regions close to what used to be the Aral Sea, infant mortality rates are apparently in excess of 100 per 1000 live births. Those children who do survive suffer a very high incidence of disease. Testing of food products in Middle Asia, as elsewhere, is neither widespread nor a regular occurrence. Thus, much of what is produced in these regions to be eaten is suspect. There have been reports that knowledgeable medical personnel in Turkmenistan, for example, refuse to eat local melons because of a perceived abuse of chemical fertilizers and the lack of testing. Such problems are not unique to this environmentally devastated region, of course. In Western Europe and North America the popularity of foodstuffs produced without chemical additives is related to the concern over the relationship between chemicals used in agriculture and personal health. The level of public awareness and concern over such issues in post-Soviet states was until recently comparatively low. But for many people recognizing the risk is one thing, finding and affording alternative foodstuffs in the current economic reality is quite another. Chernobyl' and its aftermath have certainly helped to focus attention on public health and environmental degradation.

The nuclear meltdown at Chernobyl' in late April 1986 had international as well as domestic consequences. In Lapland, reindeer herders were unable to market animals fed on lichens and moss which contained unacceptably high levels of radioactive contamination. The same problem existed in some upland English sheep farming areas. The short-term costs were enormous. It is estimated that the immediate clean-up costs of more than 1000 square kilometres around the nuclear power station, and the loss of production from affected farms, factories and Chernobyl' itself, exceeded two billion roubles in total. Since then problems and costs alike have not abated. For Belarus' alone, it is estimated that over a 30-year period Chernobyl'-related costs will be the equivalent of 32 total state 1986 budgets. The health of untold numbers of people has been affected. While the epidemiological evidence is not always conclusive as to cause of death, there is no doubt at all that the human costs of the disaster continue to climb with each passing year. More than 100,000 people have been permanently resettled, since a substantial zone immediately surrounding the Chernobyl' plant has been removed permanently from agricultural production. In total, close to one-half million hectares of agricultural land have been lost in the affected areas in Ukraine and Belarus'. Notwithstanding the Gorbachev-era spirit of *glasnost'*, the Chernobyl' disaster was not immediately reported to the Soviet public. Attributed to human error, the State Prosecutor's Office secured convictions for a number of the key personnel in the plant as well as other responsible officials in the nuclear power industry. But the resumption of operations in the remaining reactors at Chernobyl', the continuing official proclamation that it is a safe industry, and the growing interest in the post-Soviet period in further development of the nuclear generation of electricity have done little to allay public concern and criticism.

While major disasters such as the drying up of the Aral Sea, Chernobyl', nuclear radiation from 'accidents' in the Urals military industrial complex in the 1950s and from testing of nuclear devices in parts of Kazakshstan which highlight the catastrophic consequences of technology on the human condition have been the focus of much well deserved publicity, there are countless other negative environmental impacts of the industrialization drive of the Soviet era. Legal prohibitions against air pollution by industry have been in place since the 1920s. Indeed, attempts to limit industrial growth in the major cities which date from the early 1930s were related in part to the felt need to improve the quality of the urban environment through reduction of both air and water pollution. Yet over the years air and water pollution affected all major urban-industrial centres. As Fig. 10.10 indicates, there are now vast areas impacted by catastrophic proportion environmental damage, extensive sections of all major rivers which are highly contaminated, in addition to sizeable areas of nuclear disaster/radiation damage and risk. A cursory review of the distribution of some of the major industries (Figs 9.4 and 9.5, pp. 263, 265) will provide some clues as to the geography of air pollution. Industrial emissions are frequently singled out as the main source of air pollution, although by no means is industry the sole cause. From press and other scientific and technical reports there is consensus that air pollution in many industrial regions is linked to higher than average incidence of respiratory ailments of one kind or another. In most post-Soviet states, the creation of environmental protection agencies and the enactment of appropriate legislation is underway. But as in the past enforcement is weak, and will remain so until basic economic conditions improve. The situation in Russia is illustrative.

Since independence the Russian government has put in place a Ministry of Ecology and Natural Resources to carry on the work of the earlier state agencies in terms of environmental protection. There is also an inter-departmental coordinating committee intended to bring an environmental perspective to sector specific developments. The laws now in place in Russia to monitor and protect the environment are designed to operate both within the context of a market economy and the expanding authority and responsibilities of Russia's republics. Tax incentives are offered for enterprises adopting pollution control technologies, and fines are legislated for transgressions. While more encompassing and, indeed, more environmentally appropriate for the present times, the Russian legislation unfortunately shares one feature in common with the predecessor state, it is long on principles for appropriate stewardship of environment and resource development and short on effective means to enforce regulations. The fact that across Russia life expectancy is falling, that each year more people succumb to environmentally related disease, that infant mortality is on the rise, speaks to the fact that environmental legislation introduced usually requires some years for real change in ecological systems to occur. That there is still no effective implementation of the current legislation, that industrial and other pollution continues to occur at health threatening levels despite a downturn in the industrial sector, mean that the health of the population in particular and the quality of life for the average person in general will not improve for some years yet. There is little happening elsewhere in the post-Soviet scene that suggests a different and more positive relationship between environmental quality and quality of life. To date, the primacy of production over environmental protection has not been thwarted.

FURTHER READING

Adelman, D. 1994 *The 'Children of Perestroika' Come of Age: Young People of Moscow Talk About Life in the New Russia*. Armonk, NY, M.E. Sharpe.

Åhlander, A.-M.S. 1994 *Environmental Problems in the Shortage Economy: the Legacy of Soviet Environmental Policy*. Aldershot, E. Elgar.

Attwood, L. 1990 *The New Soviet Man and Woman: Sex-Role Socialization in the USSR*. Bloomington, Indiana University Press.

Cook, L.J. 1993 *The Soviet Social Contract and Why it Failed: Welfare Policy and Workers' Politics from Brezhnev to Yeltsin*. Cambridge, MA, Harvard University Press.

Feshbach, M., Friendly, A. Jr. 1992 *Ecocide in the USSR: Health and Nature under Siege*. New York, Basic Books.

Flakierski, I. 1993 *Income Inequalities in the Former Soviet Union and its Republics*. Armonk, NY, M.E. Sharpe.

Herlemann, H.G. (ed) 1987 *The Quality of Life in the Soviet Union*. Boulder, CO, Westview Press.

Jones, A. (ed) 1994 *Education and Society in the New Russia*. Armonk, NY, M.E. Sharpe.

Kampfner, J. 1994 *Inside Yeltsin's Russia: Corruption, Conflict, Capitalism*. London, Cassell.

Komarov, B., Wolfson, Z. 1993 *The Geography of Survival: Ecology in the Post-Soviet Era*. Armonk, NY, M.E. Sharpe.

Loewenhardt, J. 1995 *The Reincarnation of Russia: Struggling with the Legacy of Communism, 1990–1994*. Durham, Duke University Press.

Matthews, M. 1978 *Privilege in the Soviet Union. A Study of Elite Lifestyles Under Communism*. London, George Allen and Unwin.

Matthews, M. 1986 *Poverty in the Soviet Union: The Lifestyles of the Underprivileged in Recent Years*. Cambridge, Cambridge University Press.

McAuley, A. 1979 *Economic Welfare in the Soviet Union. Poverty, Living Standards and Inequality*. Madison, University of Wisconsin Press.

McCuen, G.E., Swanson, R.P. 1993 *Toxic Nightmare: Ecocide in the USSR & Eastern Europe*. Hudson, WI, G.E. McCuen Publications.

Millar, J.R. 1987 *Politics, Work and Daily Life in the USSR: A Survey of Former Soviet Citizens*. Cambridge, Cambridge University Press.

Moskoff, W. 1993 *Hard Times: Impoverishment and Protest in the Perestroika Years*. Armonk, NY, M.E. Sharpe.

Nelson, D.N. (ed) 1982 *Communism and the Politics Of Inequalities*. Lexington, Lexington Books.

Peterson, D.J. 1993 *Troubled Lands: the Legacy of Soviet Environmental Destruction*. Boulder, CO, Westview Press.

Posadskaya, Anastasia (ed) 1994 *Women in Russia: A New Era in Russian Feminism*. London, Verso.

Pryde, P.R. 1991 *Environmental Management in the Soviet Union*. New York, Cambridge University Press.

Pryde, P.R. (ed) 1995 *Environmental Resources and Constraints in the Former Soviet Republics*. Boulder, CO, Westview Press.

Scanlan, J.P. (ed) 1992 *Technology, Culture, and Development. The Experience of the Soviet Model*. Armonk, NY, M.E. Sharpe.

Sterling, C. 1995 *Thieves' World*. New York, Simon & Schuster.

Stewart, J.M. 1992 *The Soviet Environment: Problems, Policies, and Politics*. New York, Cambridge University Press.

Turnbull, M. 1991 *Soviet Environmental Policies and Practices: The Most Critical Investment*. Brookfield, Dartmouth.

Vaksberg, A. 1991 *The Soviet Mafia*. London, Weidenfeld & Nicolson.

11

THE POLITICS OF PLACE

'The most important thing for Russia is to survive and to keep from disintegrating any further'
(*Trud*, October 14, 1995, 1)

Ethno-national territories were the building blocks of the former Union of Soviet Socialist Republics. Each of the 53 ethno-national territories which comprised the Union had a specific status, and associated rights, in the hierarchical political-administrative structure. The 15 Soviet Socialist Republics stood at the apex of the system, in theory having all of the rights of independent states. Below them in the hierarchy were 20 Autonomous Soviet Socialist Republics, 8 Autonomous Oblasts and finally the 10 Autonomous Okrugs. While most national minorities with sizeable populations had homelands somewhere within the Union, a number were denied explicit territorial recognition. The 1.1 million strong Polish minority, like other western and southern Slavic peoples who had homelands outside the Soviet Union, were never assigned a territory within the Union. Koreans, deported from the Soviet Far East region to Middle Asia in the 1930s and numbering 439,000 in 1989, were similarly treated. The nearly 200,000 Turkic Gagauz peoples of present-day southern Moldova outnumbered many northern peoples who were assigned a homeland. Still other minorities, such as the Crimean Tatars, Volga Germans and Meshketian Turks, had their legally constituted homelands abolished by Stalin during or shortly after World War II. They were forcibly relocated to various places in Middle Asia and Siberia, where many still live. Their former homelands in Ukraine, Russia, and Georgia respectively have not been reconstituted, nor have these national minorities been encouraged to return to them. In short, the former Soviet Union was a multinational federal state, one in which nearly half of the 111 minorities identified in the 1989 census had a territorial homeland. On the map this appeared to be a tidy arrangement (see Fig. 2.5, p. 47). However, on the eve of the collapse of the Soviet Union the actual distribution of the country's many nationalities was anything but tidy.

Each of the 15 post-Soviet states made the transition from Republic of the Union to political independence territorially intact. Their borders did not change, although many are contested. Most border zones are characterized by ethnically mixed populations, a fact which serves to support competing territorial claims. Some states want territories returned which were lost during Soviet-era border adjustments. Estonia and Latvia, for instance, claim territories in Russia which were theirs during the inter-war period of independence, but which were assigned to the Russian Republic after Estonia and Latvia were forcibly brought into the Union during World War II. Inside all 15 post-Soviet states live people who are not members of the titular nationality group, people who live outside their own homeland. Within some post-Soviet states live minorities who have never had a homeland, but who are now arguing for the creation of one. Put simply, no ethno-territorial region is ethnically homogeneous. Indeed, in most homelands the titular group is a numerical minority. However, since independence migration has brought some changes in that in many ethno-national territories there has been an increase in the share of the titular population. Notwithstanding recent population movements, voluntary or forced, across the post-Soviet scene about 40 million people still live 'outside' their homelands. More than one-half are Russians. The politics of particular places are therefore very much shaped by contested claims to territory. In a few places, disputes over territory have escalated from wars of words to bloody civil wars. There is no assurance there will not be more to come.

Several examples of ethno-territorial problems in the western and southern tier states will be discussed first. While each has a unique geo-political history, grievances rooted in Soviet-era nationality policy and practice are often a common element. The discussion will then turn to the politics of place in Russia. Russia has within its borders more minorities than any other post-Soviet state. At the time of the 1989 census, for instance, non-Russians in the Russian Soviet Federative Socialist Republic comprised about 27 million of the more than 147 million total population. Amongst minorities, Tatars and Ukrainians ranked first and second in number, with 5.5 and 4.4 million members respectively. The new constitution of the Russian Federation recognizes non-Russian, ethno-national territories, many as Republics within the Federation (see Fig. 9.12, p. 280). The current federal structure does not satisfy the aspirations of all minorities, however. Some have negotiated for themselves a considerable measure of political and economic independence from the federal government in Moscow. Unilateral declarations of independence have also occurred, in the case of Chechnya with catastrophic consequences for all concerned. In addition to the sometimes heated negotiations between Moscow and the non-Russian regions over respective responsibilities and authority, there is another dimension to the politics of place in contemporary Russia. Put simply, this is the struggle for political power between the core and periphery, that is, between central government in Moscow and the regions. One of the dimensions of this struggle for power involves political representation in central government itself. Thus, the results of the December 1995 elections to the Russian State Duma will be assessed from this perspective in the final section of this chapter.

Territory and minority group self-determination in the 'near abroad'

The Russian Empire, unlike its Soviet successor, was not divided into ethno-territorial regions as a matter of government policy. Instead, in 1910 it comprised 88 *guberniyas*, or provinces. These were administrative regions, and were not intentionally correlated with the distribution of particular ethnic minorities. Many minorities were restive subjects of the Empire, some like the Poles having endured decades of overt Russification. In the early 1900s, the Bolshevik faction of the Russian Social Democratic Workers Party led by Vladimir Lenin attempted to win the support of minorities who had been caught up in the territorial expansion of the Russian Empire. The strategy was to advocate the principle of self-determination, that is, the right of each minority to a homeland, to determine its own destiny. Of course, after the revolution it was assumed that all such peoples would be enlightened enough to voluntarily join the new Union of Soviet Socialist Republics. Some like the Finns, Estonians, Latvians and Lithuanians, instead took the opportunity presented by the fall of the Russian Empire to declare their independence. Estonians, Latvians and Lithuanians, of course, only remained independent until World War II, at which time they were forcibly brought into the Union. Other declarations of independence by Ukrainians, by various peoples in the Caucasus and Middle Asia, for example, sometimes lasted only a matter of a few weeks or months before being squelched, often brutally, by the Soviet Red Army. Throughout the Caucasus, Middle Asia, the far north and Siberia in particular, indigenous peoples were incorporated into the Union and assigned territories. Elsewhere, the historical association of particular peoples with specific places was formally recognized through the creation of homelands which were then constitutionally embedded in the Union. Notwithstanding the huge gap between the constitutional rights of ethno-national territories and the reality of life under the Soviet system, for many minorities it was the first time in their history they had a legally constituted homeland. In many instances this awoke awareness of ethnic identity in terms of territory, something before not always common.

Drawing boundaries on an ethno-national basis is extremely complicated, and many, if not the majority of the borders of ethno-territories, were ultimately imposed on areas of ethnic heterogeneity. In a political system which did not tolerate open debate, which crushed any and all attempts to exercise constitutional rights of self-determination, grievances between peoples over land were suppressed. While there were periodic disturbances in the ethnic regions, for example in parts of Middle Asia, most were triggered by more general complaints, such as food shortages, as opposed to inter-ethnic territorial disputes. This changed in the late Soviet era, and especially during Gorbachev's policy of *glasnost'* which gave rise to much more open discussion of previously censored topics. The status quo on ethno-territorial issues was challenged, and the question of minority group self-determination quickly came to the fore. Ethno-territorial disputes soon began to shake the very foundation of the existing Soviet geopolitical structure. Declarations of independence by some Soviet Republics in 1989 and 1990 in fact presaged the demise of the Union. Its ultimate collapse unleashed even more conflicts over territory. As Fig. 11.1 makes clear, across the post-Soviet scene ethno-territorial problem areas abound. Not surprisingly, they are most common in regions with the greatest concentrations of different ethnic minorities, such as the Caucasus. The reasons for ethno-territorial disputes are varied. Migration of ethnic minorities into 'foreign' territories, mostly voluntary but sometimes forced, has resulted in disputes over who should be living where. Where ethno-national territories were defined by artificially imposed boundaries, disputes almost invariably have arisen since the collapse of the Soviet Union. Minorities deprived of an official homeland during the Soviet era are now regularly claiming one. Territories transferred from one jurisdiction to another by the Soviet government for political reasons are now subject to dispute between affected parties. The list of reasons for ethno-territorial conflict is lengthy. The list of possible solutions unfortunately is very short. Invariably resolution requires compromise, not confrontation. The latter, also unfortunately, is more common than the former. We will begin the discussion of examples of ethno-territorial disputes in the 'near abroad' by first examining the situation in Moldova.

Moldova

From the early nineteenth century until 1917, most of present-day Moldova was a little developed corner of the Russian Empire, the Bessarabia *guberniya*. Predominantly rural and with historic ties to Rumania, after the fall of the Russian Empire Bessarabia was annexed by Rumania in 1918. It remained a part of Rumania during the inter-war period. On the Soviet side of the border, a Moldavian Autonomous Soviet Socialist Republic was created in the early 1920s. It was a mostly non-Moldovan ethnic entity formed more to meet the needs of the Soviet political agenda of the time than it was to fulfil the aspirations of a substantial Moldovan population. For example, of its nearly 600,000 inhabitants in 1939, more than one-half was Ukrainian. Moldovans numbered only 171,000, the balance of the population comprising Russians, Jews, Germans, Bulgarians amongst others. The historic Bessarabia *guberniya* territory was acquired by the Soviet Union in the realignment of borders with Eastern Europe following World War II. A new Moldavian Soviet Socialist Republic was constituted. Its territory included the Trans-Dniester river region which had comprised part of the inter-war Moldavian Autonomous Soviet Socialist Republic, as well as some adjacent lands which had been part of

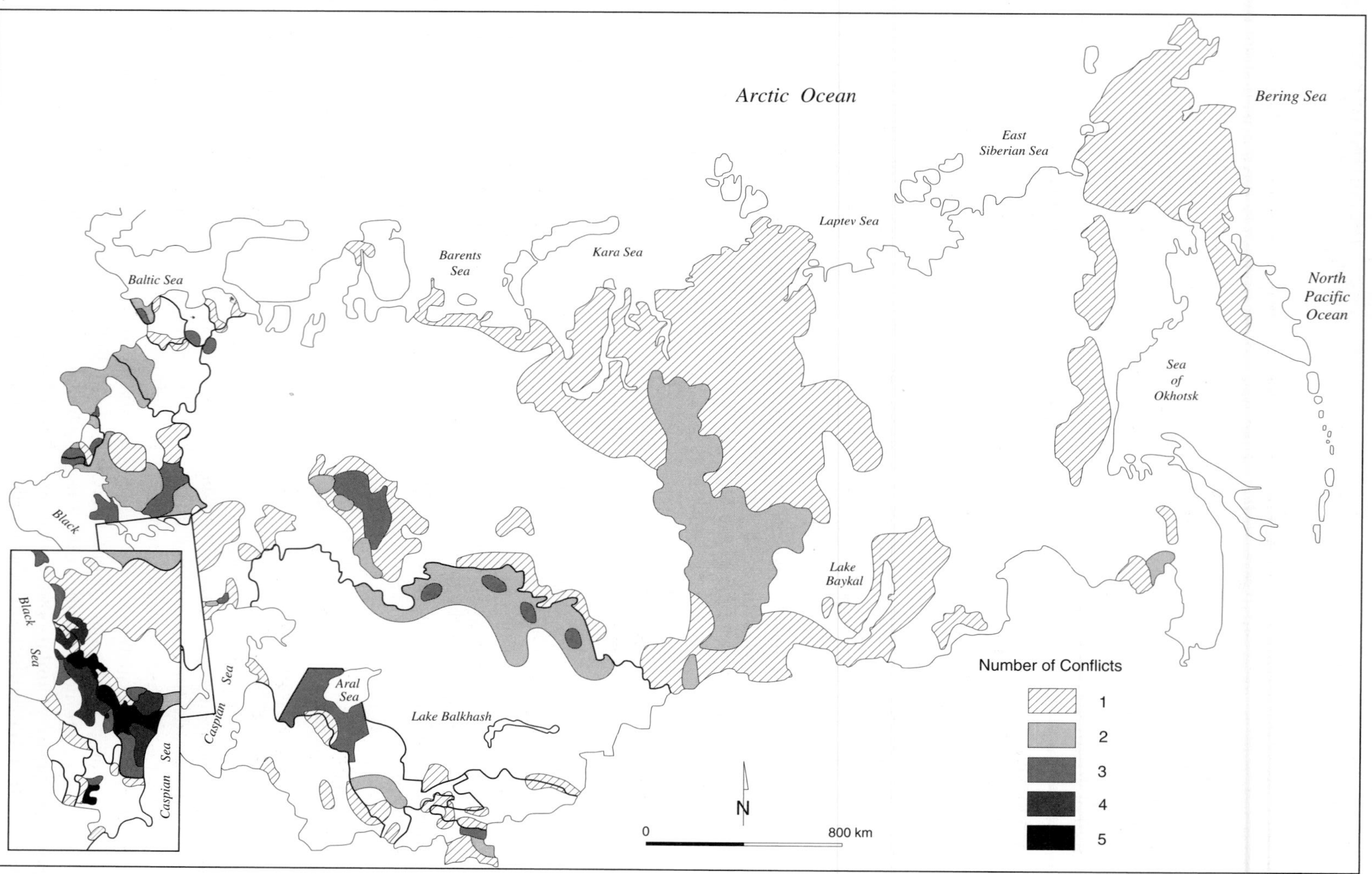

FIGURE 11.1 *Nationalities and unrest in the post-Soviet scene, 1992*
(*Source*: R.J. Kaiser, *The Geography of Nationalism in Russia and The USSR* (Princeton, NJ: Princeton University Press, 1994), 358)

Ukraine. Moldova's total population of 4.4 million in 1989 was nearly two-thirds ethnic Moldovan, a relative share little changed since the late 1950s. Russians, Ukrainians, Gagauz, Bulgarians and Jews were the principal minority nationalities in the post-Second World War era. In 1989, Russians comprised 13 per cent of the population, up slightly from the 10 per cent registered in the 1959 census. In 1989, 91.6 per cent of all Moldovans spoke Moldovan as their first language. The proportion speaking a language other than Moldovan as the first language had more than doubled since 1959, however. Of this group, the vast majority had adopted Russian as the first language. Russification was therefore a controversial issue even before Moldova gained independence in 1991.

A nationalist movement dedicated to the restoration of the Moldovan language began in 1988. The law adopted by the Moldavian Republic in early 1989 declaring that Russian was to be replaced by Moldovan as the official state language in all communication was a clear indication of the growing importance of the nationalist movement. Even though the law made some allowances for minority languages in government administration, its adoption sparked a mass strike in the city of Tiraspol in the left bank Dniester region of the Republic long dominated by Russians and Ukrainians. Amidst growing unrest across the former Soviet Union, in late 1989 a unilateral declaration of independence was issued by the government of the Moldavian Republic. Russian language schools were closed in the capital city Chisinau in 1990 and 1991 during the height of the nationalist movement. Rising Moldavian nationalism, the language issue, and a portended reunification with Rumania did little to comfort the Russian, Ukrainian or Gagauz minorities living there. Disputes over land between these minorities and the Moldovan majority resulted in calls for self-determination, this time by the disaffected minorities. All of this culminated in declarations of independence in 1990 on the part of the Gagauz, who announced the formation of an independent Republic, and on the part of the Ukrainians and Russians living in the Trans-Dniester river region (Fig. 11.2). Competing versions of history were put forward by majority and minority groups alike. Acrimonious debate escalated to more aggressive forms of behaviour, including the formation of armed partisan forces and threatened civil war. Military conflict ensued in the newly declared Trans-Dniester Republic in 1992.

The territorial integrity of the newly declared Trans-Dniester Republic was soon bolstered by the Russian army which came to the aid of what was portrayed as a beleaguered Slavic population. The army served as a buffer between Moldovan nationalist forces intent on reclaiming what they perceived to be their territory, and the dominantly Russian and Ukrainian population of the left bank, Trans-Dniester Republic, who were equally adamant as to the legitimacy of their claim to these lands. The local military leadership occasionally took decisions based on its perception of what the local and regional needs were despite contrary instructions from military headquarters in Moscow. Indeed, right wing political factions in Ukraine, but especially in Russia, made much of the need to protect Slavs from the belligerent actions of an oppressive government in Moldova. In instances when the local military command defied Moscow, it was publicly supported by nationalist elements in the post-Soviet Russian government. Within the Trans-Dniester Republic, former Communist elites used the nationality issue to their own political advantage. For example, much of Moldova's industry is concentrated in the region, and hence control of assets plays a part in the politics of the region.

By the mid-1990s hostilities had tempered, and more moderate behaviour on all sides gained momentum. The Russian government reached an agreement with the government of Moldova regarding a staged reduction in its armed forces based in the Trans-Dniester Republic. Moldova in turn has agreed to moderate its position with respect to the application of language laws, notwithstanding opposition to this move by Moldovan nationalist elements. It has also assured the populations of both the Trans-Dniester and Gagauz Republics that they will have the right to hold a free referendum concerning the will of the people should the issue of unification of Moldova and Rumania ever arise. In the Gagauz Republic, a referendum was held in early 1995 in areas in which the Gagauz peoples are concentrated in order to determine their views on being part of a special Gagauz territory recognized by Moldova. Subsequently, elections to an autonomous Gagauz government were held in a territory about one-half the size of the original homeland claimed by Gagauz nationalists. Within the Gagauz special administrative territory, different laws will prevail. For example, it has been determined that there will be three official languages, Gagauz, Moldovan and Russian. As Fig. 11.2 indicates, there are still places outside the borders of Moldova which are contested. Northern Bukovina and southern Bessarabia are part of Ukraine, but are regarded by the most hardline Moldovan nationalists as integral parts of the historic Moldovan realm. Needless to say, these claims are rejected by Ukraine. While some progress has been made in easing ethno-territorial tensions in the region, there remain places where widely variant perceptions of whose homeland they are, or should be, may yet trigger further dispute.

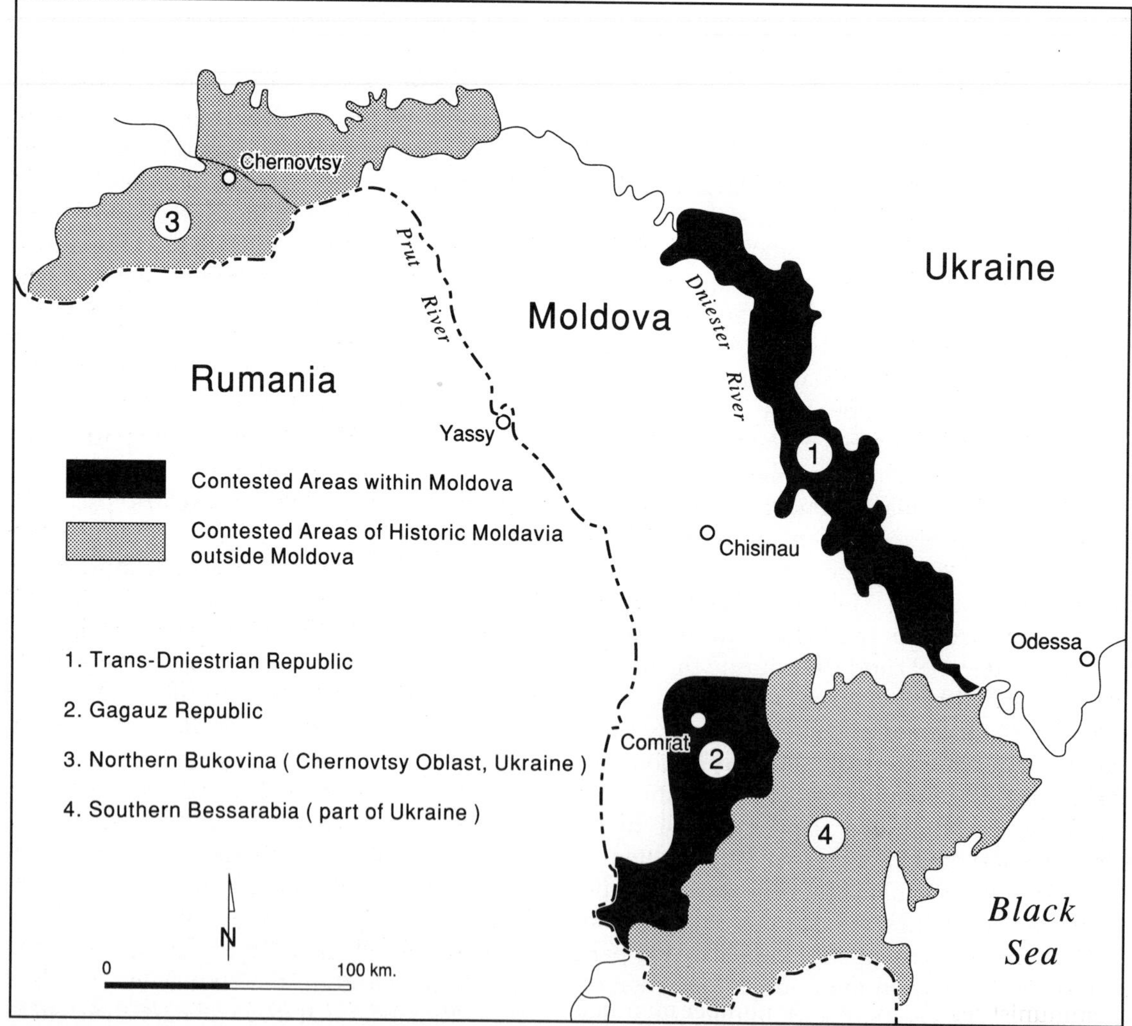

FIGURE 11.2 *National minorities in Moldova*
(*Source*: R.J. Kaiser, *The Geography of Nationalism in Russia and The USSR* (Princeton, NJ: Princeton University Press, 1994), 366)

CRIMEA

The situation in Crimea represents another type of ethno-territorial problem. Unlike Moldova where minorities are struggling to reconcile concepts of homeland with Soviet-era imposed borders, Crimean Tatars are still struggling to return to their historic homeland from which they were forcibly removed more than 50 years ago. They are having to contest their homeland claims against those of both Ukrainian and Russian interests in the peninsula. Tatars had lived in the Crimean peninsula region since the fifteenth century, albeit under the suzerainty of the Ottoman Empire. Crimean Tatars comprised a residual element of the original Mongol Horde who, under the leadership of Chingiz Khan, invaded European Russia in the early thirteenth century. The Crimean peninsula was annexed from the Ottoman Empire in the late eighteenth century by the Russian Empire, thereby ending nearly four centuries of Tatar autonomy as a Khanate. At that time, more than four-fifths of the total population of the Crimea was Tatar. The balance of the population who occupied this crossroads region comprised an array of peoples from around the Black and Mediterranean Seas. A century later, Tatars comprised barely a third of the total population owing to migration there of large numbers of Russians and some Ukrainians. In the imperial era, Crimea was part of the larger Tavrida *guberniya,* a province of the Russian Empire which extended well into south Ukraine. Crimea's southern coast, protected as it is by mountains from the continental climate of the north, offered a limited but highly attractive coastal zone of Mediterranean-like landscape and temperature. By mid-nineteenth century, it was already a favourite destination amongst the Empire's elites seeking respite from the rigours of the Russian winter. Immediately following the Russian revolution many Crimean Tatars took the opportunity to depart the new Soviet Union, as did other peoples fearful of the future under a Communist regime. A sizeable number migrated to Turkey where there is now an important Tatar minority. Notwithstanding the emigration of Crimean Tatars, the Soviet government established a Crimean Autonomous Soviet Socialist Republic in 1921. It was made a constituent part of the Russian Soviet Federative Socialist Republic. The remaining Crimean Tatars thus had a homeland, one geographically coincident with the Crimean peninsula. Promotion of Crimean Tatar cultural identity was soon initiated as part of the Soviet nationalities policy.

In 1939, the 219,000 Tatars living in the Crimean Autonomous Soviet Socialist Republic comprised barely 19 per cent of the total population. About one-half of the 1.1 million total population was Russian. As Fig. 11.3 reveals, in 1938 Tatars were concentrated in southeast Crimea. The majority of Tatars being rural, there had been little dilution of linguistic identity over the years, that is, adoption of say Russian as the first language. Indeed, even urban Tatars typically retained Tatar as the mother tongue. The roughly 60,000 Tatars classified as urban in the 1939 census were concentrated in a few small, traditionally Tatar cities away from the coast. Bakhchisaray, located midway between Simferopol' and Sevastopol', was the most important of them. Most of the larger cities throughout the Crimea were dominated by Slavs. Sevastopol', for example, was one of the most important Black Sea naval bases. As Fig. 11.3 indicates, it was a predominantly Russian city, with Ukrainian, German and Jewish minorities. Coastal resort centres such as Yalta had a similar ethnic balance. Thus, on the eve of World War II Tatars were a minority in their homeland, were predominantly rural, and lived mostly in the south and southeast parts of the peninsula. In a few short years the ethnographic map of Crimea was to be fundamentally altered, however.

During World War II it is probable that a few Crimean Tatars did collaborate with the Germans. But so too did a few members of most other minority groups directly affected by the German invasion. And after all, during the 1930s anyone suspected of nationalist leanings had been mercilessly persecuted by Soviet authorities. Nonetheless, Stalin held all Crimean Tatars responsible for the presumed actions of a few, and the whole population was subjected to brutal

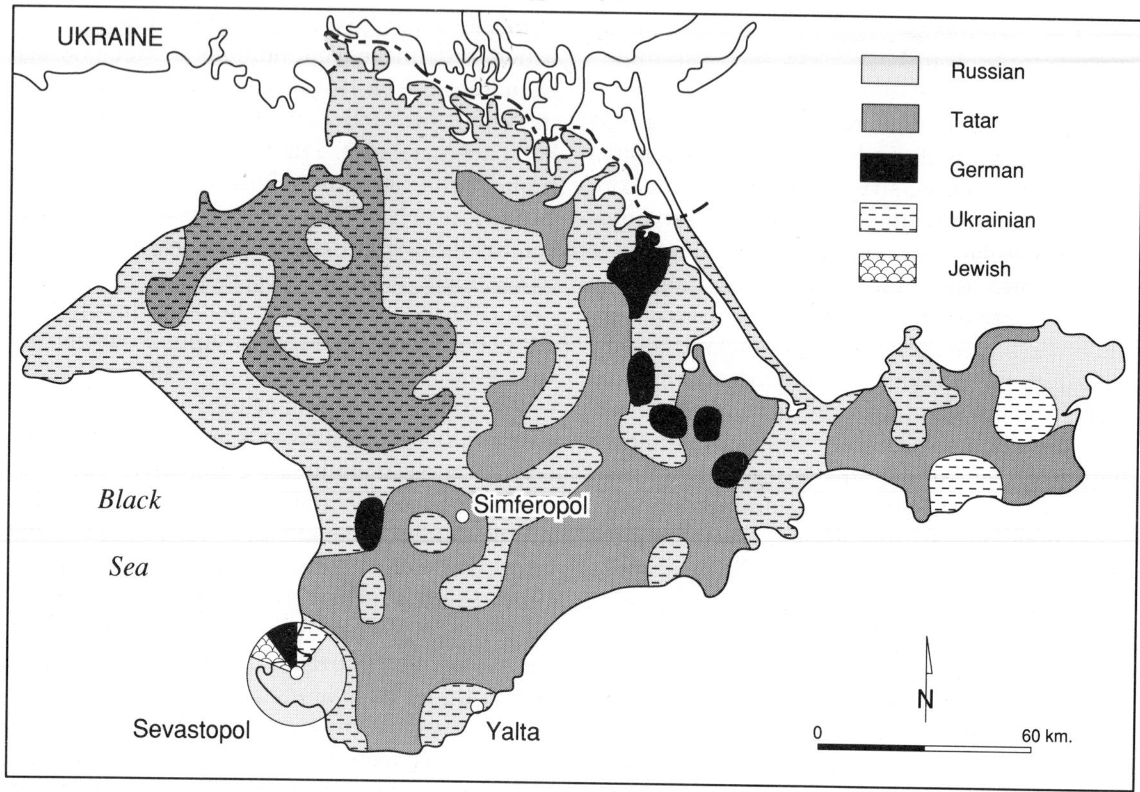

FIGURE 11.3 *Tatars in Crimea, 1938*
(*Source*: Based on S. Rapawy, *Ukraine and the Border Issues* (Washington, DC: Centre for International Research, US Bureau of the Census, 1993), Ethnographic map of Ukraine)

retribution. Following the German retreat from the Crimea in 1944, all people whose passports identified them as being of Tatar nationality, and their children, were forced from their homes, loaded into box cars with little more than the clothes on their backs, and transported en masse to Middle Asia. This was an illegal act under the Soviet constitution and civil law. A very great many Crimean Tatars did not survive, death rate estimates varying between 10 and 40 per cent of those transported, the lower figure being the official Soviet estimate. Those who survived the forced migration were settled mostly in rural Uzbekistan. According to the 1939 census, there were around 147,000 Tatars already living in Uzbekistan, about half of whom were rural. The forced migration thus added to the Tatar population in Middle Asia, but Crimean Tatars were far from well received by the locals, including some Uzbek Tatars, all of whom were already feeling the pressure of too many people for too little land.

In the Crimea, the wholesale deportation literally erased a Tatar presence from an historic homeland. Tatar homes and possessions were taken over, with government acquiescence, by new settlers brought to the Crimea, mostly Russians and Ukrainians. Crimean Tatars were officially designated an 'outlawed national minority' until this pejorative designation was finally rescinded by Nikita Khrushchev in 1956 as part of a general de-Stalinization programme. However, the Crimean Autonomous Soviet Socialist Republic, which was abolished in 1946, was not reinstated, and Crimean Tatars were not allowed to return there. Their homeland had been converted to an oblast' in 1946, that is, a non-ethnic administrative region, of the Russian Republic. In 1954, the territory was formally transferred from

Russia to Ukraine as part of the celebration of three centuries of Russian and Ukrainian union. Crimea remained an oblast′ level administrative government. At the time of the 1970 census, only 6500 people claimed Tatar as their native tongue out of a total population of more than 1.8 million in the Crimean Oblast′. Close to 574,000 Tatars lived in Uzbekistan at this date, a good number of whom were the survivors and descendants of those transported there in 1945.

While the ethnographic map was fundamentally altered by draconian, punitive government action, the Crimean Tatar homeland could not be erased from the collective memory of the people affected. In the 1960s, Crimean Tatars began petitioning the Soviet government for the right to return to their historic homeland, something for which there were already precedents. For example, thousands of Chechens also deported by Stalin, mostly to Kazakhstan, were allowed to return to their Caucasian homeland in the 1950s. But the Crimean Tatars, Volga Germans and Meshketian Turks were not permitted to do so, their homelands remaining formally erased from the map. The Tatars continued to press their case to be permitted to return to the Crimea. They used very effectively the legal rights embedded in the Soviet constitution, and various forms of passive resistance to authority, to great effect. They were able to garner thousands of signatures for petitions during an era when such manifestations of minority group solidarity put each signator at considerable potential personal risk. They organized peaceful demonstrations of several thousand people in Moscow on a number of occasions, something also entailing considerable personal risk in terms of prosecution for unlawful assembly. The petitions, demonstrations, and growing support of other people both within the Soviet Union and internationally fell on deaf ears, however. In the late 1980s, an unauthorized return migration of Tatars to Crimea began as signs of the possible disintegration of the Soviet system increased. From a few thousands in 1988, the numbers jumped dramatically as the Soviet system began to come apart at the seams. In 1990, there were reputedly 93,000 Tatars who had moved back to Crimea. They often settled illegally in rural areas in makeshift quarters since other people had long since taken ownership of family homes in villages or apartments in the cities. Following the collapse of the Soviet Union in late 1991, the number of returnees grew quickly. By 1995, Tatars numbered close to 300,000 of the roughly 2.8 million total population of Crimea. But what they returned to was a region embroiled in political turmoil over the proper place of Crimea in the post-Soviet scene, not over whether the Crimea should be reinstated as a homeland for Tatars.

Crimea's population has been preoccupied with the political future of the region in the post-Soviet period. Russians presently comprise about two-thirds of the total population(see also Fig. 4.8, p. 108), a heavily Russified Ukrainian population nearly one-quarter. Tatars are the next largest minority group. Not only is the population dominated by Russians, the Crimean peninsula is commonly perceived by Russians everywhere as being part of their homeland. Notwithstanding popular Russian sentiments, the current political reality is that Crimea is part of Ukraine. Following a referendum in early 1991, the Ukrainian government re-established a Crimean Autonomous Soviet Socialist Republic, but not to accommodate Crimean Tatars, rather to accommodate separatist tendencies amongst the majority Russian population. But this development was still not acceptable to many Russians living there. In February 1992, a new name was formally adopted, the Crimean Republic. In May 1992, a new constitution was adopted by the government of Crimea. The new constitution declared Crimea to be independent, established Crimean citizenship and set the stage for negotiations over future relations with Ukraine as an equivalent independent state, and not simply as just another lower level of government. Ukraine's response was swift, and negative. It had the real option of an effective economic blockade. Since more than nine-tenths of all energy supplies, and most of the Crimea's essential irrigation water, come from or through Ukraine, bringing the regional economy to a halt is well within the realm of possibility. Portended threat was sufficient. By September that year a number of clauses, including the declaration of independence, were rescinded by the government of Crimea. But the separatist movement did not disappear. While the Russian government was circumspect as to the future status of Crimea, especially in terms of suggestions of it becoming part of Russia, some political figures in Russia used the situation to their own advantage by arguing the need to provide protection for potentially oppressed Russian minorities in the 'near abroad'. The popular perception that Crimea was historically a part of Russia simply added credibility to the proposition.

During the Crimean independence movement which was dominated by Russians, the Crimean Tatars did not take a position regarding affiliation of the region with Russia. They want nothing less than the reinstatement of the Crimea as a Tatar homeland, of course, and are loathe to find themselves caught up in the struggle between Russian and Ukrainian interest groups. To voice the collective concerns of all Crimean Tatars, in 1991 they re-established their traditional representative body, the Kurultai, which operates as form of separate government. Each delegate to the Kurultai is elected by 1000 constituents. In 1993, the Kurultai comprised 137 members. As this is an *ad hoc* governing body which does not sit year round, a type of inner cabinet called the Mejlis is elected from the Kurultai membership to manage affairs when the larger representative body is not sitting. That such an organization could be put in place so soon after returning to such an uncertain economic and political situation is clear testimony to the coherence and solidarity of the Crimean Tatars as a minority group. It serves a variety of purposes. For example, an acute housing and employment crisis was endured with relative equanimity until a Tatar squatter settlement was forcibly removed by government authorities in 1992. This resulted in the Kurultai calling for a mass mobilization of tens of thousands of Tatars which brought most forms of land transport across Crimea to a halt. Authorities made concessions on Tatar settlements in unoccupied areas. The central issue for debate in the Kurultai continues to centre on the appropriate political strategy to reclaim Crimea as the historic Tatar homeland. During discussions in Crimea in 1993 concerning the structure of a future parliament, the Kurultai leadership argued for a special quota of seats to be set aside for themselves. The proposal was rejected out of hand by the existing government, something which prompted widespread Tatar civil unrest promoted by the Kurultai. It was finally agreed that 14 out of the 98 seats in the new government would be set aside for directly elected Crimean Tatars. However, participating in the proposed elections for 1994 was felt by some Tatars to be tantamount to recognizing the existing political authority in the Crimea. The opposing view prevailed. All 14 seats were filled in the election by candidates who were supported by the Kurultai. The elections in 1994 not only produced a new parliament with a significant Tatar presence, it produced for the first time a President of Crimea. The first President, Yuriy Meshkov, a strongly separatist, pro-Russia activist, soon found himself isolated from more moderate elements in the new government. Taking advantage of Russia's many internal ethno-territorial problems, and the fact that relations between the two countries improved in 1995, Ukraine in March that year took the unprecedented step of passing legislation abolishing the post of President of Crimea. It also took additional steps to initiate a land survey with the express purpose of finding areas for Crimean Tatars to settle. The Crimean Parliament adopted a new draft constitution in 1995 that does not explicitly mention the post of President, giving instead presidential authority to the head of parliament, thereby further isolating Meshkov. Moreover, articles in the draft constitution further dilute the earlier nationalistic claims to independence and Crimean citizenship. For instance, the draft constitution acknowledges Crimea to be an autonomous part of Ukraine, not independent of it. Ukraine now customarily refers to Crimea in official communications in more conciliatory terms as the Autonomous Republic of Crimea. But it has not given up any real authority over the region. In short, Ukraine has prevailed thus far in Crimea's struggle for real political independence. The place of the Tatars in Crimea remains unresolved.

Tatars have had many problems with which to deal. Their return has not been easy. A few have bought land and property, and are beginning to start life anew. But the vast majority have a rather precarious economic existence. The regional economy has been severely impacted both by the market reform process and the political uncertainty the independence movement and political tensions between Ukraine and Russia produced. In the latter context, control over the Black Sea fleet and associated naval installations such as those at Sevastopol' is the subject of dispute between Ukraine and Russia. The rusting Black Sea Fleet is in any event now more a liability than an asset. But the military bases did provide substantial direct and indirect employment in Crimea, however. And so too did the military-industrial complex. Most of these plants are at a standstill. Crimea's non-military-based industry which is heavily oriented to food processing is in similar straits. Factories in Crimea are customarily operating at much reduced levels, if they are operating at all. The previously important tourism sector has collapsed. Sanitorium facilities were owned by Soviet enterprises and ministries. Few now have the financial resources to maintain them. Those who used to holiday there for a modest sum now often cannot make ends meet weekly, let alone contemplate a vacation in Crimea or anywhere else. The jobs of many people living in Crimea were dependent on the year round tourists. Few now come, and thus many workers in the sector are unemployed. Organized crime controls many enterprises where jobs might be had, for if economic stability returns to Crimea there is enormous scope for enrichment from land development and tourism. For returning Tatars, jobs in any sector of the economy were not to be had easily in any event. They have in many instances returned to the land. Perhaps a majority of them remain unemployed. While economically the situation for Tatars in Crimea is problematic, they have made progress in the shaping political landscape, and have some hope that Ukraine's intervention may yet bring land for them to settle. But Crimean Tatars have not yet regained legal recognition of their historic homeland, nor are they likely to do so in the near future.

The Caucasus

Of all the regions of the former Soviet Union, the Caucasus has the dubious distinction of having the greatest number, and certainly the most bloody, of the ethno-territorial disputes (Fig. 11.1). Over the centuries, the mountainous environment has provided safe refuge for scores of ethnic groups differentiated by language, culture and religion. As Fig. 11.4 reveals, there are numerous ethno-territorial regions which are the legacy of Soviet nationalities policy. Many of these places have been embroiled in civil war since the final days of Soviet power. In the mid-1990s, Chechnya, which is reluctantly part of Russia, is perhaps the most readily recognized of the places denoted in Fig. 11.4. The situation there will be discussed in a later section. While civil war in Chechnya doubtless has cost more lives than elsewhere in the Caucasus, the sad fact is that very few corners of this region have been spared the consequences of ethno-territorial disputes. Unfortunately, there are many parallels between war-torn former Yugoslavia and this part of the post-Soviet scene.

From the early part of the nineteenth century when the Caucasus was brought into the Empire, Russia's troops spent much time and effort subduing the various minorities of the Caucasus. Indeed, in the final days of the Empire there were still pockets of resistance to imperial authority. The demise of the Russian Empire and the imposition of Soviet power did not alter this characteristic of the region. What made the most difference was the nature of the Soviet system and, especially during Stalin's time, the absolutely ruthless response of the Soviets to any overt manifestation of resistance to authority or ethno-centred nationalism. Put simply, long-standing differences between ethnic minorities in the Caucasus, and between them and the Soviet regime, were stifled. With the collapse of the system, long-standing feuds erupted, in many instances having been further inflamed by decisions taken during the early years of Soviet nationalities policy. The imposition of borders to define the many ethno-territorial homelands depicted in

FIGURE 11.4 *The Caucasus*
(*Source*: Adapted from *Atlas SSSR* (Moscow: Glavnoye Upravleniye Geodezii i Kartografii pri Sovete Ministrov SSSR, 1969), 23)

Fig. 11.4, and the assignment of some territories to particular Republic jurisdictions, left a legacy of bitterness which infects the region to this day.

The first place over which really serious fighting occurred was the region of Nagorno-Karabakh in Azerbaijan. This former Autonomous Oblast', like the Autonomous Soviet Socialist Republic of Nakhichivan, was created at the time of the 1921 treaty between the Soviet Union and Turkey (Fig. 11.4). Few of the Soviet-era boundaries imposed in the southern part of the Caucasus region correlated with the *guberniya,* or provincial, borders of the imperial Russian administrative system. Under this system, the mostly mountainous region adjoining present-day Turkey and Iran was divided into five provinces. Soviet nationalities policy, however, dictated that homelands be created. Thus, the Christian Georgians and Armenians were given homelands with the status of Soviet Socialist Republic, as was the Moslem Azeri population. But the areas depicted in Fig. 11.4 greatly simplify the complex ethnic reality of the region. The population of Nagorno-Karabakh according to the limited census data had historically been dominated by Armenians, but these data ignored the large scale seasonal migration there of Azeri herdsmen. In any event, the territory was made part of the new Azerbaijan Soviet Socialist Republic. Over the years the Armenian presence remained strong. According to the 1939 census, Armenians comprised 88 per cent of the 151,000 inhabitants. By 1970, the Armenian share was about 81 per cent of the 150,000 total population, and had dropped to 79 per cent at the end of the decade. While still a substantial majority, the political situation in Nagorno-Karabakh steadily deteriorated as Armenians were increasingly resistant to the status quo. Once again, the late 1980s Gorbachev policy of *glasnost'* permitting open discussion of previously off-limits issues brought matters to a head. The Armenian majority formally petitioned to be allowed to secede from Azerbaijan and join Armenia. The petition was rejected by the Soviet government. This unleashed pent-up anger in both Azerbaijan and Armenia against ethnic minorities from the other state. Most of these minority groups had lived peacefully with neighbours of the other nationality for generations. Nonetheless, riots, pogroms, forced migration, hardship and death were widespread. Deaths mounted rapidly, and hundreds of thousands of people were made refugees in the ensuing civil war. Armenian military intervention eventually secured Nagorno-Karabakh, and a militarized corridor from Armenian territory to Nagorno-Karabakh is maintained as a lifeline to the region. Some Armenians fleeing Azerbaijan have settled there, and former Azeri inhabitants have been forced to depart. Thus, Nagorno-Karabakh is now both larger in total population and even more Armenian ethnically. Despite years of an economic and energy blockade imposed by Azerbaijan and Turkey, Armenia has endured the hardship, substantially reforming its economy in the process. With the change in government in Azerbaijan, tensions have eased for the time being. Russian armed forces serve in a quasi-peace keeping capacity in the region and something of a stand-off now prevails. However, there remains potential for further armed conflict since in both Armenia and Azerbaijan the region is regarded as an integral part of the historic homeland.

Variations on this same theme of ethno-territorial dispute, often based on animosities predating the Soviet era, impact on other parts of the Caucasus. Within Georgia, not only has there been civil war between extreme nationalist and more moderate political factions, minority groups distinct from the Christian Georgian majority have been belligerents in civil war as well. As Fig. 11.4 indicates, within Georgia three homelands were created in the early Soviet era, the Autonomous Soviet Socialist Republics of Ajariya and Abkhaziya, and the Autonomous Oblast' of South Ossetiya. To date, only Ajariya has escaped bloodshed. But the Muslim Turkic speaking peoples of this border region have been pressing for greater autonomy. South Ossetiya was the first to be caught up in violence. Here the Muslim Ossetian peoples who comprise the majority of the population pressed for independence in the late Soviet era out of a justifiable sense that their future in a post-Soviet situation would be problematic. Georgian nationalists had long argued that the Ossetians were interlopers, having taken a part of the historic Georgian homeland, despite the fact that Ossetians made up at least two-thirds of the region's total population throughout the Soviet era. Georgians have consistently made up less than 30 per cent of the total population over the same period. Thus, the Georgian nationalist's claim to Ossetiya as part of their historic homeland is rooted in the distant past. But so too is the rebuttal by Ossetian nationalists, whose dominant presence in the region they argue can be traced back 2500 years. The South Ossetian's response to the increasingly belligerent Georgian nationalist position of the late Soviet era, like other minorities across the country, was to declare themselves independent. The adjoining North Ossetiya Autonomous Soviet Socialist Republic in Russia (see Fig. 11.4), afforded the possibility of moral and other support if not political affiliation. Georgia's official response to all of this came in December 1990. Its government passed legislation formally abolishing the South Ossetiya Autonomous Oblast', an act not in accordance with the prevailing Soviet constitution. Following the collapse of the Soviet Union in 1991, civil war broke out between Georgian nationalists forces and the counterpart in Ossetiya. Mass emigration of Ossetians over the Caucasus to North Ossetiya in particular has resulted.

In Abkhaziya, the ethnic balance is different. Georgians have long had a plurality in share of total population. For instance, Georgians comprised 30 per cent of the population in 1939; 41 per cent in 1970; and 46 per cent in 1989. Abkhazians, on the other hand, comprised 18 per cent in 1989, the same share as in 1939. But despite the share of total population they benefited during the Soviet era by virtue of policies which gave priority to Abkhazians in accessing education, jobs in government and the local economy. Georgians not surprisingly perhaps took deep exception to this policy. Abkhaziya declared independence from Georgia in 1989. An uneasy tension prevailed, until civil war erupted. Georgian nationalist forces here confronted a different situation than in South Ossetiya, however. Russia, it is claimed, has provided armaments and training for the Abkhazian nationalist forces, something that Georgia vehemently criticizes. There has long been a sizeable Russian presence in Abkhaziya, which would provide popular support for Russian involvement should ever public justification be necessary, whereas very few Russians lived in South Ossetiya. In this region the Georgian nationalist claim to territory can be, and is, based on numbers, not historic claim to land as in South Ossetiya. Not surprisingly, Abkhazian nationalists argue that their claim to the region is based on their presence there over the centuries. But whatever the real situation in the distant past, the formally thriving tourist and agricultural Abkhazian economy has been devastated by civil war. Once again, thousands of people of all nationalities have been forced to flee the region. Abkhazians, Russians and other minorities have departed to Russia, and nearly a quarter of a million Georgians to Georgia. Some observers of the situation there argue that Russia aided and abetted the Abkhazian resistance to Georgian nationalists in order to gain geopolitical advantage in the region. Eventually, the Georgian military was driven out of Abkhaziya. Whatever the original motivation, Russia in fact has emerged as the official peace keeper in the region, although by late 1995 had not yet succeeded in persuading Abkhazian authorities to permit Georgian refugees to return to their homes. But Russia has managed to leverage Georgian agreement to join the Commonwealth of Independent States, as noted in an earlier chapter, and has gained Georgian concurrence to maintain military bases on Georgian territory. Meanwhile, Abkhazian authorities are pressing for a confederal arrangement for a Georgian–Abkhaz state, something which would put them of equal footing. Georgia refuses to consider the suggestion, arguing instead for a return to the status quo ante. Georgia is using military base access and use as leverage to obtain Moscow's acquiescence for a Georgian military operation that would bring Abkhaziya back into the fold. The situation in Abkhaziya thus remains delicately poised on the brink of renewed civil war.

Tajikistan

The ethno-territorial problem in remote and mountainous Tajikistan is as complex as anywhere else, but the situation there is different because there is an international dimension to it. In addition to civil war within Tajikistan, a force of some 5000 rebels operates from Afghanistan, with the tacit approval of its government and the active support of sympathetic Islamic fundamentalists. A border guard and army of more than 25,000 men under Russian administration defend the border with Afghanistan, as well as secure the uneasy peace within the country. The politics of this remote place are related to ethnic, religious and regional differences. From the late 1930s until the outbreak of civil war in 1992, Tajikistan's population comprised three main ethnic groups – Tajiks who made up more than three-fifths of the total, Uzbeks at nearly a quarter, and about 7 per cent who were Russian, most of whom lived in the capital city, Dushanbe. Of the 380,000 Russians, at least 80 per cent had departed the country by the mid-1990s because of the unrest. The distribution of the principal ethnic minorities is presented in Fig. 11.5. As a

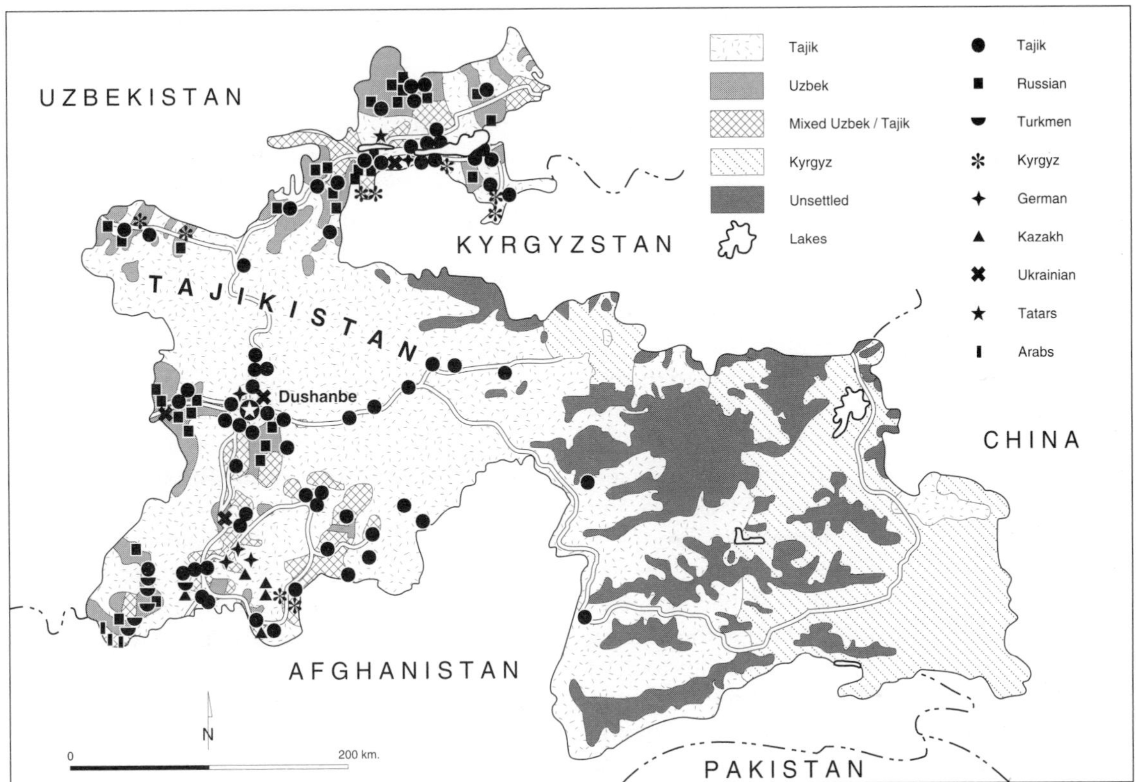

FIGURE 11.5 *Contemporary Tajikistan*
(*Source*: Adapted from *Atlas Tadzhikskoy Sovetskoy Sotsialisticheskoy Respubliki* (Dushanbe: Akademiya Nauk Tadzhikskoy SSR, 1968), 121)

close inspection indicates, the Uzbek minority was concentrated in the northern and eastern parts of Tajikistan, especially in the finger of territory which penetrates northeast into the Fergana Valley. In these largely oasis settlement regions, Tajiks and Uzbeks had lived together peacefully for centuries. Until the Soviet-era nationalities policy created separate homelands and separate histories for the various peoples of Middle Asia, Uzbeks and Tajiks both identified themselves as Sarts. Kyrgyz peoples lived mostly in the little settled, very mountainous eastern region. Around Dushanbe and to the south is a quite diverse population, as Fig. 11.5 reveals, though the total populations of particular minorities such as the Tatars, Ukrainians or Germans were typically small. Many of these people, like the Russians, have fled Tajikistan for safer havens since the civil war.

Following the collapse of the Soviet Union, a discredited Communist Party loyalist, Rakhmon Nabiev, was appointed President, by government officials whose political power base was in the northern parts of the country. He was confirmed in office in an election a short time later, an election claimed to be fraudulent by opposition forces which were very strong in the southern region. Reformist groups in Tajikistan, already bitter because their candidate lost the Presidential election, saw Nabiev as a representative of the old Party elite, of the *nomenklatura*, and a continuation of the status quo The fact that the late 1991 ban on the Communist Party was lifted in Tajikistan at the same time Nabiev was selected to take over as President widely was seen as evidence of the linkage between events. After independence in 1991, a number of opposition groups formed a coalition early the next year to press for economic reform and democracy. Demonstrations occurred, especially but not exclusively in Dushanbe, and the government responded with force. Regional differences within the country were exacerbated during the confrontation between opposition and government. In an effort to stem the unrest, President Nabiev formed a government of reconciliation which included a number of opposition leaders. Economic reform and democracy movements continued to take on elements of ethnic and regional, if not shades of religious difference. Armed partisan forces were being formed during this period. Meanwhile there was a growing Tajik nationalist movement which demanded that Tajik be adopted as the official state language. Minorities took alarm, and especially Uzbeks and Russians. Along with other ethnic groups a multi-national state was proposed. The divisions deepened. By mid-1992 civil war had erupted. Opposition forces eventually forced the Nabiev government out. Civil war in Tajikistan took at least 20,000 lives, caused several hundred thousand people to become refugees in adjoining countries, including Afghanistan, and left the country in a state of near ruin.

The civil war ended in early 1993 in large measure because of the intervention of the Russian Army. Moscow views Tajikistan as part of its geopolitical realm, even if the country is independent politically. In all other respects, it is dependent upon Russia. A new President was elected by the Tajik government in late 1992. This was Imomali Rakhmonov, a Tajik, a former collective farm chairperson from the southern region, and something of a compromise choice as leader. Since winning the Presidential election in 1994, Rakhmonov moved to curtail somewhat the long-standing dominant influence of the northern region in Tajik politics, a region in which many of the country's Uzbek minority lives. But he has also made conciliatory gestures, including the removal of Tajik as the exclusive state language from the draft constitution. Russian remains the official *lingua franca.* Rakhmonov has also taken to stressing the bonds between Uzbeks and Tajiks, rather than the differences. Rakhmonov has focused on creating a concept of Tajikistan as a homeland for all ethnic groups, not just Tajiks. He has even tried to find a place for Russians, for whom both Tajiks and Uzbeks have little sympathy, seeing them as colonizers and anti-Islamic in past action. Part of the problem in conveying the message of reconciliation is that in Tajikistan, the poorest of the former Soviet Republics, much of the predominately rural population does not read newspapers or watch television. Indeed, in parts of the remote and mountainous eastern region the state television broadcasts cannot be received. Moreover, in times of economic hardship inter-ethnic rivalries can easily come to the fore. Peace is not only maintained in Tajikistan largely through the presence of a Russian administered army, the economy is heavily subsidized by Russia. The task of economic reconstruction can only proceed if peace is maintained, and that means that Russia will continue to play a role in defending Tajikistan's southern border with Afghanistan against insurgent forces wanting to overthrow the Rakhmonov government and install a more nationalist, if not Islamic fundamentalist regime. Given Russia's own problems this is a heavy commitment. Indeed, Russia's role of peacemaker in the 'near abroad' stands in contrast to some obvious failures to maintain peace and good order within Russia itself. Chechnya is a case in point.

THE POLITICS OF PLACE IN RUSSIA

Chechens number nearly one million, the greatest majority of whom live in a north Caucasian homeland created in the early Soviet era as a result of the prevailing nationalities policy. While some minorities may have welcomed the formation of an ethno-territorial homeland by the Soviets, it is fair to say the Chechens did not. From the time Russia began its territorial expansion into the northern Caucasus in the late eighteenth century, the Sunni Muslim Chechens periodically waged Holy Wars against imperial, and Christian, Russia. They resisted the imposition of Russian political authority down to the revolution of 1917. Thus, Chechens scarcely looked upon Bolsheviks as a liberating force, to say nothing of their opinion of the formal adoption of scientific atheism as the state religion of the newly created Soviet Union. On the heels of the fall of the Russian Empire came the formation of the Gorskaya Republic, an alliance of the peoples of the north Caucasus. While Georgian, Armenian and Azerbaijan armed resistance to the Red Army collapsed in matter of weeks, it took a year for the Bolshevik forces to prevail over the resistance to them in the Gorskaya Republic of the north Caucasus, a resistance in large spearheaded by Chechens. That nearly 10,000 defenders of the Gorskaya Republic lost their lives did little to win the allegiance of Chechens in particular and the northern Caucasian peoples in general. In creating ethno-territorial regions in the north Caucasus, borders were drawn in such a manner as to create divisions where none existed before (Fig. 11.4). Clans, not nations, not territory, had been the fundamental social organizational principle across the region in the past. Territorial homelands were created with a view to dividing and conquering resistance to Soviet power. For example, in 1934 the mostly anti-Soviet Chechens were put in the same region as the less anti-Soviet Ingush peoples (Fig. 11.4). Formation of the Chechen–Ingushetyia Autonomous Soviet Socialist Republic in 1937 was scarcely adequate compensation for past transgressions. Chechens remained defiant, but paid a colossal price for doing so. Tens of thousands were liquidated in the purges initiated by Stalin against all suspected anti-Soviets in the late 1930s. After the retreat of the German invasionary forces from the north Caucasus in 1944, there was a mass deportation of some 200,000 Chechens to Kazakhstan and Siberia in particular. As in the case of the Crimean Tatars, some Chechens were accused of collaboration, the whole population was held responsible, and an enormous number died during the mass deportation. Many Ingush were similarly treated. The Chechen–Ingushetiya Autonomous Soviet Socialist Republic was abolished. But wherever Chechens ended up they mostly remained a people apart, retaining their use of Chechen as the first language, and invariably marrying within the Chechen community. Again like the Crimean Tatars and other outlawed peoples, with the de-Stalinization initiated by Khrushchev in 1957, the Chechens were 'rehabilitated'. But unlike the Crimean Tatar homeland situation, in 1957 the Chechen–Ingushetiya Autonomous Soviet Socialist Republic was restored, and Chechens were allowed to return. What they returned to was a region in which newcomers had arrived, in many instances taking over the former homes and possessions of Chechens. The borders of the homeland had also been adjusted and two districts from Stavropol' Kray were shifted to the Chechen part of the reconstituted Republic. The population of the two districts was then dominated by Terek Cossacks.

The Chechens' close attachment to homeland was accentuated in consequence of their suffering while in exile, and as unwelcome returnees in Chechnya itself. Between the late 1950s and the early 1990s, Chechen demographics assured them a growing share of the population in their region. Indeed, during the 1980s there was an absolute as well as relative reduction in the number of Russians living in the Chechen homeland, despite the importance of the oil industry in the region which had traditionally attracted a substantial Russian labour force. While the involvement of Russians in the economy declined, in the all important Communist Party hierarchy the hands on the levers power remained Russian almost to the end. Only in 1985 under Gorbachev was a Chechen appointed First Secretary, for decades a customary concession in almost all other Republics. This symbolic gesture intended to build allegiance to Moscow backfired, and simply fed the already strong nationalist movement instead. Inward looking and territorial, the Chechens were increasingly dominant in their homeland. Developments during the final years of the Soviet era portended new possibilities for self-determination. When the opportunity finally arose, Chechens wasted no time in unilaterally declaring their complete independence.

The National Chechen Congress began demanding independence in 1990. This was a nationalist organization which better represented Chechen political aspirations than did the Supreme Soviet of the official government, the Chechen-Ingushetiya Autonomous Soviet Socialist Republic. The principal advocate of complete separation of Chechnya, as the Chechen homeland was called, was a General in the Soviet Army, Dzhokar Dudaev, who had quickly emerged as a leader in the National Chechen Congress. During the abortive coup of August 1991 in Moscow, Dudaev was unequivocal in supporting Boris Yeltsin in his defence of democracy. He was also instrumental in overthrowing the discredited Republic Supreme Soviet. Once started, the independence movement in Chechnya gathered momentum, and soon the relationship between Russia, of which Chechnya was still formally a part, quickly soured. Dudaev called elections to a new government in Chechnya for late October 1991. He won handily, and on the strength of his 85 per cent majority support declared unilateral and complete independence. Russia under Boris Yeltsin reacted with a declaration of emergency and dispatched troops to Chechnya to reinstate the previous leadership of the Supreme Soviet as a provisional government. The tactic did not have public support in Russia and certainly not in Chechnya, despite Moscow's attempt to characterize the election of Dudaev as a massive fraud. Almost immediately Yeltsin issued a decree recalling the army. Russian troops were entirely removed from Chechnya in the summer of 1992. Garrisons and military supplies, including armaments, were left behind. These were immediately commandeered by the Chechen military under Dudaev. An economic blockade was introduced by Russia instead of armed intervention in the region. But it was for the most part ineffective, and gave sustenance to the further development of black market transactions and criminal activity which affected ordinary citizens more than the Chechen resistance movement. From the middle of 1992 until December 1994 Chechnya was tolerated by Russia, despite its continued aversion to recognize in any way Moscow's authority. Chechnya refused to sign any agreements, did not participate in the various referenda, did not participate in the December 1993 elections to the new Russian State Duma, and did not involve itself in the discussions and vote leading up to a new constitution for the Russian Federation. Moscow refused to accede to Chechnya's secessionist demands, and proceeded as if it were still a part of the Russian Federation, albeit one led by a misguided if not illegally elected government headed by Dudaev. Moscow's support of a provisional government in Chechnya was ineffectual since it lacked any real measure of popular support within Chechnya. By December 1994, President Yeltsin issued a decree ordering the Russian army back into Chechnya to bring order to the wayward region. This triggered a continuing civil war that has laid waste to much of Chechnya, cost tens of thousands of lives, but has not resulted in the defeat of the Chechen resistance movement.

The war in Chechnya has been a disaster for all concerned. Initiated by Yeltsin's December 1994 Presidential decree, it did not have the support of the State Duma which on two occasions later on formally voted against it. Planned to be just a brief military operation by its proponents in Moscow, the war in Chechnya has been anything but brief. The war has been a public relations disaster for both Yeltsin and the Russian army, and a catastrophe for Chechnya. The Russian army command structure was ineffectual, the troops were ill-trained for the task at hand, and the civilian casualties have been huge. In the early part of 1995, Russian troops moved into the capital, Grozny (Fig. 11.4). But months of resistance by Chechen urban guerrilla forces and bombardment by the Russian military reduced much of the city to rubble. A once thriving city of 400,000, its population was reduced by half. The fortunate ones have managed to escape, joining the stream of refugees pouring into adjoining Ingushetiya, Dagestan, and Stavropol' Kray. Thousands died during the battle for control over Grozny. A ceasefire was negotiated in the summer, after the Chechen armed forces retreated from the city into the southern parts of Chechnya. Mountainous and difficult to pursue conventional military manoeuvres, many of the towns and much of the countryside in this region remained firmly in Chechen control. In late 1995, the ceasefire negotiated in the summer was near breakdown. From this mountainous bastion, Chechnya guerrillas continue to wreak havoc. Much of the war has been accorded live coverage on Russian television. It is enormously unpopular amongst the general public.

Russia has attempted to gain support in Chechnya by undertaking to rebuild the now shattered regional economy. With most of Chechnya's economy devastated, several hundred thousand people are unemployed. Chechnya was an oil-producing and -refining centre. Much of this infrastructure lies in ruin. The scale of the reconstruction cost is staggering, and for a Russian economy already weakened by market reform, probably impossibly expensive in the near-term future. The sheer cost of reconstruction feeds the sentiment in Russia that sees the departure of Chechnya as a positive development, both economically and politically. With Moscow's active participation, a former speaker of the Chechen parliament has been appointed the Chair of a Government of National Revival. But this essentially puppet government lacks popular support, and in any event does not address the political reality of the resistance movement which still commands widespread endorsement. By the end of 1995, conservative estimates put the human cost of the Chechnya civil war at more than 20,000 lives. Dudaev was killed in early 1996, but politics in Chechnya are still not controlled by Russia. Perhaps the only positive note about the current situation is that independence movements elsewhere in Russia's north Caucasus region, including the call for a Terek Cossack homeland, have been tempered somewhat because of what has happened to Chechnya.

Throughout the rest of Russia resistance to the power of the centre on the part of ethnic regions is also characteristic. There have been some ethno-territorial conflicts as well (see Fig. 11.1). Tensions between minorities have been exacerbated in some regions because of the influx of refugees, which has put additional strain on regional and local economies in already difficult times. Virtually all of Russia's many ethno-territorial regions have declared independence at some point in the last few years, and not a few of the non-ethnic regions have done so as well. The issue of independence, however, tends to be couched in terms of general principles concerning greater political autonomy, rather than concrete steps to secede from the Federation, as in the case of Chechnya. In the realm of actions rather than words, most attention is directed toward negotiating greater control over the regional economy. The issue, as noted earlier, is control over revenue, and its allocation.

Since independence Russia's regions have attempted to gain some economic advantage at the expense of the centre. Many have been quite successful, and amongst this group the resource-rich ethno-territorial regions stand out. One of the first tactics adopted by the regions in dealing with Moscow was to form a number of strategic alliances based on existing, or newly created, organizations. Some of the more important regional associations as of late 1992 are depicted in Fig. 11.6. While both the membership and the nature of regional associations is in a constant state of flux, what they represent is a snapshot of the constant jockeying for a comparative advantage in dealing with Moscow on the part of various regionally based special interest groups. In order to gain political support for federal government or Presidential initiatives, since independence Moscow has been continuously engaged in bilateral and multilateral negotiations with the regions. Negotiations are usually over money. For example, since the beginning of 1992 the tax sharing with the regions has steadily increased at the expense of the centre. Yet many regions were even more dependent on federal transfer payments in late 1995 than they were in early 1992. Put simply, many of the already economically advantaged regions have gained additional income while those with limited resource wealth and crumbling industrial economies have nothing much to tax and hence are worse off. What Moscow has given up varies according to the perceived importance of the region, the region's resource base, and its persistence at the negotiating table. A few examples will suffice. The Kaliningrad enclave which has strong historical ties to Germany was made a special tax-free zone in the hope that this would attract substantial foreign investment (see Fig. 9.12, p. 280). It was not a successful initiative and has since been revoked by Moscow despite the fact that this was ostensibly a ten-year agreement with Kaliningrad authorities. By the end of 1995, Moscow had signed separate treaties with

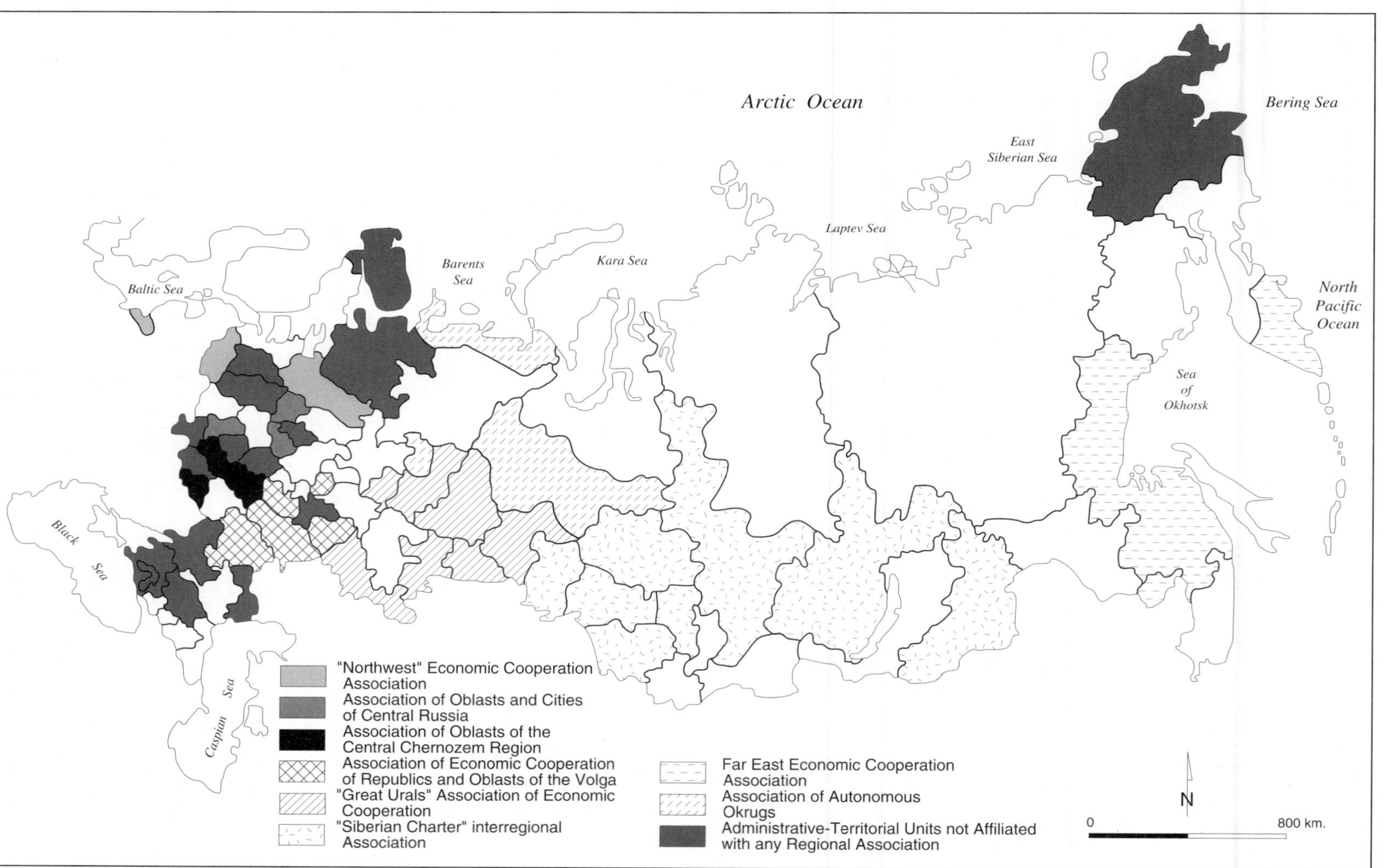

FIGURE 11.6 *Regional associations in Russia, 1992*
(*Source*: N.V. Petrov, S.S. Mikheyev, L.V. Smirnyagin, summarized and annotated by A. Bond, 'Russia's Regional Associations in Decline', *Post-Soviet Geography*, Vol. 34, No. 1, 1993, 65

eight of the Federation's republics. These treaties are all different in terms of demarcating the respective authorities of the federal and republican governments. Treaties have been concluded with Tatarstan, Bashkortostan, Kabardino-Balkariya, North Ossetiya, Sakha, Buryatiya, Udmurtiya and Tuva (see Fig. 9.12, p. 280). Negotiations with these republics covered a wide array of jurisdictions and issues. For example, the treaty signed in October 1995 with the Udmurtiya Republic located in the Urals region covers, amongst other things, the ownership and development rights for forest and petroleum resources, environmental policy and safety, defence sector industries operation, law and order, and inter-bank relations. Udmurtiya has acquired complete autonomy in development, taxation and royalty rights in forest and petroleum resource sectors. Critics of the devolution of federal responsibility for resource development and taxation argue that the process is inconsistent, and geared mostly to fixing 'squeaky wheels', a process that is divisive in a federation and therefore politically dangerous. Taken to its logical conclusion, they argue, the federal government will have no tax revenue sources left. Negotiators claim that giving up taxation responsibilities and turning more discretion over to the regions adds to their autonomy and fosters economic development, while at the same time diminishing the financial obligations of the federal government. There is no question, however, that the regions are not all equal in the negotiation process, and that social justice objectives, such as reducing regional income disparities, have been jettisoned for short-term political ones. Since elections to the State Duma were called for December 1995, just two years after the last election, the intervening period was an especially busy one in terms of special treaties with Russia's Republics in particular. The outcomes reflect the politics of particular places.

Developments in the Republic of Bashkortostan reveal clearly how local elites can acquire real decision-making authority without incurring the wrath of the central government, and without falling into the dead end of civil war which has destroyed so much of Chechnya. Bashkortostan is located in the Urals region (see Fig. 9.12, p. 280). Its total population in 1995 was a shade over four million. Of this number, barely one-fifth was Bashkir, the indigenous population. Russians comprised nearly two-fifths of the total, and there was a sizeable Tatar minority as well (28 per cent). Although historically a minority in their own homeland, few Bashkirs have become strong nationalists since the collapse of the Soviet Union. Most Bashkirs are seemingly comfortable with the political status quo, preferring instead to wrest control over the economy from the centre. The economy of Bashkortostan remains functionally tied to that of Russia and its regions. It has a substantial oil refining capacity, but only very limited domestic production and a limited market. Its refineries are geared to processing oil piped in from Siberia, and to a lesser degree from Tatarstan to the northwest, and delivering the products to the industrial regions in the Urals and central European Russia.

Since 1991, Bashkortostan has steadily, but carefully, widened its jurisdictional responsibility. It has done so as a result of negotiating with Moscow from a position of growing political strength and astute manoeuvring. Typically, President Emomali Rakhimov, a long-standing regional leader during the Soviet era, former Chairman of the Supreme Soviet of the Bashkiriya Autonomous Soviet Socialist Republic, and President since being elected to the post in December 1993, has let neighbouring Tatarstan initiate negotiations for the devolution of control from Moscow, and then come in afterwards raising the ante. His strategy has been extremely successful. In December 1993, Bashkortostan's parliament, which Rakhimov had refused to disband despite the September decree issued by President Yeltsin abolishing all local and regional governments and calling for new elections, passed legislation which claimed for the Republic complete jurisdiction over all natural resources as part of its declaration of sovereignty. The federal government lacked the resolve to try and overturn the decree. Indeed, in early 1994 Bashkortostan signed a bilateral treaty with the Russian Federation, once again on the heels of a similar treaty signed by Russia with Tatarstan. President Rakhimov claimed, also once again, that he had negotiated an even better 'deal' with Moscow than did Tatarstan. The Bashkortostan Republic was acknowledged as the owner of natural resources and won the right to develop them in its own interests. Ownership of the extensive oil refining facilities and infrastructure also passed from Russia to Bashkortostan. Russia ended up retaining ownership of the money-losing defence sector industries. The tax revenue sharing agreement is also decidedly favourable for Bashkortostan. It can also determine its own foreign policy and trade priorities. For once, the old and bankrupt Soviet-era dictum of being 'master of one's own house' had a considerable measure of validity. But who are the masters?

Local elections in Bashkortostan in early 1995 brought overwhelming support for candidates of different parties who were supported by President Rakhimov. Rakhimov's strategy of economic rather than political self-determination enjoys widespread support even though he does not head a monolithic political organization. Rather he has managed to consolidate the political support of the government leaders at the regional and local levels within Bashkortostan, as well as that of most of the key people in the private sector. Many of course are not ethnic Bashkirs. Instead of pushing overtly for political separation from the Russian Federation, Rakhimov, like the leaders of many regions, has lobbied for control over the economy, especially over natural resources, royalties from their development, and a larger share of general taxation revenues. But by no means is this a truly democratic movement. As elsewhere in Russia's increasingly autonomous regions, former members of the Communist Party ruling elite have garnered for themselves control over the economy, and in the process often have become wealthy. The politics of power in Bashkortostan are not overtly ethnically based, however. Rakhimov has continued to appoint people to positions of authority and power less on an ethnic basis and more because of presumed loyalty. He has assumed more direct control over the media, and has squelched opposition voices by economic means such as denying access to government-owned printing enterprises. Elections are influenced by gerrymandering, since the 70 per cent of the population residing in the cities can elect only one-third of the deputies to the Bashkortostan parliament. To be sure, more Bashkirs live in the country than in the city, but in general all rural inhabitants, Bashkir, Tatar or Russian, tend to be more conservative and more likely to support the status quo. Growing economic independence has meant that Bashkortostan is less reliant on transfer of funds from Moscow for what are formally federal responsibilities, but for which Moscow often has no money. Control over taxation revenues negotiated by Rakhimov has enabled the Bashkortostan government to step in and make up the difference when federal transfer payments were deficient. Such developments have obvious importance in generating political support for the current leadership. But from the national perspective, it is evident that by negotiating away ownership of infrastructure, industries, and natural resources, by negotiating away control over taxation revenues and foreign trade prerogatives, and by formally recognizing Bashkir citizenship, the Russian central government is less able to be relevant to the people of this region. This process has been repeated across all of Russia. The more such agreements, the more centrifugal tendencies in the Russian Federation will increase in consequence. If there were not enough pressures on the central government, the December 1995 elections to the State Duma unquestionably added to them.

The campaigning which preceded the State Duma election provided ample opportunity for candidates and voters alike to voice widely shared frustration over a variety of issues – Russia's place in the world, Chechnya, crime and corruption, poverty. The latter issue certainly figured very prominently in the list of complaints. While candidates running for office on a market reform platform could point out that during 1995 inflation was comparatively low, that the foreign trade account was in surplus and rising, that in a number of areas there were signs that the worst adjustments in jettisoning the planned economy for a system more akin to a market economy were now over, for most people signs were not enough. They want improvement in their lives today, not tomorrow. Voters took the opportunity to vent their frustrations when filling out the ballot. One-half of the 450 seats in Russia's State Duma are filled by means of proportional representation based on the votes on ballots listing all registered parties. A party must win at least 5 per cent of the total vote in order to qualify for a seat. The other half of the seats is filled by direct election to single-member electoral districts on a 'first past the post' basis. This provides a route into the State Duma for independents, or for those who do not make the cut-off for the proportional representation party seats in the State Duma. The multitude of parties and blocs on the current political scene in Russia makes it very difficult for any one of them to gain an absolute majority of seats given these electoral procedures.

In the December 1995 elections, only four parties out of the 43 listed on the ballot succeeded in crossing the 5 per cent threshold of votes cast. This was half the number making the cut-off in the elections to the first State Duma in December 1993. The four parties were the Communist Party of the Russian Federation (KPRF) led by Gennady Zyuganov, Liberal Democratic Party of Russia (LDPR) led by Vladimir Zhirinovsky, Our Home is Russia (NDR) led by the Prime Minister of Russia, Viktor Chernomyrdin, and Yabloko headed by Grigory Yavlinsky. Their shares of the vote were 22, 11, 10 and 7 per cent, respectively. NDR was formed in 1995 largely at the insistence of President Yeltsin. Broadly speaking, the KPRF and LDPR are conservative, if not reactionary in outlook. The NDR and Yabloko are moderate and reformist in philosophy. Amongst the previously incumbent reformist parties which did not make the 5 per cent threshold was Russia's Democratic Choice (DVR) headed by Yegor Gaidar. DVR had about a fifth of the vote in 1993, second only to the LDPR, but only 4 per cent in 1995. Gaidar is closely associated with the Yeltsin market reform drive initiated in 1992. Some of the other previously incumbent parties on the 1995 election ballot which did not make the 5 per cent threshold include the Agrarian Party of Russia, Women of Russia, and the Party of Russian Unity and Consensus. Voters sent one part of a message in terms of the fate of DVR in particular. The other part of the message is related to the change in the fortunes of the Communist Party. The KPRF received a shade less than 10 per cent of the vote in the December 1993 election. However, the KPRF's dramatic jump in popular support is less indication of a wish to return to the old order than it is a reflection of general weariness of a substantial segment of the population with the market reform process and its negative impact on the conditions of their daily lives. The rise in popular support for the KPRF parallels the experience of Communists throughout the former Soviet east bloc countries where the dismantling of the command economy and the market reform process began. From Bulgaria, to the Czech Republic, to Poland, Communists and former Communists are in power. In these countries, Communists being in power is not a step back in time however. Instead, resuscitated Communists are the present-day leading edge reformists. Moreover, they come to power with years of experience in the art of politics and governance. In Russia, by way of contrast, Gennady Zyuganov, the KPRF leader, presents his Communist party in whichever light he thinks the audience wants to see it, but his is not a Party of new ideas. And so far as can be determined, it is far from reformist. That demographically the Communists are disproportionately supported by older people is not too surprising given the appeal the Party made during the election for the restoration of the social safety net, so much a part of the old Soviet social contract with the people. It is also supported by a large segment of the hard pressed, if not disgraced, Russian military whose performance in Chechnya has raised serious questions as to its capabilities. Yet it should not be assumed that the KPRF is solidly anti-market reform either. Mostly it advocates slowing the tempo of change and protecting those hardest hit by the transformation which has occurred since independence. Privatization would not be reversed on a large scale, if at all. Indeed, not a few candidates standing for the KPRF in the December 1995 election were entrepreneurs and business people.

Russian voters are not unsophisticated. They have sent a message, but one that is subject to interpretation. Most of the 43 parties listed on the ballot did not make it into the State Duma, and those that did got quite different levels of support compared to 1993. For instance, the ultra-nationalist Liberal Democratic Party of Russia (LDPR), led by the increasing scandalous Vladimir Zhirinovsky, saw its popular support halved in 1995. Voter appeal for the LDPR topped the party list in the 1993 election to the State Duma, spanned most age groups and occupations, and was supported across most of Russia's regions. The performance of its leader since then has been an embarrassment to many ordinary Russians, and this is clearly reflected in its plummeting performance at the polls. Some observers have suggested that voters often intentionally split their vote. For example, it is possible for a voter to support the KPRF on the party list run-off ballot, but support a market reformer on the single candidate constituency ballot, or vice versa. Whatever the interpretations of individual voter behaviour might be, the overall outcome is clear – the reformist element in the State Duma elected in 1995 has been weakened considerably. The inability of the reformist parties to present a unified opposition to the Communists and ultra-nationalists during the election has been costly.

The political make-up of the State Duma elected in 1995 portends some challenges for the reformist movement and the President. The KPRF will have the most members, 158, according to the final preliminary results based on the 225 single member electoral constituencies and the 225 proportional representation party seats. NDR, LDPR, and Yabloko follow with 54, 51 and 45 seats respectively. The rigidly conservative, Communist-based Agrarian Party of Russia, while not gaining seats in the proportional representation run-off did manage to elected 20 members in the 'first past the post' single representative constituencies. So too did independents who number 77. A host of small parties had a total of 45 members elected. What might be broadly labelled the anti-reformist or at least 'slow the pace of reform' parties, collectively can muster more than one-half of the votes in the State Duma, assuming they agree on an issue. But the State Duma is only one part of government since adoption of the new Russian Constitution in 1993. The real political power is still in the hands of the President of Russia, as described in Chapter 2 (see Fig. 2.9, p. 56). The President nominates, amongst others, the Prime Minister and Deputy Prime Minister, both key positions. The President may remove the government without obtaining first the consent of the State Duma. The State Duma cannot initiate draft laws pertaining to budget matters without the approval of the President. Laws passed by the State Duma must be ratified by the 178 member Federal Assembly to which each of Russia's 89 regions sends two members. Most owe some allegiance to the President. If the Communists, or ultra-nationalists, were to win the Presidential elections scheduled for June 1996, then politics in Russia could change dramatically. But most observers do not foresee such an eventuality, arguing that the average Russian voter knows full well how to ensure that dissatisfaction with the status quo is registered without bringing further upheaval to the whole country. Even if a Communist were elected President, radical change is most unlikely. It is conventional wisdom that the market reform process has gone too far, is too broadly supported, and just too important to reverse. Amongst the many challenges ahead, the regional nature of the vote for the main parties also figures prominently. There are important differences in the results of the 1995 State Duma elections between those areas benefiting from economic reform and in many of the regions where the present economic situation is extremely difficult, and the future is less than auspicious.

As the data in Fig. 11.7 indicate, support for the KPRF is very widespread, but nonetheless regionally variable. It is strong everywhere save for a broad swath of territory in central European Russia and in some parts of Siberia. Amongst the ethnic territories such as those of the north Caucasus and mid-Volga regions where the agricultural economy is dominant and the Party apparatchiks of the Soviet era still hold sway, in the more impoverished agricultural-based regions of central and northwest European Russia, and in some industrial regions where Soviet-era command economies were heavily skewed to the military-industrial complex and which since

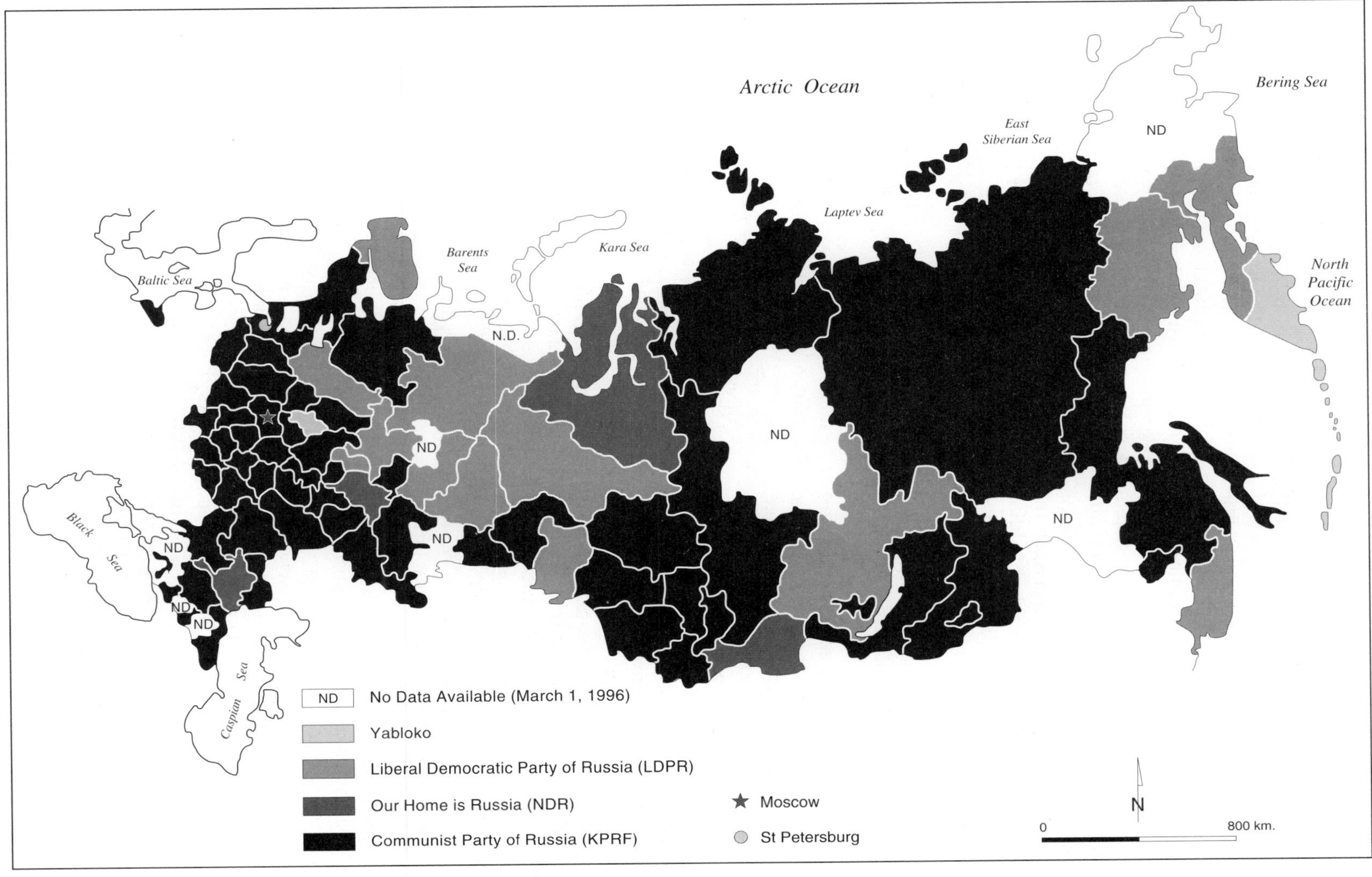

FIGURE 11.7 *State Duma election results, December 17, 1995: Party with the largest share of votes amongst the four parties which exceeded the 5 per cent threshold, by region (preliminary final results as of March 1, 1996).*
(*Sources*: *Kommersant*, No. 48, 26 December 1995, 34; *World Wide Web, http://users.aimnet.com./~ksyrah/ekskurs/lpist2.html*)

independence have been devastated by its collapse, support for the KPRF was very evident. Support for the pro-reform NDR of Prime Minister Viktor Chernomyrdin was especially strong in Moscow, where it captured about one-fifth of the vote. Tatarstan and Chechnya were also areas of NDR support, though the voter turnout in the latter place was comparatively low and support for the NDR probably reflects Moscow's special effort to have the election proceed in the face of resistance to it on the part of the Chechen nationalist movement. The reformist Yabloko party did well in St Petersburg and Kamchatka. The LDPR did well in the Far East, especially Primor Kray and Magadan Oblast', and took Murmansk as well (see Fig. 9.12, p. 280). What Fig. 11.7 underscores is that the major places of economic development, particularly Moscow and St Petersburg which tend to dominate the news, are not Russia. Russia is increasingly a country of important regional differences, and the results of the State Duma elections in 1995 outlined in Fig. 11.7 simply underscores this dimension of its geography.

In the coming years there will be many good reasons to watch carefully developments across Russia and the post-Soviet scene. Few parts of the world have undergone such profound economic, social and political change in so short a time, and impacted so many people. Public expectations of the new political economy were high, and often unrealistic. As yet only a small share of the population in any post-Soviet state has really seen significant improvement in their quality of life. For most people, conditions of daily life and labour are worse than before. As elections occur across the post-Soviet scene, voter dissatisfaction is evident in many places. But that more or less free elections do take place in these newly independent states is itself no small achievement. Democratic traditions were totally absent in the histories of many peoples now living in post-Soviet states. The legacy of the Soviet system permeates nearly all aspects of life, from values, to economic structures, to ethnic conflict, to environmental degradation. The pace of change in the post-Soviet era is not everywhere the same. The direction of change, however, is more similar than it is different. What has been presented in this book can only touch on some of the complexities of the post-Soviet scene. Of necessity, the scene described has been sketched in broad brush-fashion and with few strokes. There is much detail that can be filled in. If what has been presented has raised awareness, has stimulated more interest in this always fascinating part of the world, then its purpose will have been served.

Further reading

Beissenger, M., Hajda, L. (eds) 1987 *The Nationalities Factor in Soviet Society and Politics: Current Trends and Future Prospects*. Boulder, CO, Westview Press.

Blum, D.W. (ed) 1994 *Russia's Future: Consolidation or Disintegration?* Boulder, CO, Westview Press.

Buzgalin, A.V., Kolganov, A. 1994 *Bloody October in Moscow: Political Repression in the Name of Reform*. New York, Monthly Review Press. Translated by Renfrey Clarke.

Carrere d'Encausse, H. 1994 *The End of the Soviet Empire: The Triumph of the Nations*. New York, A New Republic Book, Basic Books.

Chorbian, L., Donabedian, P., Mutafian, C. 1994 *The Caucasian Knot. The History and Geo-Politics of Nagorno-Karabagh*. London, Zed Books.

Conquest, R. (ed) 1987 *The Last Empire: Nationality and the Soviet Future*. Stanford, CA, Hoover Institution Press.

Devlin, J. 1995 *The Rise of the Russian Democrats: The Causes and Consequences of the Elite Revolution*. Aldershot, E. Elgar.

Ehrhart, H.-G., Kreikmeyer, A., Zagorski, A.V. (eds) 1995 *Crisis Management in the CIS: Whither Russia?* Baden-Baden, Nomos.

Eikelman, D.F. (ed) 1993 *Russia's Muslim Frontiers: New Directions in Cross-Cultural Analysis*. Bloomington, IN, Indiana University Press.

Fish, M.S. 1995 *Democracy from Scratch: Opposition and Regime in the New Russian Revolution*. Princeton, NJ, Princeton University Press.

Frazer, G., Lancelle, G. 1994 *Zhirinovsky: the Little Black Book. Making Sense of the Senseless*. London, Penguin.

Friedgut, T.H., Hahn, J.W. (eds) 1994 *Local Power and Post-Soviet Politics*. Armonk, NY, M.E. Sharpe.

Galeotti, M. 1995 *The Age of Anxiety: Security and Politics in Soviet and Post-Soviet Russia*. London, Longman.

Gross, J.-A. 1992 *Muslims in Central Asia: Expressions of Identity and Change*. Durham, Duke University Press.

Hill, F., Jewett, P. 1994 *'Back in the USSR': Russia's Intervention in the Internal Affairs of the Former Soviet Republics and the Implications for United States Policy Toward Russia*. Cambridge, MA, Strengthening Democratic Institutions Project, John F. Kennedy School of Government, Harvard University.

Kaiser, R.J. 1994 *The Geography of Nationalism in Russia and the USSR*. Princeton, NJ, Princeton University Press.

Kampfner, J. 1994 *Inside Yeltsin's Russia: Corruption, Conflict, Capitalism*. London, Cassell.

Lapidus, G.W. (ed) 1995 *The New Russia: Troubled Transformation*. Oxford, Westview Press.

Lewin, M. 1995 *Russia – USSR – Russia: the Drive and Drift of a Superstate*. New York, New Press,

Lewis, R.A. (ed) 1992 *Geographic Perspectives on Soviet Central Asia*. New York, Routledge.

Malik, H. (ed) 1994 *Central Asia: Its Strategic Importance and Future Prospects*. New York, St. Martin's Press.

Motyl, A.J. 1987 *Will the Non-Russian's Rebel? State Ethnicity and Stability in the USSR*. Ithaca, NY, Cornell University Press.

Motyl, A.J. 1993 *The Post-Soviet Nations: Perspectives on the Demise of the USSR*. New York, Columbia University Press.

Saikal, A., Maley, W. (eds) 1995 *Russia in Search of its Future*. Cambridge, Cambridge University Press.

Starr, F. (ed) 1994 *The Legacy of History in Russia and the New States of Eurasia*. Armonk, NY, M.E. Sharpe.

White, S., Gill, G., Slider, D. 1993 *The Politics of Transition: Shaping a Post-Soviet Future*. Cambridge, Cambridge University Press.

Wilson, A. 1994 *The Crimean Tatars*. London, International Alert.

Yanov, A. 1988 *The Russian Challenge. The USSR and the Year 2000*. Cambridge, Cambridge University Press.

Yergin, D., Gustafson,T. 1993 *Russia 2010 – and What it Means for the World*. New York, Random House.

INDEX